Fairlight Audio Post with

DAVINCI RESOLVE 16

by Mary Plummer

Fairlight Audio Post with Davinci Resolve 16

Mary Plummer

Blackmagic Design

www.blackmagicdesign.com

To report errors, please send a note to training@blackmagicdesign.com.

Series Editor: Patricia Montesion

Editors: Bob Lindstrom and Dan Foster

Cover Design: Blackmagic Design

ISBN: 978-1-7342279-4-9 - Fairlight Audio in Resolve 16 - English

Contents

Foreword

Welcome to **Fairlight Audio Post with DaVinci Resolve 16**

I think one of the most exciting things about DaVinci Resolve 16 is that it brings together audio post, editing, color correction, and visual effects in the same software application! The Fairlight page in DaVinci Resolve gives you all the tools you need for feature film-quality audio post-production and finishing. From advanced surround sound mixing to recording, sound design, and more, Fairlight is a complete professional digital audio workstation built right into your editor. That means you'll be able to switch between creative tasks without having to export or translate files between different applications!

Best of all, DaVinci Resolve 16 is absolutely free! We've made sure that the free version of DaVinci Resolve actually has more features than any other paid editing system. That's because at Blackmagic Design we believe everybody should have the tools to create professional, Hollywood caliber content without having to spend thousands of dollars.

I hope you'll enjoy using DaVinci Resolve 16 and we can't wait to see the amazing work you produce!

Grant Petty
Blackmagic Design

Who This Book is For

This hands-on training guide is designed for DaVinci Resolve editors, audio newcomers, and experienced audio professionals who want to create, enhance, and mix soundtracks in the Fairlight page. Each lesson takes you step by step through practical real-world projects such as a feature film trailer and scenes from a science fiction narrative film. You'll start out building a powerful trailer soundtrack, and then move on to recording voiceover and ADR tracks, as well as editing and repairing dialogue. Next, you'll explore sound design secrets to create original sound effects, layer sound and use FairlightFX plug-ins to add depth and dimension to your tracks. Finally, you'll step through advanced mixing techniques to balance, sweeten, pan, automate, bounce, and deliver your finished soundtrack. Download DaVinci Resolve 16 software for free at: blackmagicdesign.com.

Getting Started

Welcome to **Fairlight Audio Post with DaVinci Resolve 16**, an official Blackmagic Design certified training book that teaches professionals and students the art of sound design, recording, editing, sweetening, and mixing, as well as how to get the most out of audio tracks created and edited in DaVinci Resolve 16. Beginning audio editors and assistants will find clear workflow-driven lessons, while seasoned audio professionals will quickly grasp Fairlight's user-friendly tools for realizing incredible sound.

As you step through the lessons, you'll gain experience with Fairlight's powerful new features: ADR tools, external sound libraries, video and audio scrollers, fixed-playhead playback, clip normalization, built-in native audio plug-ins, and more! Best of all, you no longer need to send projects out to another audio application because DaVinci Resolve 16 puts professional audio post tools just a click away from your editing tools and timeline.

This guide blends practical, hands-on exercises with the real-world "craft" of audio post-production such as sound editing, dialogue editing, dialogue repair and replacement, sound design, sound effects editing, and mixing. As you step through Resolve 16's real-world audio post-production tools and workflows, you'll discover new techniques for whatever tasks you take on. You'll go deeply into audio editing and mixing in the Resolve's Fairlight page to explore tricks and techniques used by professional audio editors, sound designers, and engineers to enhance the soundtracks in your projects.

After completing this book, you are encouraged to take the 50-question online proficiency exam to receive a Certificate of Completion from Blackmagic Design. The link to the exam is located at the end of this book.

About DaVinci Resolve 16

DaVinci Resolve is the world's fastest growing and most advanced editing software. It also has a long history of being the world's most trusted application for color correction. DaVinci Resolve 16 includes a complete digital audio workstation with professional recording, audio editing, and mixing tools that enable you to complete projects using only one piece of software!

What You Will Learn

In these lessons, you'll work with multiple projects and timelines to learn both fundamental and advanced *practical* techniques used in several audio production genres. You'll acquire real-world skills that you can apply to real-world productions.

Lesson 1

Plunges you into the middle of an exciting movie trailer's soundtrack as you explore the Fairlight page interface and tools and finish building the soundtrack.

Lesson 2

Utilizes three practical recording techniques to develop voiceover tracks; perform automated dialogue replacement (ADR); and use Resolve's built-in oscillator, to record noise and tones that you can apply to multiple creative sound design and sound effects tasks.

Lesson 3

Explores Resolve 16's unique audio track layers. You will edit a composite of the best takes from a voiceover recording and create crossfades between music clips in the same track.

Lessons 4, 5, and 6

These lessons focus on dialogue tracks and the jobs of a dialogue editor. In Lesson 4, you'll create a checkerboard edit that splits each character's dialogue clips to a separate track, and balance the clip levels within each track. In Lesson 5, you'll apply common dialogue repair and replacement techniques to reduce plosives and other unwanted sounds, and then stitch together a room tone track to create a seamless dialogue scene. In Lesson 6, you'll use advanced FairlightFX dialogue repair tools such as the De-Hummer, De-Esser, and Noise Reduction plug-ins—as well as the gate dynamics processor in the Mixer to remove hum, sibilance, and noise.

Lessons 7 and 8

These lessons focus on sound design for enhancing a soundtrack. In Lesson 7, you'll use Pitch and Chorus plug-ins to change a human voiceover into a processed computer voice, and you'll experiment with time-based plug-ins to add depth and dimension to Foley sound effects. In Lesson 8, you'll create original sound effects from oscillator recordings to simulate a heart monitor, sci-fi computer warble, and rotor blades, and produce a multilayered computer sound effect.

Lessons 9 and 10

These lessons demystify the art of mixing and sweetening your soundtrack and cover the tools and techniques you'll use to polish and complete a mix.

Lesson 11

Steps you through bouncing stereo and 5.1 timelines to create stems. You'll also explore a finishing workflow for panning and linking six mono tracks into a single 5.1 surround track. Then, you'll move to the Delivery page to render your final stems and mix to meet various delivery standards.

The Blackmagic Design Learning Series

Blackmagic Design publishes several official certification books as part of the Blackmagic Design Learning Series. They include:

- The Beginner's Guide to DaVinci Resolve 16
- Advanced Editing with DaVinci Resolve 16
- Color Correction with DaVinci Resolve 16
- Fairlight Audio Post with DaVinci Resolve 16
- Fusion Visual Effects with DaVinci Resolve 16
- And more to come

Whether you want to learn more advanced editing techniques, color grading, or visual effects, certified training has a learning path for you.

After completing this book, you are encouraged to take a one-hour, 50-question online proficiency exam to receive a certificate of completion from Blackmagic Design. The link to this exam is located at the end of this book.

For more information on additional books in this series and Blackmagic Design certification training, visit www.blackmagicdesign.com/products/davinciresolve/training.

System Requirements

This book teaches Resolve 16 for macOS and Windows. If you have an older version of DaVinci Resolve, you must upgrade to the current version to follow along with the lessons. Fortunately, DaVinci Resolve 16 is a free upgrade from previous versions of DaVinci Resolve.

Downloading DaVinci Resolve 16

You can download the free version of DaVinci Resolve 16 from the Blackmagic Design website:

1. Open a web browser on your macOS, Windows, or Linux computer.
2. In the address field of your web browser, enter **www.blackmagicdesign.com/products/davinciresolve**.
3. On the DaVinci Resolve landing page, click the Download button when it appears.
4. Follow the installation instructions to complete the installation.

When you have completed the software installation, follow the instructions in the following section to download the content for this book.

Acquiring the Lesson Files

You must download three zipped R16 audio post lesson files that contain the media files you'll use while performing the exercises in this book. After you download and save the compressed files to your hard disk, extract the zipped files and copy them to a single R16 audio post lessons folder that you create in your Documents folder.

To Download and Install the Three Lesson Files:

When you are ready to download the three lesson files, follow these steps:

1. Connect to the Internet and navigate to: www.blackmagicdesign.com/dvres/audio-post-resolve16-pt1.

   The download will begin immediately.

   The **R16 audio post pt1.zip** file is 533.3 MB in size, and depending on your Internet connection, should take roughly five minutes to download to your computer using a standard broadband connection.
2. Download the second part of the lesson files by navigating to www.blackmagicdesign.com/dvres/audio-post-resolve16-pt2.

   The **R16 audio post lessons pt2.zip** file is 4.48 GB in size and should take roughly 12 minutes to download to your computer using a standard broadband connection.
3. Download the third part of the lesson files by navigating to www.blackmagicdesign.com/dvres/audio-post-resolve16-pt3.

   The **R16 audio post additional media.zip** file is approximately 685.9 MB in size and should take roughly five minutes to download to your computer using a standard broadband connection.
4. After downloading all three of the zip files to your computer, open your Downloads folder, and double-click the three zip files to unzip them (if your computer doesn't unzip the file automatically).
5. In the Documents folder, create a new folder named **R16 audio post lessons**.

6 Move the unzipped **R16 audio post lessons pt1**, **R16 audio post lessons pt2**, and **R16 audio post additional media** folders to your Documents > "R16 audio post lessons" folder.

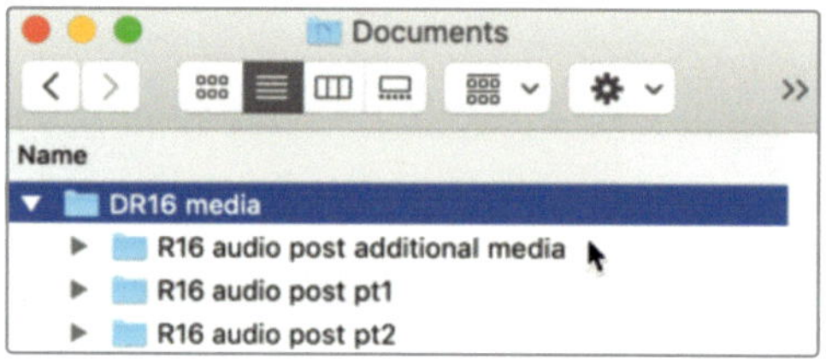

Your media lessons and media are ready to go.

Creating a New Disk Database for the Sound Library

Before you start Lesson 1, you'll need to create a new disk database for the "Sound FX Sampler for Fairlight" folder that was included with the additional media for this book. Although you can create a disk database at any time, if you create a database for your sound effects now, you'll be able to stay on task in Lesson 1 and simply add the library sound effects to your database

You can use either disk or PostgreSQL databases for your sound libraries. For this book, you'll create and use a Disk database.

To create a new Disk database:

1 Click the button at the upper-left of the Projects window to open the Databases sidebar.

2 At the bottom of the sidebar, click the New Database button.

3 When the New Database window appears, click the Create tab.

4 Leave Type set to Disk.

5 In the remaining fields, do the following:

- In the Name field, type **soundfx**.
- Click in the Location field, and use the file system navigation dialog box to locate the Documents folder.
- In the Documents folder, click New Folder. In the New Folder dialog box, type **Sound FX**. Click Create.

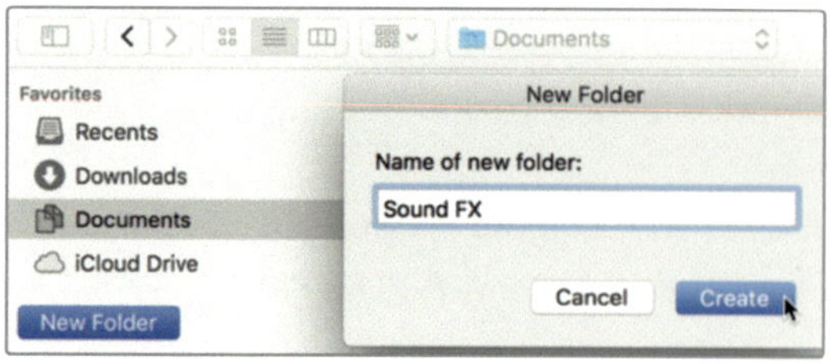

6 Select the new Sound FX folder, if necessary, and click Open to create the new database in the Sound FX folder.

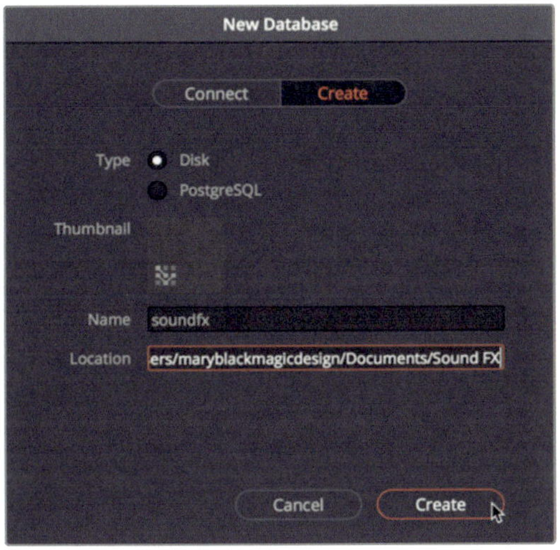

7 In the New Database dialog box, click Create.

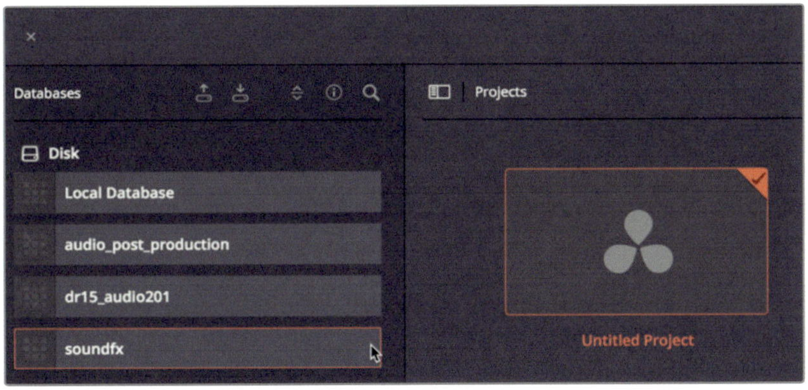

The new disk database appears in the Disk database section of the Databases sidebar. The new soundfx database that you created will be used for indexing sound effects in Lesson 1.

8 In the Project Manager, open the local disk database that you wish to use for the lesson projects.

You are now ready to begin Lesson 1, "Building a Soundtrack."

Acknowledgments

We would like to thank the following individuals for their contributions of media used throughout the book:

- Steven Esparza for the *Girl on Wave* trailer. *Girl on Wave* is a feature length documentary film produced and directed by Steven Esparza, and written by Steven Esparza and Sarah Hauser. Property of ECHOHOUSE films, www.echohousefilms.com.
- Nguyen-Ahn Nguyen for *Hyperlight*, a short film produced and directed by Nguyen-Ahn Nguyen. Written by Nicholas Billon, based on an original idea by Nguyen-Ahn Nguyen and Simran Dewan. Property of Nguyen-Ahn Nguyen.

Lesson 1

Building a Soundtrack

Audio post-production is more than simply adjusting volume levels and mixing tracks. In fact, as you embark on your journey through this book and your own projects keep your eye on the big picture. That is, the visual and audio components of a project are of equal importance, and it takes only a few seconds of audio—good or bad—for someone to form an opinion for your entire production. And here's the kicker: The audience doesn't care how you got there; they just want the experience to be awesome. So, it is up to you as the audio editor, sound designer, supervising sound editor, re-recording mixer, or all of the above to deliver a professional quality soundtrack regardless of your budget, format, length, deliverables, or distribution methods.

Time

This lesson takes approximately 65 minutes to complete.

Goals

Luckily, the Fairlight page in DaVinci Resolve 16 includes all the tools you'll need to transform your basic production sound into a fully realized soundtrack from start to finish.

In this lesson, you'll step into the role of the audio editor at a busy trailer post-production house where you use the Fairlight page to finish building an exciting one-minute trailer soundtrack for the documentary film *Girl on Wave*, directed by Steven Esparza of EchoHouse Films. Along the way, you will explore the interface, build tracks, edit and trim audio clips, and balance the volume levels while taking ongoing directions from the "clients" to follow their creative lead as inspiration strikes. Let's get started.

NOTE This project assumes you have downloaded the media for this book. If you are unsure, return to the "Getting Started" section of this book and follow the download instructions in "Acquiring the Lesson Files."

Opening and Playing a Project

In this scenario, the original audio editor has called in sick, and it's up to you to finish it within the hour for the clients who are down the hall grabbing a double latte. This lesson will introduce you to the primary interface elements and tools you need as you go.

Let's begin by opening the project with most of the trailer soundtrack already created so you can get to know the Fairlight audio timeline and playback controls. Once you are comfortable with navigation and playback, you'll use additional interface panels and tools to finish the project.

1 In the Project Manager, right-click and choose Restore Project Archive. Select R16 audio post lessons > R16 audio post pt1 > lesson 01 build soundtrack.dra, and click Open.

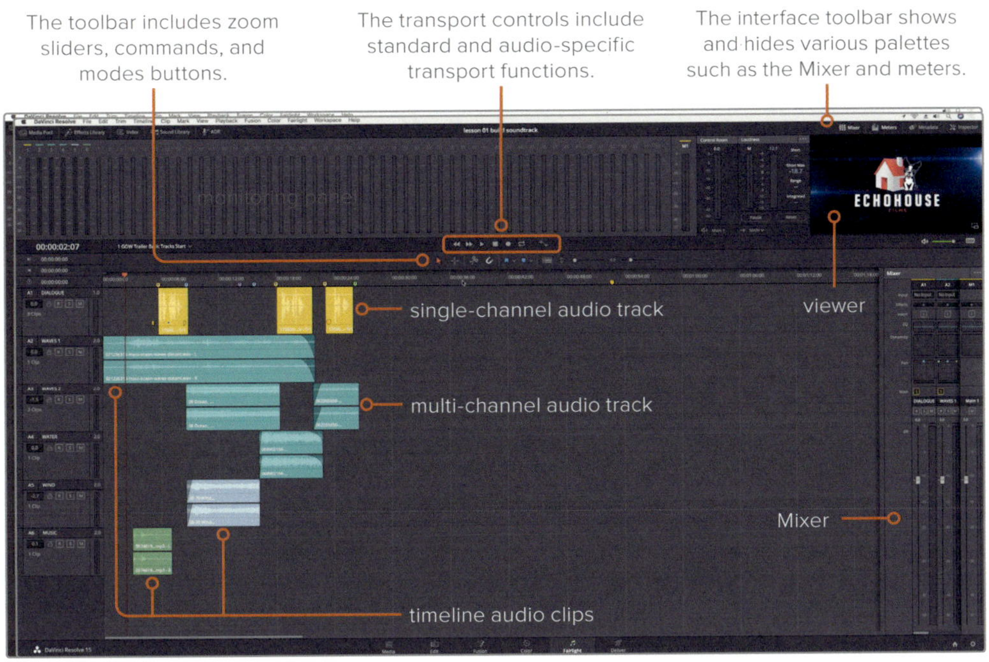

The default Fairlight page layout includes the timeline, monitoring panel, and mixer. The timeline is optimized for audio mixing and uses single channel or multichannel tracks.

The project opens in the Fairlight page with the audio timeline, monitoring panel, and Mixer visible. The 1 GOW Trailer Basic Tracks Start timeline includes six colorful audio tracks. The track colors were added for this lesson to help organize each track's contents ranging from dialogue to sound effects and music.

NOTE If your project does not open to the Fairlight page, or additional panels are visible, feel free to click the Fairlight button at the bottom of the screen to go to that page. To reset your user interface, at the top of the page, choose Workspace > Reset UI Layout.

2 In the upper-left corner of the timeline window, click the timelines pop-up menu to see all the project's timelines. In the list of timelines, choose 4 GOW Trailer Mix.

This timeline has only one stereo track containing the finished mix of the trailer. The corresponding frame of video is displayed in the viewer in the upper-right corner of the interface.

To move the playhead to a specific timecode, you can use the timeline ruler and corresponding timecode display as a guide.

> **TIP** You can right-click the timecode display to choose a different timecode format including SubFrame, Samples, and film Feet and Frames.

3 Press the Home key, or click the Timecode ruler at the beginning of the timeline, to move the playhead to that position (00:00:00:00).

The transport controls above the timeline toolbar are the standard buttons that you'll find on any professional audio recording and playback device.

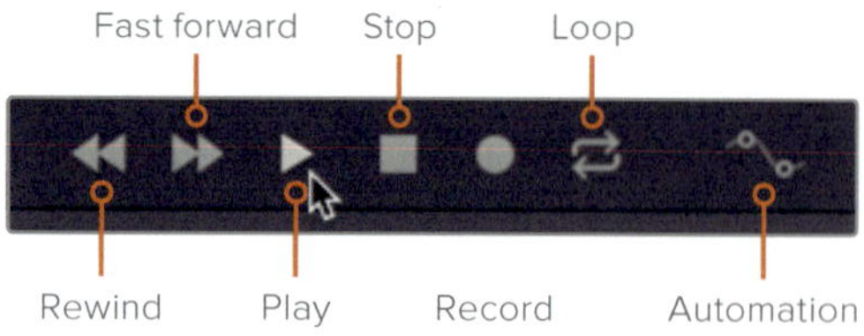

4 In the transport controls, click the Play button, or press the Spacebar, to start playback. During playback, watch the meters and video playback in the monitoring panel at the top of the window and the meters in the Mixer to the right side of the screen.

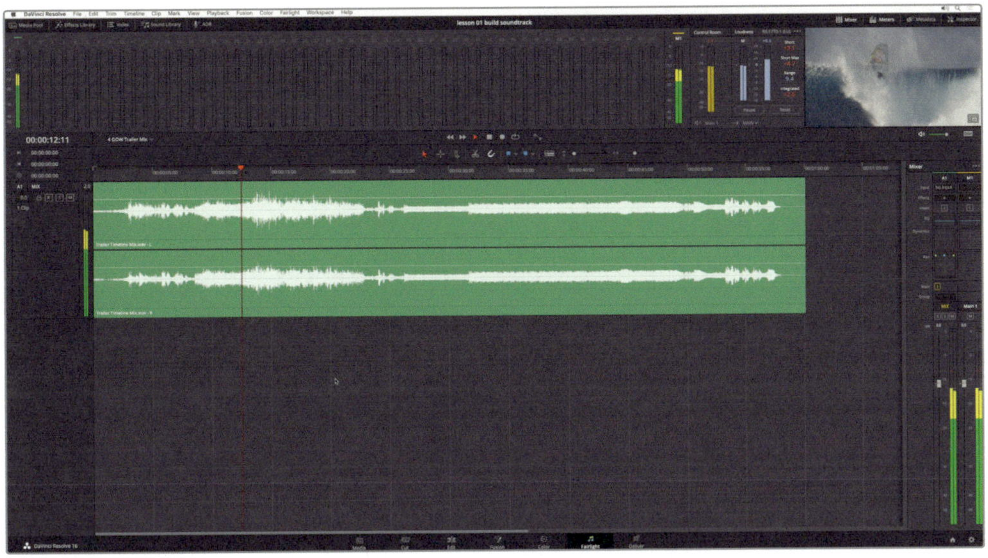

As the name suggests, the monitoring panel displays all the audio and video contents of the current timeline. The Mixer shows a large channel strip for each track, as well as the main output.

If you are gripping the edge of your desk with white knuckles and thinking that there are too many meters and buttons to keep track of, you can relax. The Fairlight page is as user-friendly as it is powerful. At any time, you can simplify the interface to show just those tools you need as you need them.

5 Press the Spacebar to stop playback.

6 In the Interface toolbar at the upper-right of the screen, click the Mixer button to hide the Mixer.

You'll work with the Mixer and monitoring panels extensively in later lessons. For now, let's simplify the interface by expanding the viewer as a separate window and hiding the meters.

TIP Pressing the Home and End keys will move you to the start or end of a timeline, respectively. Some Mac keyboards do not have Home and End keys. When using one of those keyboards, press Fn-Left Arrow to move to the start of the timeline and Fn-Right Arrow to move to the end.

Monitoring Video Only

In a traditional audio-post workflow, an audio editor will need a separate reference video to use as a guide when building the soundtrack. In Resolve, you do not need a reference video because the Fairlight page includes a handy viewer that shows the timeline's video. As you have already discovered, the viewer is located to the far-right side of the monitoring panel. You can easily expand the viewer as a separate floating window.

1 In the lower-right corner of the viewer, click the Expand button to turn the viewer into a floating window.

The viewer appears in the middle of the interface. Now, you can drag the top of the viewer to move it, and drag its edges or corners to resize it.

2 Click the Meters button to hide the monitoring panel.

3 Drag the viewer down to the bottom of the empty tracks area in the timeline. Then, drag the upper-right corner of the viewer to resize it to about twice its original size. When you are finished, the viewer should fit nicely in the lower half of the timeline.

To return the viewer to the monitoring panel, you can click the Dock button in the upper-right corner of the viewer, or press Return. For now, let's leave the viewer open as a separate window.

NOTE Feel free to resize or move the viewer at any time as you work through these exercises.

If you have a Blackmagic DeckLink card or UltraStudio, you can also view the video on a second video display.

Evaluating Tracks during Playback

Actively listening to and evaluating tracks during playback is a necessary step toward building powerful soundtracks. It involves listening to tracks played together, independently, and in various combinations to determine if they are working as a whole.

Normally, you would not be able to fast-forward to the end of a project and listen to its mix ahead of time. However, because this is a training book, and you happen to have the mix already in your timeline, this is a wonderful opportunity to see and hear what you are striving to accomplish before you build the tracks in this lesson. In this exercise, you'll listen to the finished mix once, and then open the starting version of the timeline and evaluate the tracks.

1 Press the Home key to move the playhead to the beginning of the timeline. Then play the timeline once to see and hear the finished trailer.

   Now that you have heard the mix, let's go back to the starting timeline.

2 In the timeline pop-up menu, choose 1 GOW Trailer Basic Tracks Start.

   To evaluate the tracks, you can use the controls in the track headers. At the left of each track, the header shows each track's name, number, and control buttons.

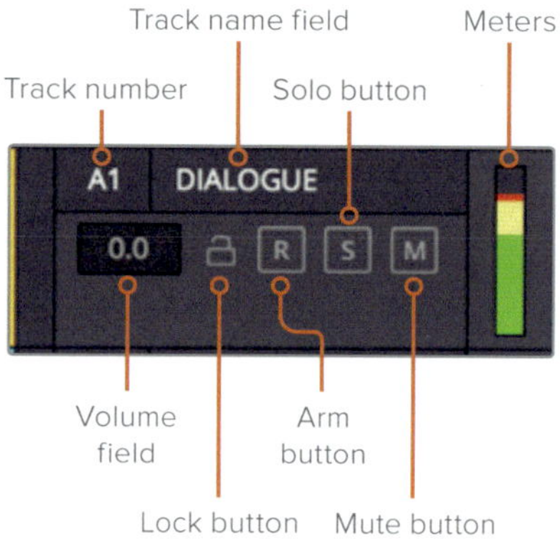

3 Move the playhead to the beginning of the timeline.

4 Click any empty space on the A1 DIALOGUE track header to select that track.

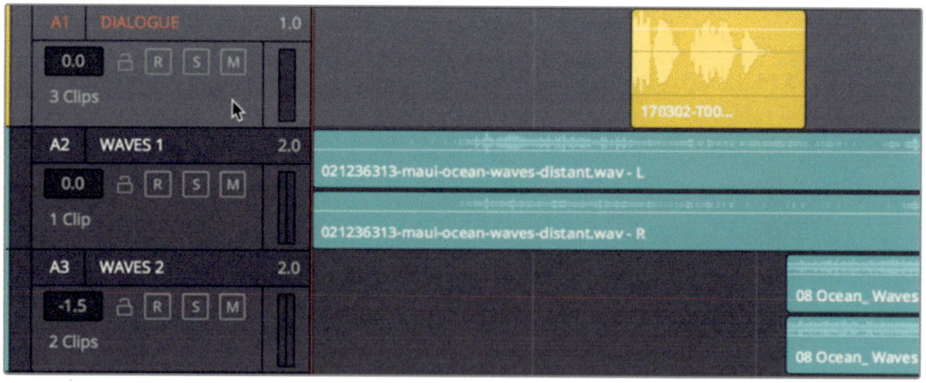

Notice that when a track header is selected, the name and number turn red, and the track brightens.

5 Click any empty space in the timeline to deselect the A1 track.

> **NOTE** Track selection is important when you are recording or editing using shortcuts. You can evaluate tracks without selecting them; however, it is always a good idea to be aware of which tracks or clips are selected as you work.

Now let's try using the Solo and Mute buttons on several tracks. Solo isolates a track by temporarily muting all other tracks. Mute silences a track until it is unmuted. You can click these buttons to toggle each control on and off during playback.

6 Solo the A2 WAVES 1 track, and then mute the A5 WIND track.

7 Start playback from the beginning of the timeline. At around 00:00:12:00, unsolo the A2 track. Then, solo the A1 DIALOGUE track and listen to the last two dialogue clips. When you are finished, stop playback.

As you can hear, the Mute and Solo buttons are an easy way to selectively listen to specific tracks during playback. You can also swipe them on or off to quickly solo multiple tracks at once. Let's leave the A5 track muted and try swiping the Solo buttons on the other sound effects tracks.

Your goal is to evaluate the sound effects tracks to see how well they enhance the impact of the visuals. First, you'll listen only to the waves and water sound effects (FX) on tracks A2, A3, and A4. Then, you'll play the timeline again while listening to all of the tracks and evaluate the A5 WIND track with the other tracks.

8 Unsolo the A1 track. Start playback from the beginning of the timeline. Starting from the Solo button on the A2 track, drag downward across the Solo buttons on the A3 and A4 tracks. When you are finished playing the clips, stop playback.

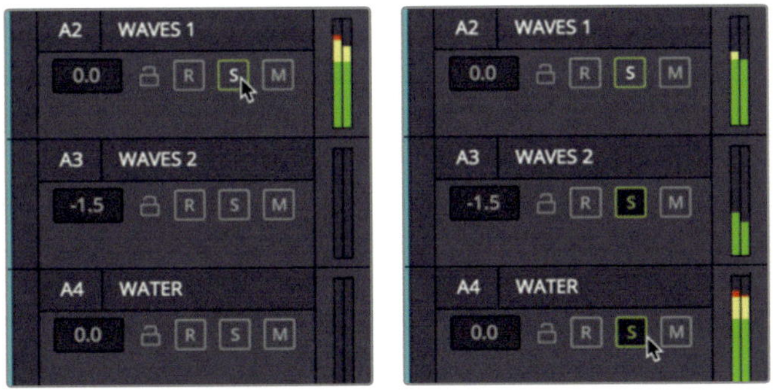

9 Swipe off the Solo buttons on the A2, A3, and A4 tracks.

10 Mute the A1 track.

Now it's time to evaluate the section of the timeline that includes the wind effect. Instead of starting playback at the beginning of the timeline, let's start closer to the action.

11 Click the timeline ruler between the yellow and teal markers (at around 00:00:07:10).

12 Start playback, and click the Mute button on the A5 WIND track to toggle the track off and on several times to hear the soundtrack without and with the wind sound. When the wind clip is finished and the viewer shows underwater footage, stop playback.

13 Unmute the A1 and A5 tracks.

14 To replay the exact same section of the timeline, choose Playback > Play Again, or press Option-L (macOS), Alt-L (Windows).

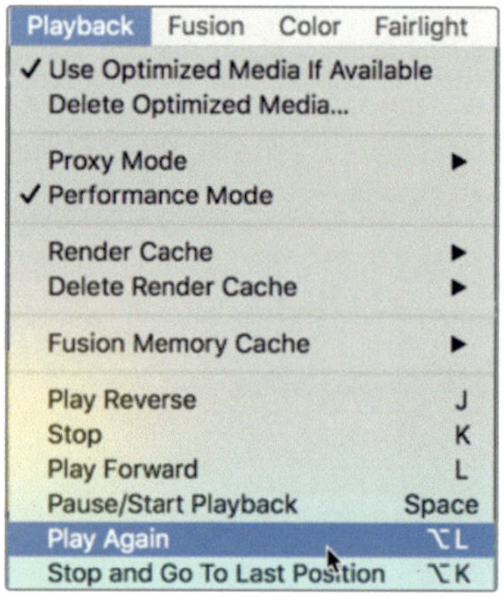

15 Stop playback when the wind clip is finished.

By now, you should have formed an opinion about the effectiveness of the various tracks and the wind effect, specifically. For the purposes of this lesson, you will keep it. Here is another truth when you are working with motion picture sound: reel does not equal real. In other words, sound effects used in soundtracks are much larger than real life.

Take this trailer as an example. Not only was a wind effect present in the actual trailer's soundtrack, but the sound of a tornado was included to really intensify the sense of danger. As you work through these lessons, you'll learn that sound is cumulative, and a little bit goes a long way. Even so, sometimes during key dramatic moments, more is better, and on occasion too much is best.

Sound designers routinely mix all kinds of sounds to create more dramatic effects such as adding a lion's roar to the Death Star explosion in *Star Wars IV: A New Hope*, or in this case, violent winds to intensify the sound of strong waves.

By the way, the clients returned with their double lattes while you were evaluating the tracks, and they are anxious to see how the second half of the trailer works with the music. What music? The exciting trailer music is conveniently located in the Media Pool with the other audio files.

Next, you'll explore the Media Pool to preview, mark, and edit additional audio clips into the timeline—starting with the music.

Previewing Audio Clips in the Media Pool

The Media Pool contains all the media and timelines in your current project. In this exercise, you'll select and preview several audio clips and add them to the timeline. Whenever you load a clip into the preview player, you can preview the entire clip from its first available frame to the last.

1 In the upper-left corner of the Interface toolbar, click the Media Pool button to open the Media Pool.

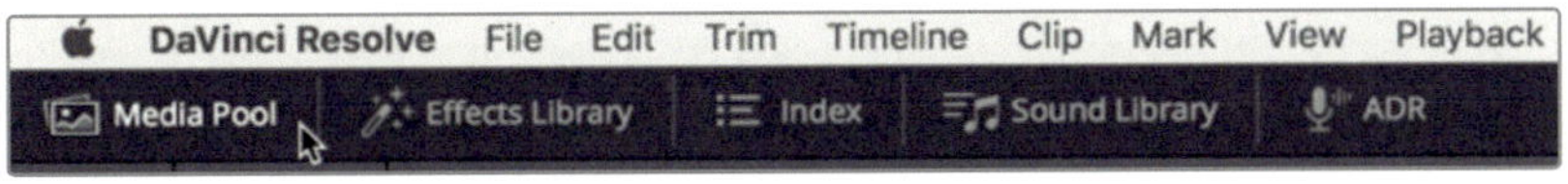

The Media Pool opens on the left side of the Fairlight page. For the simplicity of this first lesson, the media for this project was organized into three bins: Video Clips, Audio Clips, and Timelines.

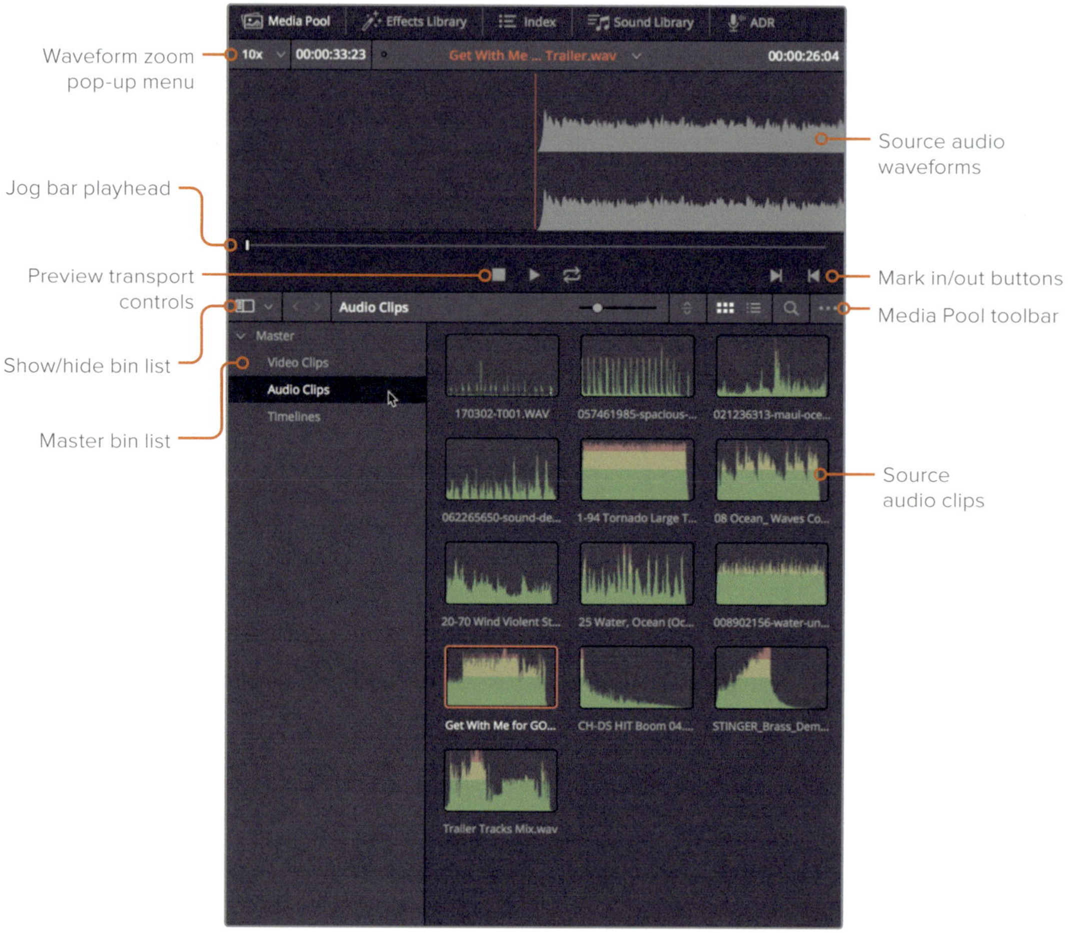

2 In the Media Pool bin list, click the Audio Clips bin to view its contents.

You can view source clips in either list or icon view. The default is icon view. If you look carefully, you'll see that each audio clip's icon shows a graphical representation of the waveform.

3 Select the **Get With Me for GOW Trailer** clip to load it into the preview player at the top of the Media Pool.

4 Press the Home key to move the playhead to the beginning of the clip in the preview player. Then, click the Play button, or press the Spacebar, to preview the clip. When you are finished, click stop, or press the Spacebar.

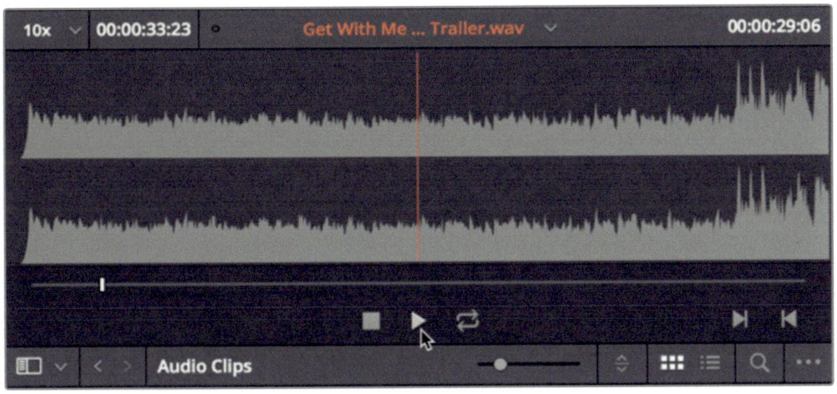

To see more information about your media files, let's change to list view.

5 In the Media Pool toolbar, click the List View button.

List view is handy for reading clip names and sorting clips by a specific column. In this case, let's use the list view to find and preview another type of music clip called a *stinger*. Stingers are single notes or chords–often strings–that are inserted to add tension and suspense.

6 Select and preview the clip **STINGER**. While you listen to the clip, try to imagine where it might go in the soundtrack.

So, what did you think of the stinger? The clients love it! Out of context, the stinger may seem like overkill, but once it is placed in the timeline and the levels are balanced with the other tracks, you'll understand the power of the stinger. Any idea where to place it? Stay tuned to find out.

You'll add these clips previewed to the timeline in a few minutes. First, let's look at the Markers index and the timeline markers that you'll use as a guide for adding the clips.

Working with Markers

Markers are often used in audio post-production as a guide to setting music cues, marking sound effects, or identifying elements such as dialogue problems that need to be addressed. In this exercise, you'll use existing markers to aid navigation in the Timeline and Markers index. When you become comfortable with markers, you can use them as a guide when adding clips to the timeline in the next section. Let's start by looking at the Markers Index.

1 In the upper-left corner of the Interface toolbar, click the Index button to open the Index panel. Then, at the top of the Index panel, click the Markers button to show the Markers index.

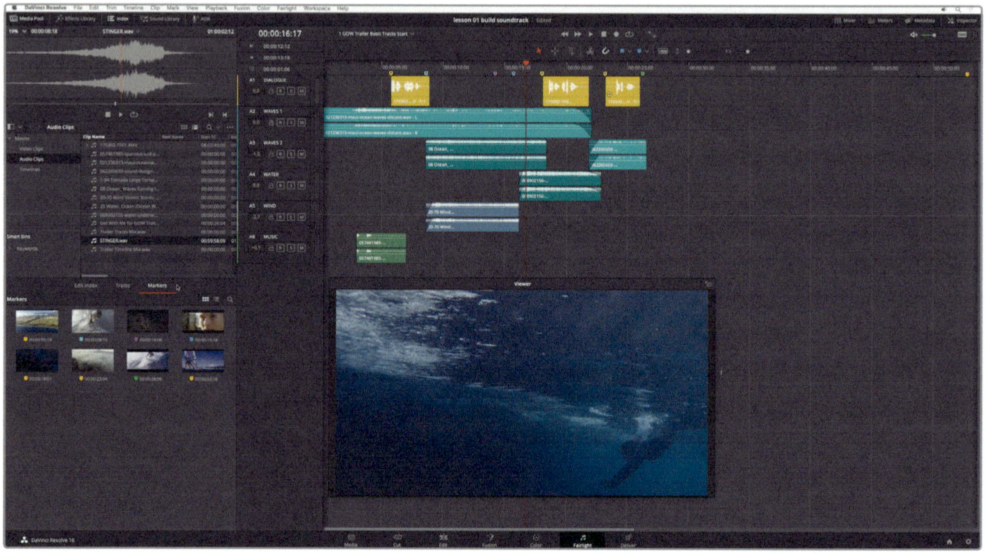

The Markers index opens in the lower-left corner of the page below the Media Pool. Here you can see the current timeline markers and a corresponding thumbnail frame for each marker. Just as with the Media Pool, you can view the markers in either icon or list view.

2 In the Markers index toolbar, click the List View button to change the index to list view.

In list view, you can see columns that show each marker's number, name, and other information. You can use the maker thumbnail icons to move the playhead to the selected marker. Let's try it.

3 Double-click the thumbnail icon for Marker 7 to move the playhead to that frame.

Here is where you will add the first music clip. Let's take a moment to streamline the information in the Markers index columns.

4 Right-click the column headers, and deselect all the checkboxes except #, Frame, Name, Color, and Notes.

Now you can clearly see the Name, Color, and Notes for each marker in the list.

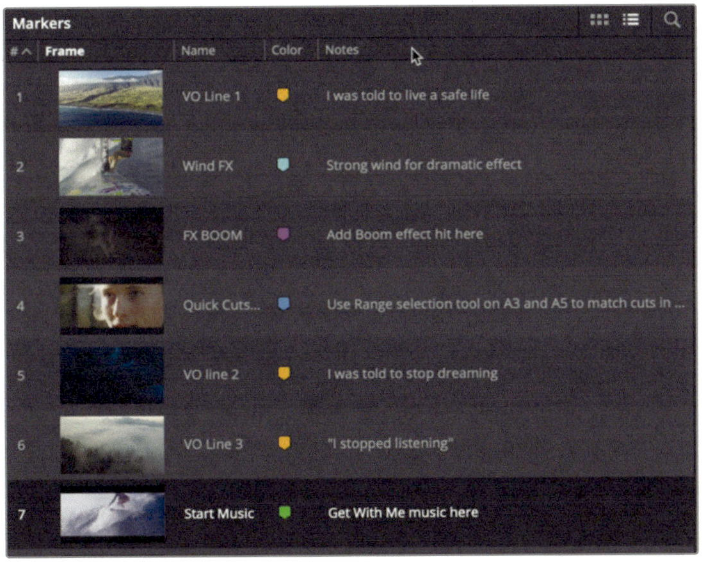

NOTE You can sort the marker list by any column header.

To navigate between markers in the timeline, you can hold down the Shift key and press the Up Arrow or Down Arrow keys. Pressing Shift-Down Arrow moves down the timeline to the next marker. Pressing Shift-Up Arrow moves the playhead to the previous marker.

5 Click anywhere in the timeline to make it active. Then press Shift-Down Arrow to move to the next marker.

6 In the timeline, double-click the yellow marker beneath the playhead to open the Markers panel, and enter the information as follows:

- In the Name field, type **VO line 4**.
- In the Notes field, type **My name is Sarah Hauser, and this is my story.**

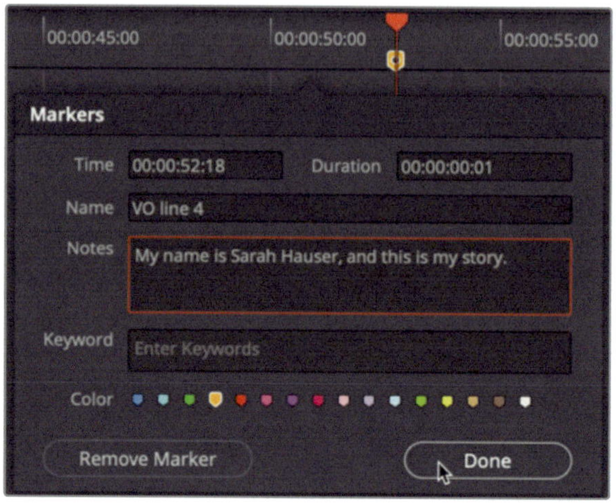

7 Click Done.

The Markers index updates to show the new marker 8 information.

8 In the timeline, press Shift-Up Arrow to move the playhead to the previous marker, or in the index, double-click marker 7.

In the next exercise, you'll add a music clip to the timeline using this marker as a guide.

Adding Audio Clips to the Timeline

Once you find a clip that you want to work with, and you know where you want to place it in the timeline, you need only drag it from the Media Pool to the timeline. You can add clips to an existing track or create a new track by dragging a clip to the empty space below the current tracks. In this exercise, you'll add the **Get With Me...** clip to the MUSIC track, and create a new track for the stinger.

The preview player includes a handy pop-up menu that shows the ten most recently accessed clips.

1 At the top of the Media Pool, click the pop-up menu arrow, and select **Get With Me for GOW Trailer.wav.**

2 Drag the clip from the waveforms at the top of the preview player to the playhead position in the A6 MUSIC track.

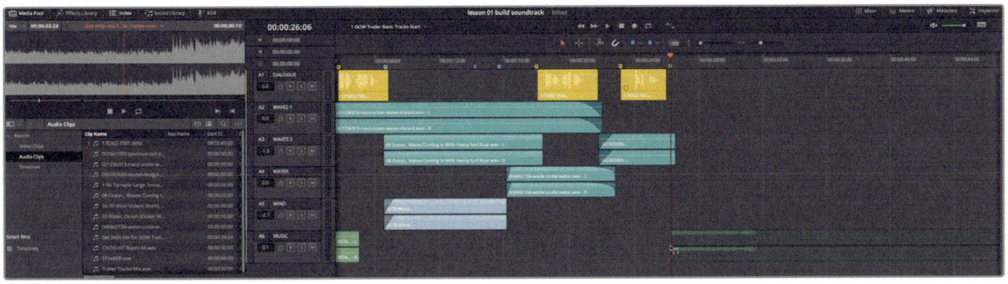

The clip appears in the track once you release it.

3 Resize and move the viewer to the lower-right corner of the timeline to make room for the next music clip.

It's time to add the stinger to a new track. You could place it in the A6 track with the other music clips, but because there's no limit to the number of tracks you can use in your soundtrack, let's create a new one just for the stinger.

As for placing it, your clients suggest that you place it right before she falls, or right after she falls, or perhaps under the close-up shot of her face while she reflects on the fall. In light of the clients' indecision, let's try starting it with the dramatic wind FX.

Move the playhead to the teal-colored Wind FX marker (00:00:08:15).

This time you'll drag the stinger clip from the Media Pool list to the empty space below the timeline tracks. By doing so, a new track is created for the source audio clip. When you drag the clip to the empty space in the timeline, you will see a translucent image of the new clip in the new track before you release the clip.

4 Drag the **STINGER** clip from the Media Pool list to the playhead position (the Wind FX marker) below the A6 track.

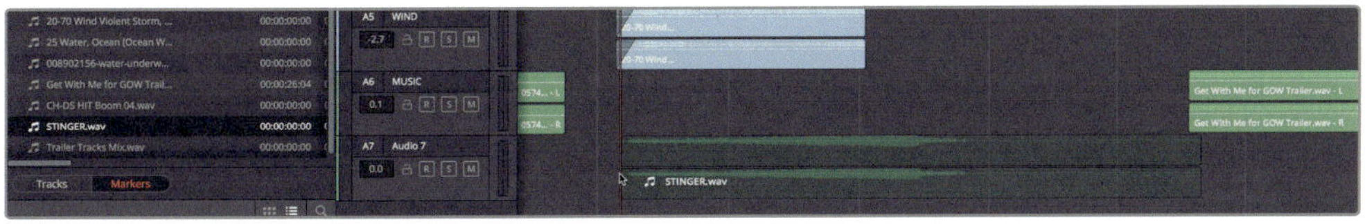

A new A7 track appears along with the stinger clip.

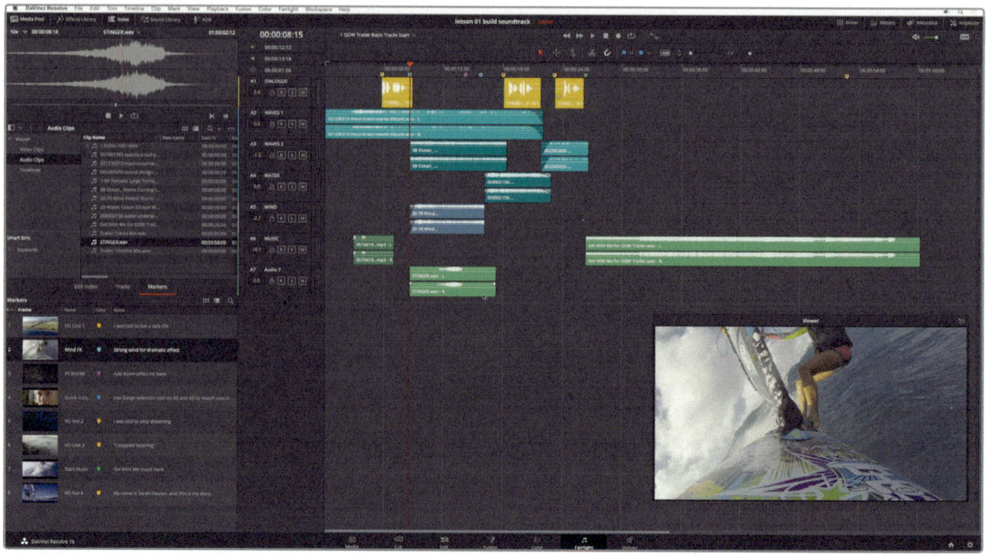

5 Play the timeline from the beginning and listen to the new music clips in the soundtrack.

The soundtrack is coming along nicely. As you just heard, the right music choices make a big difference. In this case, the stinger adds a dramatic element to the action sequence and the theme song carries the entire second half of the trailer without the presence of any other sound effects.

Naming and Moving Tracks

Descriptively naming and organizing your tracks can save you time later in your mixing process. In this exercise, you'll name the A7 track and move it above the other music track.

1 Double-click the A7 track name field, which is currently named Audio 7.

2 Double-click the current track name within name field to select it. Type **STING**, and press Return to rename it.

To move a track up or down, you can use the Right-click shortcut menu in the track header.

3 Right-click the A7 track header, and choose Move Track Up.

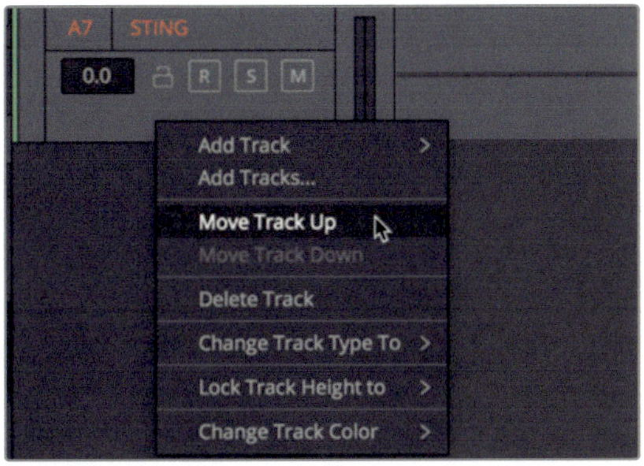

The STING track moves above the MUSIC track while the track numbers remain in order.

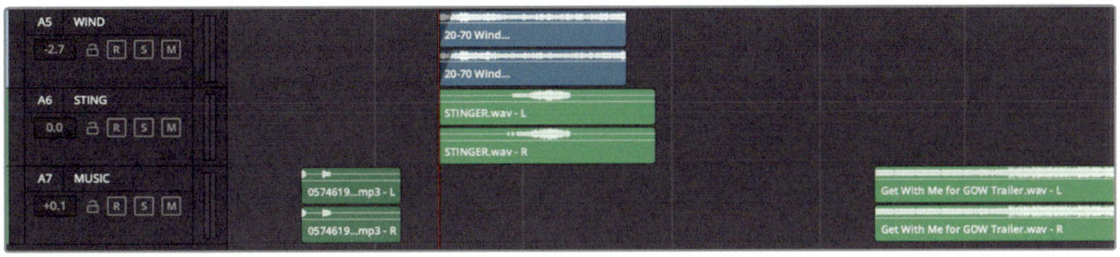

Next, you'll apply all of the skills you've learned so far to preview and edit the last voiceover clip.

Marking and Editing Select Portions of a Clip

So far you have previewed and added whole audio clips. However, in audio post-production, you'll often work with long clips comprising multiple takes. When you are working with a clip that includes more content than you need, you can place markers on the clip to label specific sections. You can also mark a useful clip range by placing in and out points. In this exercise, you'll use both methods to mark the clip.

Production audio files often have complicated numeric names based on the recording metadata. Although you could rename the clips, or use metadata as a reference as to their content, it's a good idea to keep the original names when the clips are edited into the timeline. As luck would have it, the voiceover file for this timeline has a numeric name. Instead of hunting for the clip you need, let's use one of the voiceover clips already in the timeline to find the original clip in the Media Pool.

1 In the A1 DIALOGUE track, right-click one of the yellow clips, and choose Find in Media Pool.

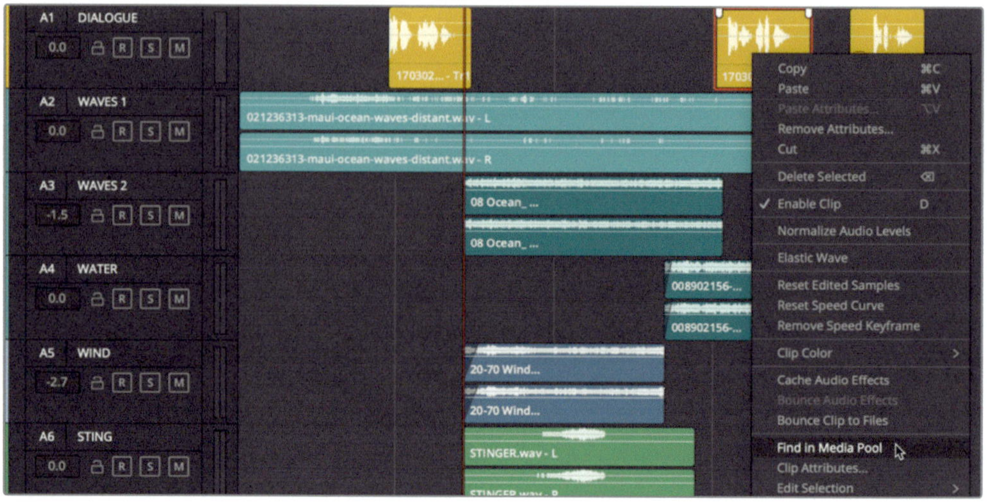

The source clip is selected in the Media Pool and source viewer. Notice the yellow markers in the preview player jog bar. You can navigate through clip markers in the preview player just as you can with timeline markers. They also appear in the Clip Name list.

2 Click the arrow at the left side of the clip's icon to show the markers in the Media Pool list.

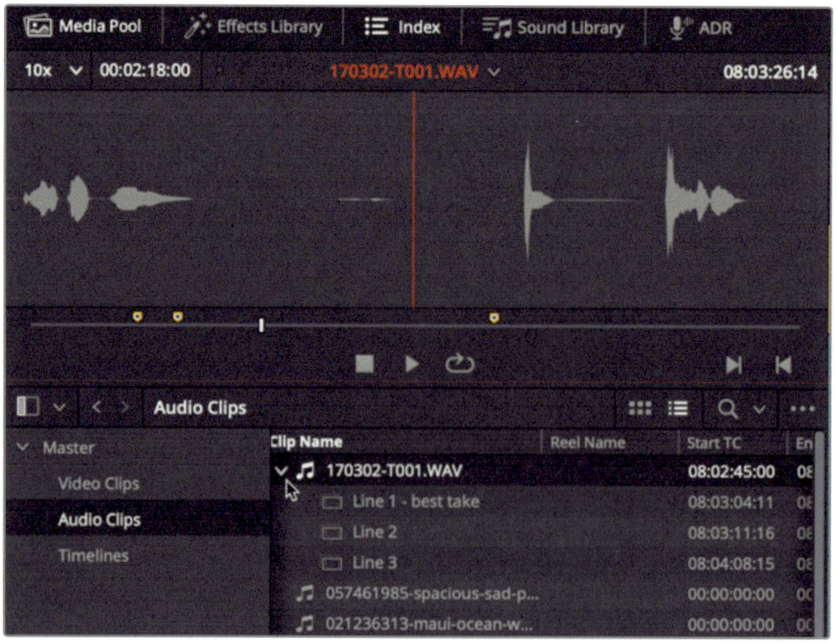

3 In the list, click the marker named Line 3 to move the preview player playhead to that marker.

> **NOTE** You can press Shift-Up Arrow or Shift-Down Arrow to move from marker to marker in the preview player. Also, when the playhead moves to a marker, the marker information appears in the preview player.

Your goal is to choose the best take for Line 4 and mark it. Let's change the zoom level in the preview player to see more of the clip's audio, and then use the jog bar to find the last line of voiceover.

4 In the upper-left corner of the source viewer, in the zoom menu, choose 1x.

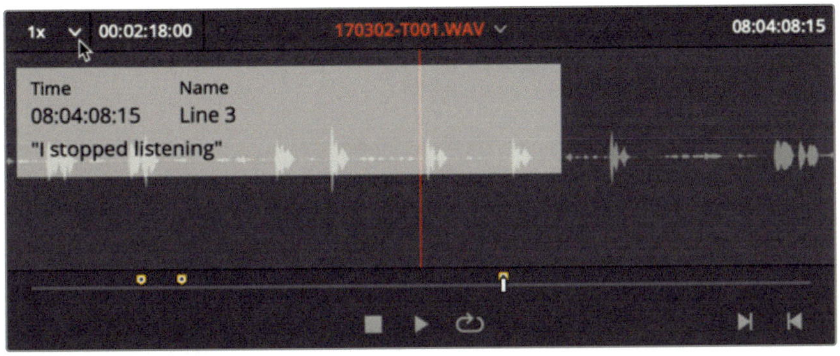

At this zoom level, you can clearly see the multiple voiceover phrases over time. Each phrase has a distinctive appearance in the waveform.

5 Press End to move the playhead to the last frame of the source clip.

Chances are that the last sets of peaks represent the three takes of the last voiceover line.

6 Drag the white jog bar to the left until the source playhead is placed before the first set of peaks. (It's around 08:04:35:00 in the source timecode field in the upper-right corner of the source viewer.)

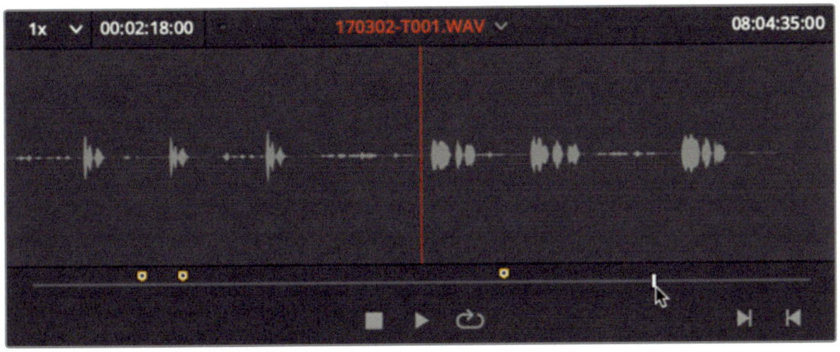

7 Press M to set a marker on the source audio clip. Press M again to open the Markers dialog. Then, in the Markers dialog, type **Line 4**. In the Notes field, type **"My name is Sarah Hauser and this is my story."** Finally, change the marker color to yellow. When you are finished, click Done.

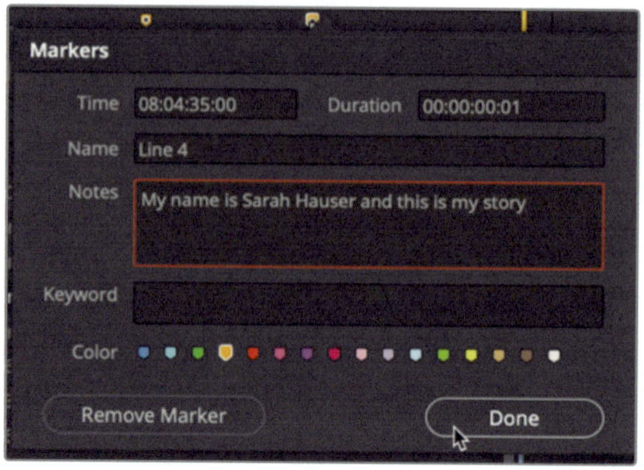

Now that you've set a marker for the last section of voiceover, you can preview the audio. To simplify this process, let's use a new set of shortcuts to quickly scrub through the audio without using your mouse.

Scrubbing with the JKL Keys

Using the JKL keys allows you to quickly scrub through clips in both the timeline and preview player. Pressing the L key plays forward, pressing the J key plays backward, and pressing the K key stops playback. Tapping the L or J keys multiple times will speed up playback. Holding down L or J will play one frame at a time and stop playing when you release the key. Holding down K while tapping L or J will play forward or backward one frame at a time, whereas pressing K together with L or J will play in slow motion. Let's give these keys and combinations a try as you preview the Line 4 voiceover takes.

1 Press L to play forward through the first take of Line 4.

2 Press K when you hear the voice of director Stephen Esparaza giving Sarah feedback after her first take.

3 Press the JKL keys to preview the last two takes. When you are finished, move the playhead to the beginning of your favorite take. If you aren't sure which one to choose, use take 2 at 08:04:43:00.

Marking In and Out Points

In the lower-right corner of the preview player, you'll see a set of buttons to mark in and out points. The keyboard shortcuts for placing these marks are I and O.

1 Click the Mark In button, or press I, to mark the in point at the beginning of the take.

2 Hold down L to scrub through the take. When you are finished, release the L key.

3 Click the Mark Out button, or press O, to mark an out point at the end of the take.

   With the in and out points marked, you're ready to edit that section of the clip to the timeline.

4 In the timeline, click the timeline ruler above the last yellow marker to move the playhead to that frame.

5 In the preview player, drag the marked clip's waveform from the viewer to the playhead position in the A1 DIALOGUE track.

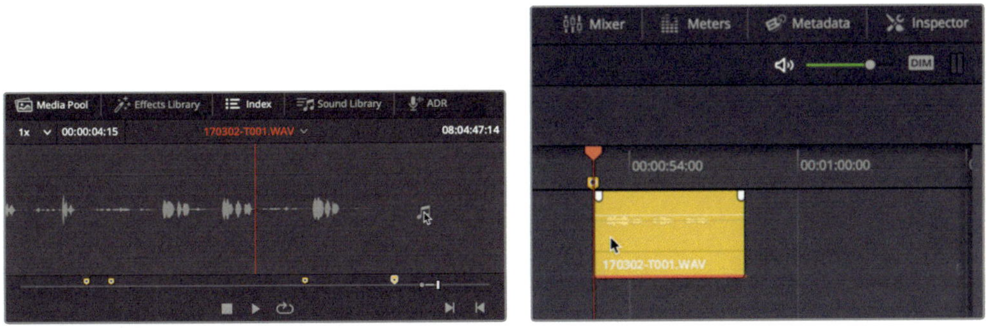

6 Press the JKL keys to play the clip in the timeline.

You may have immediately noticed that the music is so loud you can't hear the last line of voiceover. No worries, you can easily fix that in the Inspector.

Changing Clip Volume Levels in the Inspector

The Inspector gives you easy access to controls for any selected clip. While you will be using the Inspector in many ways throughout this book, let's currently focus on the clip volume controls. Also, this is a good time to streamline the interface and close any panels or windows that you don't need at present.

1 To the left of the Interface toolbar, click the Media Pool and Index buttons to hide those panels.

2 To the right of the Interface toolbar, click the Inspector button to show the Inspector.

> **NOTE** From this point, you'll simply be instructed to show or hide an interface panel during exercises.

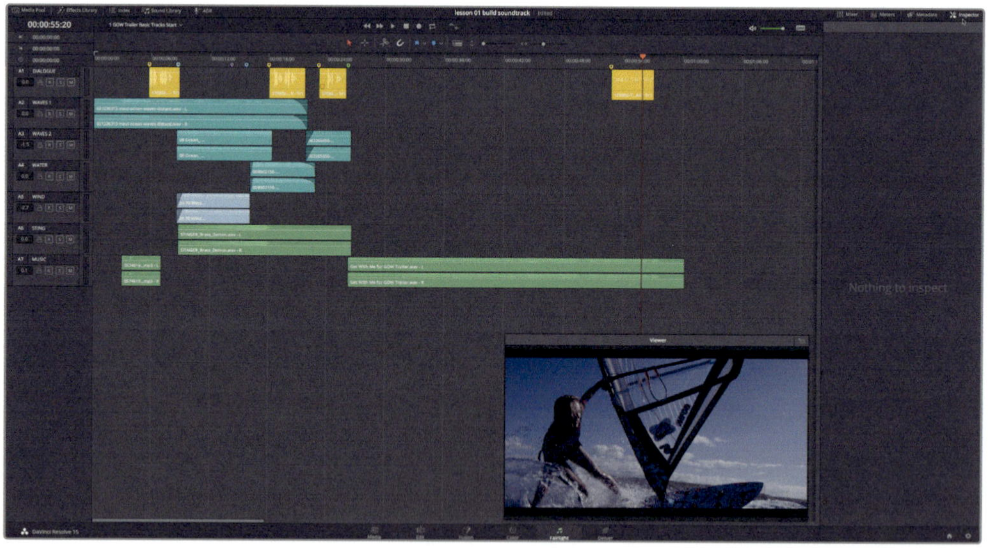

The Inspector indicates that it has nothing to inspect.

3 In the A1 track, select the fourth yellow clip to see the clip's controls in the Inspector.

The Inspector includes controls for Clip Volume, Clip Pan, Clip Pitch, and a four-band Clip Equalizer. In this exercise, you'll use the Clip Volume controls. You can see that the current volume level for the selected clip is 0.00. This does not mean that the clip volume is actually at 0 decibels (dB). Instead, the Clip Volume field indicates the amount of volume change being applied to the original source level. In this case, it is no change at all.

4 In the A1 track, select the second yellow clip to see that clip's information in the Inspector.

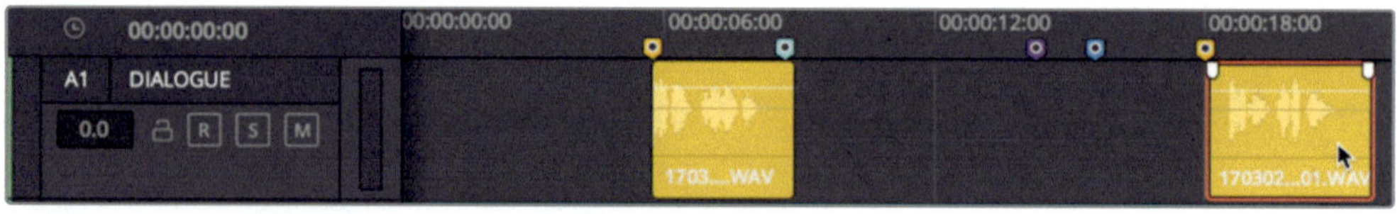

According to the Clip Volume field, this clip's volume level is raised by 10.00 dB.

5 Select the first clip in the A1 track to see that clip's level (11.20 dB).

You can change the Clip Volume in the Inspector by dragging the Volume slider, typing a new number in the Volume field, or dragging left or right on the field.

6 Select the fourth clip in the A1 track, and in the Inspector, raise the Clip Volume level between 10.00 and 12.00. You'll finesse this level later.

7 Play the fourth clip in the timeline and listen to the new volume level with the music.

The music is still dominant in this section. So then, which is the priority: the music or the voiceover? Before you decide, let's learn how to hide all but the dialogue and music tracks, and use the zoom sliders to adjust the size of the remaining tracks, as needed.

Showing, Hiding, and Zooming In and Out of Tracks

The more tracks in your timeline, the more control you'll need over their heights and visibility. To hide tracks in the timeline, you'll use the Tracks index, which shares a panel with the Marker index.

You'll now work within the tracks index to hide most of the tracks while you focus on just the dialogue and music tracks. Because you don't need the viewer for the next few exercises, let's hide It as well.

1 In the upper-right corner of the viewer, click the Return button to return the viewer to the monitoring panel.

> **NOTE** In future steps, you will simply be instructed to show or hide the viewer.

2 Show the index, and then click the Tracks tab to show the Tracklist.

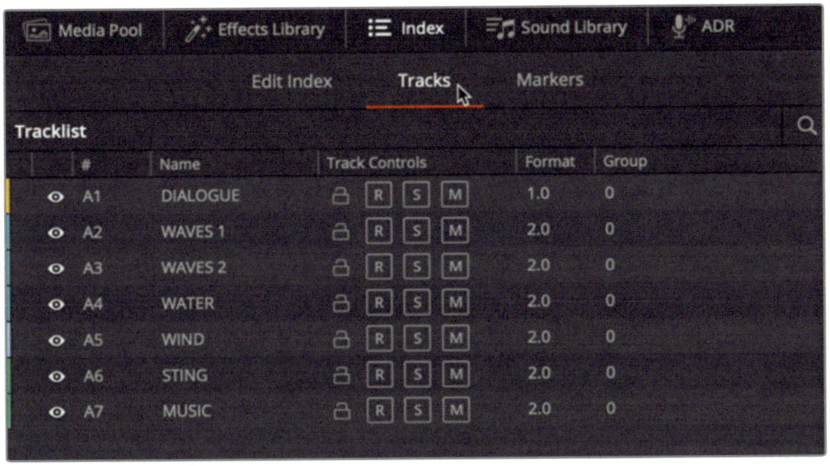

Here you can see a row for each track. To the far left of the Tracklist, you can see visibility controls represented by eye icons. You can click these controls off and on to hide or show a track. Or, you can click and swipe multiple contiguous tracks off and on at once.

> **NOTE** If you are using DaVinci Resolve 16.2 or higher, you can also swipe on or off track controls, including Track Lock, Record Enable, Solo, and Mute. Additionally, the new, improved Tracks index lets you double-click to rename tracks and drag and drop to reorder tracks, right in the tracks list!

3 In the Tracklist, with your pointer over the visibility control for track A2, drag down over the visibility controls for tracks A3 through A6 to hide them.

Those tracks are no longer visible in the timeline; however, their audio contents will continue to play normally.

Next, you'll resize the tracks using the zoom sliders that are located in the toolbar above the timeline.

4 Drag the vertical zoom slider to the right to increase the height of the timeline tracks.

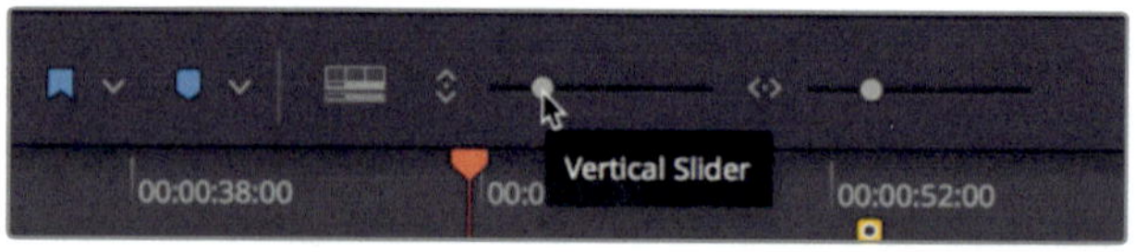

5 Drag the horizontal zoom slider to the right to zoom in to the waveform at the playhead position.

6 Press Shift-Z to fit the timeline clips horizontally in the visible timeline area.

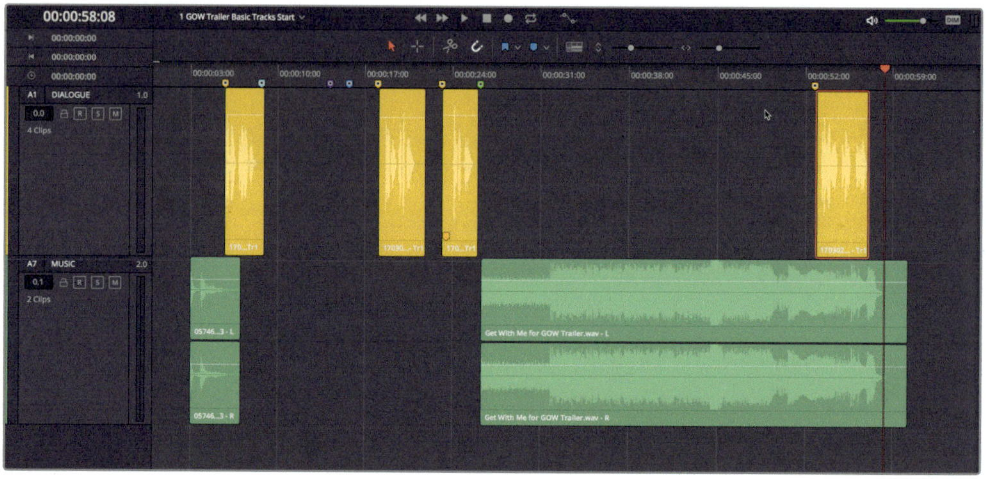

TIP You can use the zoom sliders in the toolbar to zoom in or out of the timeline vertically or horizontally. Vertical zooming is based on track selection, whereas horizontal zooming centers on the playhead. Press Shift-Z to toggle between zoom to fit horizontally and the previous zoom level.

Now that you have a focused view of the dialogue and music tracks, let's adjust the music track levels to accommodate the voiceover or vise versa.

Changing Clip Levels in the Timeline

We are back to the question of which is the priority: dialogue or music? This is a good time to mention the golden rule of audio post-production, also known as the "Dialogue is King" rule. In short, spoken words such as voiceover, narration, or dialogue take priority over all other audio tracks. This rule also applies to a music score, when the actors' lines take priority; action scenes, when characters' shouting must be heard over helicopters and explosions; and even documentary films, when subjects are talking in a noisy location.

If the voiceover track always takes priority, what happens to the music? In this project, music drives the entire second half of this trailer's soundtrack. Turning down the entire music clip to accommodate one dialogue line near the end is not a viable option. The solution is to reduce, or dip, the music clip's volume level only during the voiceover clip. This technique is also called *ducking*. First, let's adjust the overall clip level in the timeline, and then add keyframes to change the track levels as needed.

Each timeline audio clip includes a volume level curve that looks like a white horizontal line across the upper third of the clip. To adjust the volume level of a timeline clip, you can drag the clip's volume level curve up or down.

1 In the **Get With Me** clip, drag down the volume curve to reduce the clip's volume level.

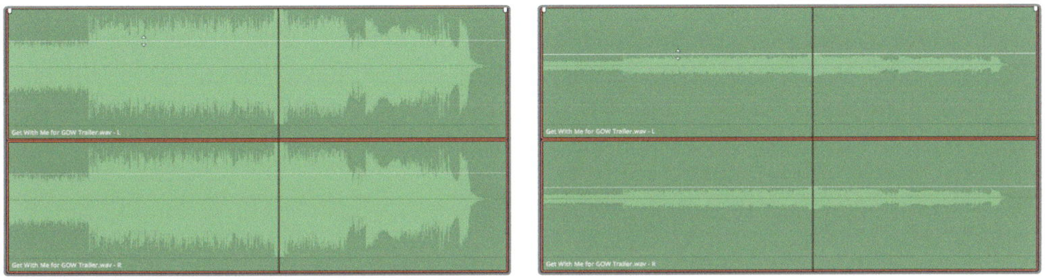

2 Double-click the volume curve to reset it to its original level.

To bend the volume curve so that the volume level changes over time, you can Option-click (macOS) or Alt-click (Windows) the curve to add keyframes.

Let's use the voiceover clip on the A1 track as a guide for setting the keyframes.

3 In the timeline's **Get With Me** clip, Option-click (macOS) or Alt-click (Windows) the volume curve at approximately the first and last frame of the fourth yellow clip in the dialogue track.

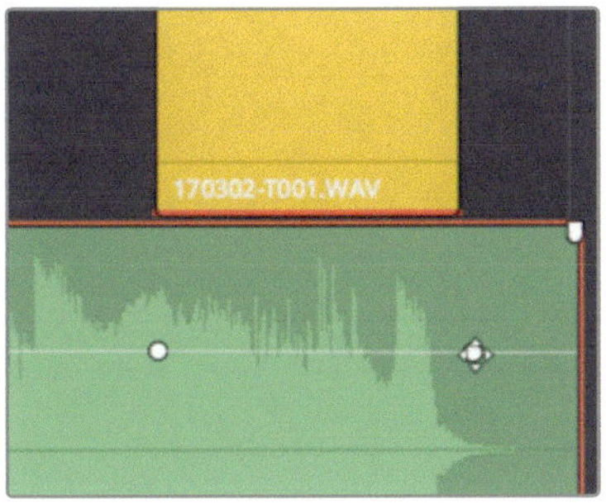

4 Place another keyframe before the first one and after the last so you have a pair of keyframes on either side of the yellow dialogue clip.

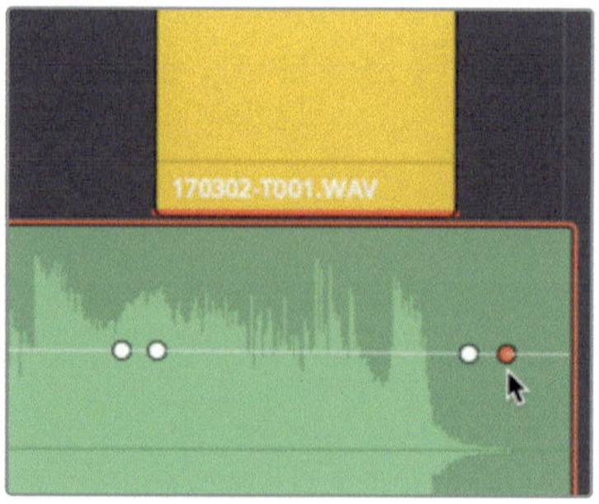

Now you can raise or lower the portion of the music clip that is between the keyframes, and in the Inspector, see the change in volume level. The Inspector will show the volume level of the selected clip at the playhead position.

5 Show the Inspector. Move the playhead over the middle of the last dialogue clip. Drag down on the music clip's volume level curve at the playhead position. When you release the mouse button, the level will appear in the Inspector. Dip the volume level to somewhere between -8.00 and -10.00 in the Clip Volume field.

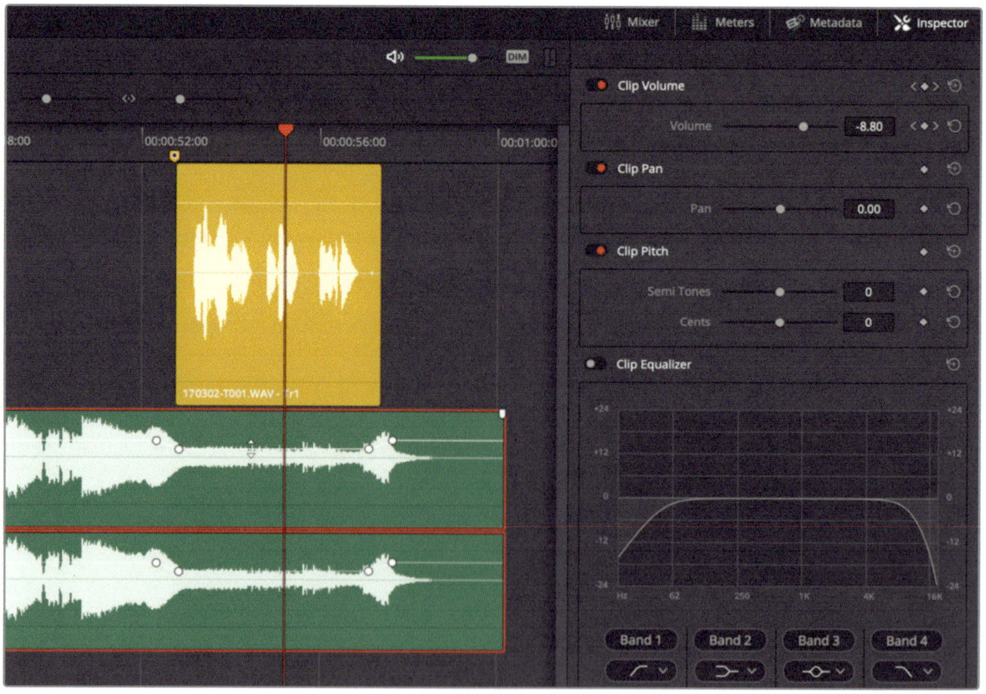

6 Play that section of the timeline to hear the dipped music.

The dialogue and music now work much better together. If the audio transition sounds too abrupt, you can increase the space between your pair of keyframes, and make the initial music fade down more gradually. The trick to moving keyframes for timing is to hold down the Shift key as you drag to constrain the movement to a horizontal axis. This takes a little practice, so take your time; and if you make any unintended level changes, you can choose Edit > Undo and try again.

7 Shift-drag the first keyframe to the left to double the original distance between your initial keyframes. Feel free to increase or decrease the distance to suit yourself.

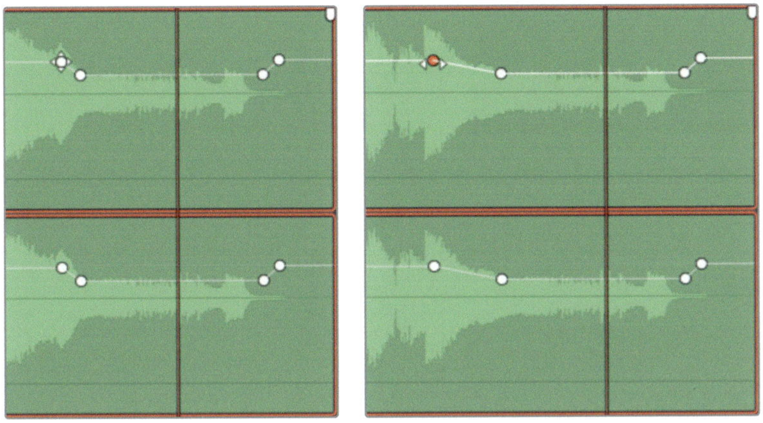

8 Play the dipped music section one more time to make sure you are satisfied with the results.

9 When you are finished, do the following:

- Hide the Inspector.
- Swipe on visibility for all of the tracks in the Tracks index.
- Hide the Index.
- Use the Vertical Zoom slider to fit all of the tracks vertically in the timeline.
- Press Shift-Z to fit all of the clips horizontally in the timeline.

Well done! You've dipped your music clip level to accommodate the voiceover. Next, you'll create a new track and use the Sound Library to add another audio clip to the timeline.

Manually Creating a New Track

You can create a single track or multiple tracks simultaneously from within the timeline. To do so, you'll create a new sound effects track below the A5 WIND track.

1 Right-click the A5 WIND track header, and choose Add Tracks.

In the Add Tracks dialog, you can determine how many tracks you add and where they are placed.

2 Set the Number of tracks to 1; set Insert Position to Below WIND; and set the Audio Track Type to Stereo.

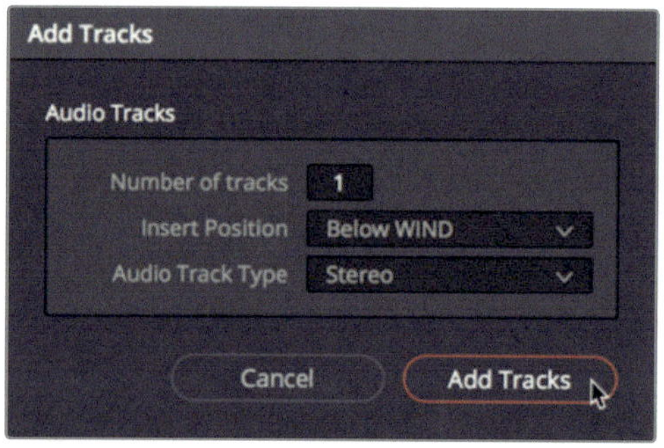

3 Click the Add Tracks button.

A new track appears Below the A5 WIND track.

4 Name the new A6 track **IMPACT FX**.

Now, let's use the Sound Library to find and audition an effects clip.

Adding Sound Effects to the Sound Library

Many editors and audio professionals have vast libraries of project-ready sound effects and music that are often stored on servers or separated hard drives. You can access your own sound collections in the Fairlight page within the Sound Library panel.

NOTE This exercise assumes that you have followed along with the "Getting Started" section of this book and downloaded the SFX Sampler for Fairlight folder and created a new **sound fx** database. If not, please return to the "Getting Started" section of this book and follow the instructions in "Setting Up Your Sound Effects Database."

In this exercise, you'll add the downloaded SFX Sampler for Fairlight folder contents to your database.

1 In the Interface toolbar, click the Sound Library button to show the Sound Library.

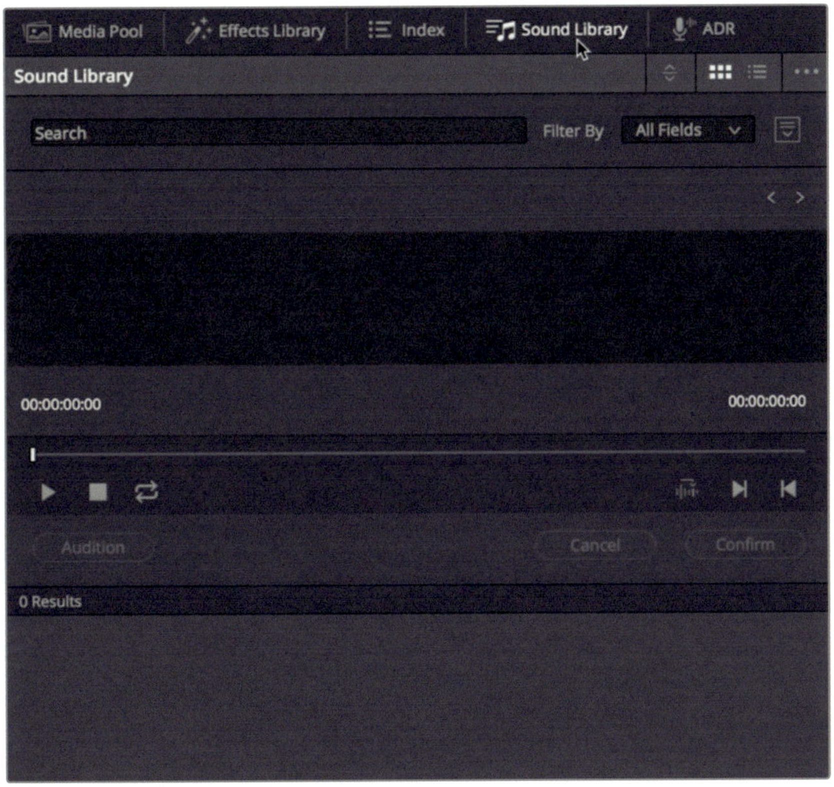

2 Click the Database button to the right of the search field to show the Database pop-up menu.

3 In the Database pop-up menu, choose sound_fx.

Once you have selected a database in the menu, you can add sounds. In this case, you'll add the SFX Sampler for Fairlight folder.

4 In the upper-right corner of the Sound Library, click the Options menu (three dots), and choose Add Library, or click the Add Library button in the middle of the Sound Library panel.

5 In the browser, select the SFX Sampler for Fairlight folder, and click Open.

You will see a brief progress dialog while the clips are scanned and indexed. When finished, a dialog indicates that the scan was successful.

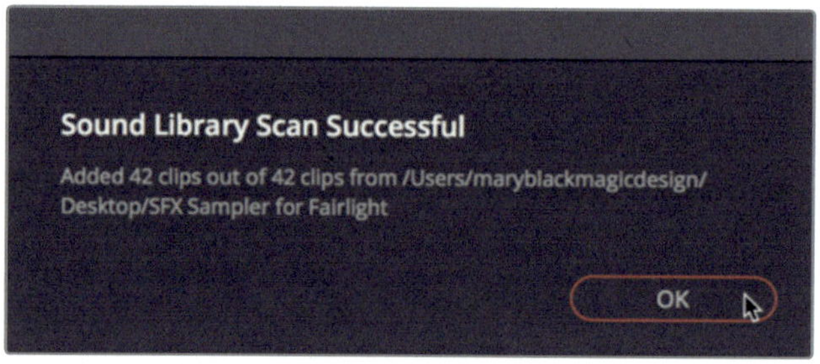

6 Click OK.

Your sample sound library is ready to search. Next, you'll determine where you want to add a sound effect in the timeline.

Auditioning Sound Effects in the Timeline

One of the great things about working with the Sound Library is that you can access thousands of sounds in multiple databases and audition them directly in the timeline, thereby hearing them in the context of the other tracks without committing to their use until you are sure they work. All you need to do is select a track and move the playhead to the position where you want the sound to start.

In the following exercise, while watching the viewer, you'll move the playhead to a specific timeline position based on the onscreen action. Then, you'll select the A6 IMPACT FX track, and search for a sound effect to audition. Once you find the right sound effect, you can commit within the Sound Library without needing to drag the clip into the timeline.

For this example, the clients want you to find an impact sound effect to accentuate the action when Sarah falls from her board. At the moment, the stinger that you added earlier is dominating the soundtrack during her fall; so, let's mute the stinger while you audition the impact effects.

1 Show the viewer and move it to the empty space in the upper-right corner of the timeline.

2 Mute the A7 STING track.

3 Play the timeline from the beginning, and stop playback when Sarah falls from the board at the top of the high wave at approximately 00:00:12:05.

4 Select the A6 IMPACT FX track.

   Now, you can find a sound effect to audition in the Sound Library.

5 In the Sound Library search field, type **impact**.

   A list of the impact sounds appears in the results list.

6 Select any of the impact sounds from the list to see its waveform in the Sound Library preview player.

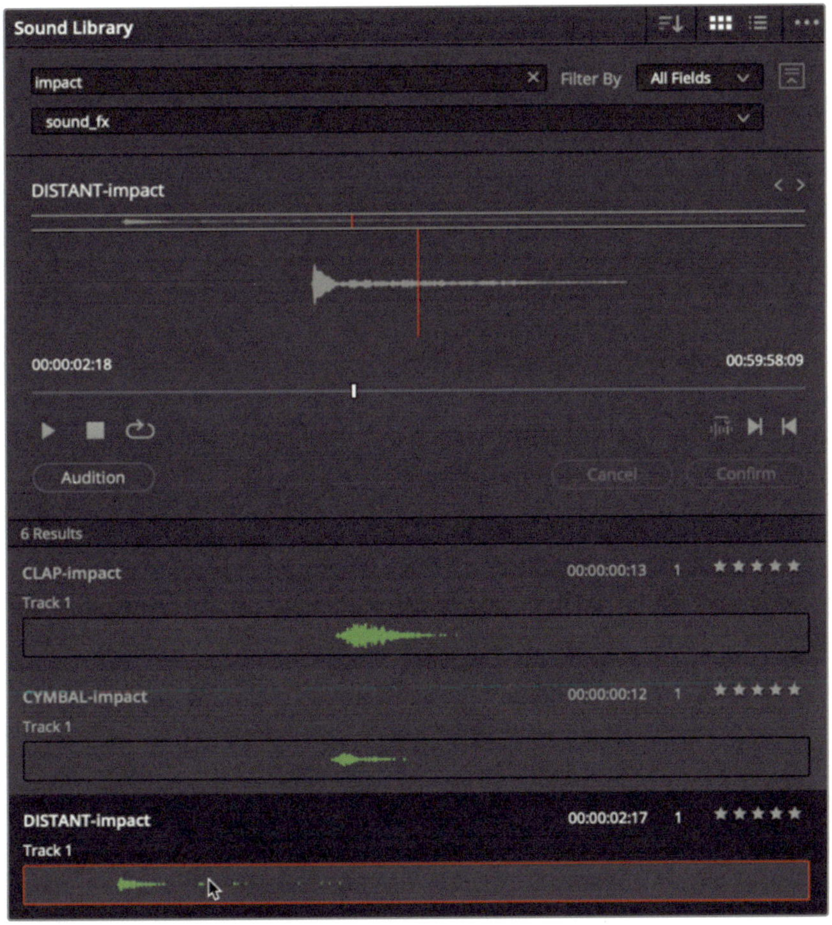

The process of selecting and previewing sounds in the Sound Library is very similar to working in the Media Pool source viewer.

7 In the Sound Library preview player, click the Play button, or press the Spacebar to play the sound effect.

All of these impact sound effects are exciting—but don't fit this trailer.

Let's locate a specific sound effect for this timeline.

8 In the Sound Library search field, type **soft**.

Four sounds, including SOFT-HIT, appear in the list.

9 Preview the SOFT-HIT sound.

This sound effect has just the right level of drama and impact for the surf action without distracting from the overall trailer soundtrack.

10 In the Sound Library, click the Audition button.

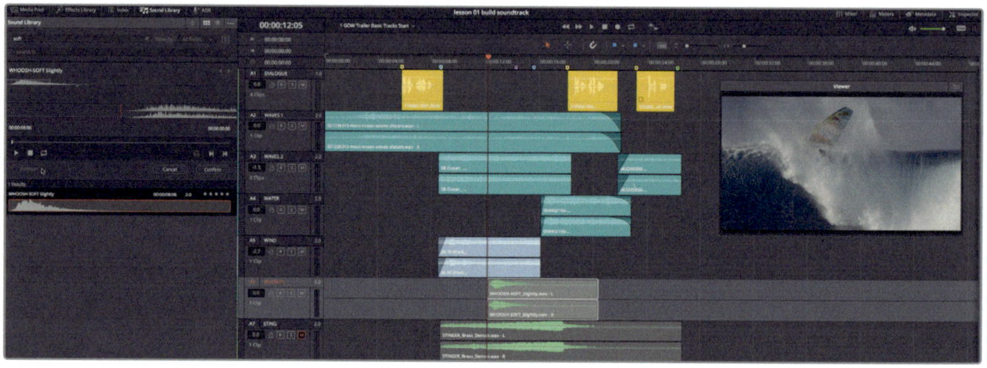

The sound effect appears in the timeline ready to play.

11 In the timeline, move the playhead back to the teal marker and play the timeline. Stop playback when you hear the voiceover.

The sound effect works pretty well and adds one more dramatic element to the sound design.

12 In the Sound Library, click the Confirm button to retain the sound effect in the timeline.

> **NOTE** Clicking the Cancel button removes the previous sound effect and gives you the opportunity to audition another audio clip in the same place.

13 Hide the Sound Library.

Now that you know how to use the Sound Library, you can audition your own sound collections in your projects.

Moving and Trimming a Clip in the Timeline

In your timeline, dialogue may be king; but in the real world, your client is all powerful, followed by the director and the sound designer or supervising sound editor (often the same person). For this project, the clients like your soundtrack choices, but they want you to move the soft-hit impact effect to the first frame of Sarah's face as she reflects on the fall. Your response, as always, should be something like, "Sure, great idea," or "Of course, I'll get right on it." Then, you make the change and share the results. You can choose your battles later when something really doesn't work. In this case, the clients' change will be awesome.

To move clips in the timeline, you simply drag them to a new location. Because you are using the first frame of a video clip as a guide, let's show the video track in the timeline. You'll find the options to show or hide additional timeline elements in the Timeline View Options menu located between the markers and zoom sliders in the toolbar.

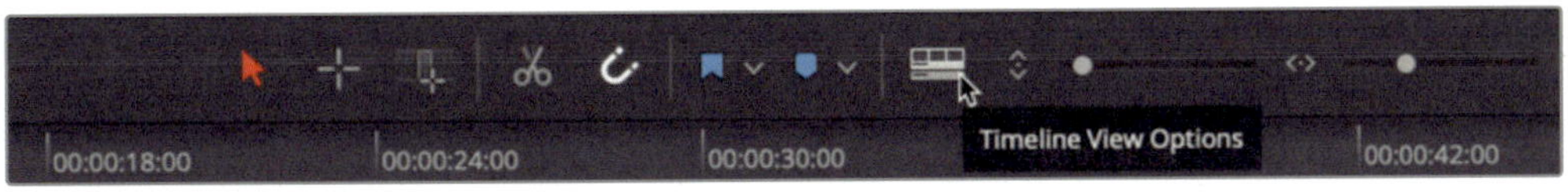

1 In the Timeline View Options menu, click the video tracks icon. Then, click the timeline to hide the menu.

The video tracks from the Edit page appear at the top of the timeline. You can now use those individual video clips in the track as a guide when placing your audio clip.

2 Move the playhead to the purple marker in the timeline (00:00:14:06). In the video track, you'll see that the playhead is also at the first frame of Sarah's clip.

3 Drag the **SOFT-HIT** clip in the A6 track to the right so it starts at the playhead position.

4 Play the timeline from the teal marker to the second yellow marker to hear how the sound effect works in its new position. Press Option-L (macOS) or Alt-L (Windows) to replay that section.

As promised, the effect works even better here and adds to the dramatic flash cuts that happen next. Did you notice that the end of the sound effect is silent? Because of that, you may think the effect is longer than it actually is. Let's take a moment to trim the tail end of the sound effect.

To trim a timeline audio clip, you can drag an edge of the clip toward the right or left. While trimming, you'll see a translucent waveform image of the entire audio file that you can use as a guide as you trim.

5 Drag the tail (right edge) of the **SOFT-HIT** clip to the left until it reaches the end of the actual waveform.

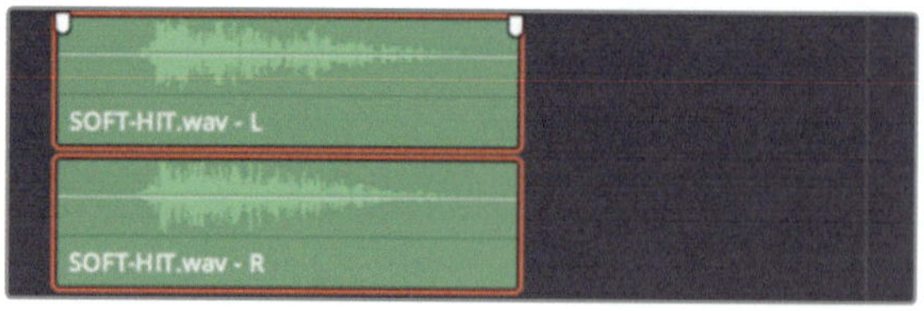

Let's remove the silence from the end of the stinger clip, as well.

6 Trim the tail of the stinger clip in the A7 track.

The clients are happy, but, as clients will, they want you just one-more-thing to enhance the drama of Sarah's reflection moment. They want you to create silent gaps in the soundtrack to match the flash cuts on the video track. And, of course, your answer is "Sure!"

Selecting and Deleting a Range in the Timeline

To select and delete a specific range in the timeline, you'll use the Range Selection tool. So far, you've been working in standard editing mode using the Selection tool to select, move, and trim clips in the timeline. To go into range selection mode, press the R key. (Incidentally, if you haven't guessed already, the shortcut for entering standard editing mode is A for arrow.)

In this exercise, you'll find the gaps in the video track and use the Range Selection tool to select and delete those sections of audio in the timeline. Along the way, you'll use additional techniques to zoom and navigate in the timeline.

1 Press R, or in the toolbar, click the Range Selection Mode button.

2 Move the playhead to the blue marker at 00:00:15:15.

3 Drag the horizontal zoom slider, or press Command-= (equals sign) in macOS or Ctrl-= (equals sign) in Windows to zoom in to the playhead until you can clearly see the gaps in the video track. Drag the scroller at the bottom of the timeline as necessary to move the gaps to the middle of the timeline.

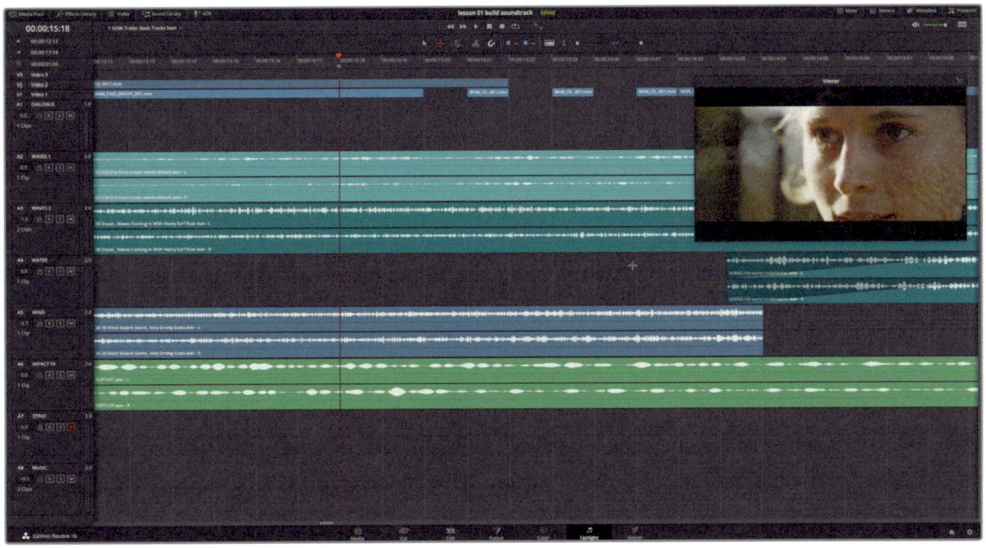

Your goal is to use the Range Selection tool to drag a selection that is the same width as each gap and includes the teal and blue clips on the A2 through A5 tracks.

> **TIP** When dragging a specific range with the Range Selection tool, drag the specific amount first—in this case, the gap—and then extend the range in the other direction to include all of the necessary clips.

4 Using the Range Selection tool, drag from left to right directly below the gap in the video track, and then drag down until you include the clip in the A5 WIND track.

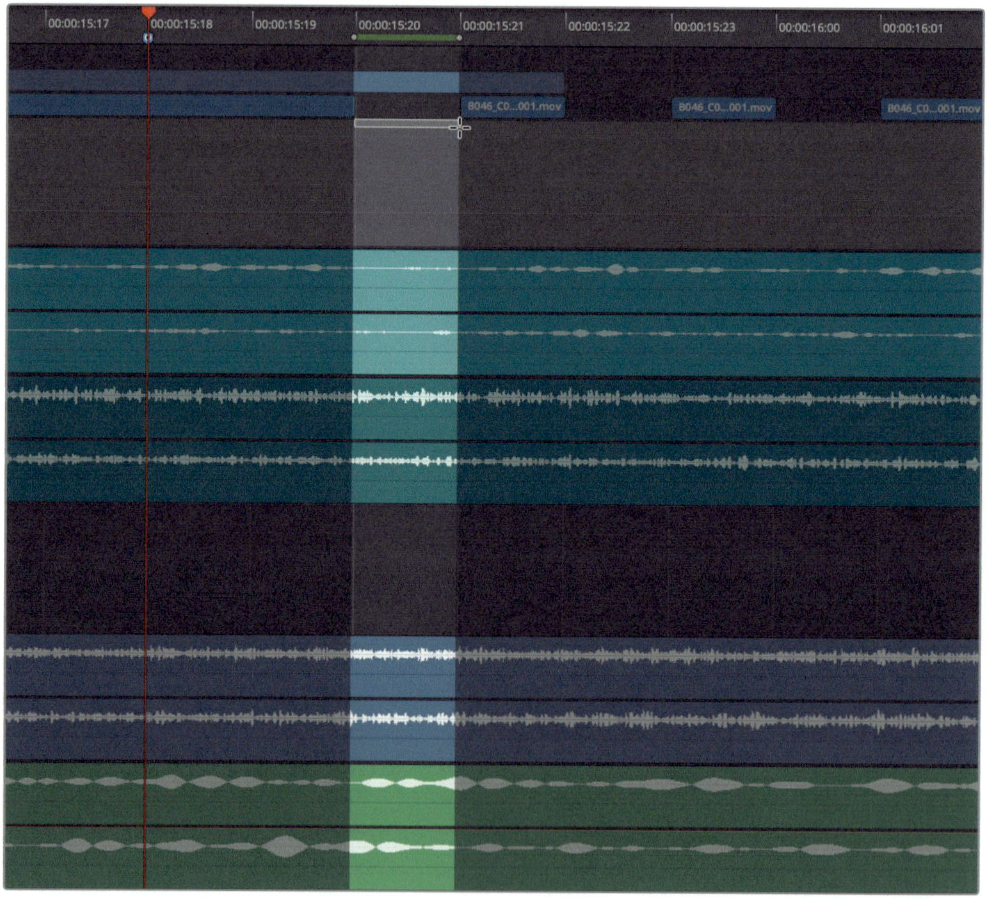

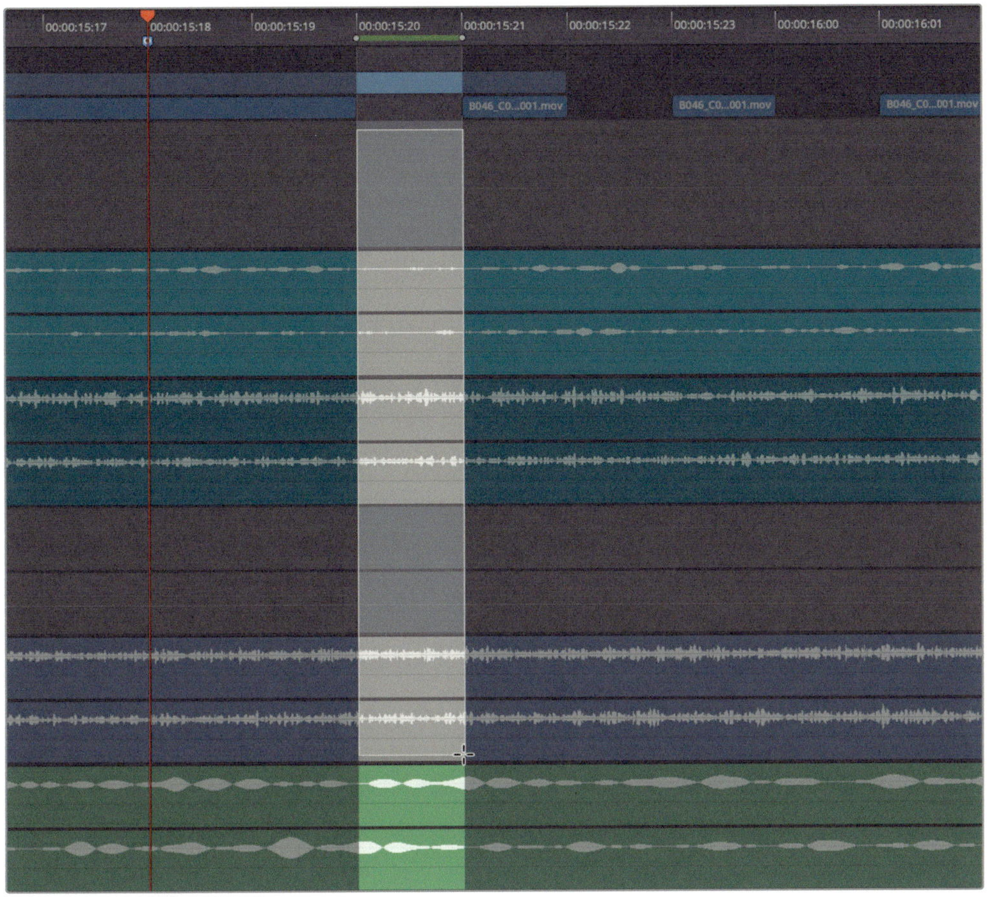

The green bar in the ruler above the timeline indicates the in and out points of the range. You can also see the range's in and out points in the timecode display. The sections of the selected clips that are within the range appear in gray. You can now delete that selected range.

5 Press Delete.

6 Repeat steps 4 and 5 to delete the sound effects beneath the other two gaps.

7 When you are finished, deselect the current range. Then, press A to return to the standard selection mode.

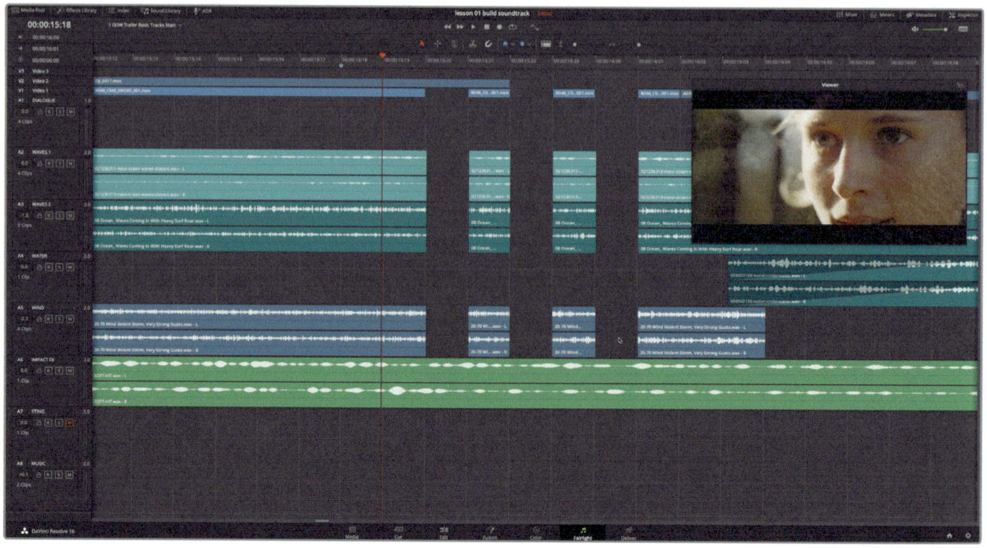

8 Press Shift-Z to display all of the clips horizontally in the timeline.

9 Play the timeline from the purple marker to hear how your edits work with the other tracks.

Wow! The gaps in the sound effects tracks sound awesome. Now let's see what happens when you add the stinger to the mix.

10 Unmute the A7 STING track and play the timeline from the beginning.

Overall, the soundtrack is really working. However, the stinger is stepping all over the audio gaps and the soft-hit effect. Luckily, audio isn't an all-or-nothing medium. You may still be able to keep the stinger by lowering the volume level—a lot. Then again, you may love the stinger and want to keep it as it is. Ultimately, you have to trust your own ears and creative judgement.

Balancing Track Levels

The last step in completing this soundtrack is to balance the multiple track levels. You can adjust track levels quickly in the track header or in the Mixer. In this exercise, you'll first tackle the volume level of the A7 STING track using the volume level field in the track header. Then, in the Mixer, you'll adjust the other track levels.

1 In the A7 STING track header, locate the volume level field, which currently reads 0.0.

Like the clip level field in the Inspector, the volume level field in the track header indicates the amount of change applied to the original default track level.

You could try to change the level now while the playhead is stopped. However, if a track's level is competing with other tracks, it's best to adjust the level while listening to it in the context of the other tracks.

2 Move the playhead to the teal marker, and start playback.

3 As you listen to the stinger build along with the other tracks, drag downward on the A7 track volume field to lower the level until it balances with the other tracks. Repeat this step until you are happy with the level. If you aren't sure what level to us, try -12.

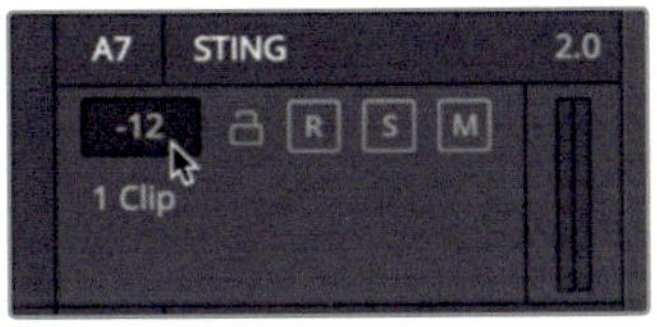

4 Close the viewer. Click the Mixer button to show the Mixer.

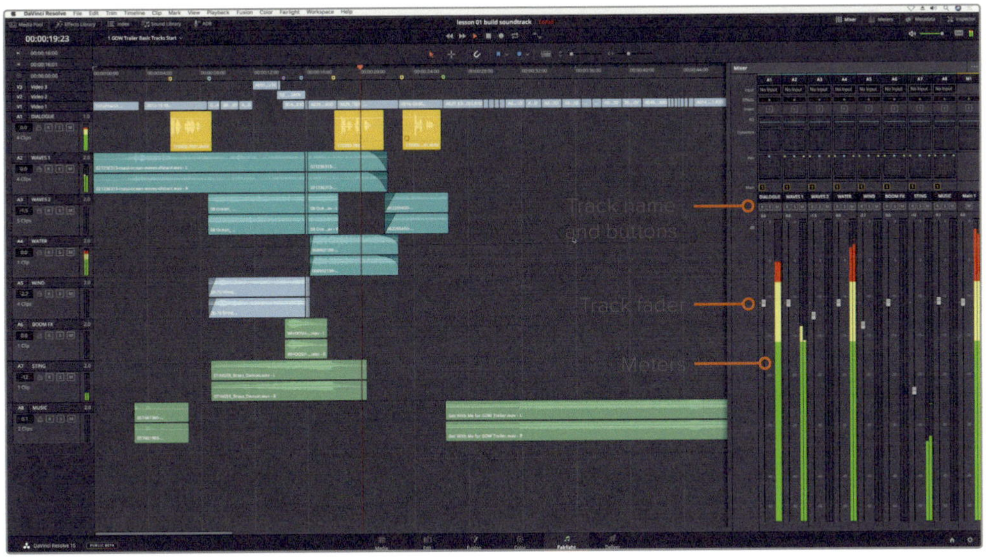

At a glance, you can see that the Fairlight Mixer is loaded with controls, especially at the top where you'll find an Effects slot, EQ, Dynamics, and Pan controls. (You will be working more extensively with each of these later in the book.)

For this lesson, let's focus on the track faders for balancing the volume levels of the tracks. Notice that the fader in the A7 channel strip reflects the volume change that you applied in the track header.

5 In the timeline, select the A1 DIALOGUE track header.

The A1 DIALOGUE channel strip in the Mixer brightens (just like the corresponding A1 track in the timeline) to indicate it is selected.

If you look at the track names in the Mixer, you'll see that their capital letters make them easy to read. This is why many audio professionals use all capital letters and short names or initials in audio track names.

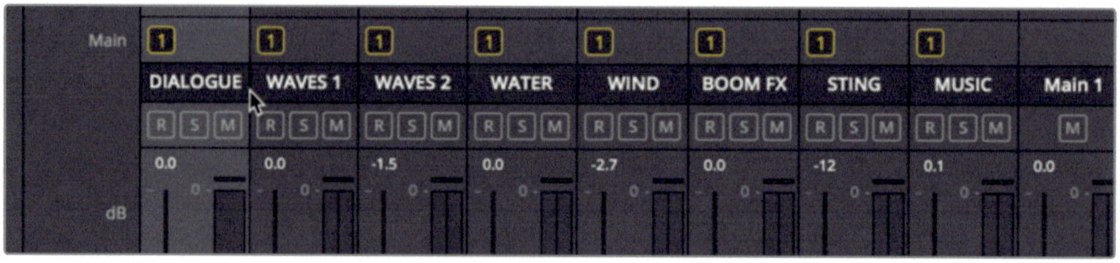

You already balanced the levels of the clips in the A1 Dialogue track, so let's assume those levels are good. Specific target levels and delivery requirements will come later. For now, this is your opportunity to use the most powerful audio input and processing tool known to man—your ears—to balance the levels to your own taste.

6 Start playback from the beginning, close your eyes, and just listen to the timeline until the "Get with me" song starts.

   Did any sounds seem out of place or too loud?

7 In the Interface toolbar, click the Meters button to show the monitoring panel and viewer.

8 Play the timeline again and watch the video while listening to the tracks.

9 Adjust the faders on the channel strips to suit yourself.

10 When you are finished, stop playback and hide the Mixer.

Congratulations! The clients are almost happy. Yes, almost happy. Your clients have another tiny request. They are wondering if you can add an echo to the soft hit on the IMPACT FX track so that it enhances the flash cuts. Sure!

Adding a FairlightFX Plug-In to a Clip

DaVinci Resolve 16 includes a full set of audio plug-ins called FairlightFX that can be added to any clip or track. To become familiar with them, you'll add the Echo plug-in to the SOFT-HIT clip. and adjust its parameters in the FX interface or in the Inspector.

First, let's solo the A6 track and set in and out points on either side of the soft-hit clip to make playback of that timeline clip easier.

1 Solo the A6 IMPACT FX track.

2 Move the playhead to the beginning of the **SOFT-HIT** clip, and press I to set an in point. Move the playhead to the end of the clip, and press O to set an out point.

The portion of the timeline within the in and out points highlights.

3 In the Interface toolbar, click the Effects Library button.

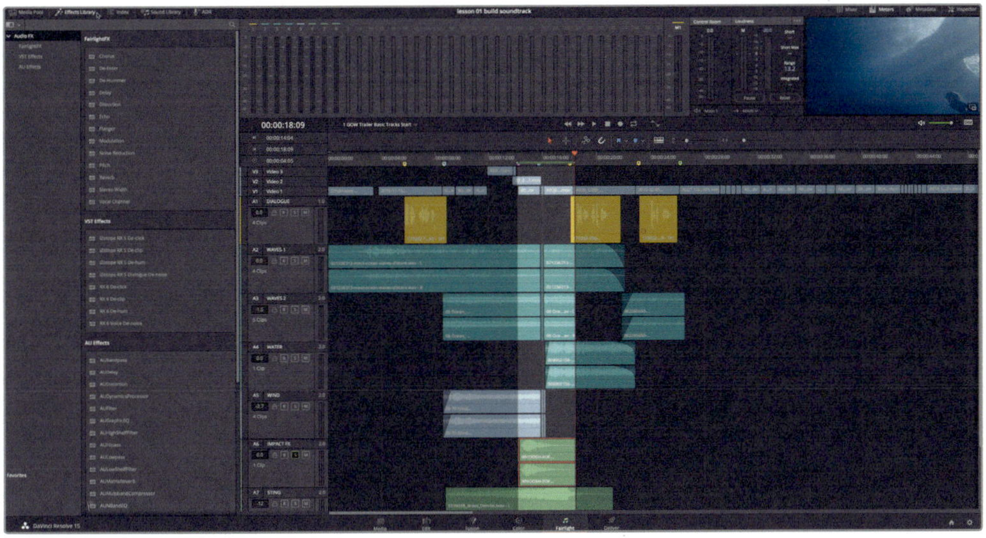

The Effects Library shows the FairlightFX at the top, followed by any VST Effects or Audio Unit effects installed on your system.

4 In the FairlightFX list, drag the Echo effect to the **SOFT-HIT** clip in the timeline.

The Echo controls panel opens. Here you can adjust several FX controls. For this exercise, let's play the clip with the default Echo settings, and use one of the Echo presets to change the parameters.

5 Press Option-/ (forward slash) in macOS or Alt-/ (forward slash) in Windows to play the timeline from the in to the out points.

You can really hear the Echo effect on the clip. Now let's apply one of the presets.

6 In the upper-left of the Echo controls panel, click the Preset menu to see the available presets. Choose Fast Ping Pong.

7 Play the clip and listen to it with the Fast Ping Pong preset applied.

As you can hear, the Fast Ping Pong preset has a subtler echo effect than the default preset but adds some nice depth to the sound along with a fast-moving delay back and forth (ping pong) between the left and right channels.

8 Close the Echo Controls panel.

If you need to see the controls again, you'll find them in the Inspector.

9 In the timeline, select the SOFT-HIT clip, if necessary. Show the Inspector.

Parameter controls for the Echo effect appear at the bottom of the Inspector. You'll also see a trash icon to delete the effect and a Customize button to show the controls window.

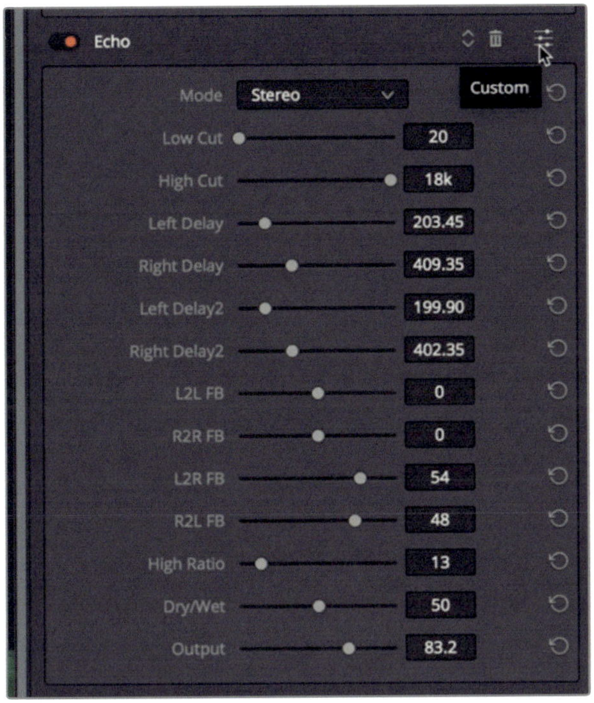

10 In the Echo controls in the Inspector, click the Customize button to open the Echo controls panel.

11 Feel free to experiment with the different Echo presets and plug-in controls. When you are finished, close the Echo controls window.

12 Stop playback, if necessary.

13 Choose Mark > Clear In and Out, or press Option-X (macOS) or Alt-X (Windows) to clear the in and out points.

14 Unsolo the A6 track. Hide the Effects Library and the Inspector.

15 In the Timeline View Options menu, turn off the Video track.

16 Play the timeline and listen to the entire soundtrack.

> **NOTE** If you didn't finish all of the steps in this lesson, open the timeline 3 GOW Trailer Basic Tracks Finished to hear the finished soundtrack.

Great job! You not only worked your way through most of the Fairlight interface, but you also finished the trailer soundtrack and learned some real-world audio techniques.

More Drama Please!

As luck would have it, the clients were happy until they held a preview screening of the trailer and the audience consensus was that they wanted to feel more inherent danger during the action sequence. So, you'll need to apply your new skills to please the client and save the soundtrack in this self-guided exercise.

Your goal is to really step up the element of danger in the action sequence by adding the violent winds sound from a tornado effect. Fortunately, that effect is already in the Media Pool. All you need to do is create a new track, find the sound effect, preview it, mark the part you want to use, and add it to the timeline in the same position as the Wind clip. Don't forget to name the new track and check the volume levels when you're done. If you have any other creative impulses for improving the tracks or changing the mix, go for it and have fun with it!

Lesson Review

1 In the default layout, which Fairlight panels are showing with the timeline? Choose all that apply.

a) Mixer

b) Media Pool

c) Index

d) Monitoring panel (meters)

2 True or False? The viewer in the Fairlight page shows only the selected video tracks in the Edit page.

3 Where in the Fairlight page can you go to open an existing timeline? Choose all that apply.

a) Metadata

b) Media Pool

c) Timelines pop-up menu

d) Index

e) Timeline tab

4 How do you separate the viewer from the monitoring panel so that you can use it as a floating window? Choose all that apply.

a) Double-click the viewer.

b) Option-click (macOS) or Alt-click (Windows) the viewer.

c) Command-click (macOS) or Ctrl-click (Windows) the viewer.

d) Click the Expand button in the viewer.

e) Click the Viewer button below the transport controls.

5 True or False? When zooming horizontally in the timeline with the zoom slider or shortcuts, the selected clips (or track) are always centered in the timeline.

6 What happens when you drag a two-channel stereo clip from the Media Pool to the empty space in the timeline below the existing tracks?

a) The clip splits into left and right channels and appears in two new mono tracks.

b) The clip appears in a new two-channel stereo track in the timeline.

c) Nothing happens because you can't drag a clip to the empty space in the timeline below the existing tracks.

7 Where can you change the volume level of individual timeline clips?

a) On the clip in the timeline

b) Mixer

c) Inspector

d) Preview player

e) Media Pool

Answers

1 a or d

2 False

3 b or c

4 d

5 False. Horizontal zooming centers on the playhead, while vertical zooming centers the selected tracks.

6 b

7 a or c. You can change the volume level of individual timeline clips by using either the volume curve overlay on the clip, or the clip volume slider in the Inspector.

Lesson 2

Recording Voiceover and ADR

Audio post-production often requires recording new audio files to fill out the soundtrack. These tracks can range from scratch voiceover (VO) to automatic dialogue replacement (ADR) to Foley sound effects or multitrack music recordings. The number and type of tracks that you record depend entirely on your audio interface and type of project.

You can easily record your own tracks in DaVinci Resolve 16 in the Fairlight page timeline. All you need to do is set up a microphone, patch the microphone input to a track, arm the track, and start recording.

Time

This lesson takes approximately 45 minutes to complete.

Goals

In this lesson, you'll set up and perform three types of audio recordings. First, you'll record a simple voiceover track for the voice of an intelligent computer. Then, you'll use the new ADR tools to set up and record automatic dialogue replacement in the same scene. Finally, you'll patch and record Resolve's built-in oscillator for sound effects.

NOTE The exercises in this lesson build from the tools and skills that you learned in the previous lesson. If you skipped ahead to this lesson, you may need to look back at the previous lesson as a guide for skills already covered.

Setting Up Your Microphone

Before you can start recording, you first need to set up your microphone. If you don't have a separate microphone, you can use a built-in microphone in your computer, if available. When setting up new audio hardware with your system, it's a good idea to first close DaVinci Resolve; otherwise, the newly attached hardware may not be recognized. Once your audio interface and microphone have been connected and recognized by your computer, they should also be available when you open Resolve.

MORE INFO For more specific information for compatible audio interfaces including Fairlight, MIDI and MADI interfaces, Fairlight Accelerator card, or additional monitors, please refer to the DaVinci Resolve 16 documentation or the DaVinci Resolve 16 Configuration Guide.pdf available at http://documents.blackmagicdesign.com /DaVinciResolve/20180407-79c607/DaVinci_Resolve_16_Configuration_Guide.pdf

1. Close DaVinci Resolve, if necessary.
2. Attach a microphone to your computer or audio interface.

3 In your computer's preferences, configure your microphone (or the built-in mic) as the audio input device.

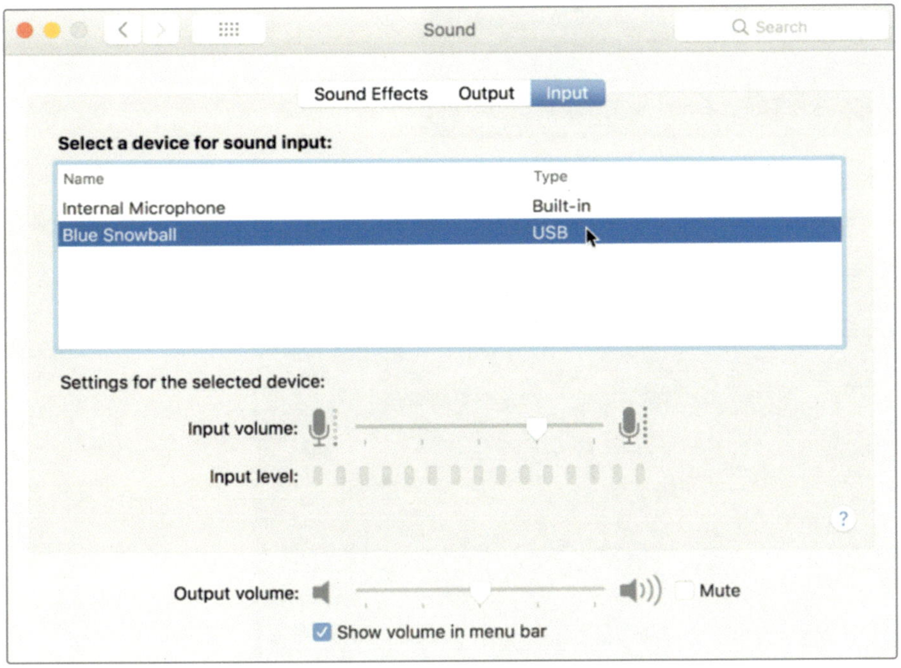

Once the microphone is set up, you can use it in Resolve. Next, you'll open the project that you will use for the recording exercises.

Preparing the Project

In this exercise, you will restore the Hyperlight Audio Post project that contains the scenes, timelines, and media that you'll need to complete Lessons 2 through 11. Then, in the Media Pool, go to the Lesson Timelines bin to see the timeline organization for this project, and open the first timeline.

1 In the Project Manager, right-click and choose Restore Project Archive.

2 Select R16 audio post lessons > R16 audio post pt2 > Hyperlight Audio Post.dra, and click Open. Then double-click the Hyperlight Audio Post project to open it.

The project opens in the Fairlight page with the 2 Recording VO Start timeline visible.

This is a good time to save a working copy of the project so that you can keep the unaltered original project for later reference.

3 Choose File > Save Current Project As.

4 In the name field, type your initials after the project name.

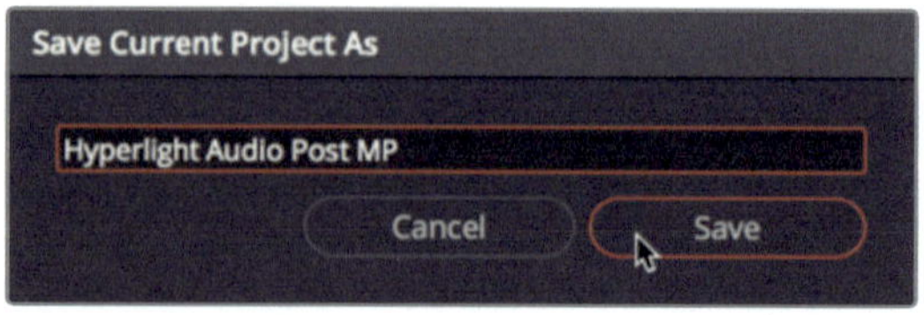

5 Click Save.

6 Show the Media Pool. In the Media Pool bin list, select Lesson Timelines > Lesson 02.

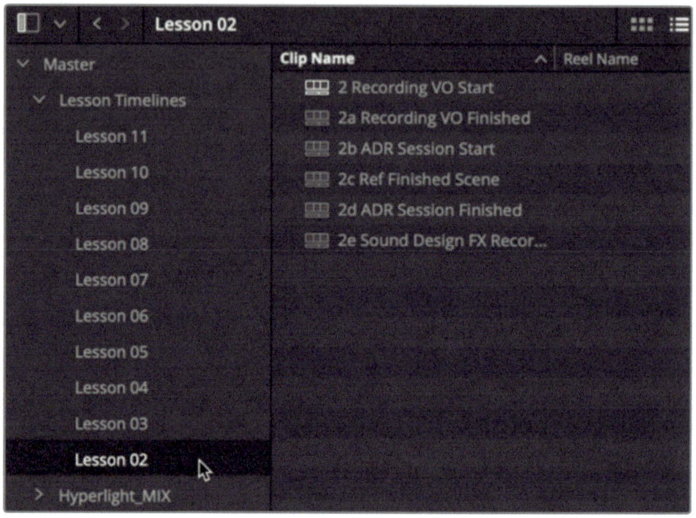

Here you will see all the timelines for this lesson. Each lesson includes sequential timelines starting with the lesson number. In this case, the first timeline is 2 Recording VO Start, which should already be open. You'll work with this timeline shortly. For now, let's open a different timeline that will serve as a reference for the finished scene.

7 Open the timeline, 2c Ref Finished Scene.

For subsequent exercises, you'll simply be instructed to open a specific timeline.

> **NOTE** You can also access any of the lesson timelines for this book in the Timelines smart bin at the bottom of the Media Pool bins list, if smart timelines is enabled in User Settings, or in the Timeline pop-up menu.

Now that you've restored and saved the project, and opened the reference timeline, let's take a few minutes to learn more about the voiceover part that you will record.

Previewing the Scene for Context

You are about to embark on a real-world audio post-production journey through several scenes from the short film, *Hyperlight*, directed by Nguyen-Anh Nguyen. Along the way, you'll use professional tools and features in the Fairlight page to perform various audio post jobs as you turn raw production sound into a fully mixed, stereo, and 5.1 surround soundtrack.

Hyperlight is a futuristic science fiction short film that takes place in space. Two main characters, Emiliana and Philip, accidentally cross a space–time threshold during hyper-space travel and end up encountering themselves. Only one version of each person can survive.

The voiceover you will record is that of the ship's computer, ADA. Chances are you've seen a film or television show—such as *2001: A Space Odyssey* or *Star Trek*—that features a talking computer throughout the ship.

In this exercise, you'll experience your own time travel as you play through several versions of this scene to hear the original production sound, a voiceover test, and the voice used in the final mix.

Let's start by streamlining the interface.

1. Hide the Media Pool and Mixer, if necessary.
2. Resize and move the viewer to the empty space to the right of the timeline. Hide the Monitoring panel (Meters).

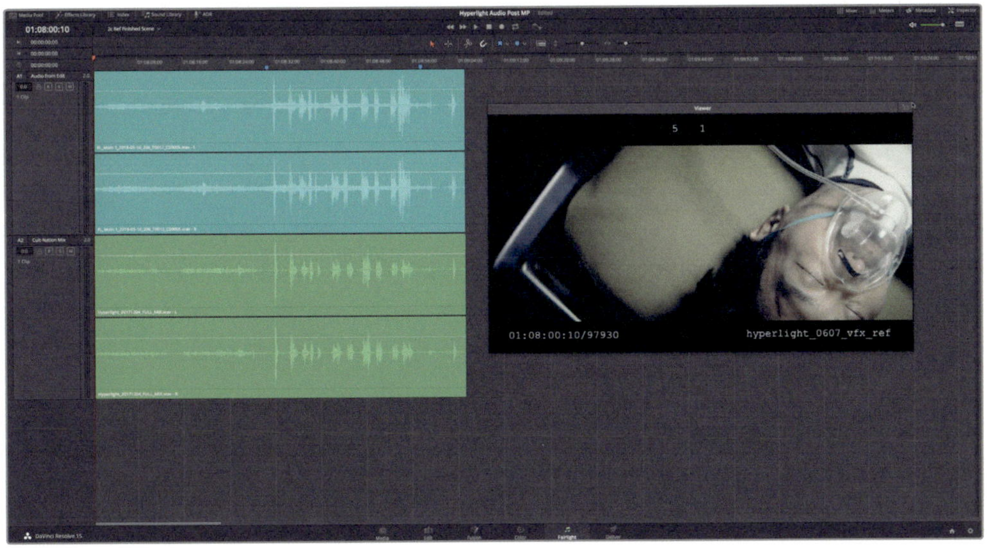

This timeline includes an early version of the soundtrack from editorial along with the final mix.

Track A1 contains the kind of sparse soundtrack that is provided by the editing department when a scene is picture locked. You'll work with these tracks shortly.

3 Solo track A1, and play the scene from the beginning. During playback, pay close attention to the sound of the computer's voice.

4 Un-solo the A1 track, and solo the A2 track.

The A2 track contains the professional stereo mix for this section of the scene.

5 Play the scene and listen to the fully developed soundtrack, including the computer's voice. The computer dialogue is between the blue timeline markers, so feel free to just play from marker to marker.

Wow, what a difference a professional sound designer, composer, and re-recording mixer can make to the soundtrack! You probably noticed that the computer's voice sounds a lot like a smart phone's interactive voice assistant in the first example, and more emotive and human in the final mix.

Now, let's go back to an earlier version of this timeline and listen to the original production dialogue recording with the scratch computer voiceover.

6 Open the 2 Recording VO Start timeline. Press Shift-Z to horizontally fit the timeline clips in the timeline.

7 Resize and move the viewer to the lower-center of the timeline below the timeline clips.

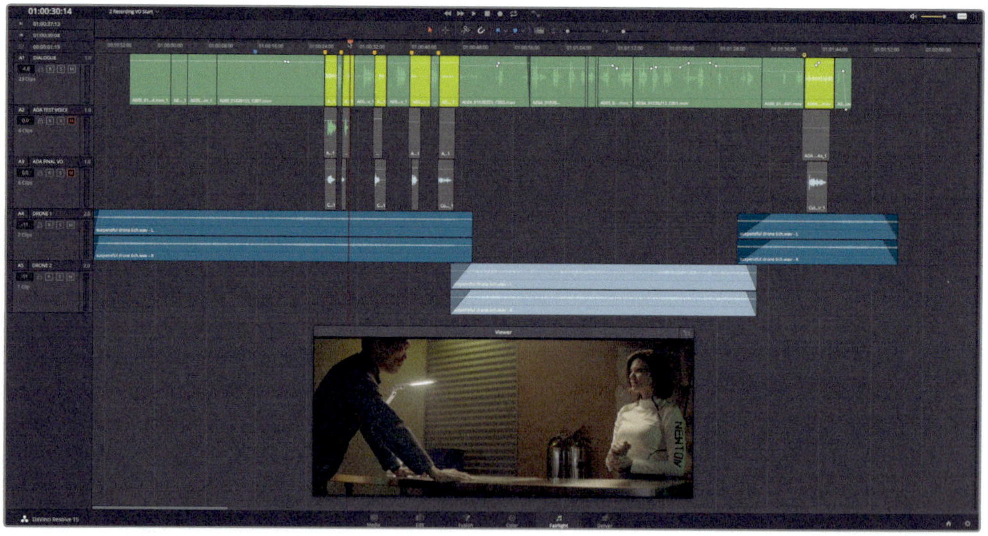

The timeline includes five audio tracks. All the dialogue is in the A1 track. The computer voice clips are colored lime green.

Tracks A2 and A3 contain alternate versions of the computer voiceover and are currently muted.

Tracks A4 and A5 contain drone sound effects that are used to temporarily fill the gaps between dialogue lines instead of using ambient sound and temp music.

NOTE Because this is a simple dialogue scene of two people talking in a room, it doesn't require a lot of additional sound effects tracks to see if the edit and performances are working. In many cases, the editorial department can cut and lock the picture with only one dialogue track and some temporary music or ambient sound to fill the gaps. In this scene, the drone sound effects set the mood well enough to pass it along to the sound department for finishing. Of course, you will transform the soundtrack in subsequent lessons. For now, you'll focus on re-recording the computer voice.

8 Move the playhead to the blue marker at 1:00:15:15, if necessary. Play the scene and listen to the production dialogue with the original scratch computer voice.

Clearly, the voice recorded during production was meant as a placeholder to cue the talent. Now that you've heard the original scratch version of the computer voice, let's select and disable those clips.

Selecting and Disabling Clips

In the previous lesson, you used the Mute and Solo buttons to control which tracks were audible during playback. Sometimes, you may have a need to mute specific clips within a track. In that case, instead of using the Mute button in the track header, you can disable the clip. In Resolve, you can disable both video and audio clips with a single keystroke. To do so, you select one or more clips and press D (for disable). This keyboard shortcut works as a toggle to disable or enable the selected clips. Let's try it out on the scratch computer voiceover clips in the A1 track.

1 Move the playhead to the third lime green clip in the A1 track.

2 Press Command-= (equals sign) in macOS or Ctrl-= (equals sign) in Windows to zoom horizontally in to the timeline clips.

The timeline zooms horizontally with the playhead in the center of the timeline. The first six lime green clips should now be easier to see and select.

3 Select the first lime green colored clip in the A1 track, and press D.

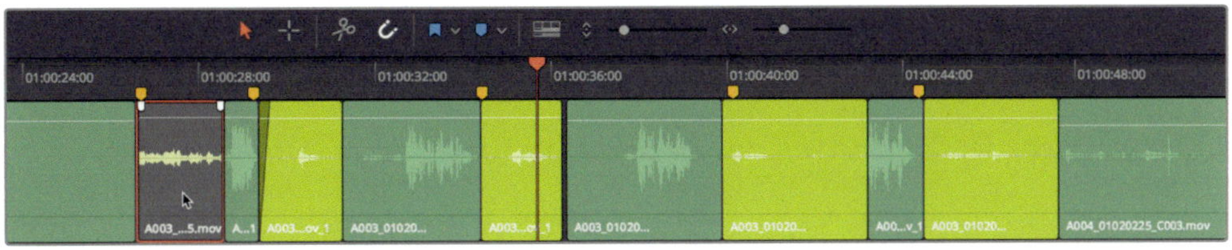

The clip turns gray to indicate that it is disabled. Let's disable the rest of the tracks at once.

4 Select the second lime green clip, then Command-click (macOS) or Ctrl-click (Windows) the next three lime green clips to select them all.

5 Press D to disable the selected clips.

With these scratch production computer dialogue clips disabled, you can selectively listen to some of the other computer voice options and even record your own.

Changing Playback Behavior

DaVinci Resolve 16 includes several playback features that you can use to simplify your timeline playback. In this exercise, you'll use the Stop and Go to Last Position feature to play back a specific part of the timeline based on the starting playback position. You can enable this feature in the Playback menu or from the transport controls.

1 Move the playhead to 01:00:22:00, which is right before Philip says, “ADA,” for the first time.

2 In the transport controls, right-click the Stop button, and choose Stop and Go To Last Position.

3 Un-mute the A2 track, and start playback. Stop playback by pressing the Spacebar just after ADA says, “I can't explain this discrepancy.”

When you stop playback, the playhead returns to the starting position (01:00:22:00).

What did you think of the computer voice this time? This version of the voice was created by an actual computer reading the script lines. You probably noticed that some of the lines have changed since the original production script. Now listen to the A3 track to hear the final version of the voiceover. Of course, this final version includes some reverb and other processing effects to make it sound more like a computer voice. You'll work more with those effects in a later lesson.

4 Mute track A2 and un-mute track A3. Play the scene again to hear the final computer voice in context. When you are finished, stop playback.

This time the computer voice was obviously recorded by a human with more subtle inflection and intuitive responses. Now that you are familiar with the computer voiceover part, you're ready to set up your timeline for recording.

Choosing a Location for New Audio Recordings

When you record audio files in the Fairlight page, you simultaneously create new clips in the timeline and Media Pool, as well as storing new media on your hard disk. Files are stored in the Capture location on your system, while the corresponding clips will appear in the selected bin in the Media Pool. In this exercise, you'll create a new bin for the recorded clips and locate where the new files will be stored on your storage disk.

1 Show the Media Pool.

2 In the Hyperlight Audio Post bin list, right-click the Hyperlight Audio Post bin, and choose Add Bin.

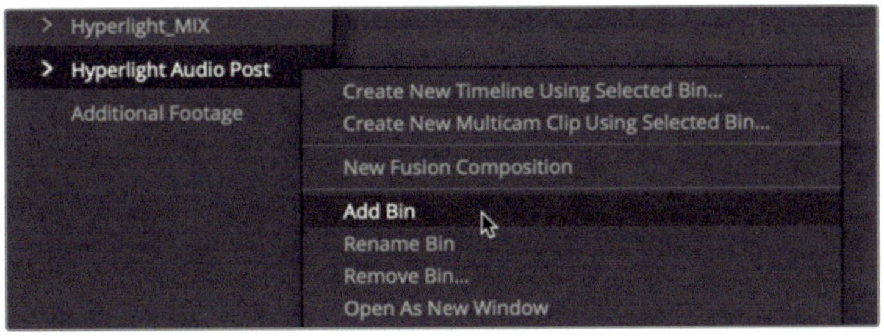

The new bin appears in the bin list with the name field selected and ready for you to type in a new name.

3 In the new Bin's name field, type **Recordings**, and press Return.

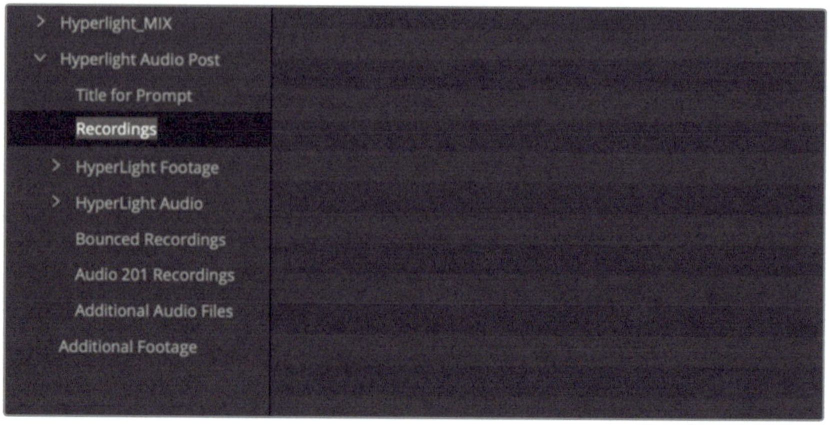

You have created a bin for the recorded clips in your project, and now you'll choose a location to store the new files. For this project, you can use the default capture settings; however, it's a good idea to know where the default capture location is and how to change it. You can specify the location of your recorded audio files in the Capture and Playback panel of the Project Settings.

4 Press Shift-9 to open the Project Settings window.

5 In the Project Settings sidebar, click Capture and Playback to show the Capture and Playback panel.

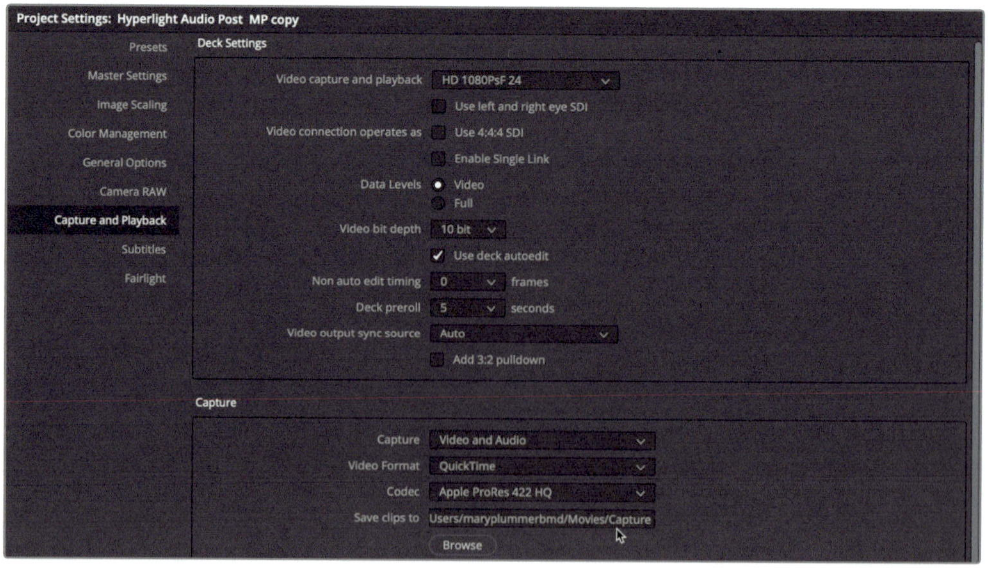

6 At the top of the Capture settings, ensure that the Capture pop-up menu is set for Video and Audio. If you reset the user interface in Lesson 1, you should be all set. If necessary, change the Capture pop-up menu from Video to Video and Audio.

7 Below the Capture pop-up menu, find the "Save clips to" field.

This location is where newly recorded files will be stored. The default capture location is the Users Movies > Capture folder. For the purposes of this project, though, you can keep the current settings.

If you want to save the recording to a different location, click Browse, and choose that new location now. Keep in mind that the new location will apply to all newly captured video and audio media.

> **TIP** It's always a good idea to check the capture location before recording audio or capturing video files. Also, if you share a workstation and change the capture location, be sure to either change it back, or coordinate with the other users so that you can all keep track of your media as you go.

8 Close the Project Settings window.

The last steps before recording are to create a new track, patch the microphone input to the track and arm it for recording.

Creating and Patching Tracks for Recording

You create a track for recording just as you would for editing. The important thing to consider is whether you want to record in stereo or mono. In most cases, the preferred track format for dialogue is mono, especially when you are working with voiceover or narration.

For this recording, you'll create a new mono track and use the Input settings in the Mixer to patch the microphone to that new track.

1 Right-click the A1 track header area, and choose Add Track > Mono.

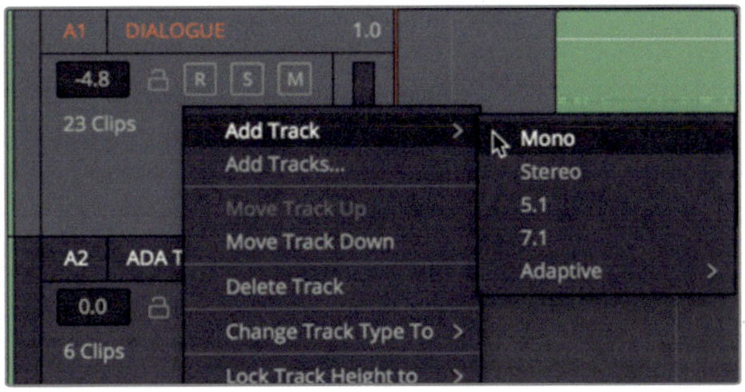

A new mono track named Audio 6 appears at the bottom of the timeline below the A5 DRONE 2 track.

2 Name the A6 track **Scratch VO**.

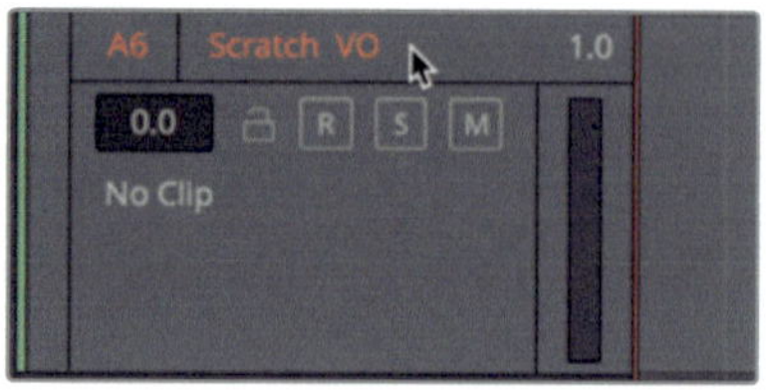

What Is a "Scratch" Voiceover?

A "scratch" voiceover is a temporary recording that is used as a placeholder for editing, timing, and early approval processes prior to recording (and paying) the voice artist to perform the final voiceover. Usually, the voiceover script will change several times during post-production while the edit and soundtrack are refined. In audio post-production, scratch voiceovers are often recorded for public address announcements, movie trailer voiceovers, and dialogue involving phone conversations.

Now you must set the input for this track. To patch a track's input, you can choose Fairlight > Patch Input/Output, or click the Input settings in the Mixer. Let's use the latter method for this recording.

1 Show the Mixer.

2 Select the A6 track, if necessary, and ensure that the Mixer is expanded to show the A6 channel strip.

   Input settings are displayed at the top of each channel strip just below the track number assignment. Tracks without an assigned input will display No Input.

3 At the top of the A6 channel strip, in the Input pop-up menu, choose Input.

The Patch Input/Output window opens. This window displays the source inputs to the left and the destination tracks to the right.

4 Select the first channel for your microphone as the source, and select the Scratch VO track as the destination.

NOTE If you were recording with a stereo microphone, or recording multiple microphone sources to multiple tracks, you would patch each track's audio input channels accordingly.

5 Click Patch to set your input patch.

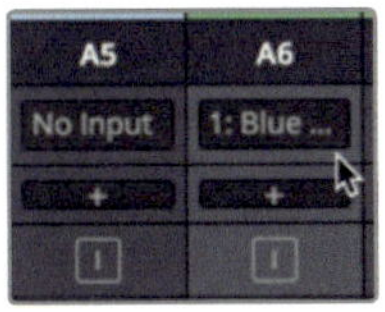

The Patch Input/Output window and the A6 channel strip in the Mixer update to show that the microphone channel you selected is patched to the A6 track.

6 Close the Patch Input/Output window.

Your track is created and patched, so now you can arm it for recording. The Arm button, labeled "R" for record enable, is available in both the track header and the track's channel strip in the Mixer.

7 In the A6 track header, click the Arm button (R).

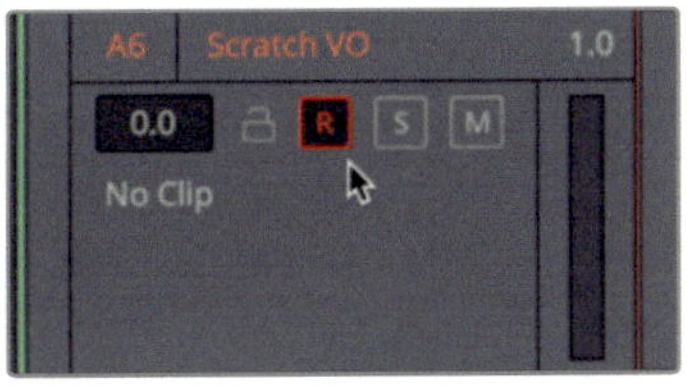

The Arm buttons in both the track header and channel strip turn red to indicate that the track is armed for recording.

8 Hide the Mixer, and mute the A3 track.

Muting tracks prevents them from playing. In this case, let's also turn off visibility for the A2 and A3 tracks, which will make it easier to focus on the track you are recording.

9 Show the Tracks Index. Then, in the Tracklist, turn off visibility for the A2 and A3 tracks.

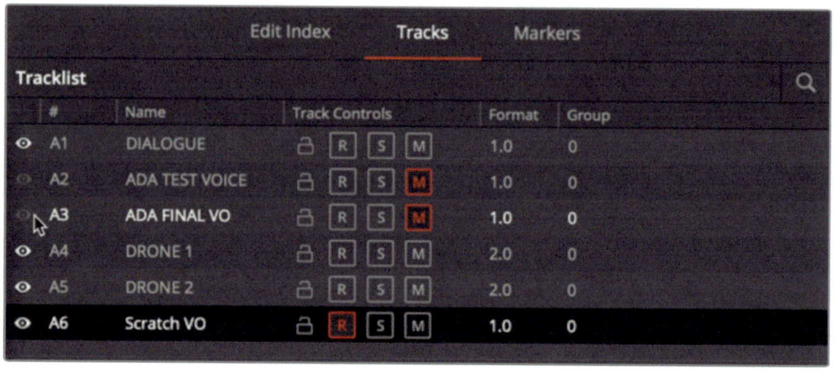

> **NOTE** If your hardware supports multitrack recording, you can patch and arm multiple tracks accordingly.

Now that the track is patched and armed, it's a good idea to take a moment to check your monitoring style.

Monitoring the Track Input

DaVinci Resolve 16 offers a variety of user-defined Input Monitoring styles with which you can hear the microphone signal before, during, and after you record. In the Fairlight menu, you have five Input Monitoring styles from which to choose, including Input, Auto, Record, Mute, and Repro.

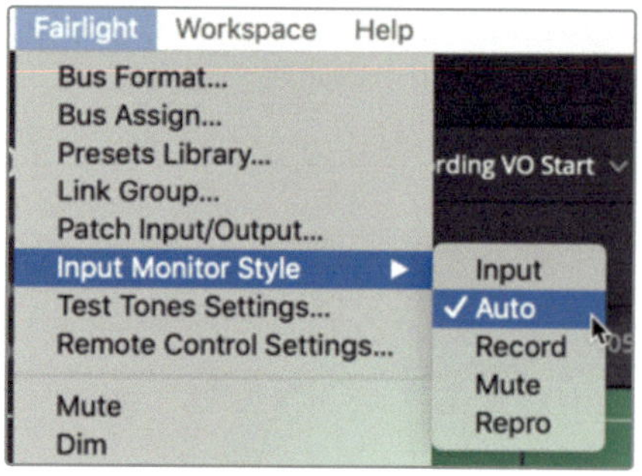

The Input style lets you hear only the live signal on input, whereas the Record style limits you to monitoring the input signal to active recording when the Record button is on. Repro monitors only the audio already recorded to the track; instead of hearing the live input, you can review what was just recorded during recording.

As you may have guessed, when you choose the Mute style, you'll hear nothing. If you think you'd never need to use the Mute style, imagine trying to record a scratch voiceover in the same room as the speakers/monitors but without using headphones.

Fortunately, the default input monitoring style, Auto, is perfect for this recording exercise because it lets you hear the live input signal from armed tracks during recording, as well as the contents of each track during playback.

1 In the Fairlight menu, choose Input Monitor Style > Auto.

2 Speak into the microphone. You should be able to hear the input signal through your headphones and/or speakers.

TIP If the microphone and monitoring speakers are in the same room, you'll need to use headphones to monitor your input during recording. If you don't have headphones available, you can mute playback during recording or use the Mute input monitoring style.

Finally, if you are using an external microphone and audio input/output interface, you may need to increase the microphone input level. You can do so in the Path Settings window in the Input pop-up menu.

3 In the Mixer, in the A6 channel strip's Input pop-up menu, choose Path Settings.

The Path Settings window opens with standard controls for adjusting the input level of audio signals for each track. Here, you can use the Mic/Instr control to adjust the Mic/Instrument level for the source patched to the A6 track.

4 If your microphone input level is low and you are using an external microphone and audio interface, click the On button for the Mic/Instr controls to enable the mic input level control. If your microphone requires phantom power, click the 48V button to enable phantom power. Then, adjust the input level, as needed.

5 When you are finished, close the Path Settings window.

6 Hide the Mixer.

> **MORE INFO** You can find detailed information about all the Input Monitoring styles and Path Settings controls in the DaVinci Resolve 16 Reference Manual, available in the Help menu.

The last thing that you need before recording is the script. Rather than rely on printed text in the book, let's show the text right in the viewer.

Using a Text Clip as a Script Prompt

One distinct advantage DaVinci Resolve's Fairlight page has over audio-only applications is that you can utilize the text tools in the Edit page without leaving the current application or timeline. If you work in television or in a large facility, you probably have access to a teleprompter system to display the script when recording voiceover. However, with DaVinci Resolve, a simple text clip in the Edit page will do the trick.

In this exercise, you'll go to the Edit page to enable a text clip already in the timeline.

1 Click the Edit button, or press Shift-4, to go to the Edit page.

2 In the Edit page timeline, select the disabled clip in the V2 Video 2 track, and press D.

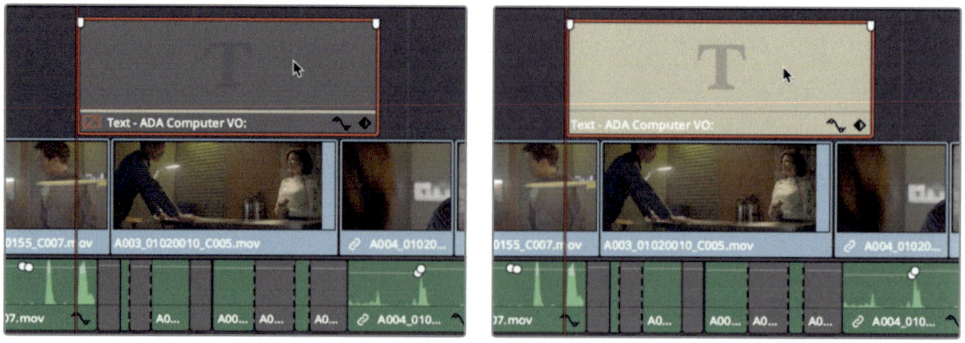

3 Press Shift-7 to return to the Fairlight page.

4 In the Timeline View Options, show the Video Tracks.

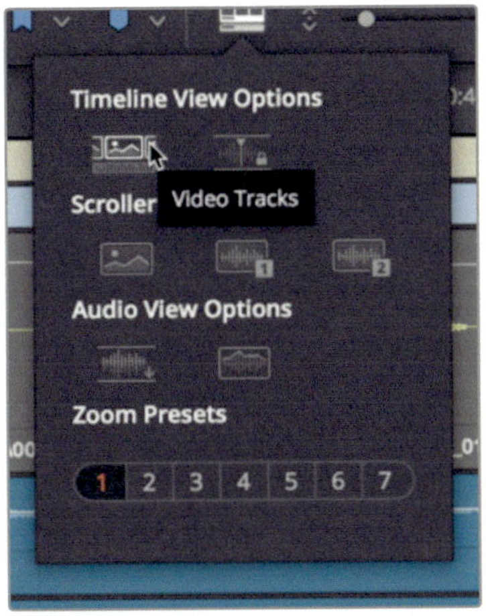

Now you can see the video tracks including the beige-colored text clip in the V2 track.

5 Move the playhead over the beginning of the text clip in the V2 Video 2 track so you can read the text in the viewer.

6 Use the scroller at the bottom of the timeline to center the section of the timeline that you'll be recording.

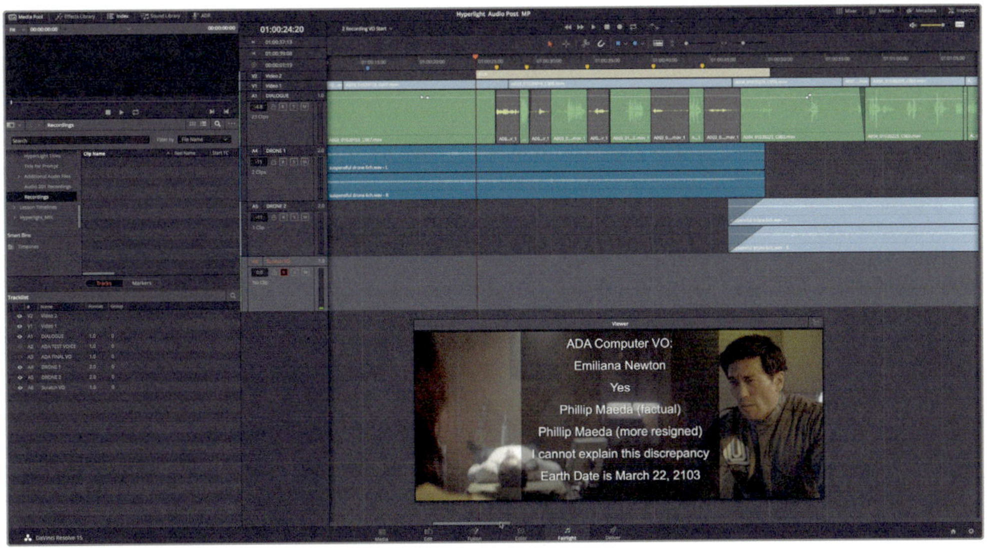

You are ready to record.

Recording in the Timeline

You have completed all the necessary steps to prepare your project for recording. Remember, you're creating a scratch voiceover track, so don't worry about the quality of your microphone or background noise. Before you start recording, it's a good idea to practice your lines and double-check that the Recordings bin is still selected.

1 If necessary, in the Hyperlight Audio Post bin, select the Recordings bin to designate it as the storage location for your new recordings.

2 Solo the A6 Scratch VO track.

For the purposes of this voiceover recording session, you will need to read through the lines sequentially, one after another. You'll separate the different lines and move them into position in the next lesson. Keep in mind that the final voiceover line comes at the end of the scene.

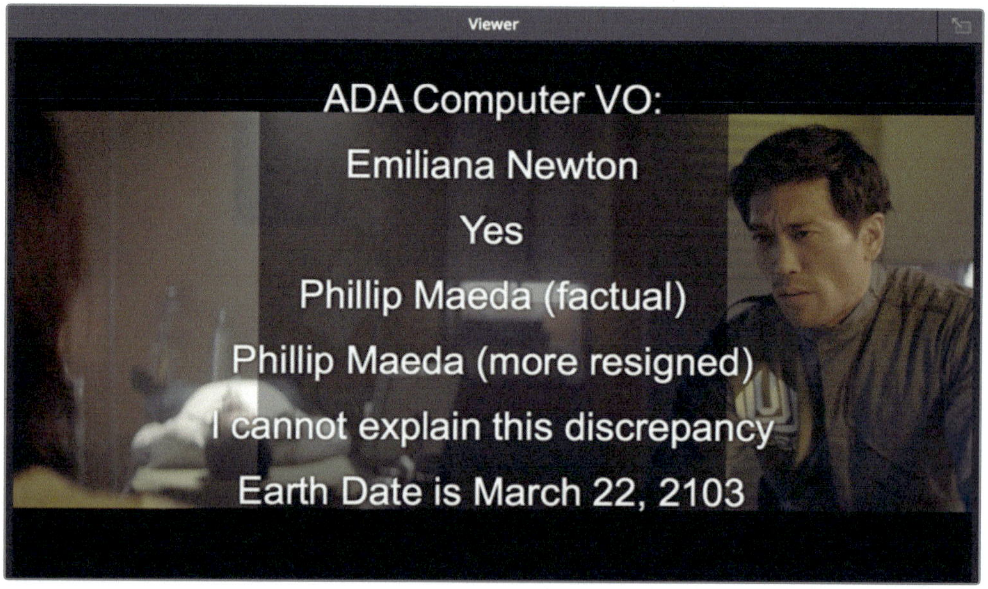

3 Practice reading your voiceover lines.

4 Make sure that the A6 track's Arm button is on (red).

Recording starts at the playhead position on the armed track.

5 Move the playhead to the beginning of the text clip in the V2 track (01:00:24:20).

You'll find the Record button in the transport controls at the top of the screen. When the playhead starts moving and drawing a red region within the armed track, you are recording.

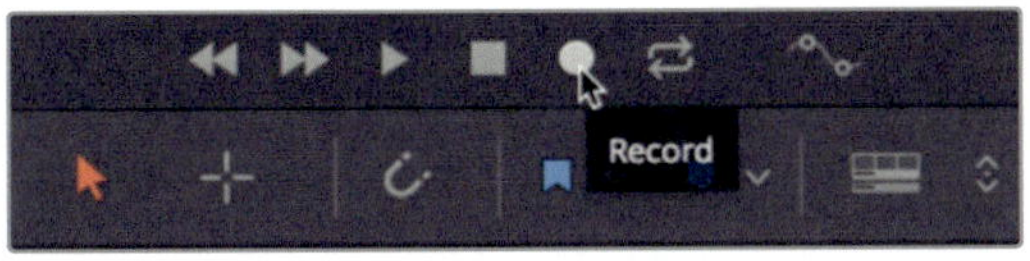

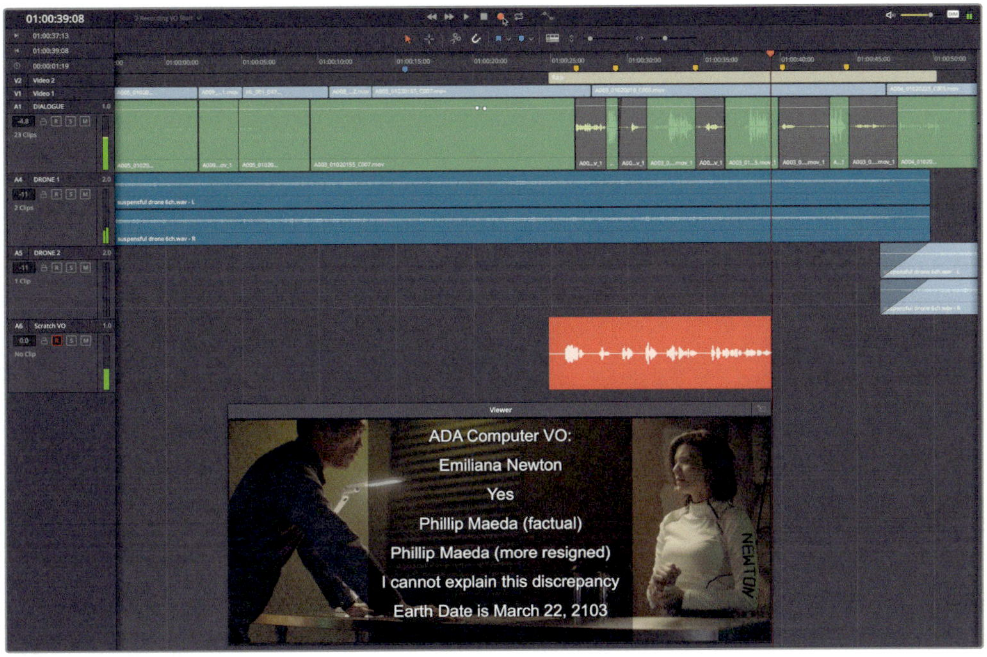

6 In the transport controls, click the Record button. Read the script into the microphone. When you are finished, press the Spacebar to stop recording.

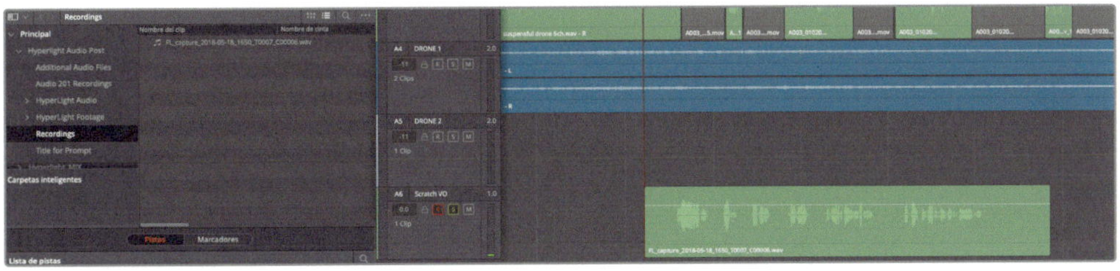

The new audio clip appears in the A6 Scratch VO track and in the Recordings bin as a .wav file. Notice that the clip name starts with FL_capture followed by the date. The FL stands for Fairlight.

7 Play the timeline to hear your recorded take.

Don't worry if your recording was rough. The purpose of this lesson is to give you the skills to record audio in the Fairlight page, not perfect your talent as a voiceover artist. That being said, if you work in corporate or commercial video, you know that recording scratch (rough) voiceover and narration tracks is all part of the job.

Let's record another take over the first one. Doing so will not overwrite the first recording; instead, the new take will be saved as a separate layer in the same track. To keep track of your takes, let's color code them as you go.

8 Right-click the first take in the A6 track, and choose Clip Color > Yellow.

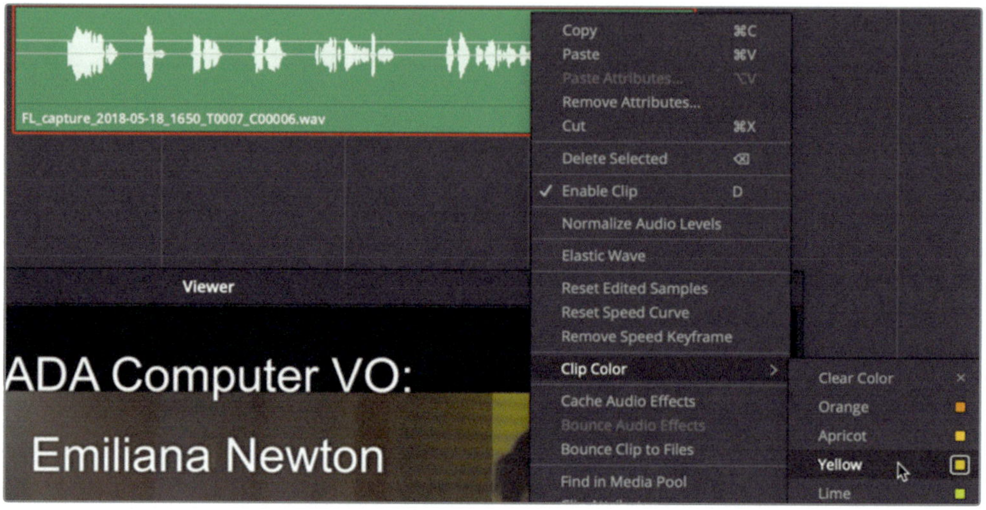

9 Record another take. When you are finished, stop playback and change the color of the second take to orange. Listen to the second recorded take.

If your second take is shorter than the first, you'll see where the second recording overlaps the first take in the track.

Now that you have recorded several full takes, you can try recording a partial take of just the fifth phrase, "I can't explain this discrepancy."

10 In the Playback menu, choose Stop and Go To Last Position, or Press Option-K (macOS) or Alt-K (Windows), to turn off the playback mode.

11 Press the J K L keys, using the waveform as a guide, to move the playhead to the beginning of the fifth phrase.

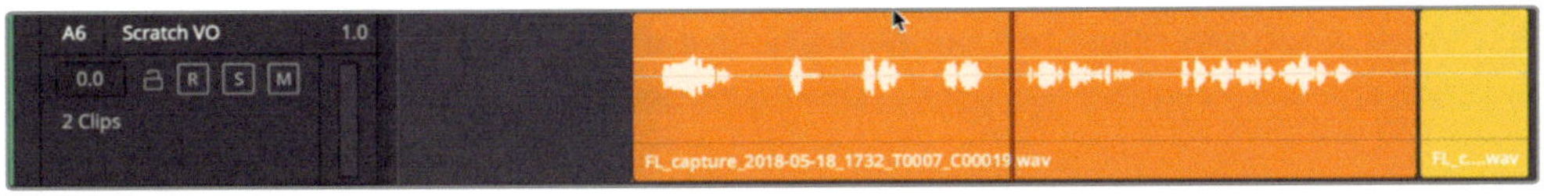

This time, when you record the third take, vary the line slightly so the computer says "can't" instead of "cannot."

12 Click the Record button, and record the entire sentence: "I can't explain this discrepancy." Click the Stop button, or press the Spacebar, to stop recording.

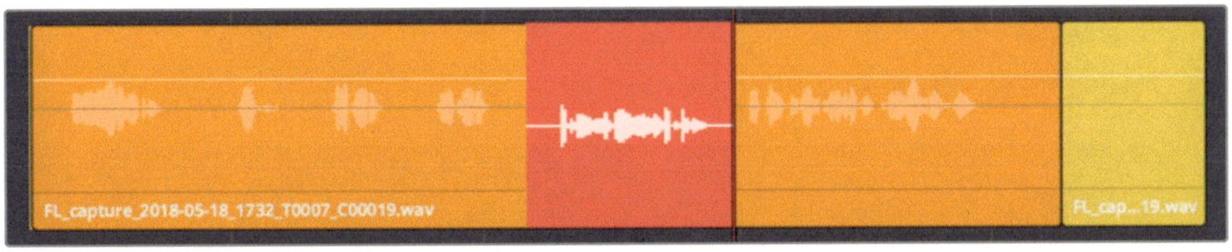

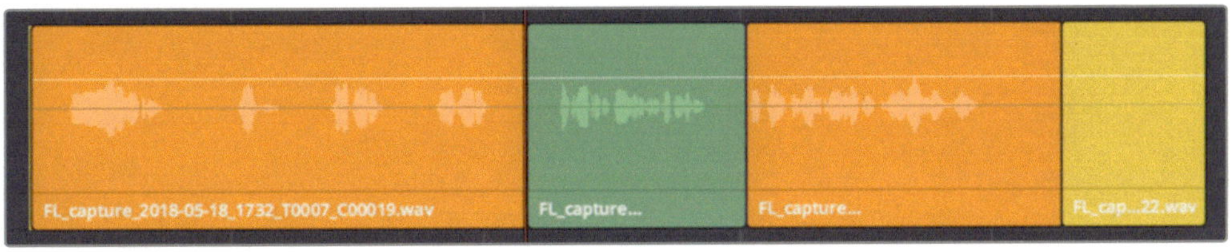

13 Change the partial take color to apricot.

14 Un-solo and disarm the A6 track. Notice that the A6 track header shows that it contains three clips.

15 Save your project.

Now that you have recorded several takes in the same track, let's show the audio track layers so you can see all of your takes at once.

Showing Audio Track Layers

Track layers are a powerful way to record and edit multiple clips in a single track. The most important thing to remember when working with track layers is that you hear only the uppermost clip at any given time.

You'll work more extensively with audio track layers in the next lesson. For now, let's show the track layers so you can see your separate recorded takes.

NOTE If you didn't record any takes, feel free to open the 2a Recording VO Finished timeline and follow along.

1 Hide the Media Pool and Index. Move the viewer to the lower-right corner of the timeline.

2 Choose View > Show Audio Track Layers.

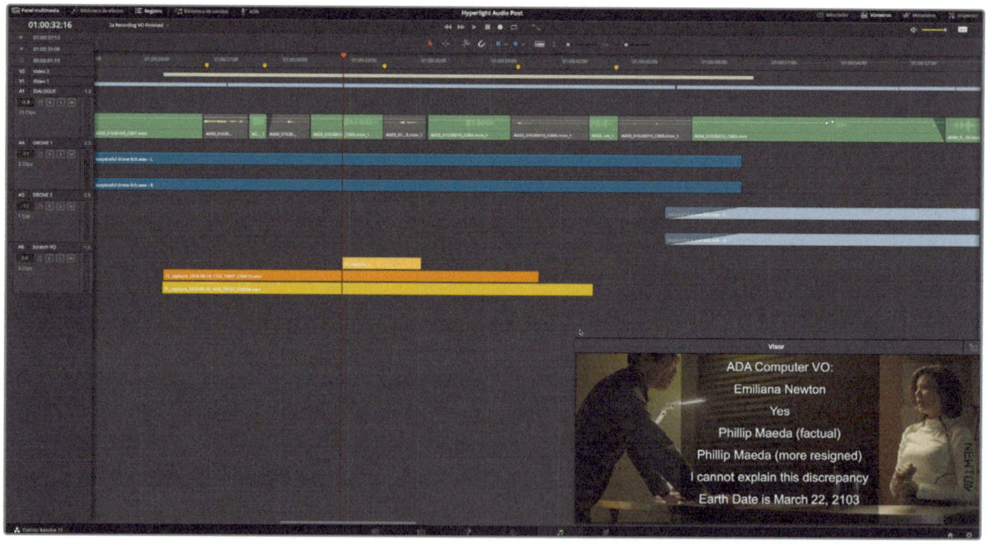

Each clip has its own layer, and an empty, clip-sized space appears in each track above the clips. The height of each clip and layer is based on the number of clips in the track.

Track A1 is a mono track with a single layer of audio clips. Tracks A2 and A3 are stereo tracks with two channels (left and right) and an empty layer above each channel. Track A6 Is also a single-channel mono track containing stacked mono clips, each on a different layer.

When starting playback, only the uppermost clip exposed to the playhead is played.

3. Solo the A6 track and Play the stacked takes.

   At the moment, you only hear take 2 (orange) and take 3 (apricot).

4. Un-solo the A6 track. Choose View > Show Audio Track Layers to deselect that option.

   You have successfully set up, patched, armed, and recorded a scratch voiceover track in the Fairlight timeline. In the next lesson, you'll further explore the power of the audio track layers to select, play, and edit takes, as well as create crossfades within a track.

NOTE If you are using a Fairlight Desktop Audio Editor or Fairlight Audio Editor panel in a Fairlight console, you'll have additional options for manually recording in the timeline. These options are available in the Audio Editor's Record mode toolset, and include: Record Here, Record Range, Record Clip, Record Again, Punch In Again, and Record Head.

What Is ADR?

If a dialogue recording can't be used because it is damaged, noisy, or unclear, it must be replaced with audio from other takes or be re-recorded. The process of re-recording production dialogue is called automated dialogue replacement (ADR), or looping. ADR sessions usually require the original actor to re-record the dialogue to improve the sound quality or reflect script changes. ADR sessions are also commonly used for dubbing films in other languages, recording dialogue for animated films, and performing Foley sound effects. The term "looping" refers to the mechanics of recording dialogue replacement that involves playing, rehearsing, and recording the same dialogue cue over and over again.

NOTE Although most of the lessons in this book follow a real-world workflow, the following set of ADR exercises is slightly out of order. Dialogue replacement usually happens at the end of the dialogue editing process, which you'll encounter in Lesson 4. However, because this lesson is dedicated to recording, it made sense to demonstrate recording dialogue replacement in this context.

Setting Up an ADR Session

DaVinci Resolve 16 includes a full ADR toolset in the Fairlight page. In this exercise, you'll open a new timeline for ADR recording and use the ADR Setup panel to customize your session.

1 Open the 2b ADR Session Start timeline.

This is a longer version of the scene you worked with in the previous lesson. As you can see, each character's dialogue clips have been separated to different tracks. This is part of the dialogue editing process. Also, the A1 ADR track was created for the replacement dialogue recordings.

2 Play the scene once to familiarize yourself with the content and how it has evolved. When you are finished, stop playback, and move the playhead to the beginning of the timeline.

Now that you have listened to the scene, let's locate the dialogue lines that you will replace. They are identified with markers.

3 Press Shift-Down Arrow to move the playhead to the first pink marker. Then, zoom in horizontally so the pairs of pink and yellow timeline markers are near the center of the ruler and the clips in that section are easy to see.

In a few minutes, you will record replacement dialogue for either Emiliana's line, "Identify the person on the medlab," or Philip's line, "ADA how is that possible?" Let's play them once, and then go to the ADR tools to set up your recording session.

4 Play the timeline from the first pink marker to the last yellow marker.

5 In the interface toolbar, click the ADR button to show the ADR tools. Click the Setup button at the top of the ADR tools to show ADR setup panel.

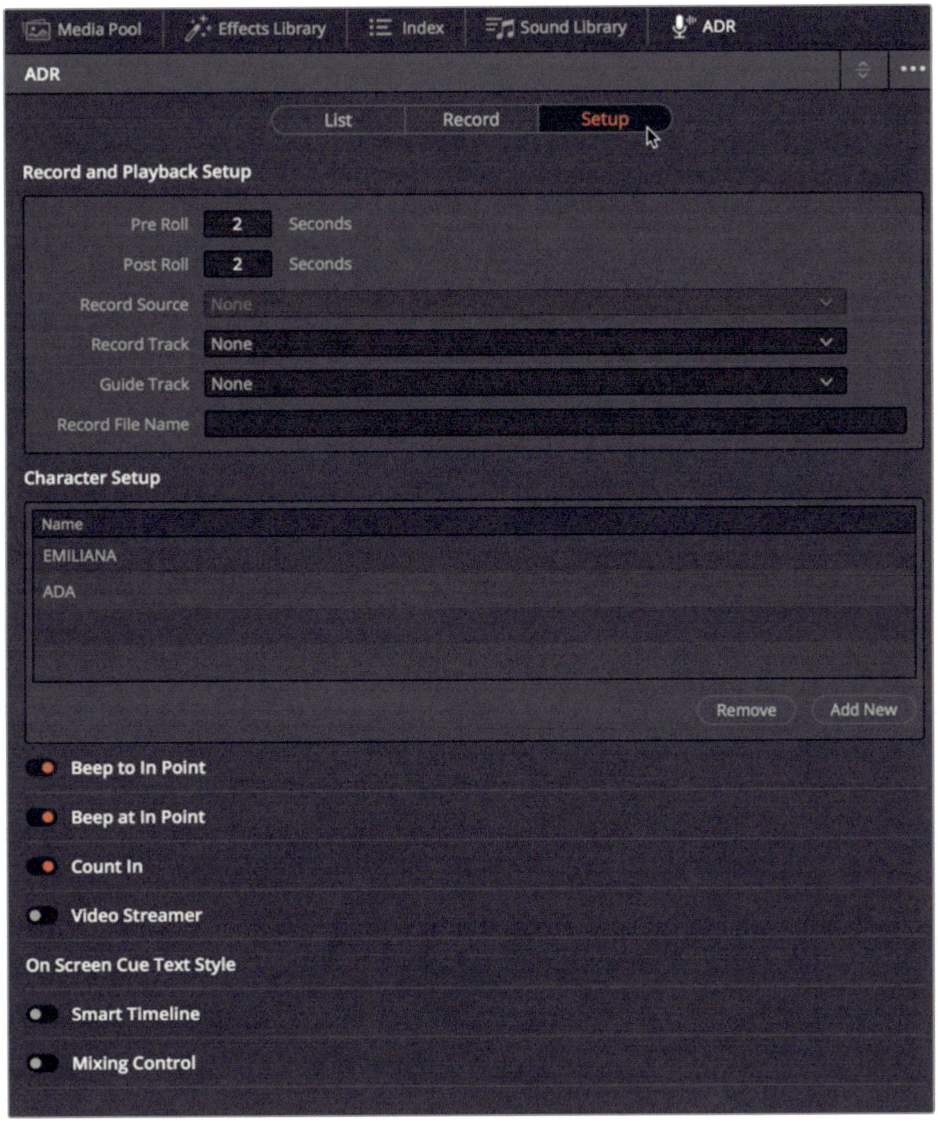

The setup panel is self-explanatory, so let's just walk through the setup for this session, starting at the top with the Record and Playback Setup controls.

The Pre Roll and Post Roll controls let you determine how much time you have before and after a cue is recorded. Think of pre roll as allowing time for a count-in before recording begins.

6 Set the Pre Roll to 6 Seconds, and the Post Roll to 2 Seconds.

Notice that the Record Source menu is empty. To choose a source, you first need to choose the Record Track.

7 Choose Record Track > ADR. Then, in the Record Source pop-up menu, choose your microphone.

The Guide Track is the track with the original dialogue that you will listen to as you record the replacement.

8 Choose Guide Track > EMILIANA.

9 In the Record File Name field, type your initials followed by **ADR Session 1**.

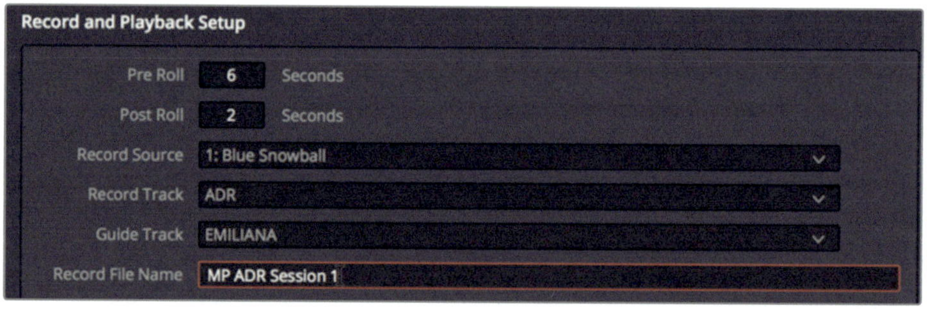

Next, let's move down to the Character Setup area in the center of the ADR Setup panel, and add a new character.

10 Click the Add New button. In the selected Character name field, type **PHILIP**. When you are finished, press Return.

When working with professional talent in an ADR session, you'll find that actors each have unique personal preferences for working with cues. Some like to hear a series of beeps before they speak. Some prefer a visual count-in, or *streamer*, and no beeps, whereas others want absolutely no visuals and only a single beep right before they speak. Fortunately, the Fairlight ADR setup panel includes a variety of audio and visual cues that you can mix and match to suit your sessions and your talent. In addition to switching the different audio and visual cues on or off, you can also double-click the setting name to reveal the full controls for that element.

NOTE If you are using a Fairlight Desktop Audio Editor or Fairlight Audio Editor panel in a Fairlight console, you'll have additional options for manually recording in the timeline. These options are available in the Audio Editor's Record mode toolset, and include: Record Here, Record Range, Record Clip, Record Again, Punch In Again, and Record Head.

11 Click the following controls to select them: Beep to In Point, Beep at In Point, Count In, Video Streamer, and Smart Timeline, as shown.

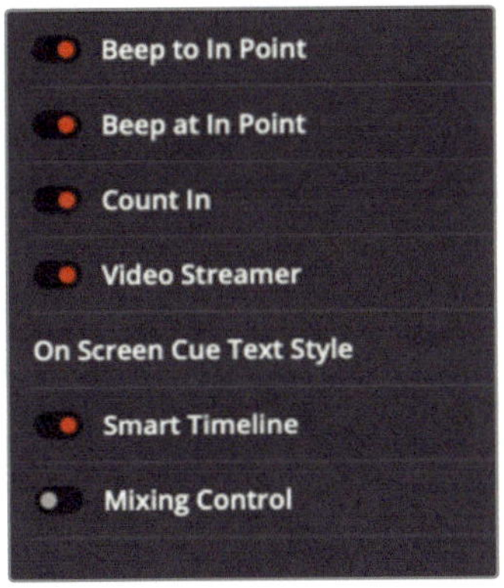

12 Double-click setting titles for the Beep to In Point and Beep at In Point settings to show the controls for those settings.

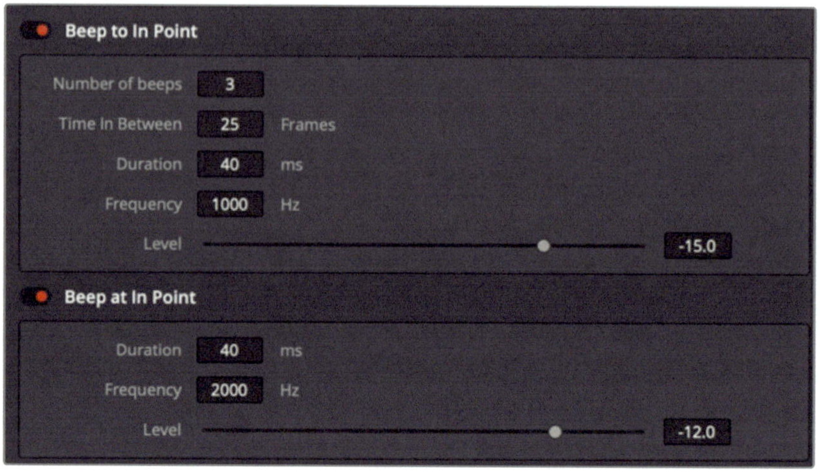

These Beep settings offer two types of audio cues to help the talent prepare to record their replacement dialogue lines. If you look closely at the settings for each, you will see that the Beep to In Point is set to emit a -15 dB, 1000 Hz beep three times before the In Point for the recorded ADR Cue, while the Beep at In Point will emit a single 2000 Hz beep at -12 dB at the In Point where the recording begins. The slight difference in frequency and volume level between the two types of beeps gives the actor an easy-to-follow “one, two, three, go!” during rehearsal and recording.

To utilize the Beep audible cues, you first need to create a track and assign the input to the Beeps in the built-in Oscillator available via the Test Tones Settings in the Patch Input/Output window and Fairlight menu. You'll work more with the Test Tones Settings later in this lesson. For this exercise, the Test Tones Settings Beeps have already been assigned to the A9 ADR BEEP track in the timeline.

13 Double-click the Count-In header to reveal the Count-In settings. Then, change the Count From field to 5. Leave the to field set to 0 (zero).

14 Double-click the On Screen Cue Text Style header to reveal the settings.

Here you can change the Font, Size, and other style elements for the onscreen text cues.

15 Change the Font to Arial Black and the size to 50.

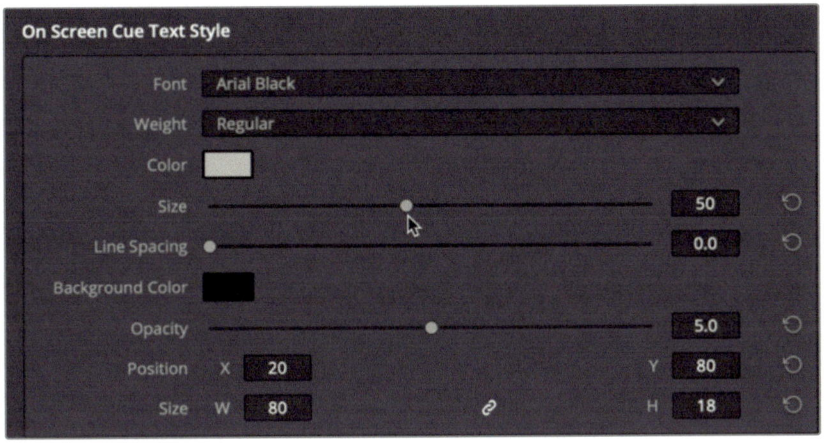

16 Option-double-click any of the setup headers to hide all the opened Setup settings panels.

The ADR session is set up. Next, you'll set up the cues in the List panel.

NOTE You can go back to the ADR Settings panel anytime to modify your settings while you work.

Manually Creating ADR Cues

You can import an entire ADR cue list or manually create new cues as you go. In this exercise, you'll create two new cues right in the timeline. To manually create cues, you simply mark an in point and out point for the cue in the timeline and click the New Cue button in the ADR List panel. You can also add the text and character to the cue. Let's try it.

1. Click the List button to show the ADR List panel.

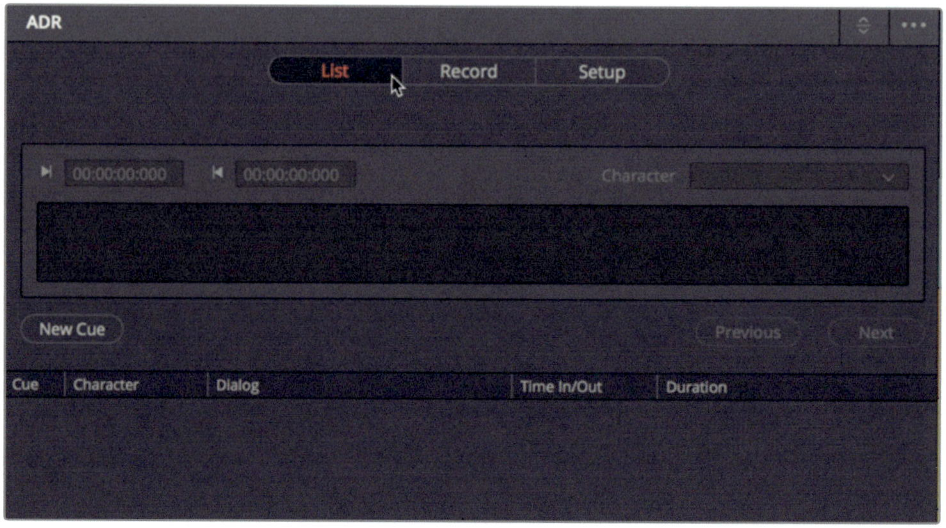

2. Move the playhead to the first pink marker. Press I to set an in point.
3. Press Shift-Down Arrow to move the playhead to the second pink marker. Press O to set an out point.
4. In the ADR list panel, click New Cue.

Cue 1 appears in the list. When a cue is selected in this list, you can modify its information in the cue editing controls above the list. Let's set the character for the selected cue.

5 Select the cue in the list, if necessary. Then, in the cue editing controls, in the Character pop-up menu, choose EMILIANA.

The only element missing now is the dialogue.

6 In the dialogue field above the cue list, type **Identify the person on the medlab**.

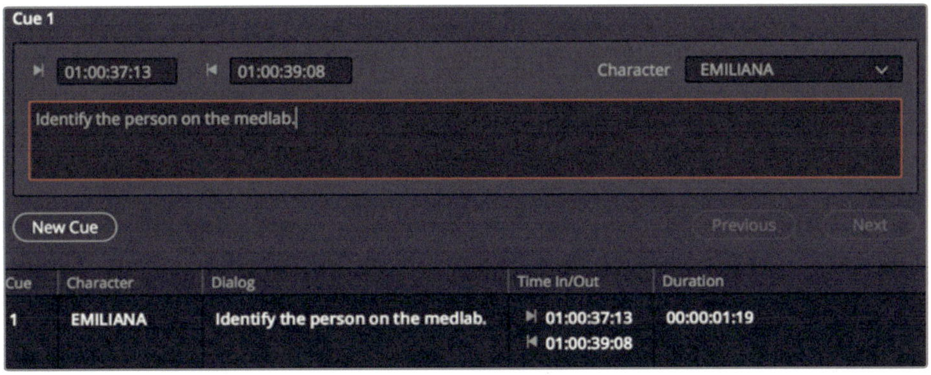

The text appears in the dialogue field in the cue list as you type. Now that you have created one cue, you are ready to create the next one.

7 Choose Mark > Clear In and Out or press Option-X (macOS) or Alt-X (Windows) to clear the in and out points. Then, in the timeline, use the yellow markers to mark the in and out points.

8 Click New Cue to create Cue 2. Change the character to Philip. Finally, type the dialogue, **ADA how is that possible?**

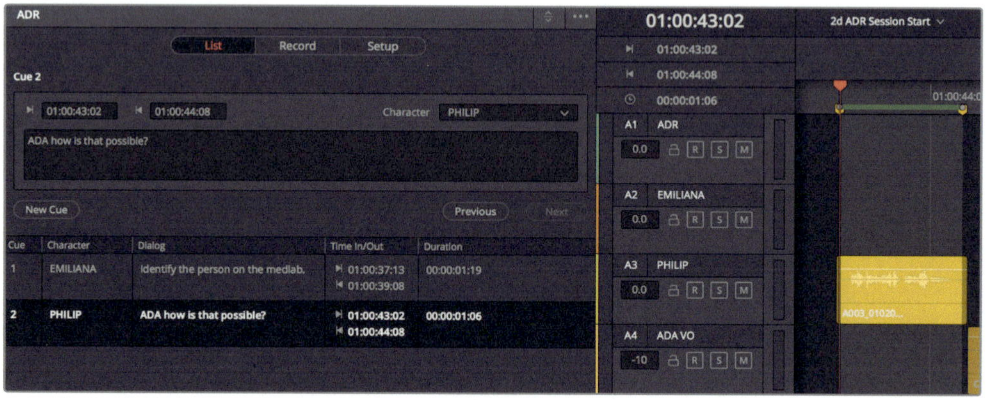

As with all recordings, you also need to select a bin for them in the Media Pool; otherwise, they will appear only in the Master bin.

9 Show the Media Pool, and select the Recordings bin. Then, hide the Media Pool and reopen the ADR panel.

10 Press Option-X to clear the in and out points in the timeline.

Your ADR session is ready for recording.

Recording ADR Cues in the Timeline

Once you have set up your ADR session, you can simply go to the ADR Record panel to record. However, your talent will probably need to rehearse a few times before recording. In that case, you'll use the Rehearse button located to the far-left of the Record panel controls. In this exercise, you will open the ADR Record panel and select a cue to record. Then, after rehearsing the cue, you'll record three takes.

1 Click the Record panel button to go to the ADR Record panel.

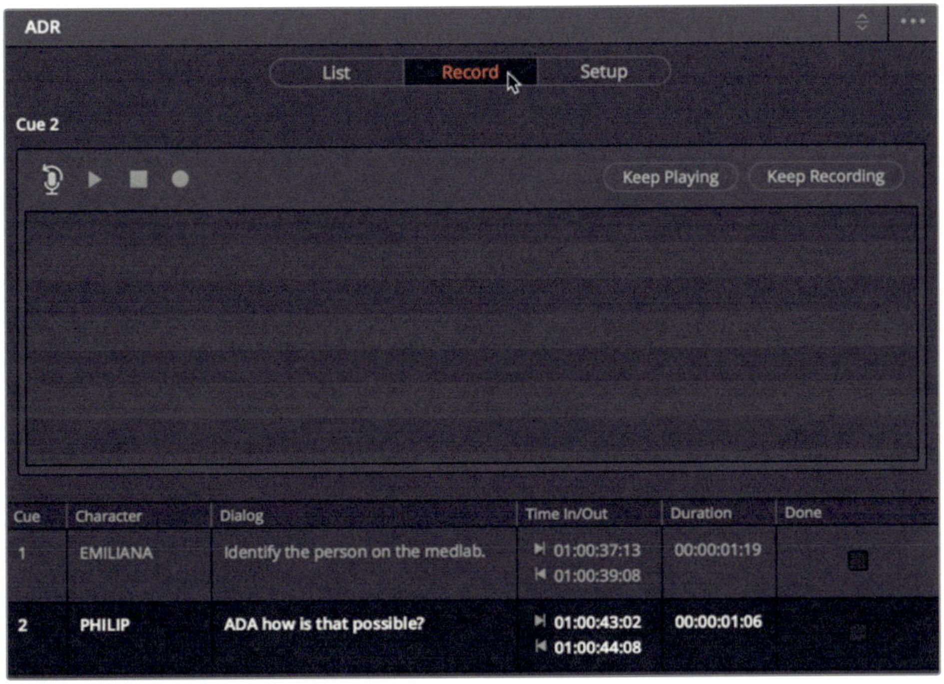

2 In the Record panel cue list, select one of the two cues to record.

Notice that the playhead moves to the in point of the selected cue.

3 On the A1 ADR track, if necessary, click the Arm button to enable the track for recording.

4 Zoom in to the cue in the timeline, as necessary, and resize the viewer so you can clearly see the talent speaking on-screen. As you resize the viewer, keep in mind that the only tracks you need to see during recording are A1, A2, and A3.

5 In the Record panel, click the Rehearse button.

You will hear the audible cues in your headphones and see the visual cues in the viewer during the rehearsal.

Chances are your first rehearsal was a little rough. No worries, you can rehearse again. Just remember, when recording replacement dialogue in sync with the picture, you will need patience and many takes to get it right.

6 Rehearse the cue again, or as many times as you need. Don't worry about getting it perfect. When you are ready to record, move on to the next step.

7 Click the Record button and record the cue.

A newly recorded region, which is the exact length as your cued in and out points, appears in the A1 ADR track.

8 In the Record panel, click the Play button to audition the recorded cue.

9 Rate your performance by clicking the stars to the right of the take in the list.

10 Repeat step 7 so that you have at least three takes.

The takes are recorded in different layers in the same track and listed in the Record panel.

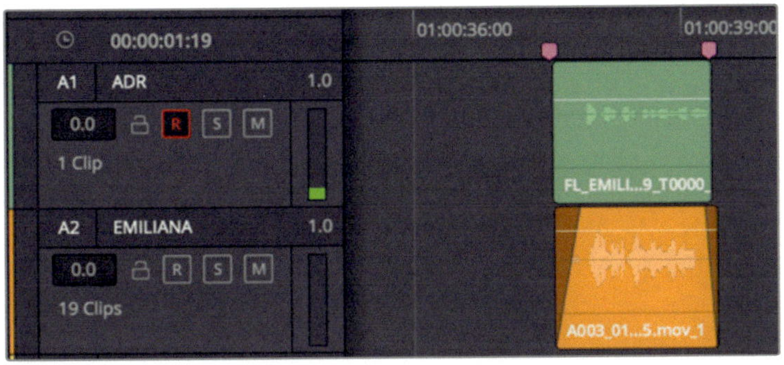

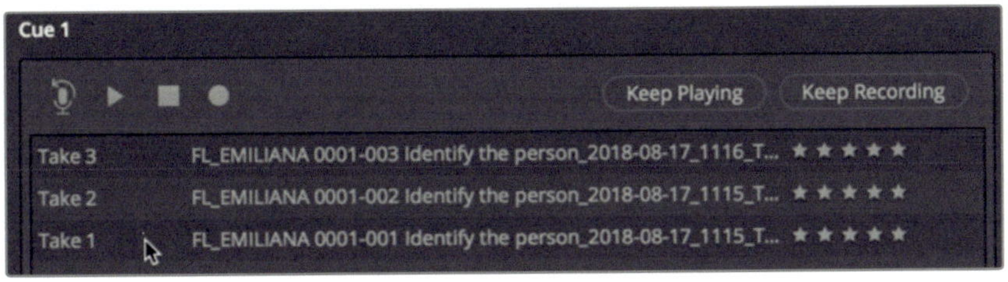

NOTE If you look carefully at the names of the recordings in the list, you'll see that they include useful information: (from left to right) FL (Fairlight) followed by the character's name, take number, cue dialog, and the date.

To see the clips for all of the takes in your timeline, you need to show the audio track layers.

11 Choose View > Show Audio Track Layers. Adjust the track zoom, as necessary, to clearly see the multiple takes in the A1 track.

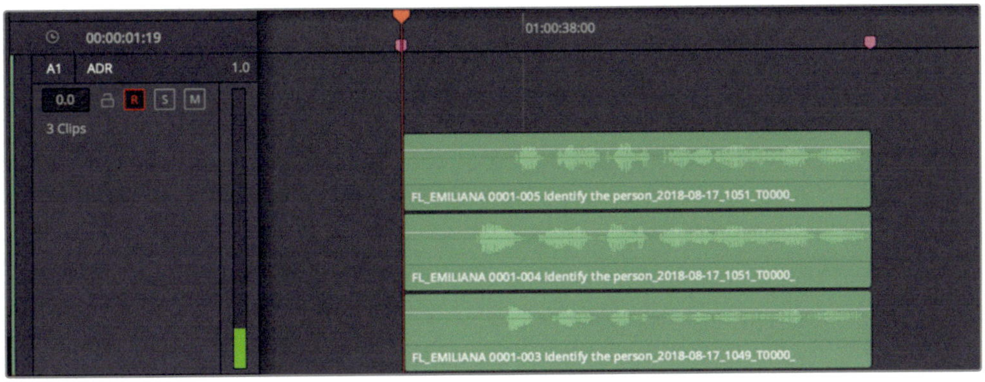

All of your ADR takes are in the timeline and ready to use in your soundtrack. This exercise was just an opportunity to see your takes in the track. You'll work more with audio track layers in the next lesson.

12 Hide the audio track layers.

13 When you are finished, select the box in the far right column to mark the cue as done. Then, deselect the Arm button on the A1 track.

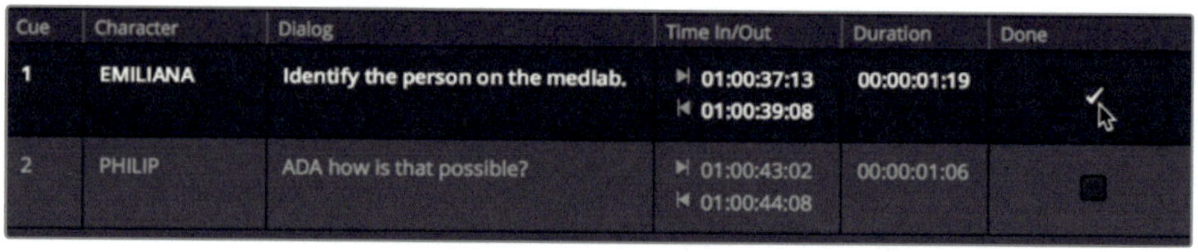

NOTE If you enjoyed recording ADR and want another go at it, feel free to practice by recording the other cue or additional takes.

Importing an ADR Cue List

Not only can you create your own cues in the timeline, you can also import a comma separated value (CSV) file containing an ADR cue list. In this exercise, you'll import a CSV file with the dialogue cues for the entire scene.

1 Hide the viewer.

2 Click the List panel button to go to the ADR list panel.

You'll find the options to import and export cues lists in the ADR panel options menu (three dots) located in the upper-right corner of the ADR panel.

3 In the ADR panel options menu, choose Import Cue List.

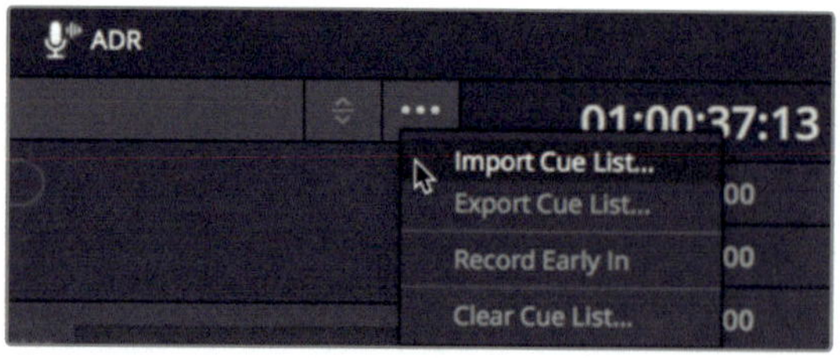

The .csv list opens into the ADR Setup window. Here you can verify that the columns are in the correct order for ADR recording: In Point, Out Point, Character, Dialog.

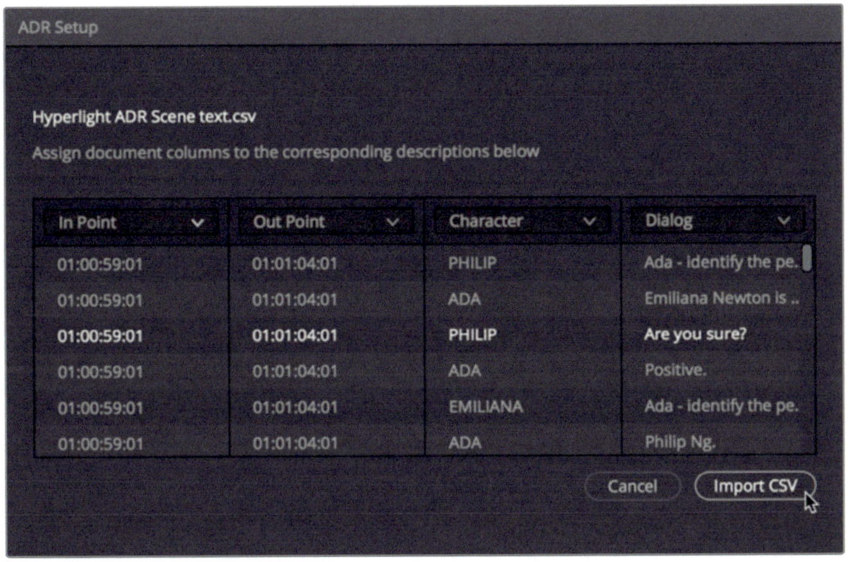

> **NOTE** If the column contents do not match the column header in the ADR Setup dialog, you can use the menu at the top of the column to choose the correct header.

4 Click Import CSV.

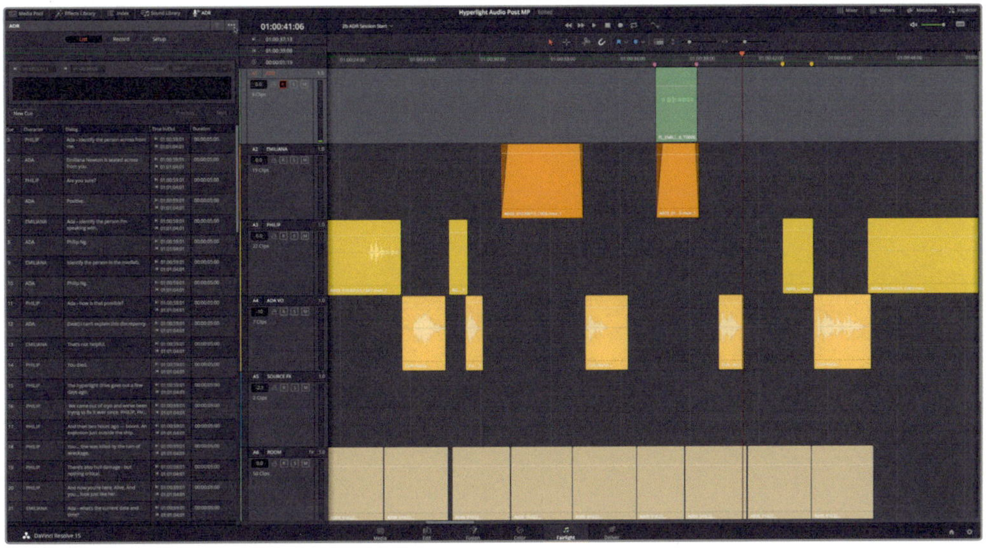

The ADR cue list is populated with a full set of ADR cues for the entire scene. You can create, edit, and export CSV files in most spreadsheet programs.

5 Hide the ADR panel. Hide the Video tracks, if showing.

You never know when you might need to record some ADR for your own projects. As you have just seen, DaVinci Resolve 16 makes this process easy to set up and use at any time. ADR recording techniques also work great for recording Foley sound effects to match on-screen action.

Recording the Built-In System Generator

Fairlight includes a System Generator that you can use to generate sounds such as white noise, beeps for ADR audible cues, or tones that you can transform into sound effects. In this last recording exercise, you will create a new timeline, patch the System Generator, and record two different sounds that you can later use for sound design.

First, let's create a new timeline and track.

1 Show the Media Pool, and create a new bin named My Timelines.

   This bin is where you can save any new timelines that you create while working through the various exercises.

2 Press Command-N (macOS) or Ctrl-N (Windows) to create a new timeline.

3 In the New Timeline dialog, in the Timeline Name field, type **Sound Design FX**. Leave the default settings as they are. Click Create.

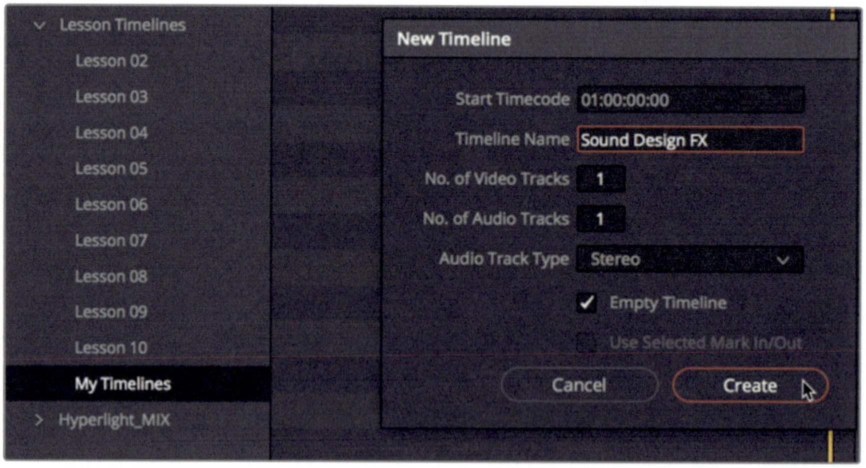

A new, empty timeline opens with one stereo audio track.

4 Name the A1 track **GEN FX**.

Now that the timeline is created, let's take a moment to select the Recordings bin in the Media Pool. Keep in mind, if you ever forget to select a specific bin for your recordings and no bin is selected, they will appear in the selected bin, or under the Master bin. You can also move them to another bin after you are finished recording.

5 In the Media Pool, select the Recordings bin. Hide the Media Pool.

You can use the Patch Input/Output window to patch the built-in System Generator to the track just as you would a microphone or other external audio device. Previously, you opened the Patch Input/Output panel from the Mixer. This time you'll do so within the Fairlight page.

6 Choose Fairlight > Patch Input/Output.

The Source Audio Input offers numerous options from which to choose, so you'll need to scroll down in the Audio Inputs menu to find the System Generator.

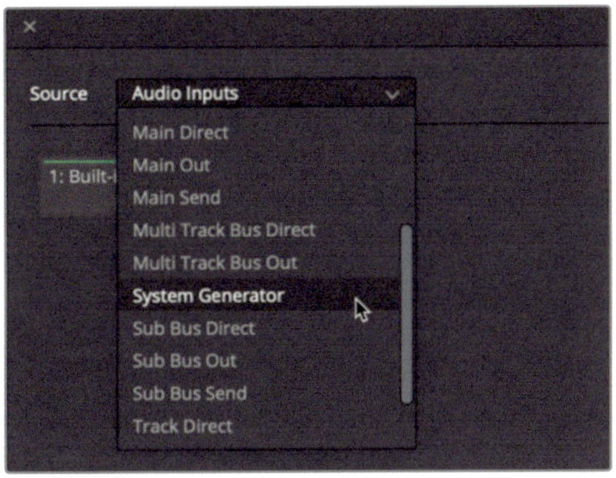

7 Choose Source > System Generator.

The System Generator can generate three types of sound that you can patch and record. Let's start with the Noise generator.

8 Click Noise as the System Generator Source. Then, choose Destination > Track Input. Select both channels of the GEN FX track for the destination because this is a stereo track.

9 Click Patch, and close the Patch Input/Output window.

WARNING! Do not arm the A1 track yet. You just patched a noise generator to the track, so if you armed the track, you would immediately hear it just as though it were a live microphone. Let's first take a moment to find and use the master playback volume controls.

Controlling the Playback Volume Level

Whether you are recording, building tracks, or mixing your final project, it's always a good idea to know how to control your master playback monitoring levels. In the upper-right corner of the timeline, you'll find the master playback volume controls. These controls do not affect the levels of your main output or individual tracks. However, they are incredibly useful for quickly adjusting the playback levels in your speaker system, computer speaker, or headphones.

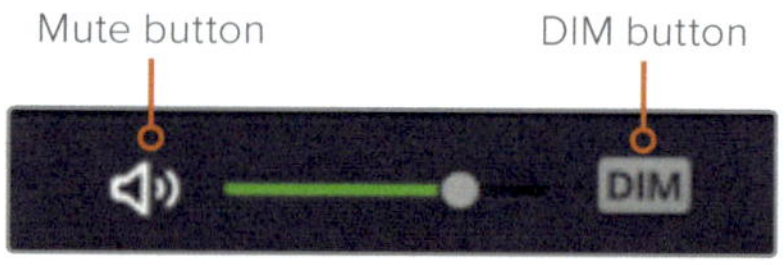

1 Click the DIM button to instantly lower the playback volume by 20 dB.

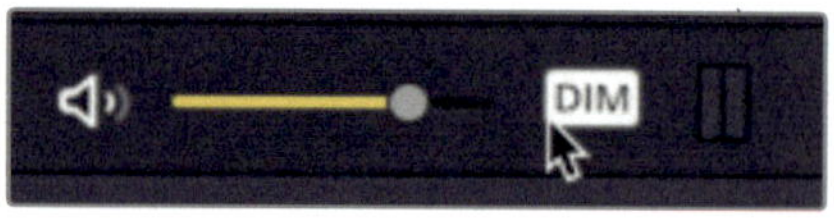

The playback volume fader turns yellow to indicate that the DIM button is on.

NOTE The ability to quickly lower the playback volume is very useful when you are in the middle of a session and someone asks you a question and you need to answer, or as in this case, to protect your ears from the incoming noise generator until you are prepared.

2 Arm the A1 track.

As promised, you can really hear that noise generator through your headphones or speakers, even with the playback dimmed by 20 dB. So, how loud is the System Generator source? The System Generator generates sound at -15 dB. This is also the average dialogue level when you start editing dialogue and mixing.

3 In the playback controls, click the Mute button to mute playback.

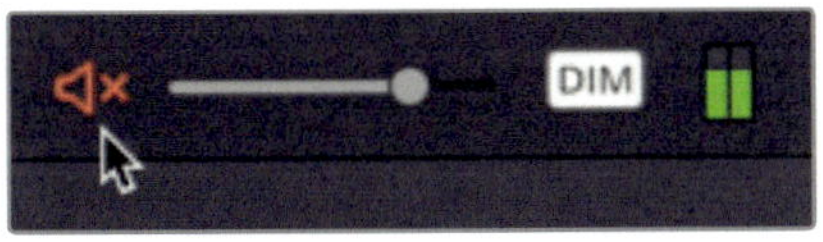

You can combine the playback controls to go from DIM to mute and back whenever necessary.

Changing the Test Tone Generator Settings

Resolve's built-in System Generator settings are available in the Fairlight menu. In this exercise, you'll open the Test Tone settings for the System Generator and experiment with some of its controls.

1 Choose Fairlight > Test Tone Settings to open the Test Tone Settings window.

2 Unmute the playback volume controls. Leave the DIM button turned on.

Feel free to mute the playback volume at any time, and vice versa.

In the Noise controls (located in the lower-right corner of the Oscillator Settings window), you can change the noise type from White to Pink.

3 If necessary, lower the Tone Level until it reads -15. Click the Pink button to change from white to pink noise. Click back and forth between the two types of noise to compare them. When you are finished, leave the Noise set to Pink.

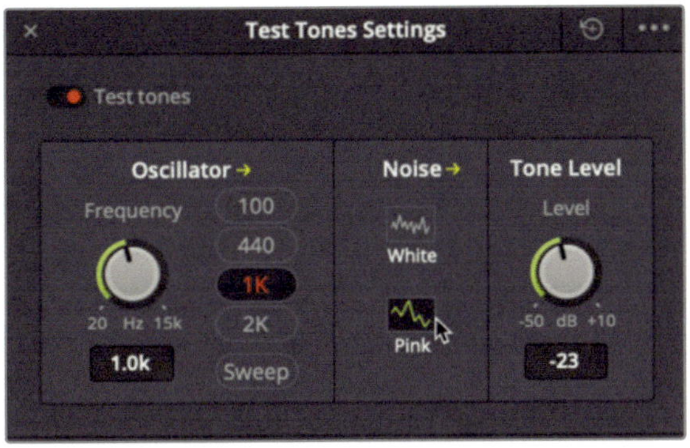

NOTE Both white and pink noise include all the frequencies that the human ear can hear. The difference is that white noise has equal energy per frequency, similar to white light that contains all the wavelengths of the visible spectrum at equal intensity. Pink noise is filtered white noise that has equal energy per octave. With pink noise, the intensity decreases as the frequency increases, which is how humans perceive sound; therefore, it's easier on the ears.

Now let's record a bit of that noise to use later.

4 Click the Record button to record about 10 seconds of pink noise. When you are finished, click the Stop button, or press the Spacebar.

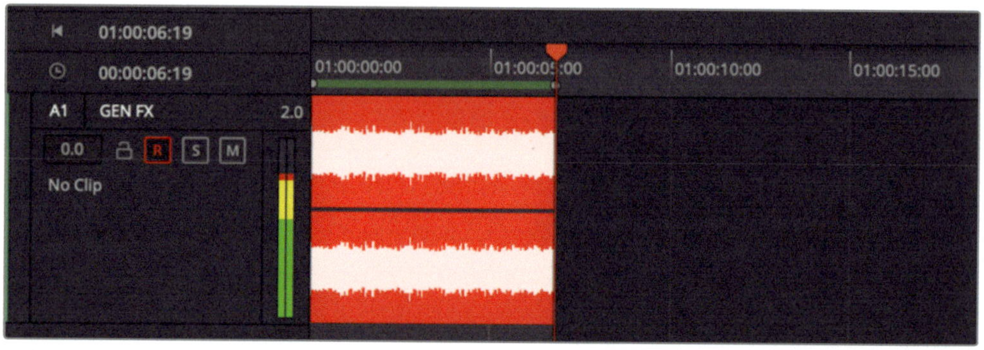

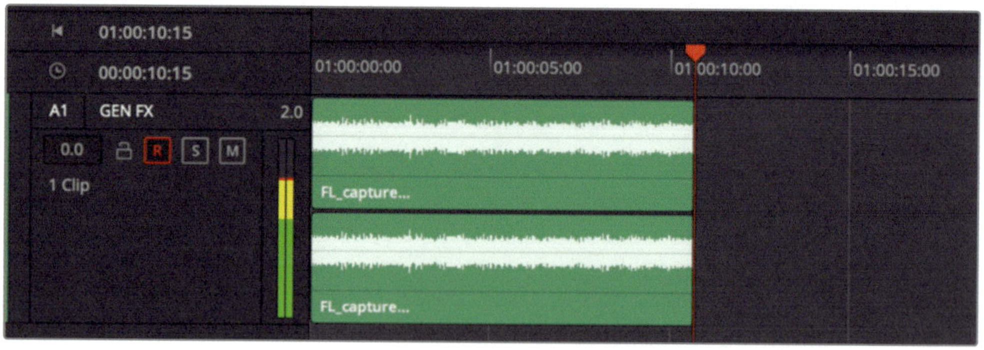

Now let's record a tone that you can control using the oscillator. To do so, you'll need to change the patch input setting. This time, let's open the Patch Input/Output window right from the Test Tone Settings window.

5 Click the options menu (...) in the Test Tones Setting window and choose > Patch Input/Output.

6 In the Patch Intput/Output window, click the Un-Patch button to unpatch the Noise generator.

7 In the Source controls, click the Osc button. Then, in the Destintion controls, select both channels of the GEN FL track. When you are finished, click Patch.

8 Close the Patch Input/Output window.

The default oscillator tone should sound familiar. That is the 1 kHz tone used for 2-pops and for setting up audio monitor levels.

9 In the Oscillator Settings, drag the Frequency knob to the left to lower the frequency, and drag to the right to raise the frequency.

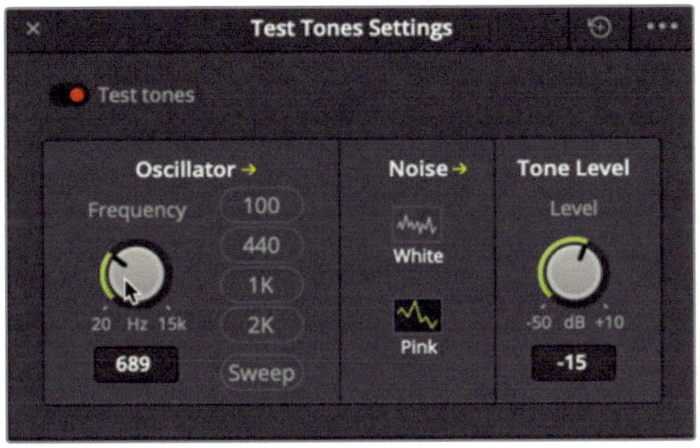

10 Click the buttons to the right of the Frequency knob to switch the Oscillator to that frequency preset.

> **TIP** You can double-click the knobs in the Test Tone Settings to reset them to the default settings.

11 Disarm the A1 track and save your project.

Although you may not need to record an oscillator sound every day, it is useful to know where it is and how to record it for your projects.

You have successfully executed three different types of recording in the Fairlight page. In the next lesson, you'll be working with audio track layers.

Practicing More FX

Because your oscillator is patched and handy, and the track is set up, let's record a few more sounds. In the following list, you will see several sounds you can record using the oscillator. When you are finished, be sure to disarm the track and save your project. Also, as you go, it's a good idea to create markers over each type of sound that you record in the timeline so you can easily identify them later. While you are at it, change the clip colors, as well. If your ears are sensitive to some of the frequencies, you can always record the oscillator with the timeline playback dimmed or even muted.

Please record approximately 10 seconds of each of the following oscillator sounds. Leave at least one second between recordings in the track.

- 100 Hz
- 200 Hz
- 440 Hz
- 1000 Hz or 1 kHz
- 2000 Hz or 2 kHz
- Ascending frequencies from 20 Hz to 10 kHz, by dragging the knob while recording
- Descending frequencies from 10 kHz to 20 Hz
- White noise

When you are finished, disarm the track, close the Oscillator Settings window, turn off the DIM button, and save your project.

NOTE If you'd like to see a finished version of the recorded frequencies, open the timeline 2e Sound Design FX Recorded.

Lesson Review

1 After connecting a microphone to your computer, what two steps are necessary before you can record to a specific track?

a) Patch the audio source to the track, and select the track.

b) Select the track, and Shift-click the Record button.

c) Patch the audio source to the track, and arm the track.

d) Select the track, and click the Record button in the track header or Mixer.

2 What keyboard shortcut can you use to disable or enable a selected clip in the timeline?

a) Shift-R

b) Shift-D

c) E

d) D

e) Shift-D

3 What happens to a recorded clip in the timeline when you record another take in the same track?

a) The new take overwrites the previous take.

b) You create a punch-in recording.

c) The new take is recorded in an audio track layer above the previous take.

d) The new take is recorded in an audio track layer below the previous take.

e) The new take temporarily replaces the previous take, but all takes can be recovered by choosing the Restore Takes option in the right-click menu.

4 Where do you set up an ADR session?

a) In the Media Pool Preview player

b) In the Setup tab on the ADR panel

c) In the Cue Index tab on the ADR panel

d) In the Sound Library, ADR Cue tab

5 True or False? To record beeps, noise, or tones, you must first open the Test Tones Settings Bin in the Media Pool.

6 Identify the two types of noise available in the Test Tones Settings window? (Pick two.)

a) White noise

b) Brown noise

c) Pink noise

d) Noise print

e) Ambient noise

7 What file format can you use to import and set up an ADR Cue list?

a) .TXT

b) .XML

c) .CSV

d) .CUE

8 What creates the beeps, noise, and tones that you can generate on the Fairlight page?

a) Velociraptor

b) Noiseinator

c) Noise FX plug-in

d) Oscillator

e) Beep generator

Answers

1. c. To record a connected microphone to a specific track, you must patch the audio source to the track in the Patch Input/Output window and arm the track.
2. d
3. c
4. b
5. False. To record beeps, noise, and tones, you must first patch the Test Tone Settings as the audio source in the Patch Input/Output window. The Test Tone Settings window is in the Fairlight menu.
6. a and c
7. c
8. d

Lesson 3

Working with Audio Track Layers

Audio track layers are a powerful device for editing multiple takes together to realize the best performance. You can stack multiple sound effects or music selections in several layers in the same track, enabling you to experiment with alternative versions, and then store those versions in lower layers for later access. In this lesson you'll apply audio track layers to two different timelines. First, you'll split and edit together the best takes in the scratch voiceover track from the previous lesson. Next, you'll use audio track layers to create crossfades between music clips in the same track.

Time

This lesson takes approximately 30 minutes to complete.

Goals

Preparing the Project

For the first exercise, you will use the pre-recorded VO in the Recording VO Finished timeline.

1 Open DaVinci Resolve 16, if necessary.

2 In the Fairlight page, show the Media Pool.

3 Open the "2a Recording VO Finished" timeline.

Your recording is in the "2 Recording VO Start" timeline, and you'll work with it toward the end of the lesson. For now, you'll edit a pre-recorded version so that it matches the steps and screenshots in this book.

Let's duplicate the timeline to create a version you can use for this lesson.

4 In the Lesson Timelines bin, select the "2a Recording VO Finished" timeline, and choose Edit > Duplicate Timeline, or right-click the timeline and choose Duplicate Timeline.

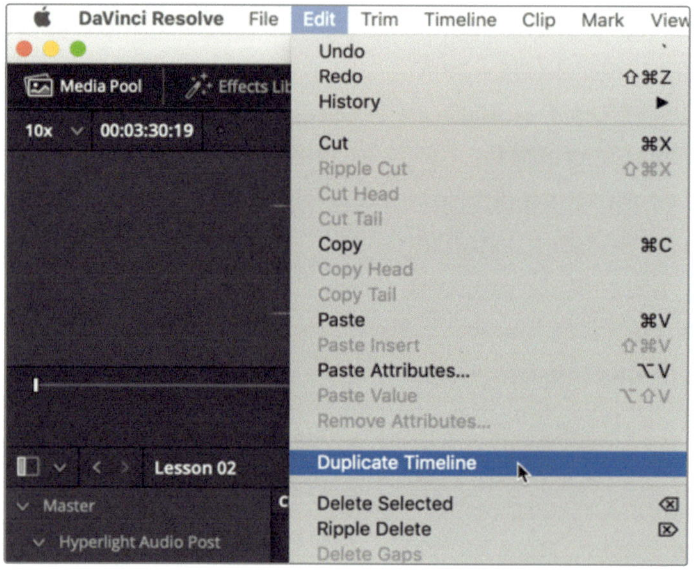

The duplicate timeline is directly below the original in the Lesson Timelines bin. Now, you can give it a new name.

5 Double-click the duplicate timeline's name (2a Recording VO Finished copy) and type 3 Audio Layers Editing. Press Return.

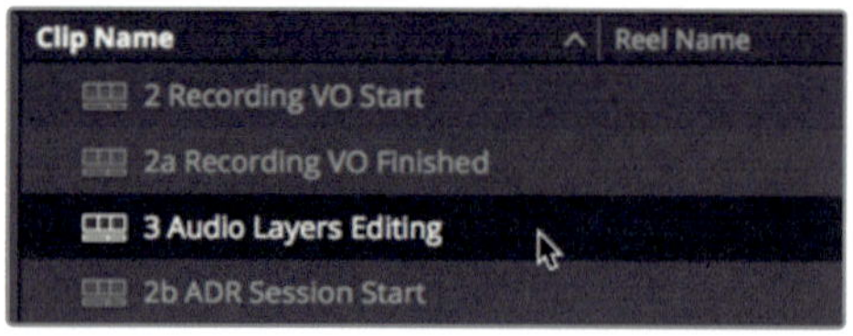

NOTE Another method for duplicating timelines is to use standard Copy and Paste available in the Edit menu or with standard keyboard shortcuts.

Because you will use this duplicate timeline in Lesson 3, let's move it to the Lesson 03 bin.

6 Drag the 3 Audio Layers Editing timeline from the Lesson 02 bin to the Lesson 03 bin. Then, select the Lesson 03 bin to see the timelines for this lesson, including the one you just added to the bin.

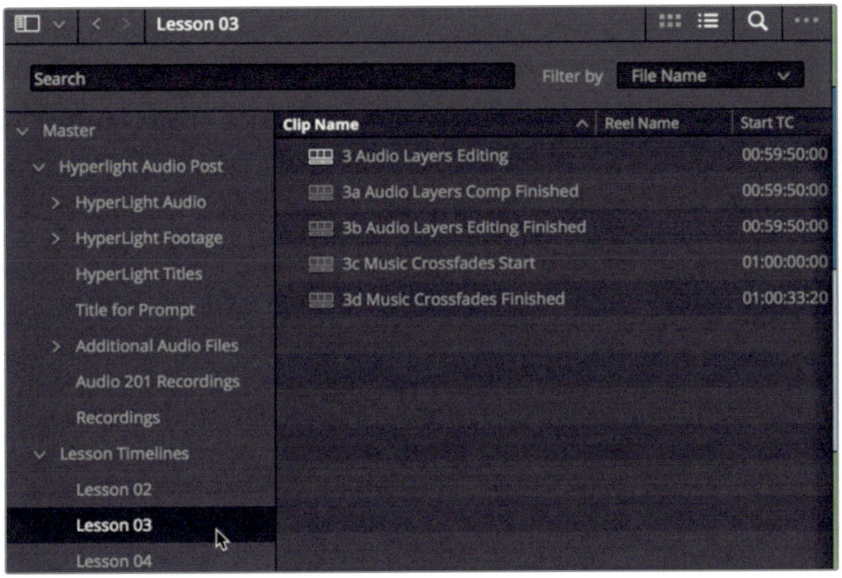

Excellent. The timeline is duplicated and moved to the correct lesson bin. Next, let's disable the text clip in the Edit page.

7 Press Shift-4 to open the Edit page. In the V2 track, select the text clip, and press D to disable it.

8 Press Shift-7 to open the Fairlight page.

9 Save the project.

The timeline is open and ready for action. Before you dive into the audio track layers, this is a good time to streamline the interface for editing.

Streamlining the Fairlight Interface for Audio Editing

One of the great features of editing audio in the Fairlight page is the large amount of user-interface real estate dedicated to the timeline. In fact, with only a few clicks, you can focus your entire screen on your tracks and clips, such as the A6 Scratch VO track. In this exercise, you'll hide all the tracks and panels that you won't be working with and resize the A6 track to maximize your view of the voiceover clips that you will edit.

1. Solo the A6 Scratch VO track.
2. Hide the Media Pool, and hide the viewer, if necessary.
3. Show the tracks index.

   You turned off the visibility of tracks A2 and A3 in the previous lesson.

   Let's turn off the visibility for the other audio tracks except for A6.

4. In the Index Tracklist, turn off visibility for the A1, A4, and A5 tracks. When you are finished, only the A6 Scratch VO audio track should be visible in the timeline.

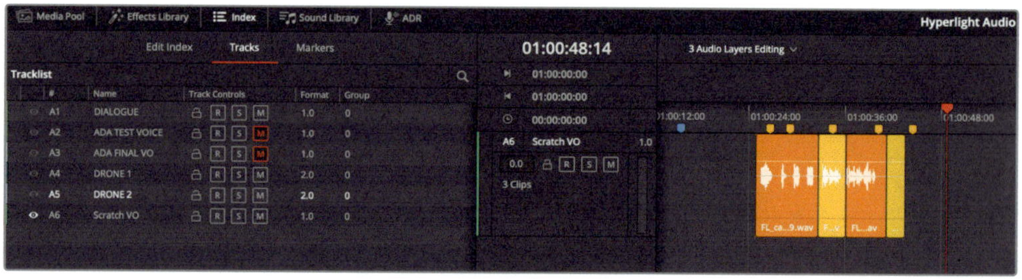

5. Hide the Index.

   Next, let's move the playhead to the middle of the voiceover recording, and zoom in horizontally several times. Remember that zooming with the View menu or keyboard shortcuts always centers the playhead and zooms horizontally from that position.

6. In the timeline, move the playhead over the third yellow marker. Choose View > Zoom > Zoom In, or press the Zoom In keyboard shortcut several times, to make the clips in the A6 track wider and easier to work with.

   You can use the Vertical zoom slider to change track height or resize tracks individually right in the track's header.

7 Position the mouse pointer over the bottom of the A6 track header and wait until it becomes a resize arrow. Drag down on the A6 track header to enlarge the A6 track. You can make the track's height as tall as you'd like within the timeline window.

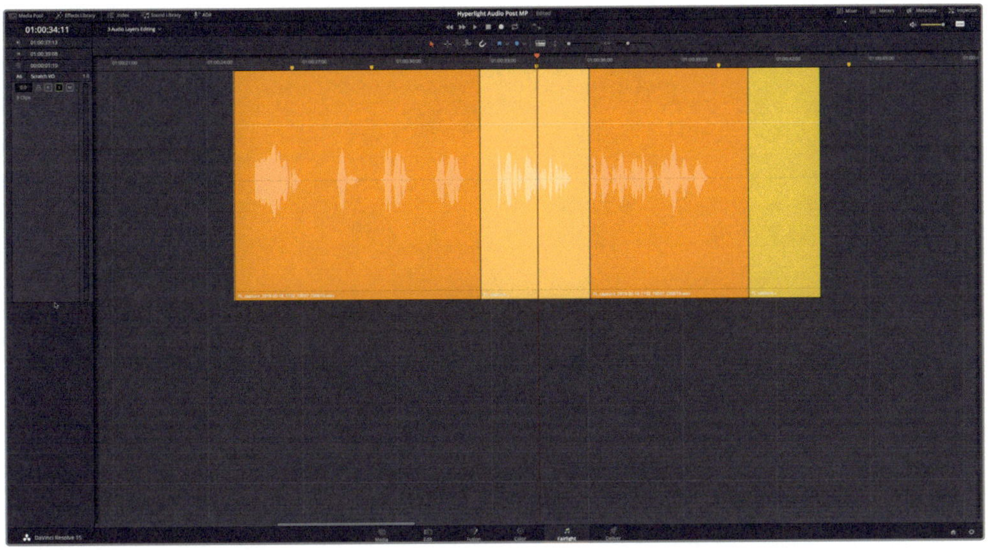

8 Choose View > Show Audio Track Layers to see the layers within the A6 track.

Now, you can clearly see the audio track layers to edit the multiple voiceover takes. Obviously, you wouldn't have to hide all the other tracks and increase the size of the clips and tracks, but you certainly can if want to, and that interface flexibility is one advantage of working on audio in the Fairlight page.

NOTE If for any reason the layered clip colors do not match the image shown, feel free to right-click the clips in each layer and choose a new color.

Moving Clips to Different Audio Layers

When you are working with audio track layers, an empty clip-sized space will always be present at the top of the track so you will have room to drag clips up or down to change their priority. When you drag a clip to the top layer, the other clips shuffle down in order. Let's try it.

1 In the A6 Scratch VO track, drag the yellow clip in the lowest layer upward to the empty space at the top of the track.

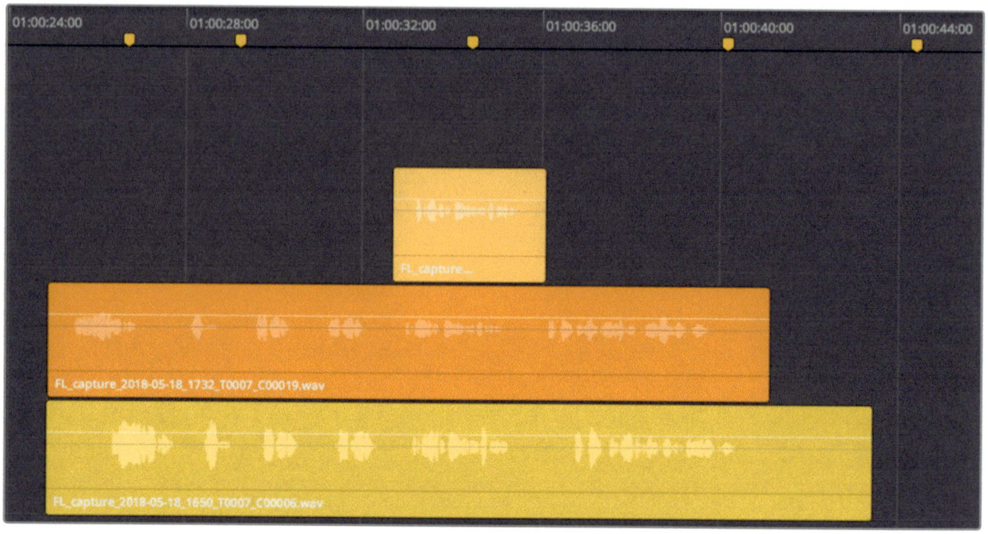

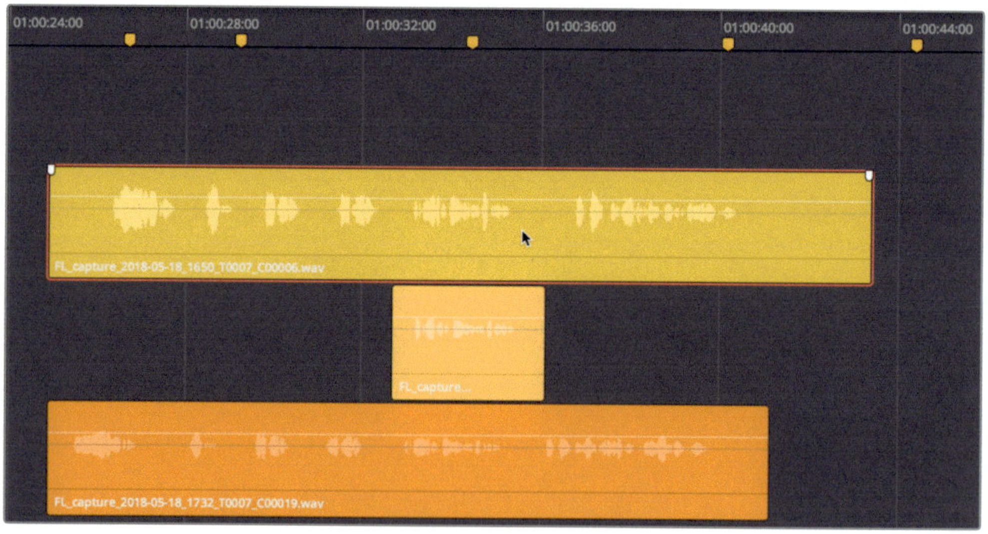

2 Start playback to hear the uppermost clip (yellow) in the A6 track.

This technique is really useful for selecting takes because you can move clips to different layers during playback.

3 Continue playback, and drag the lowest clip (orange) to the top position.

4 Continue playback, and drag the short clip in the lowest position (apricot) to the top position. Stop playback. Press Home (or Fn-Left Arrow on a laptop computer) to move the playhead to the beginning of the timeline.

As you can see, it's easy to move clips between layers, even during playback. Next, you'll divide the clips into segments and assemble the best performance.

Aligning and Splitting Audio Clips

One advantage that audio editing has over video editing is that you can cut together the best phrases, words, or even syllables without the audience being aware of your edits. Although you'll work more extensively with precision dialogue editing in Lesson 4, "Working with Dialogue Tracks," for now, you'll focus on splitting long voiceover clips into phrases and selecting the best takes.

In video editing, you use a blade tool or a razor keyboard shortcut to place an edit within a clip. The resulting edit is referred to as a *through edit*. Because through edits do not lose frames at the edit point, the video will play straight through as if no edit existed.

When working with audio tracks, the same type of edit is referred to as a *split* because you split a single audio clip into multiple audio clips. The more you split the original audio clip, the more separate audio clips you'll have with which to create your finished audio composition, or *comp*.

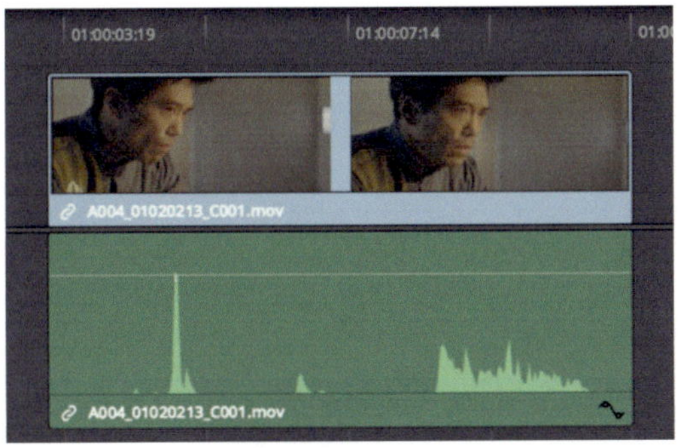

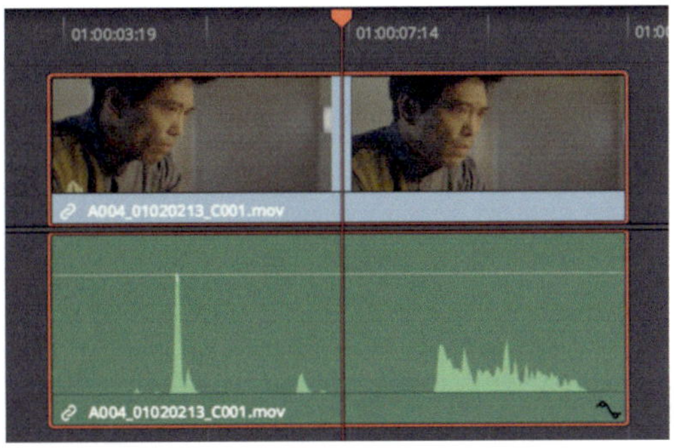

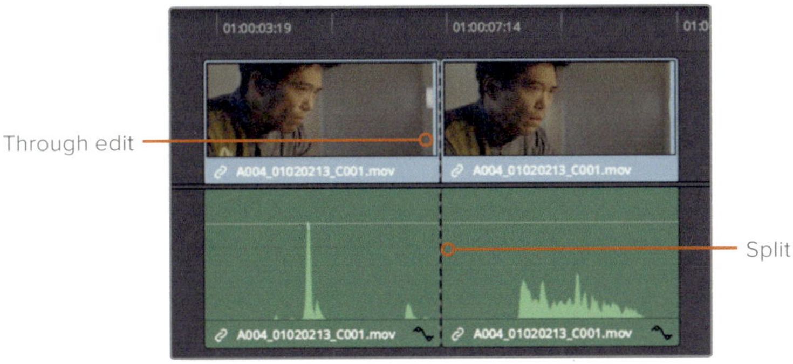

Because DaVinci Resolve supports professional-level video and audio editing, you'll find shortcuts in the Timeline menu to razor or split clips at the playhead position. Just keep in mind that the results are identical. A handy Razor tool (scissors) is also available right in the toolbar.

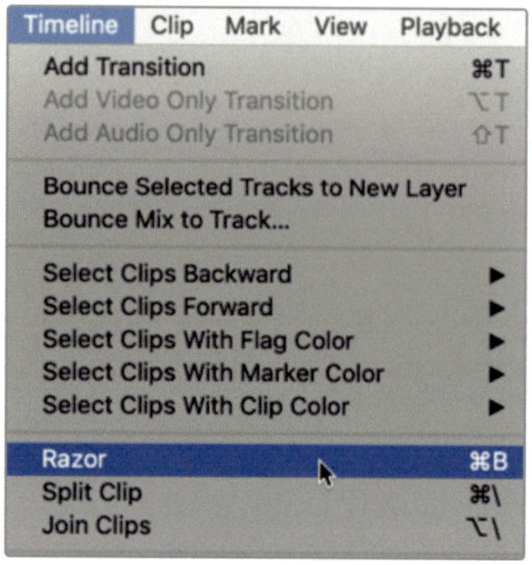

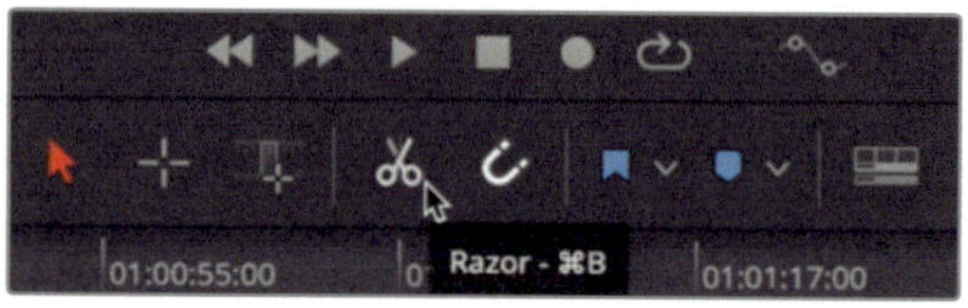

NOTE Splitting audio clips in the Fairlight page will never alter video clips in the Edit page, even if those audio and video clips are linked.

In this exercise, you'll align the stacked clips, and with one operation, split all the voiceover clips between phrases.

If you compare the waveforms between the voiceover clips in track A6, you'll see repeated patterns in each track, separated by silence (no waveform). Each waveform pattern indicates words or phrases. Although the waveforms don't indicate which words they represent, you can definitely see where and when the spoken phrases start and stop.

At a glance, the first phrase in the orange clip clearly starts earlier than the same phrase in the yellow clip. Let's move the yellow clip to the left to align the first phrase with the same phrase in the orange clip.

Rather than drag the clip into position, this is a great opportunity to use the Nudge commands available via keyboard shortcuts and in the Trim menu. Unlike manually dragging, the Nudge commands allow you to precisely move a selected clip one frame at a time to the left or right by pressing the , (comma) and . (period) keys for single-frame movement, or the Shift-, (comma) or Shift-. (period) key combinations to nudge multiple frames.

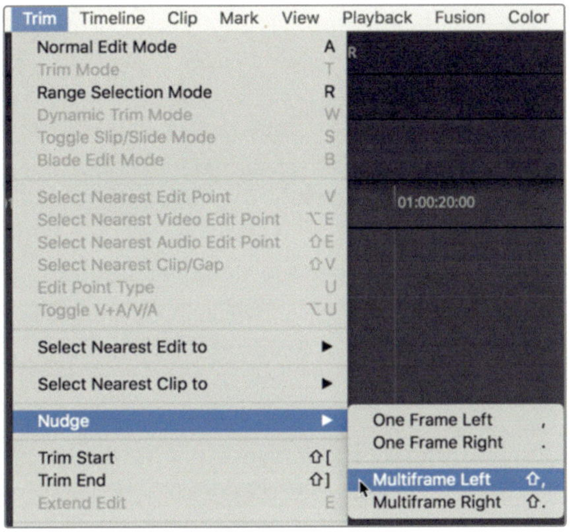

Let's use the playhead as a guide to mark the beginning of the first phrase in the orange clip, and then nudge the yellow clip into position.

1 In the second layer in the A6 track, move the playhead to the beginning of the orange clip's waveform .

2 Select the lowest clip in the A6 track (yellow). Choose Trim > Nudge > Multiframe Left, or press Shift-, (comma).

The selected clip moves five frames to the left, but still needs to move another 15 frames or so.

3 Press Shift-, (comma) three more times to move the clip 15 frames to the left. Then, using the , (comma) and . (period) keys, nudge the clip one frame left or right to line up the beginnings of the waveforms.

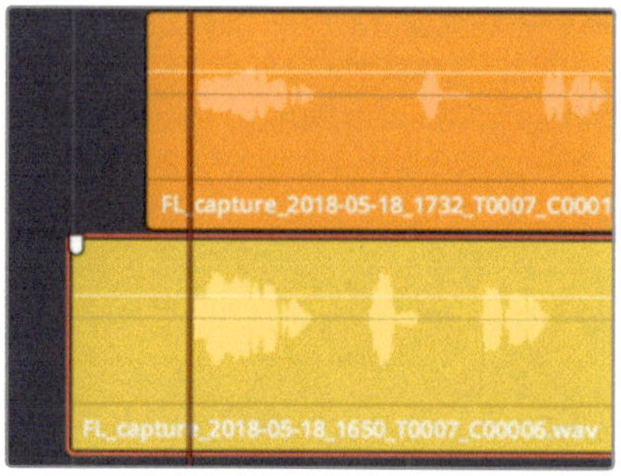

Remember, the goal in this exercise is to align the different takes for each phrase, and split them into a layered stack. Then, you'll be able to choose the best take for each line of voiceover, and move them into position.

The trick to splitting multiple takes in multiple layers at once is to line them up as you go. This first phrase is aligned. However, you probably noticed that the waveforms for the other phrases are still out of alignment. No worries. You'll align these layers one phrase at a time. After you split the first phrase from the others, you can align the next phrase, and so on.

Now you can split the clips at the end of the first phrase. To do so, you will move the playhead to the space after the first phrase, select the clips, and split them using the Razor tool.

4 Press the JKL keys to play the clips from the beginning, and stop playback after the phrase, "Emiliana Newton," at around 01:00:27:00.

5 In the A6 track, select the orange and yellow clips.

6 Click the Razor button, or Press Command-B (macOS) or Ctrl-B (Windows) to razor cut the selected clips at the playhead position.

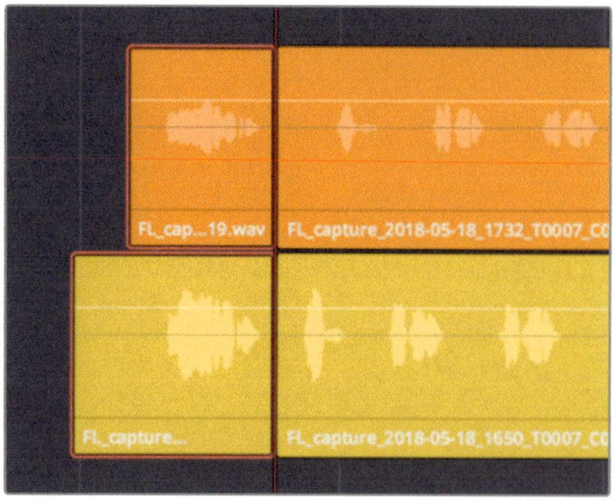

7 Press the JKL keys to play forward to the end of the next line of dialogue, "Yes."

8 In the lowest layer, select the yellow clip, and nudge it to the right until the second phrase is roughly aligned with the orange clip. The alignment of each phrase doesn't have to be perfect. If they are close, you'll still be able to split them into stacks.

9 Select the yellow and orange clips, and press Command-\ (backslash) in macOS or Ctrl-\ (backslash) in Windows to split the clips at the playhead position.

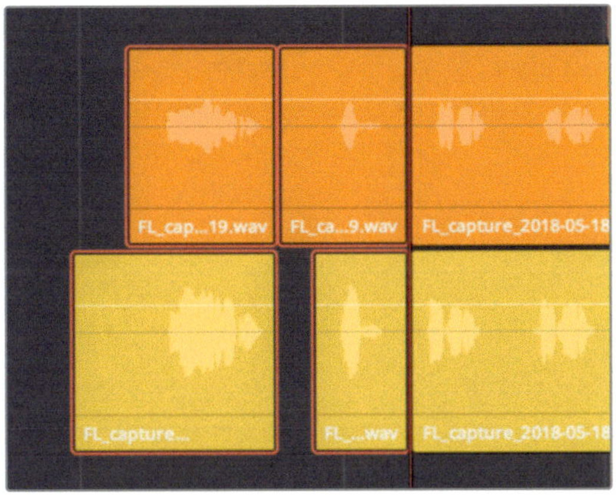

The results are the same regardless of the shortcut, razor or split, that you use. So, from this point you can choose whichever method you prefer when splitting audio clips in the timeline.

> **TIP** If you are simultaneously splitting all the audio clips beneath the playhead, you can press Command-A (macOS) or Ctrl-A (Windows) to Select All the clips on all the tracks. Doing so will select and split only those clips currently visible in the timeline. Tracks with their visibility turned off in the index will not be affected.

Splitting Clips with the Range Selection Tool

So far, you have been working in Selection mode using the familiar Selection tool (arrow pointer). In this exercise, you'll switch to Range Selection mode so that you can simultaneously select and split a range of clips.

The waveforms line up fairly well for the next three phrases and then they drift out of alignment for the last phrase. The drift is due to slight timing differences in the voiceover performance in each take.

In Range Selection mode, you can quickly separate the remaining phrases, and align them before selecting the best takes.

1 Press R, or click the Range Selection tool, to switch the timeline to Range Selection mode.

You've already split the clips at the beginning of the third phrase, "Philip Maeda," so you needn't select that phrase. Instead, you'll use the Range Selection tool to select the fourth phrase, which also happens to be "Philip Maeda." After you drag a selection using the Range Selection tool, you can drag or nudge that selection to a new location. For the first example, you will drag the selection down slightly to jiggle it into place, which will also split all clips within the selection at the range in and out points.

2 Drag the Range Selection tool from just above the orange clip at the beginning of the fourth phrase to below the yellow clip just after the fourth phrase. Make sure that the range selection includes the entire phrase on both clips. Then drag down slightly on the lowest clip in the selection to split the clips at the range in and out points.

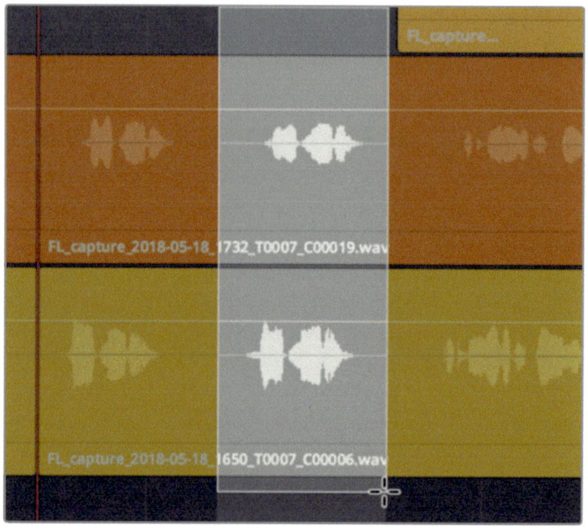

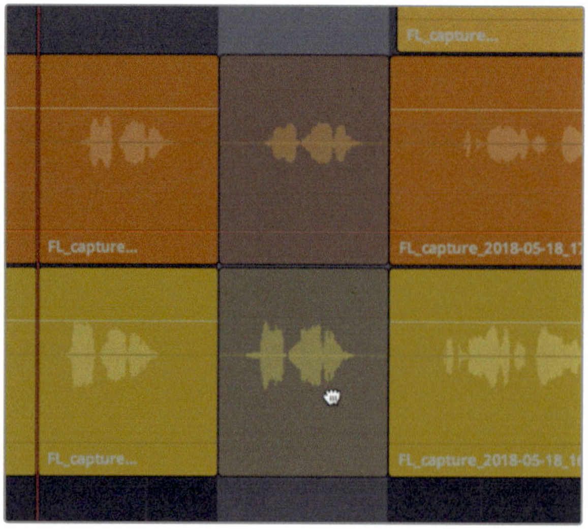

3 Click any empty track space above or below the clips to deselect the current range.

NOTE If any of your clips moved to different layers, it was because they were overlapping other clips in the same layer. If that occurs, you can press Command-Z (macOS) or Ctrl-Z (Windows) to undo the move. The clips will remain split even after you undo the previous move. If you repeat the undo shortcut, you will undo the split.

4 Drag a range selection around the last phrase in the orange and yellow clips when ADA says, “Earth date is March 22, 2103.” Then, press . (period) to nudge the selection one frame to the right to create the split.

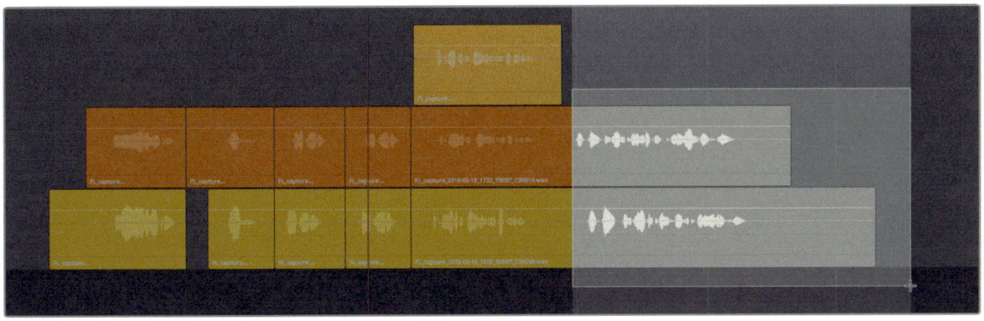

5 Press A to return to Selection mode and deselect the clips.

6 Trim the tail of the last yellow clip so it matches the length of the last orange clip.

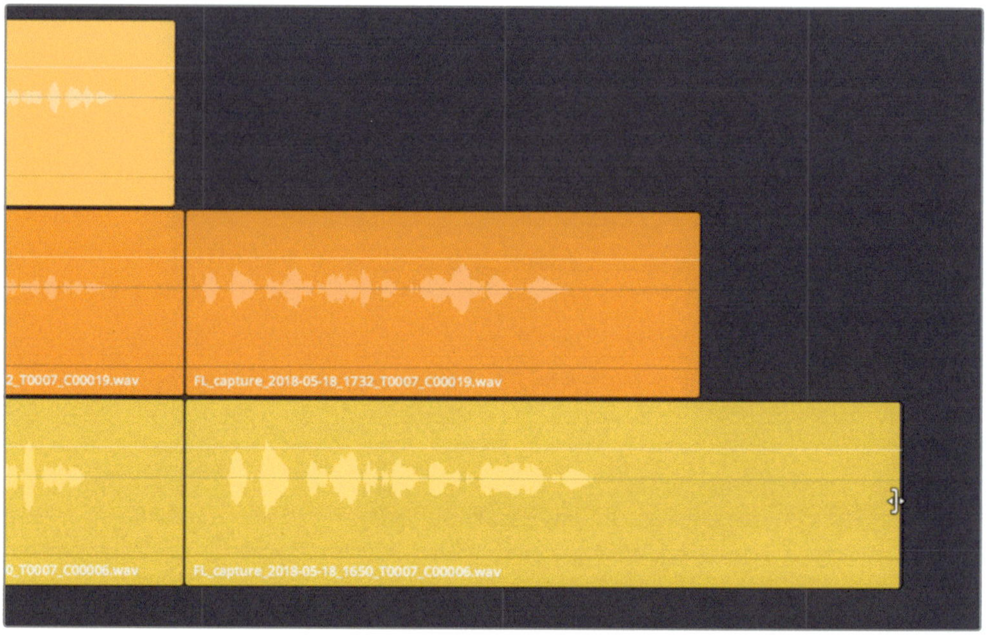

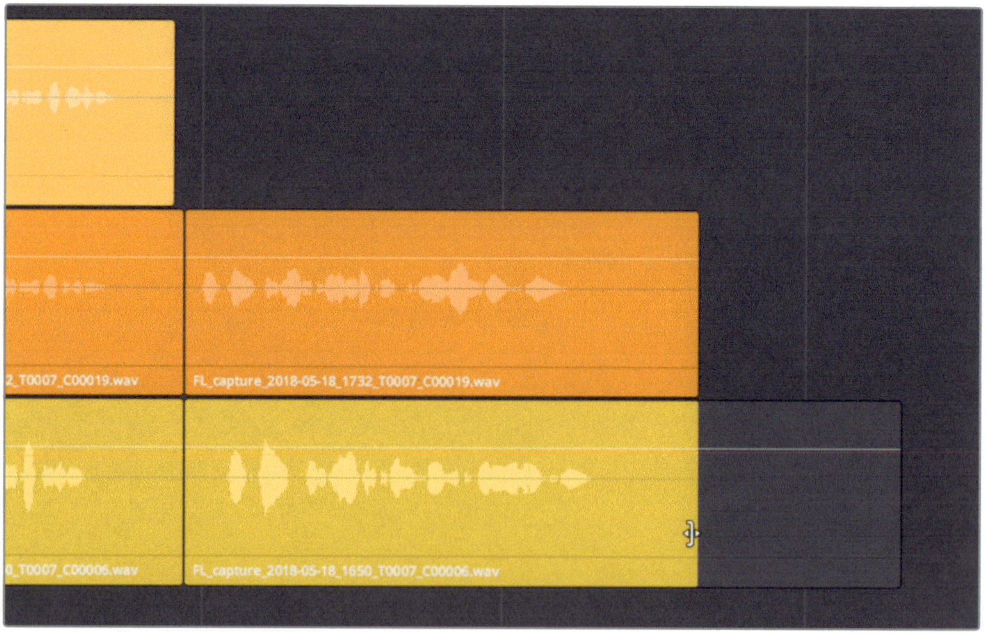

7 Trim the head of the first yellow clip to match the length of the first orange clip. Deselect all the clips.

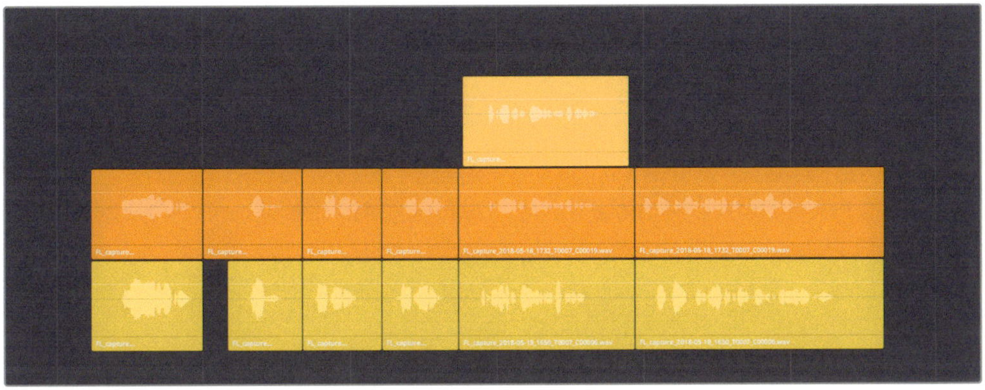

The clips are split and stacked and ready to swap between layers to piece together a composite voiceover track.

Building a Voiceover Composite Track

By definition, a composite is made up of several parts, or elements. Composite images comprise multiple visual layers. Composite audio tracks refer to audio tracks containing the best takes that have been edited together seamlessly, as in the case of this finished ADA voiceover track.

In this exercise, you'll finish building the voiceover composite by moving the clips into position in the timeline, selecting the best take for each section, and moving it to the uppermost layer. Once you are finished moving and selecting takes, you can hide the audio track layers so that only the finished composite (comp) is visible in the timeline. You already have the skills to perform this section, so this exercise will guide you through the first stack of clips, then leave you on your own to complete the rest. As you do so, you can practice using many of the Fairlight page shortcuts and techniques that you have learned in the previous lessons.

1 Press Shift-Z to zoom the timeline horizontally to fit all the timeline clips on-screen, including the clips on hidden tracks.

   You should see six yellow markers in the timeline. Each of these represents the starting position of a corresponding voiceover clip. To keep things simple, you'll work from right to left, moving the last phrase to the corresponding marker, and so on.

You'll want to make sure the snapping control is turned on so that the clips you move will snap to the marker's position. When snapping is turned on, the playhead, markers, and clips snap into alignment as you drag them toward each other. In this case, the head of the clip you are dragging will snap to the closest marker as you drag. Press N to toggle snapping on and off.

2 In the toolbar, make sure that the Snapping button (which looks like a magnet) is turned on (white).

3 Using the Selection tool (arrow), drag a selection around the last stack of voiceover clips to select them. Then, drag the clips to the right to the last yellow marker.

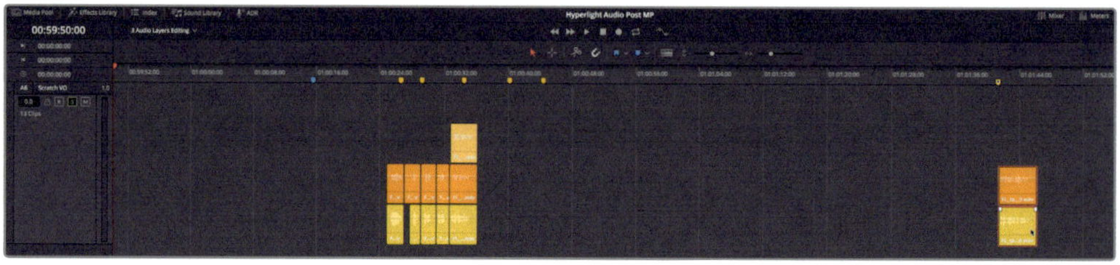

TIP Dragging audio clips from the area below the waveform and volume line avoids accidentally changing the clip's volume levels as you drag.

4 Working from right to left, select the next stack of clips, and move them to the last available yellow marker.

5 Repeat step 4. As you move the clips to the markers, feel free either to drag them or use the nudge keyboard shortcuts. When you are finished, deselect all clips.

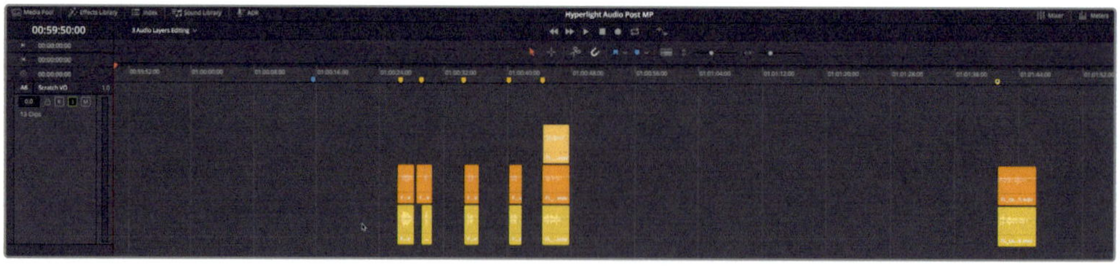

Next, you'll show the other dialogue tracks in the index so you can choose the best take in the context of the other tracks.

6 Show the tracks index, and click the visibility icon for the A1 DIALOGUE track.

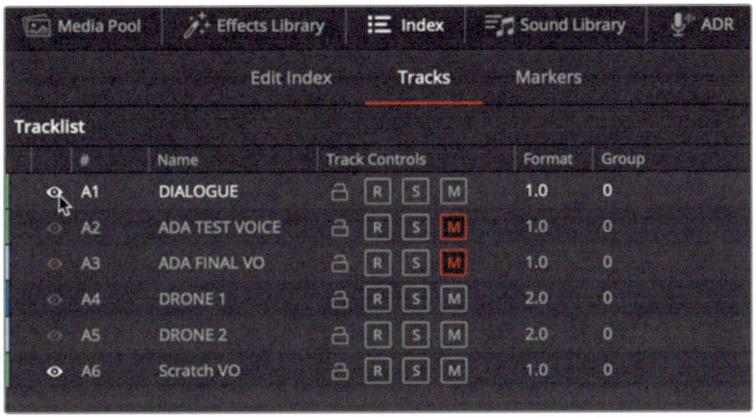

Track A1 DIALOGUE is once again visible in the timeline. Because the timeline is still showing track layers, you'll also see an empty space above the dialogue clips in the A1 track.

7 In the Index, solo the A1 track. Hide the index.

8 At the end of the timeline, select the lime green clip, and press D to disable it.

Both the A1 and A6 tracks should be soloed.

9 Show the viewer. Resize and place the viewer in the lower-left corner of the timeline window so it doesn't cover any of the clips in the A6 track.

10 Play the timeline from the blue marker through the fifth voiceover clip, "I can't explain this discrepancy."

The clips line up pretty well with the production dialogue track. However, the first line, "Emiliana Newton," seems to start a little late. Instead of aligning the beginning of the clips with the yellow marker, you need to align the beginning of the waveforms.

11 Move the playhead to the first yellow marker.

12 Select the first two voiceover clips. Nudge the clips to the left until the waveform starts at the playhead position (first yellow marker).

13 Play the first group of clips again to see if the timing is better. Nudge the first clips until you are happy with the timing of ADA's response to Philip's first question.

Now that you have heard the first five voiceover clips in context, and the first stack is in position, you can focus on the takes one at a time and move the best take in each stack to the uppermost track layer. To simplify this process, you'll use the looped playback feature.

NOTE If you didn't follow all the previous exercises, feel free to open the timeline 3a Audio Layers Comp Finished to catch up.

Using Looped Playback to Select Takes

Earlier, you moved clips to different track layers during playback. In this exercise, you'll mark in and out points in the timeline to determine the range that you'll play back. Then, you'll turn on looped playback to play the range over and over while setting in and out points in the timeline. Let's start by setting a playback range that includes Phillip's first two questions and ADA's responses.

1 Press the JKL keys to move the playhead just before Phillip asks his first question, "identify the person in front of me" (1:00:24:21). Press I to mark an in point.

2 Use JKL to play through the timeline until after the second stack of clips when ADA says "Yes." Press O to mark an out point (01:00:31:00).

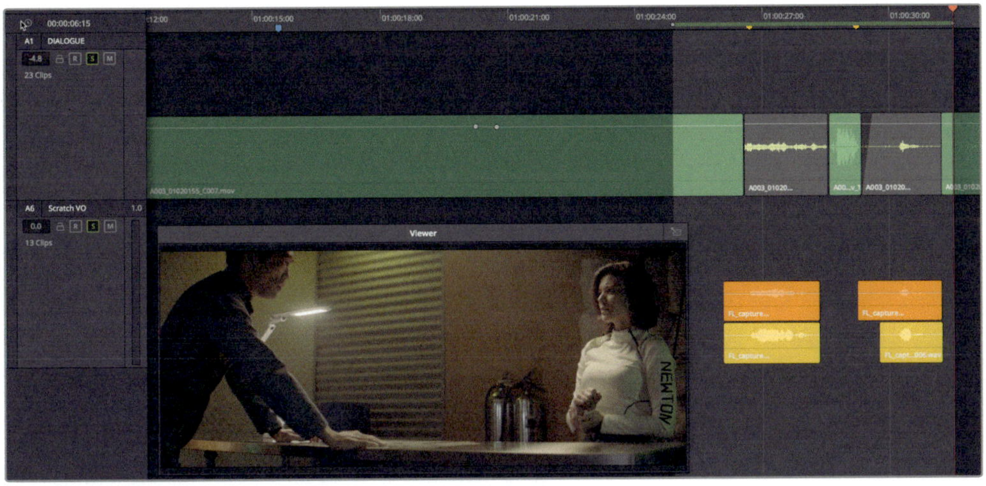

3 In the playback controls, click the Loop button, or press Command-/ (slash) in macOS or Ctrl-/ (slash) in Windows, to turn on looped playback.

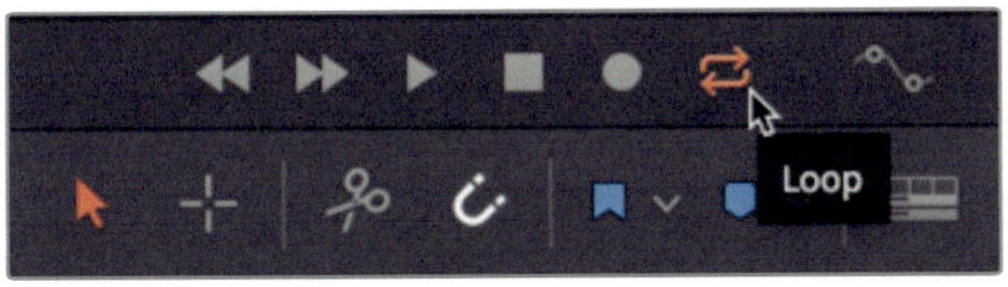

4 Choose Playback > Play Around/To >Play In to Out, or press Option-/ (slash) in macOS or Alt-/ (slash) in Windows, to start looped playback between the in and out points.

Keep in mind that you can later adjust the volume levels of the clips and apply effects to make the voiceover track sound more like a computer. For now, just go with your guts and your ears and choose a best take for the first two voiceover sections. Don't worry if you can't decide; the other take will still be there in the lower layer so can always change your mind later.

5 During looped playback, listen to the takes in the uppermost clip (orange) and drag the lower clips up one at a time (yellow) to audition each of them in context. Continue moving between the clips until you choose a best take.

You can also set in and out points in the timeline using the Range Selection tool.

6 Press R to switch to Range Selection mode. Move to the next two stacks of clips. Drag a range selection that includes the production dialogue in the A1 track and the next two stacks of clips in the A6 track. Press A to return to Selection mode.

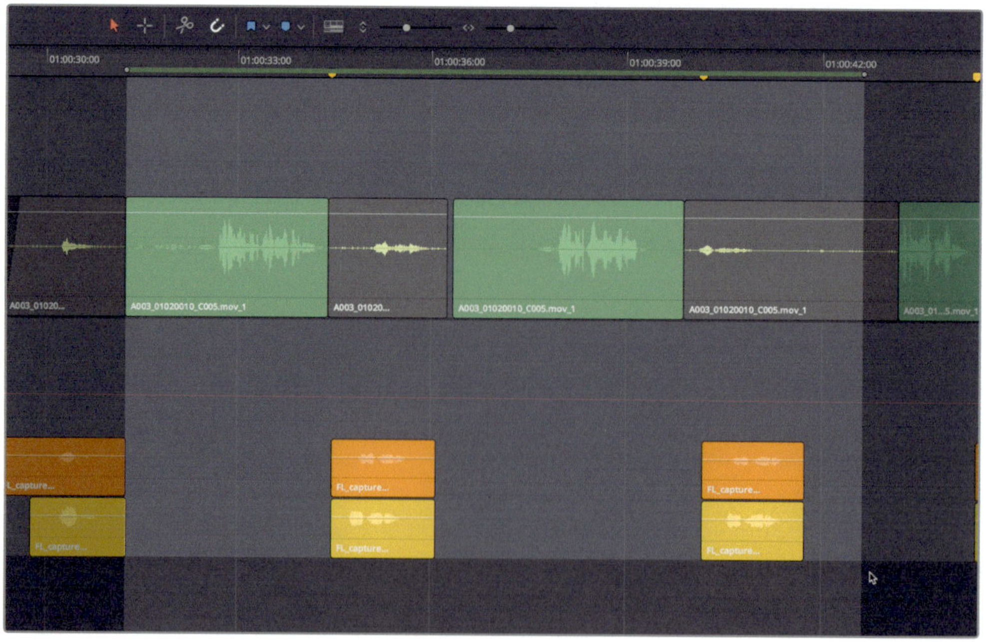

7 Start looped playback, and use the Selection tool (press A) to swap the positions of the takes during playback. When you've identified the best take, place it in the uppermost layer. Nudge the position of the clips, as necessary, to best fit the production dialogue.

8 Repeat step 6 for each of the remaining stacks of takes in the A6 track. Don't forget the last stack of clips near the end of the scene.

9 When you have finished, choose Mark > Clear In and Out, or press Option-X (macOS) or Alt-X (Windows).

10 Choose View >Show Audio Track Layers to deselect that view option and hide the layers.

11 In the Tracks Index, turn on the visibility for the A4 and A5 tracks. Then use the Vertical slider to resize the height of your tracks.

12 Press Shift-Z to zoom the tracks to fit the current timeline space.

13 Un-solo the A1 and A6 tracks.

It's always a good idea to group similar tracks together such as dialogue tracks, sound effects, and music tracks. To reorder tracks, you can right-click the track header and choose Move Track Up. With DaVinci Resolve 16.2, you can also drag and drop tracks to reorder them in the Index.

14 Right-click the A6 track header, and choose Move Track Up. Repeat this step until the Scratch VO track is located directly beneath the A1 DIALOGUE track.

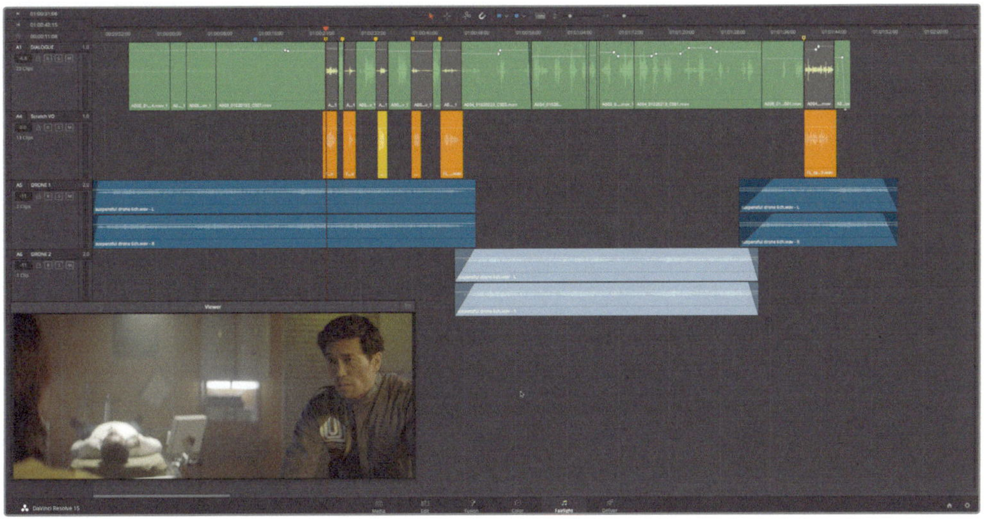

Notice that the Scratch VO track is now track A4 because it is the fourth audio track from the top. Remember, tracks A2 and A3 are still hidden.

15 Play through the timeline once from start to finish to hear the entire scene with the new voiceover clips in place.

16 Save the project.

In the next lesson, you'll work more extensively with dialogue editing. First, you'll use audio track layers to overlap and crossfade some music clips.

Preparing the Project

In the next series of exercises, you'll work with a different scene from *Hyperlight*. The music selections that you'll use are from the actual soundtrack, composed by Mathieu Lafontaine. You'll start by opening the timeline and watching the scene to get a feel for the context and existing music track.

1 Open the "3c Music Crossfade Start" timeline.

2 Play the timeline from the beginning and listen to the music in the context of the scene, a part of the movie's dramatic opening sequence.

3 Move the playhead to the beginning of the timeline.

Notice that the current frame in the timecode display does not match the timecode burn-in in the viewer.

Setting a New Start Timecode

The default start timecode for a timeline is 01:00:00:00, which you can change in the Editing User Preferences, or when creating a timeline, in the New Timeline dialog. Once a timeline contains media, you can change the starting timecode in the Media Pool as you'll do in this exercise. Also, this is a great time to show Resolve's floating timecode window, which is available in the Workspace menu.

1 Choose Workspace > Timecode Window to show the floating timecode window.

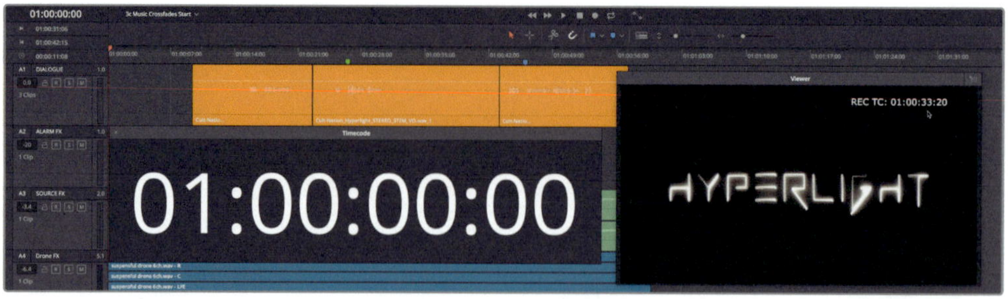

As you can see, the timecode window shows a larger version of the current playhead position, also visible in the timecode display. You can resize and move this window anywhere on the screen. In the viewer, the timecode window burn-in shows a starting timecode of 01:00:33:20.

NOTE To add a timecode data burn-in to the viewer or to your rendered output, choose Workspace > Data Burn-In, and configure the options as necessary.

2 Show the Media Pool.

3 Right-click the "3c Music Crossfades Start" timeline, and choose Timelines > Starting Timecode.

You can now enter the new timecode in the Set New Start Timecode dialog. When you type in the new number, you do not need to add the colons between pairs of numbers because they are added automatically as you type. Your goal is to type a number that matches the timecode data burn-in in the viewer: 01:00:33:20.

4 In the Set New Start Timecode dialog, type **01003320**. Click OK.

The first frame of media now starts at the new timecode. However, you'll also see an empty space in the timeline that indicates the space between the original starting timecode and the new timecode. To refresh the timeline, you need to go to another timeline and return.

5 In the Media Pool or timeline pop-up menu, open another timeline. Then, reopen the previous timeline, "3c Music Crossfades Start". Hide the Media Pool.

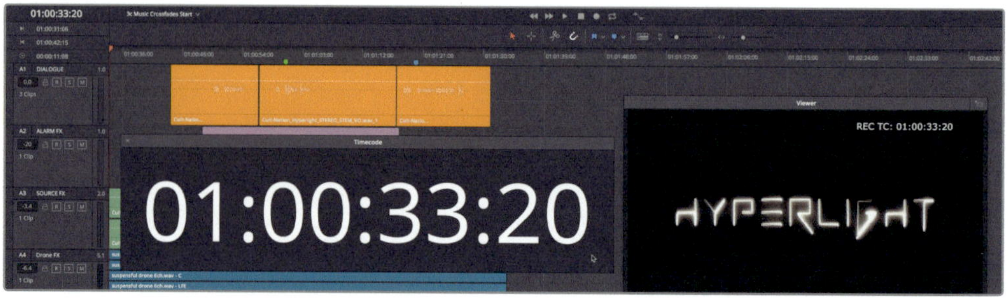

Now the starting timecode matches the window burn-in.

6 Close the timecode window, and hide the Media Pool.

Now that the timeline is set, you can commence with creating crossfades between the music selections on the A5 track.

Creating Crossfades in Audio Track Layers

You have three ways to transition between music cues in a soundtrack: You can cut from one piece to another. You can fade one clip out entirely and fade in a new selection. Or, you can fade one clip out while the other fades in, hence the term, *crossfade*.

In this exercise, you'll create crossfades between three music clips to seamlessly change music during the scene. This time, you'll work with stereo music clips that include two audio channels and an empty layer above each channel.

1 Choose View > Show Audio Track Layers.

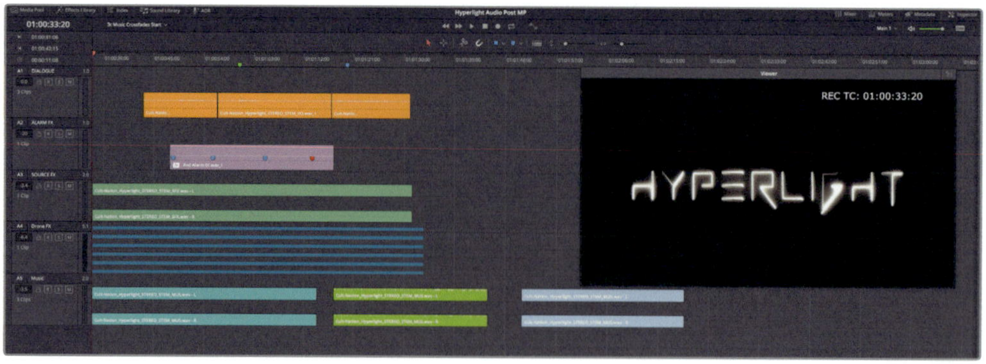

You can clearly see the empty layers above each channel in each track. Tracks A1 and A2 are mono tracks, tracks A3 and A5 are stereo tracks, and track A4 is a 5.1 surround sound track with six channels of audio.

2 Drag the vertical zoom slider all the way to the left to minimize the track heights as much as possible. Then, drag down the bottom of the A5 track header to increase the height of that track.

3 Move the viewer below the music clips, and horizontally resize the tracks until you can see all three clips in the A5 track.

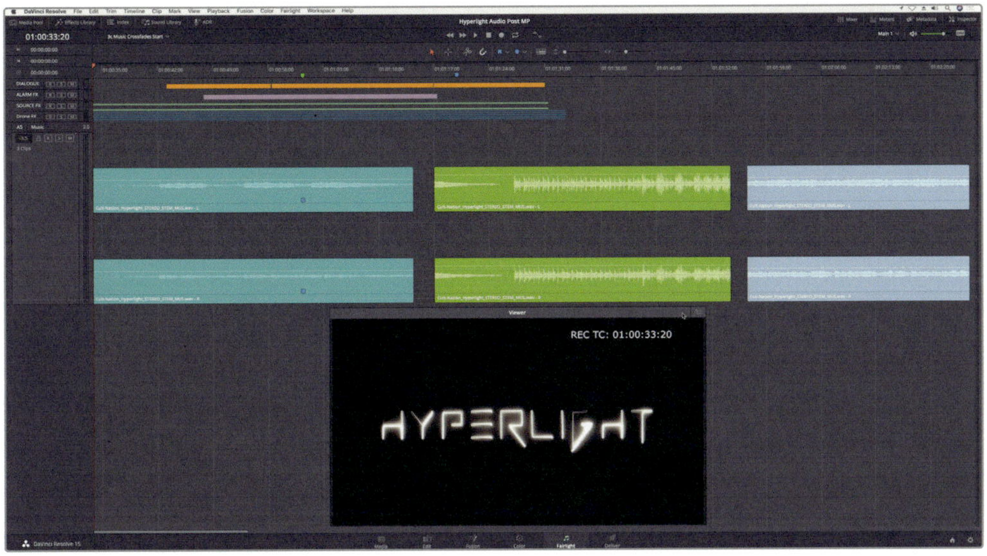

4 In the A5 track, play from the beginning of the second music clip (green).

This clip starts with a bit of strings like a stinger, and then kicks into a cool tension-inducing percussion beat. Let's trim off the bit of that stinger in the green clip. To trim a clip in audio track layers, you need to drag from the lower edge of the clip.

> **TIP** It is easier to transition from non-percussive music to percussive music because you don't have to match the tempo or key of the clips.

5 Using the waveform as a guide, in the A5 track, drag the lower-left edge of either green clip channel to the right until the clip starts with the percussive beat.

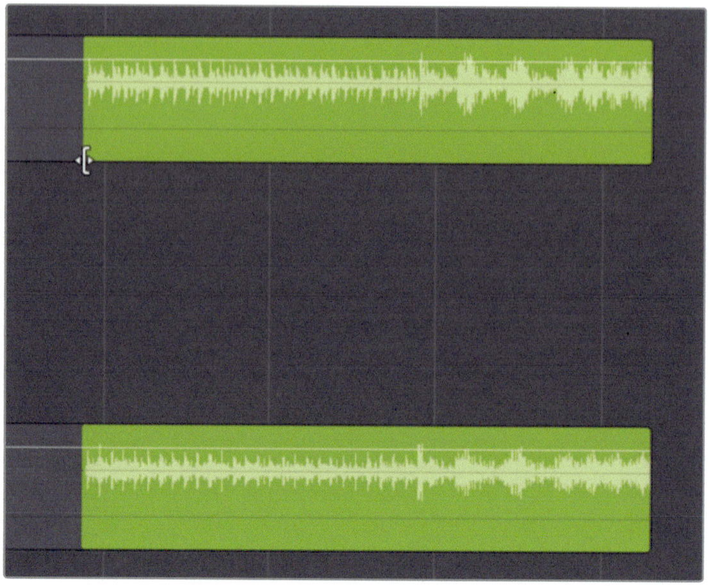

6 Drag the second music clip (green) to the left, and place it above the first clip (teal). Drag the upper music clip until it starts at the green timeline marker.

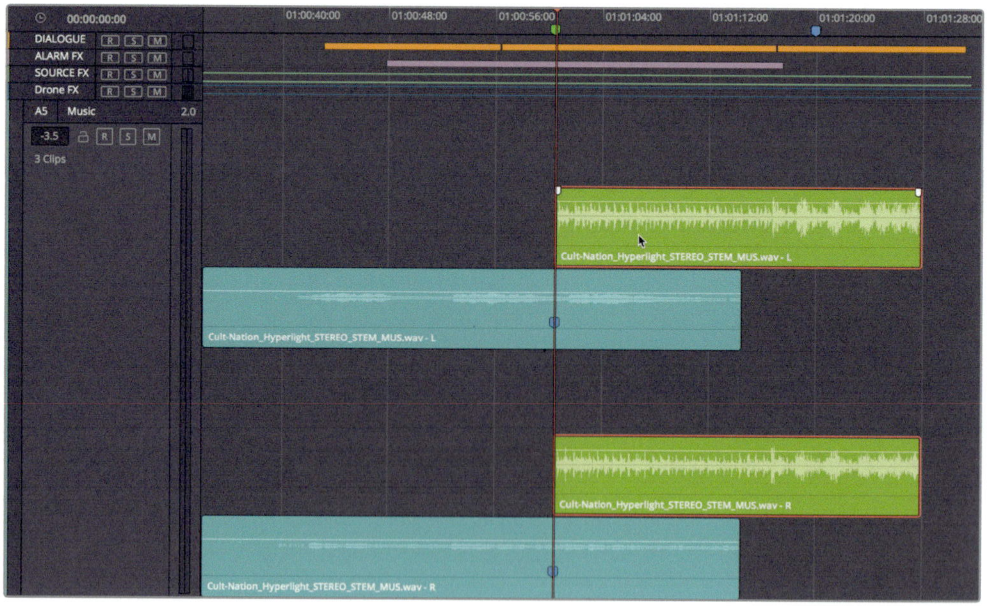

Remember, you'll hear only the uppermost layer during playback. So, you'll hear the teal clip during playback until the playhead reaches the green clip. Then, the sound will cut to the percussion music in the green clip on the second layer.

7 Play the transition between the first two music clips to hear the abrupt cut from the clip on the lower layer to the clip on the upper layer.

Sometimes, cuts between music clips work. This transition actually sounds fairly good. However, it might work even better if you smooth the transition by adding a fade-in to the upper clip (green).

Let's add a five-second fade-in to the upper clip. To measure seconds using the playhead, you can hold down Shift and press the Left or Right Arrow keys.

8 Move the playhead to the beginning of the upper music clip (green). Press Shift-Right-Arrow five times to move the playhead five seconds to the right.

9 Select the upper channel of the green clip to see its fade handle. Then, drag the fade handle to the playhead.

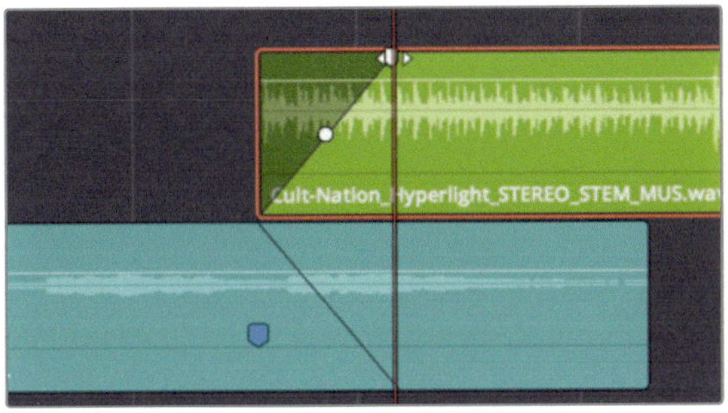

As you drag the fade handle, a mirror image of the fade appears on the lower clip to indicate a matching fade-out.

10 Play the transition between the clips to hear the musical crossfade.

It sounds great. Feel free to experiment with changing the length of the fade to hear the impact of a shorter or longer crossfade.

Now you'll add the third music clip. This could be placed above the green clip on a new layer, or on the lowest layer. The result will sound the same.

11 Drag the third music clip (blue) to the blue marker position above the other music clips.

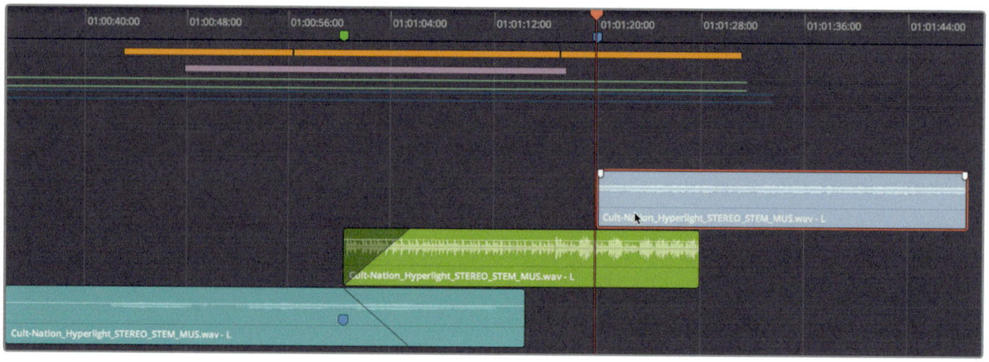

Instead of manually dragging the fade handle, you'll use the Fade In to Playhead command in the Trim menu.

12 Select the highest music clip (blue) and move the playhead to the end of the green clip.

13 Choose Trim > Fade In to Playhead.

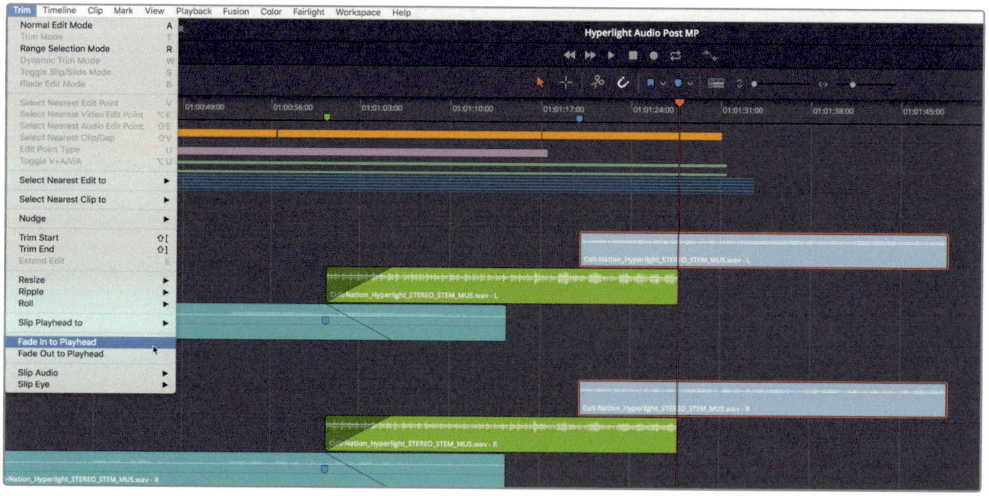

The selected clip fades in to the playhead.

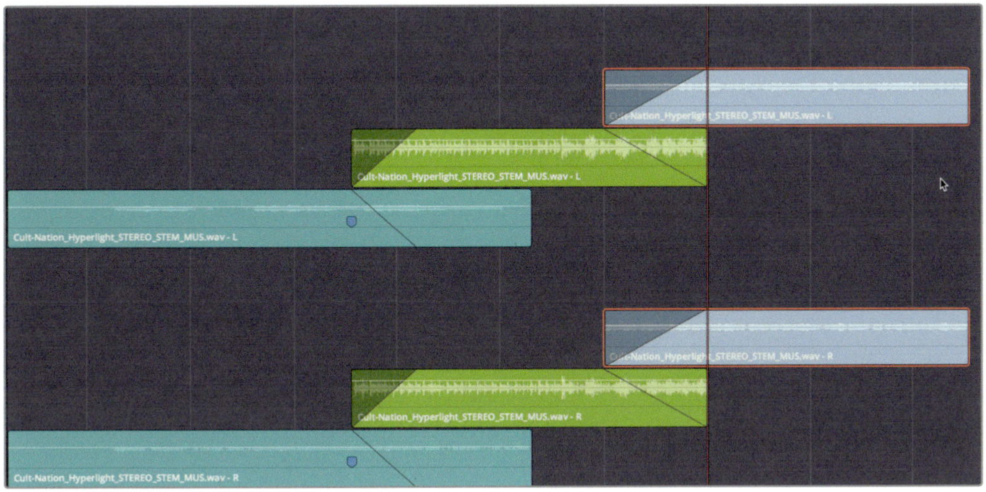

14 Play the scene from the beginning to hear the scene with additional music clips.

Now, you will move the third clip to the lowest layer and remove the fade. Instead of a fade-in, you will add a fade-out to the green clip.

15 Drag the third music clip down to the lowest layer. Drag the fade handle on the third clip to the left to remove the fade. Drag the fade handle at the end of the upper clip (green) to the left so that it starts at the beginning of the third clip.

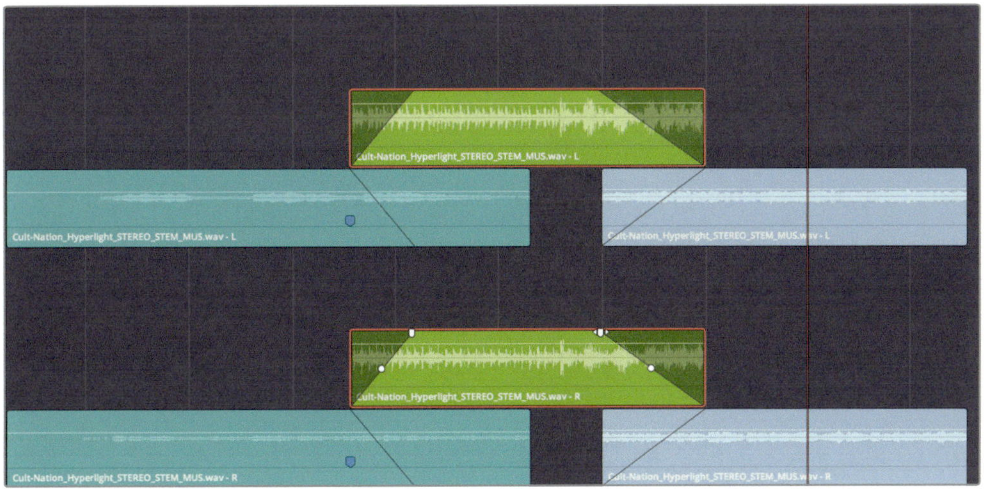

16 Play the transition between the last two music clips and listen to the crossfade.

As promised, the results are identical. So, when you apply musical crossfades between layers in your own projects, keep in mind that you can stack them whichever way you want to create the musical transitions.

NOTE The audio fade handles include a center control for changing the fade curve.

17 Choose View > Show Audio Track Layers to hide the layers.

18 Drag the horizontal and vertical zoom sliders to resize your tracks to fit the window. Press Command-Z.

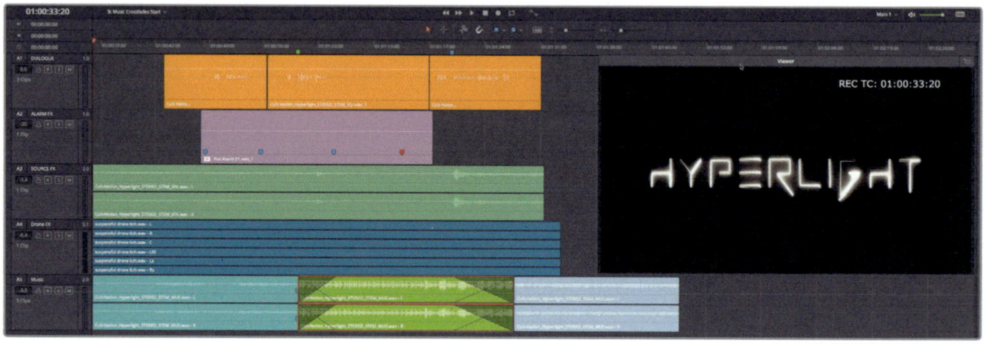

You can clearly see the crossfades between music clips in the A5 track. Creating crossfades between audio track layers is an easy and effective way to make seamless audio transitions in your soundtrack.

More Audio Layer Editing Please!

It's your turn to apply the skills that you have learned to the voiceover track that you recorded in the previous lesson. Earlier in this lesson, you stacked and edited the voiceover takes that were provided with the book media. Now, you will create a finished voiceover comp track based on your recording. Once you open the timeline, you can follow along guided by the steps in this lesson, or try it on your own without guidance. Have fun and take your time. Remember, the more you practice these Fairlight page techniques, the more proficient you'll be when working on your own projects.

Here are some loose guidelines to get you started:

- Open the "2 Recording VO Start" timeline in which you recorded the voiceover takes in the previous lesson.
- Show audio track layers.
- Color the different takes, if necessary.
- Stack and split the phrases.
- Move the phrases to the timeline markers moving right to left.
- Select the best takes and move them to the upper layer.
- Hide the audio track layers when you are finished.

Lesson Review

1 How do you move clips between layers in the same track? (Choose all that apply.)

a) Option-drag (macOS) or Alt-drag (Windows)

b) Press Shift-Up Arrow or Shift-Down Arrow.

c) Drag the clip up or down to a different layer.

d) Press , (comma) and . (period) to move the selected clip up or down to a different layer.

2 True or False? During playback, the lowest audio track layer will always be audible.

3 How do you split a selected clip at the playhead? (Choose all that apply.)

a) Click the Razor (scissors) tool.

b) Press Command-B (macOS) or Ctrl-B (Windows).

c) Press Command-W (macOS) or Ctrl-W (Windows).

d) Choose Timeline > Razor.

e) Choose Timeline > Split clip.

4 True or False? To create a crossfade between overlapping music clips in track layers, you need to add fades to the upper clip.

5 True or False? Clips on the lower audio track layers are automatically deleted when you deselect the Show Audio Track Layers option in the View menu.

Answers

1. c. You can drag clips up or down to a different layer, and you can drag or use keyboard shortcuts to nudge clips left or right in the same layer.
2. False.
3. a, b, c, d, or e. (All the answers are correct.)
4. True.
5. False. Clips in lower track layers remain in the track unless they are manually selected and deleted.

Lesson 4

Editing Dialogue Tracks

Welcome to dialogue editing, in which every word from every character counts. In this lesson, you will assume the role of a dialogue editor. As such, you will follow a real-world workflow to transform production dialogue into effective dialogue tracks. Spending extra time on the spoken words is essential to creating a polished soundtrack, regardless of the type of project, budget, or size of your crew. Along the way, you'll also explore channel mapping for dealing with multiple microphone sources, and some effective keyboard shortcuts for editing and moving clips between tracks.

Time

This lesson takes approximately 60 minutes to complete.

Goals

NOTE The exercises in this lesson build from the tools and skills that you learned in the previous lessons. If you skipped ahead to this lesson, you may need to look back at the previous lessons as a guide for skills and tools already covered.

Preparing the Project

In this exercise, you'll continue working with the scenes from Hyperlight. However, your focus will be entirely on the dialogue tracks and the words themselves. Let's start by opening the first timeline, setting up the viewer, and previewing the scene.

1. Open the 4 Multi-Channel Dialogue timeline.

   The timeline opens with three audio tracks.

2. Move and resize the viewer so that it fits in the lower-right third of the timeline.
3. Resize the tracks so you can clearly see the clips in all three tracks.

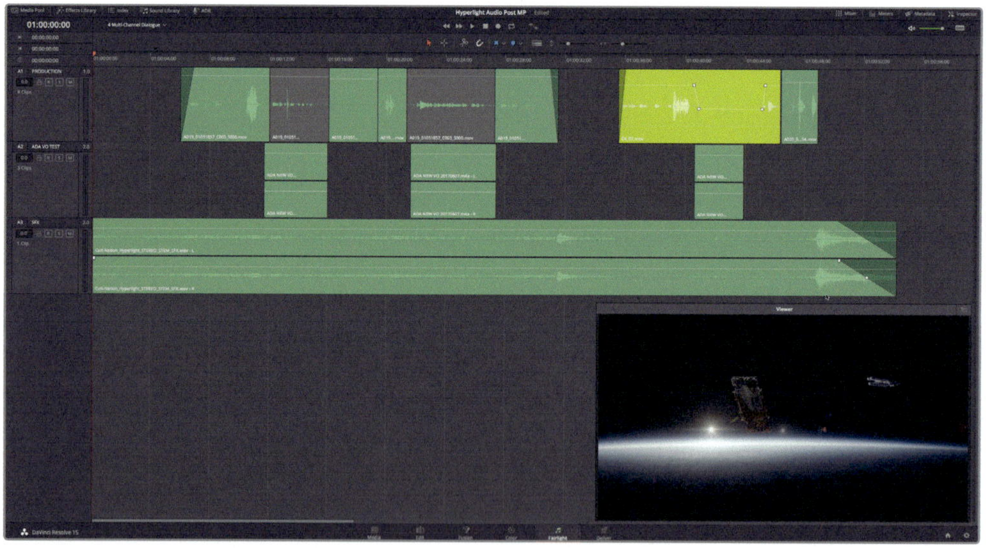

NOTE For this example, most of the color correction and visual effects have been completed. Also, for the purpose of this book, the A2 ADA VO TEST track contains the temporary ADA voice read by a personal computer, and the clip in the A3 SFX track is from the finished sound effects stem: the final stereo mix of the soundtrack's sound effects, as created by Cult-Nation Studios.

4 Play the scene and listen to the dialogue exchange between the characters Emiliana and ADA, the computer.

This timeline is an example of what a dialogue editor may receive from the editorial department as "locked picture." In other words, the picture edit is not supposed to change. Meanwhile, the visual effects team, colorists, and audio artists are free to begin work.

Of course, with DaVinci Resolve 16, all four departments can work on the same project and timelines without leaving Resolve.

What Does the Dialogue Editor Do?

As the title suggests, dialogue editors are responsible for all of the spoken words in the soundtrack, including dialogue, narration, and voiceover. In many cases, dialogue editors are also responsible for recording ADR.

However, in addition to dialogue, the dialogue editor is also responsible for all of the production sound recorded during the shoot. When the picture editor is finished with a cut, she passes it to the sound department where the dialogue editors organize the sounds from the set into appropriate tracks. They also remove unwanted "production" sounds such as camera noise, buzzing lights, and squeaky dolly wheels. The dialogue editor also ensures that every syllable of dialogue is clear and present, even if it requires wading through outtakes to find a better-sounding take. He may also swap the sound in wide shot with audio from clean close-up takes to improve clarity and presence.

The dialogue editors' "big picture" goal is to improve the clarity and consistency of each word within the context of a scene, as well as to create a seamless dialogue experience. So, in addition to cleaning up and improving spoken words, they also fill in and smooth the gaps between lines to eliminate any silence or jarring changes in ambience between cuts that could distract from the dialogue-driven performances.

NOTE The following sections on audio channel mapping and track types are crucial for properly setting up your clips and timelines for successful audio editing. In fact, about 80% of all the troubleshooting questions this author receives regarding audio editing in Resolve stem from a lack of understanding about channel mapping. The good news is that once you step through the following sections, you'll be armed with the skills needed for all future projects, whether it is dealing with multichannel audio, eliminating an unwanted channel, or simply isolating the center channel (dialogue) from a 7.1 surround timeline clip.

Evaluating Dialogue Tracks and Audio Channels

The first step in dialogue editing is to evaluate the existing tracks to determine if the audio clips and tracks have the correct channel mapping and microphone channels exposed. Basically, you need to look at the tracks and clips to make sure that they have matching channels. If not, you'll fix them. Luckily, the Fairlight page displays audio track information so that you can quickly determine how the synced audio tracks and channels are mapped in the timeline.

In the following figure, you can see all of the available audio channels in each track. The meters to the right of each track header, as well as the number above the meters, indicate each track's audio channel mapping. Additionally, in the Fairlight page, you can always see the audio channels in the timeline. The Edit page, however, only shows one channel per track, so editors may inadvertently edit an entire mono channel dialogue scene without realizing they're working with a stereo track.

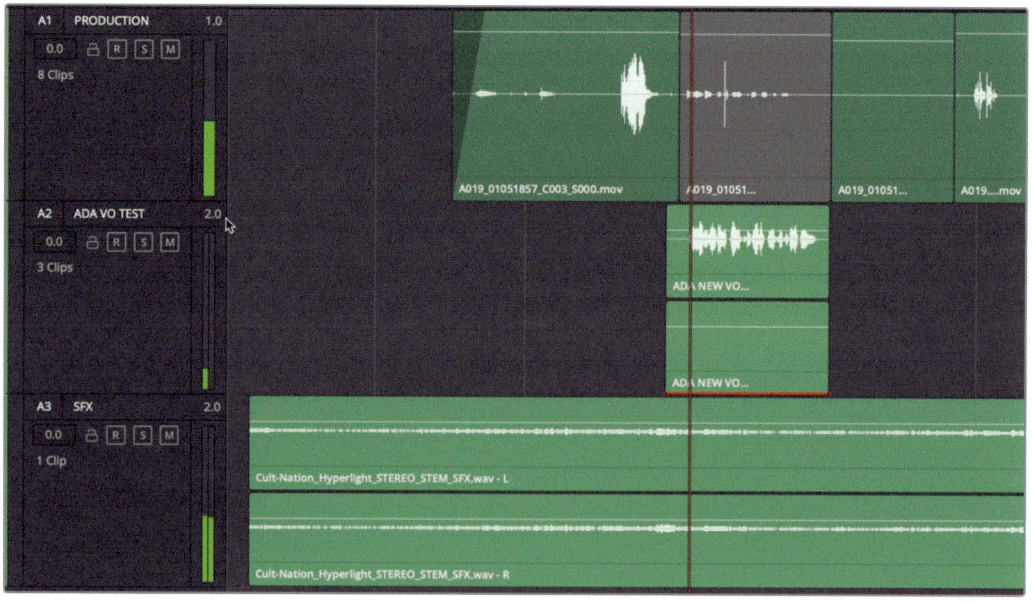

The 1.0 or 2.0 in the upper-right corner of each audio track header indicates the presence of one mono channel or two audio channels: L (left) and R (right), or stereo. The audio clips display available channels based on the track settings.

1 Count the number of audio channels visible in the A1 track.

The track header shows "1.0" to indicate the presence of one mono audio channel, and the clips in the track display only one channel of audio.

2 Count the number of audio channels in the A2 and A3 tracks. Each track should have two channels.

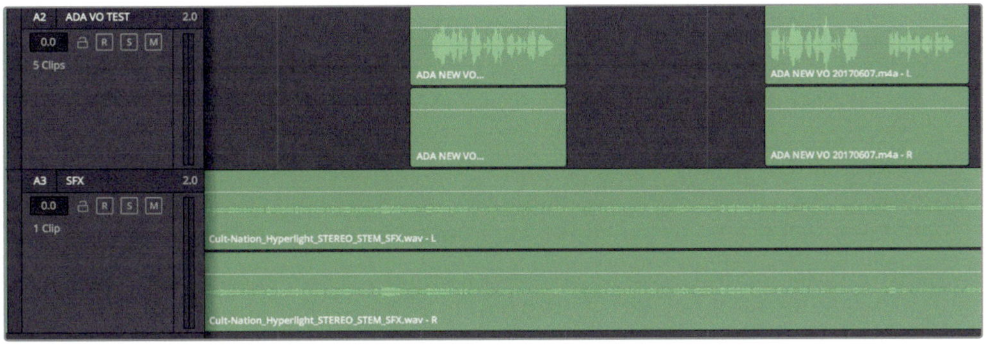

However, the waveforms reveal that one of the tracks may be mismatched. Looking closely at the clip waveforms in the A2 and A3 tracks, you'll see that the A3 track clip has waveforms in both channels, whereas the A2 track waveforms appear only in the upper channel, and the lower channel appears empty, which is an indicator that the clip does not match the track settings. Although this discrepancy may not seem like a big deal, it can later become an issue when you are editing, panning, and mixing tracks.

Because the second channel does not contain a waveform, you can change the track format from stereo to mono.

3 Right-click the A2 track header, and choose Change Track Type To > Mono.

The visible waveforms now match the track type in the A2 track, as a single channel of audio in a single channel (mono) track.

In addition to the watching for visual cues when matching audio tracks, you also must use your ears while evaluating the audio clips. In this scene, Emiliana's dialogue is clear throughout, but one of her clips sounds a little more muffled than the others, as if recorded with a different microphone.

4 Solo the A1 track. Play the last two clips, and focus on Emiliana's dialogue lines, "Plot a course to put me next to him," and, "Positive. Do it."

Although both clips sound good, the first one (colored lime green), definitely has a different sound quality. Perhaps another microphone source channel is embedded in the clip. In this case, that's exactly what you're hearing. In Resolve, a clip can have up to 24 embedded audio channels, so there may be more to this clip than meets the eye and ear. There's one way to find out. Let's compare the clip attributes for both clips.

5 Right-click the last clip in the A1 track, and choose Clip Attributes. In the Clip Attributes Audio pane, click the Source Channel pop-up menu to see the number of audio channels embedded in the clip.

The Source Channel pop-up menu reveals that the last clip contains only one channel of audio.

6 Click OK to close the Clip Attributes window.

Now let's check the lime green clip, which sounded a bit muffled.

7 Right click the second to last clip (lime green) in the A1 track, and choose Clip Attributes. Click the Source Channel pop-up menu.

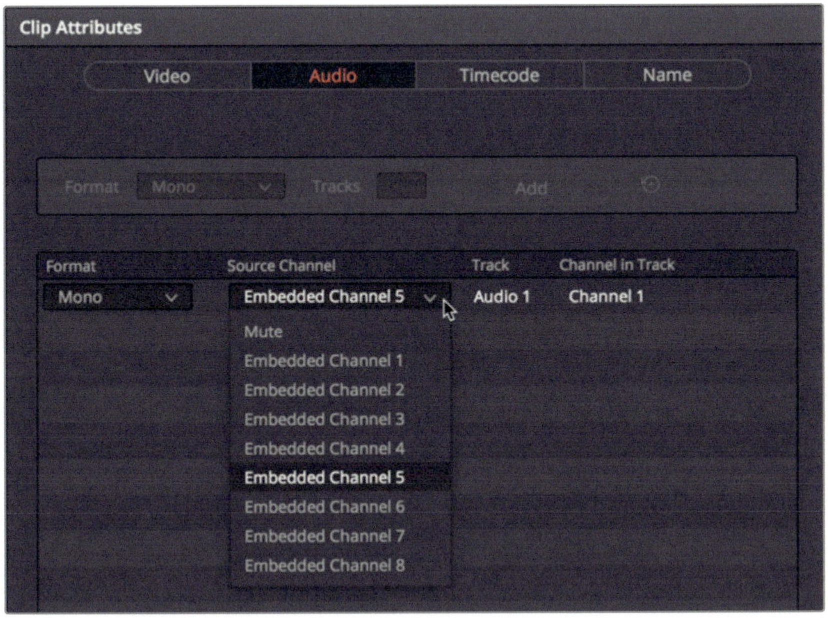

As you can clearly see, the clip format is Mono and Embedded Channel 5 is currently exposed. The pop-up menu also reveals that the clip contains eight embedded audio channels.

8 In the Clip Attributes window, click any empty space to close the Embedded Channel pop-up menu. Then, click OK to close the window.

There you have it. The clip has quite a few embedded audio channels. In fact, eight channels seems to be a little excessive for a dialogue clip. Usually, a clip with eight channels indicates a 7.1 surround sound clip. Chances are that this clip has multiple audio channels but was simply incorrectly mapped as a 7.1 surround clip when it was first synced. No worries. This audio channel mystery can easily be solved and fixed in the Media Pool.

Identifying Audio Channels

Although final dialogue tracks are usually contained in a single mono audio channel, one for each character, the original production sound was probably recorded with multiple microphones. For example, a typical dialogue recording setup would include a boom microphone and a wireless microphone on each actor. The boom microphone sound is usually the best quality and therefore will be best for use as the final track. However, sometimes the wireless microphone sound is preferable, or a combination of the two independent microphone sounds is needed to optimize dialogue clarity.

When you are editing dialogue, it is most important that you be in control of the recorded audio channels you want to use. Independent control of the channels starts at the clip level, even before the clips are edited to the timeline. Fortunately, you can always change a clip's mapping for dialogue editing.

In the following exercise, you'll open the lime green clip in the Media Pool to see the clip attributes displayed there. Then, you'll apply different channel mapping configurations to the clip to determine which is best for this scene. Finally, you'll go back to the clip in the timeline and change the mono channel to the best embedded channel.

1 Show the Media Pool.

2 In the A1 track, right-click the last clip, and choose Find in Media Pool.

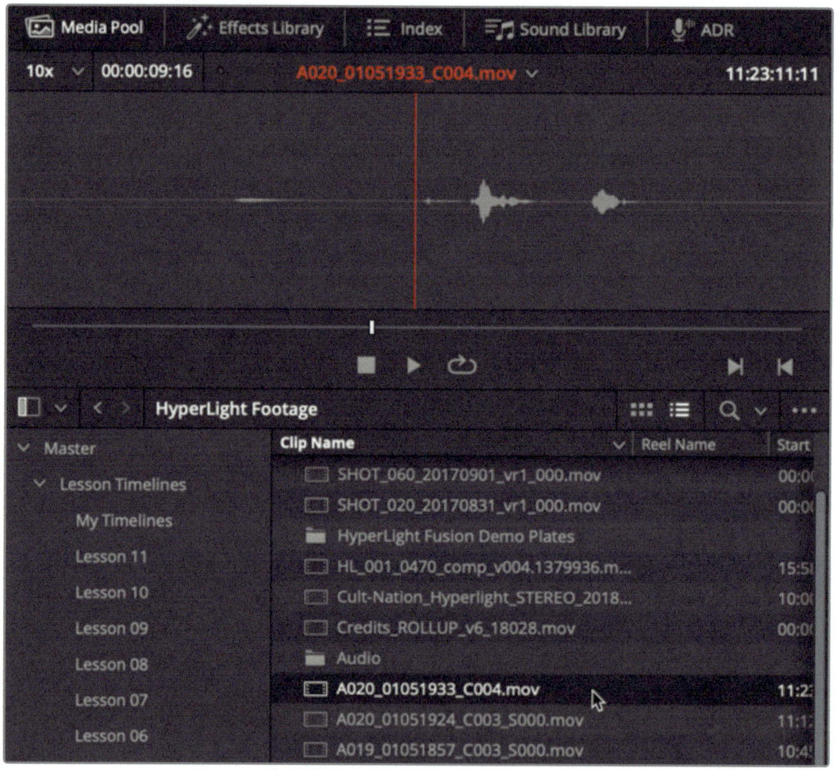

The clip is selected in the Media Pool list. In the preview player, you can also see that it has only one audio channel.

3 In the A1 track, right-click the second to last clip (lime), and choose Find in Media Pool.

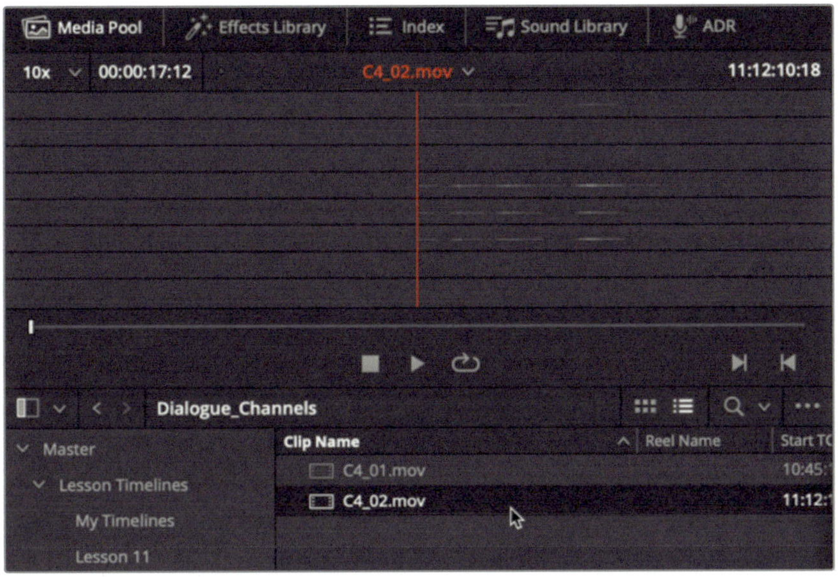

In the Media Pool preview player, you can see the eight embedded channels of audio and can also see that the clip really was incorrectly mapped as a 7.1 surround clip when it has audio only on channels 1, 4, 5, and 6. To better understand channel mapping, let's edit this clip into a new timeline.

4 In the Media Pool Dialogue Channels bin, right-click the C4_02.mov clip, and choose Create New Timeline Using Selected Clips.

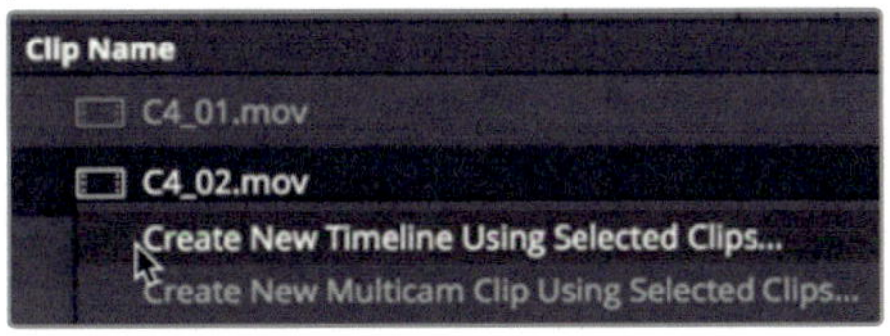

5 In the New Timeline dialog box, name the timeline **Channel Mapping**. The other default settings are fine as they are (especially Audio Track Type > Based on selected media). Click Create.

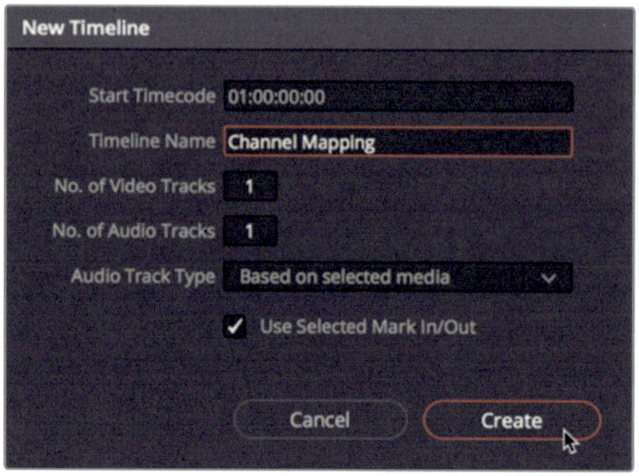

The Channel Mapping timeline opens with a single track containing the eight channels of audio.

6 Click anywhere in the timeline to make it active. Hold down Shift while scrolling up with your mouse wheel, or in the toolbar, drag the vertical zoom slider to zoom in vertically to the timeline track.

7 Hold down Option (macOS) or Alt (Windows) while scrolling up with your mouse wheel, or drag the horizontal zoom slider to zoom in horizontally.

The timeline clip confirms eight channels in a 7.1 surround track with audio waveforms visible in channels 1, 4, 5, and 6. Each waveform is slightly different, indicating that they were recorded using different microphones or represent a mix of several microphones created by the production sound mixer. The channels with larger waveforms indicate a louder, more focused microphone recording of that person's dialogue.

To evaluate and balance the levels of the dialogue tracks, you need to mute, solo, delete, raise, or lower the levels of each audio channel. However, to exert that level of control, you must map individual channels to separate tracks in the timeline. This is not possible for multichannel tracks, as you will see in the next step.

8 Move your mouse pointer over the white volume overlay on the upper channel of the clip. When the pointer changes to an up/down arrow, drag the volume overlay up to raise the volume of that channel of the clip.

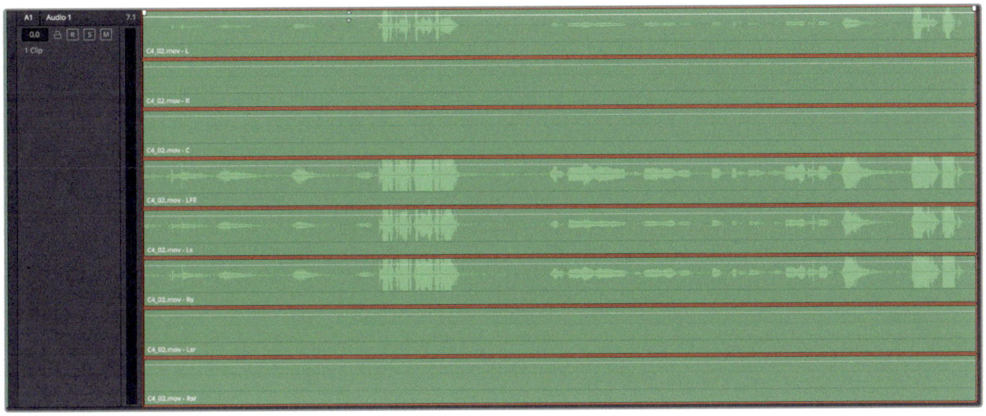

Notice that any change you make to the volume overlay on the upper channel also applies to all other channels in the clip. So, that method clearly will not work for dialogue editing because it doesn't allow you to adjust each channel independently.

9 Double-click the volume overlay to reset it, or press Command-Z (macOS) or Ctrl-Z (Windows) to undo the changes to the volume overlay.

It's easy to identify the separate channels in a timeline clip or in a source clip in the Media Pool. Next, you'll look at remapping audio clips so you can fix these discrepancies when they occur or entirely prevent them from happening in your own projects.

Remapping Audio Channels in the Timeline

You can change a track's channel configuration to match any type of audio clip, even those that don't fit standard multichannel formats like stereo and 5.1 surround. Resolve has a unique and flexible format called *adaptive tracks* that can accommodate up to 24 audio channels. In this exercise, you'll change the audio clip from an eight-channel 7.1 surround sound clip to a four-channel adaptive clip, and expose the four channels containing waveforms. Then, you'll go one step further and change both the timeline clip and track to mono.

1 In the A1 track, right-click the clip, and choose Clip Attributes. In the Clip Attributes dialog box, set Format to Adaptive 4; and set the four Source Channel pop-up menus to Embedded Channel 1, Embedded Channel 5, Embedded Channel 5, and Embedded Channel 6.

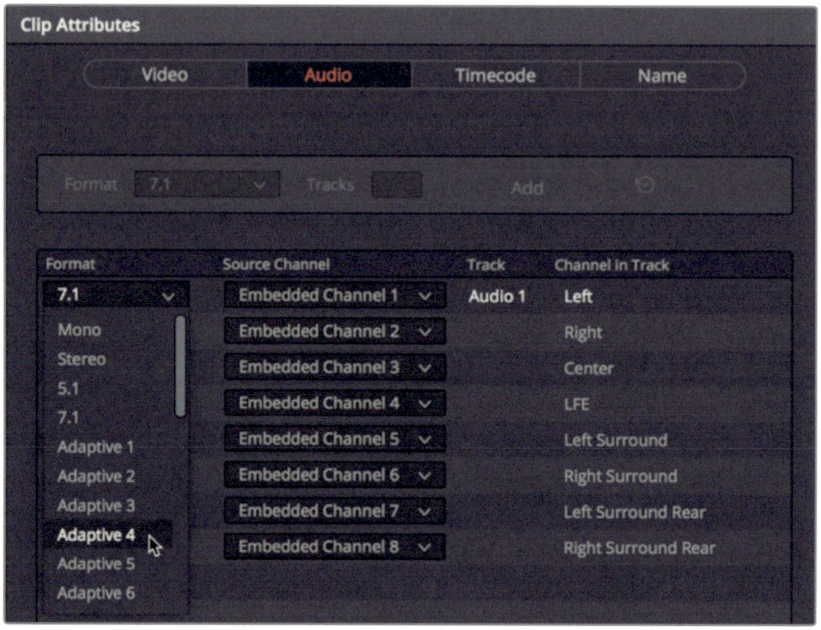

Set the Format to Adaptive 4.

Set the four source channels to 1, 4, 5 and 6.

2 Click OK.

The four-channel adaptive clip is visible in the A1 track. However, because the track is still an eight-channel 7.1 surround track, the lower four channels are empty.

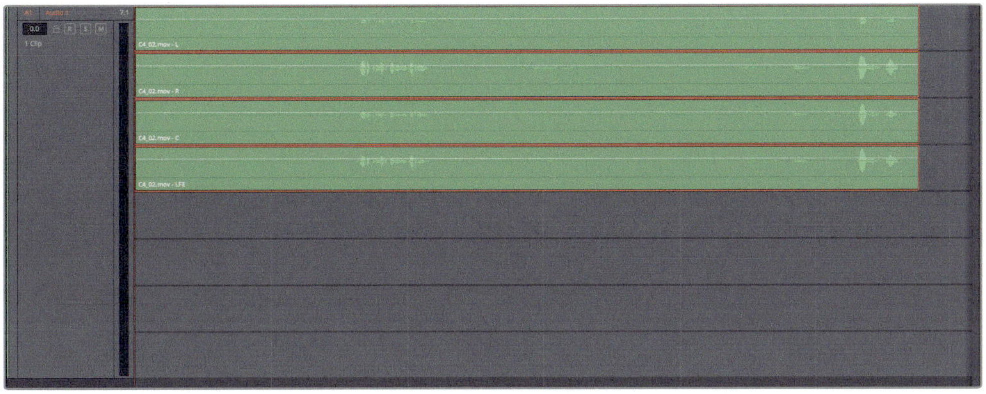

3 Right-click the A1 track header, and choose Change Track Type To > Adaptive > 4.

You finally have a 4-channel adaptive clip in a matching track.

Now you can see and hear all the sources at once while editing. That's ideal for the editor or assistant editor who is assembling selects (select takes, or best takes) and cutting together a scene.

4 Press Shift-4 to show the Edit page.

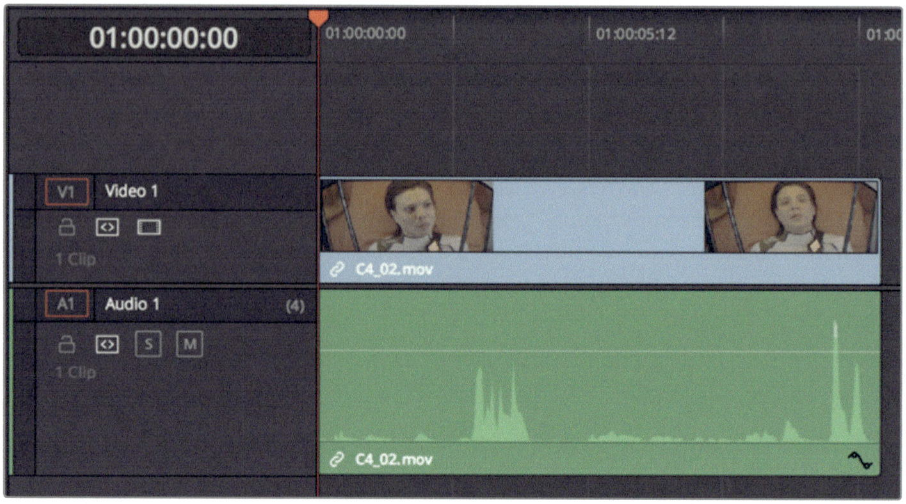

In the Edit page timeline, notice that the clip shows only one audio channel, which is easy for the editor to work with, whereas the Fairlight page always displays all the audio channels for each track.

> **TIP** In the Edit page, you can display individual channels for a clip by right-clicking any clip and choosing Display Individual Audio Channels. Also, to display audio waveforms in the Edit page, click the waveforms icon in the timeline options menu.

5 Press Shift-7 to show the Fairlight page.

6 On any of the audio channels, drag the volume overlay up or down. Double-click the volume overlay to reset it.

As a dialogue editor, you still need to gain independent control of the separate channels to narrow down your choice to the best-sounding channel. Luckily, you have plenty of options to do so.

7 Right-click the clip and choose Clip Attributes. In the Clip Attributes dialog box, change Format to Mono. In the Source Channel pop-up menu, choose a channel: 1, 4, 5, or 6.

8 Right-click the A1 track header and choose Change Track Type To > Mono. Click OK.

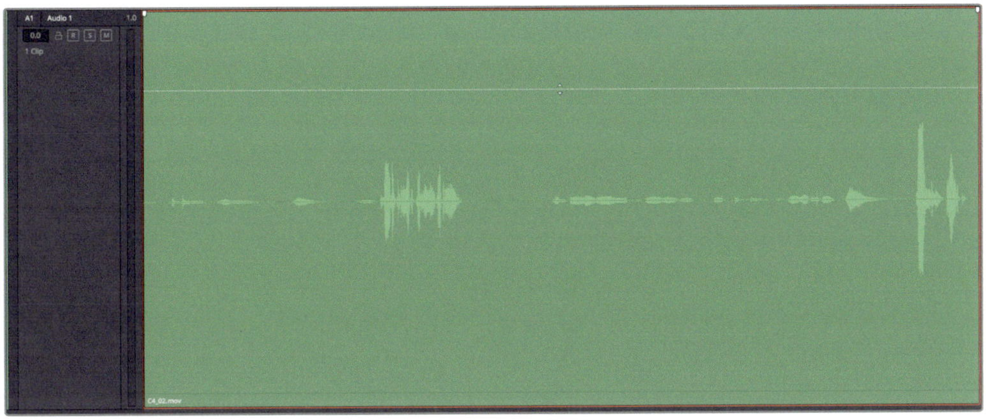

Alas, you have successfully changed the incorrectly mapped 7.1 clip into a mono clip in a mono track. You have experienced first-hand Fairlight's flexibility for transforming clips and tracks as much as needed. Of course, the looming question remains: How do you know which of the four embedded audio channels to use in the dialogue track? Sure, you could go back to the mono clip's attributes and systematically audition each mono channel. But it's much easier to evaluate and compare audio channels when you can hear them together, and solo or mute them on-the-fly for comparison. To do so, let's return to the original source clip and modify the source clip's attributes for testing those embedded audio channels.

Remapping Source Clips in the Media Pool

In this final remapping exercise, you'll return to the Media Pool to change the channel configuration of the source clip. Then, you can edit the revised clip to the timeline and evaluate the individual channels. When you're finished doing so, you'll have the skills needed to properly map, evaluate, and transform your clips and tracks in your own projects.

1 In the A1 track, right-click the clip, and choose Find in Media Pool.

2 In the Media Pool, right-click the selected C4_02.mov clip, and choose Clip Attributes.

And now for the fun part. You can use the Format and Tracks settings near the top of the window to set up and add any channel configuration or combination of channels that you'd like.

For this example, let's go a little crazy and keep the original 7.1 format with eight-channels in one track, and add an Adaptive 4 track along with four separate mono tracks, each with a different channel.

3 In Clip Attributes settings, set Format to Adaptive 4. Keep the default Tracks setting as 1, and click Add.

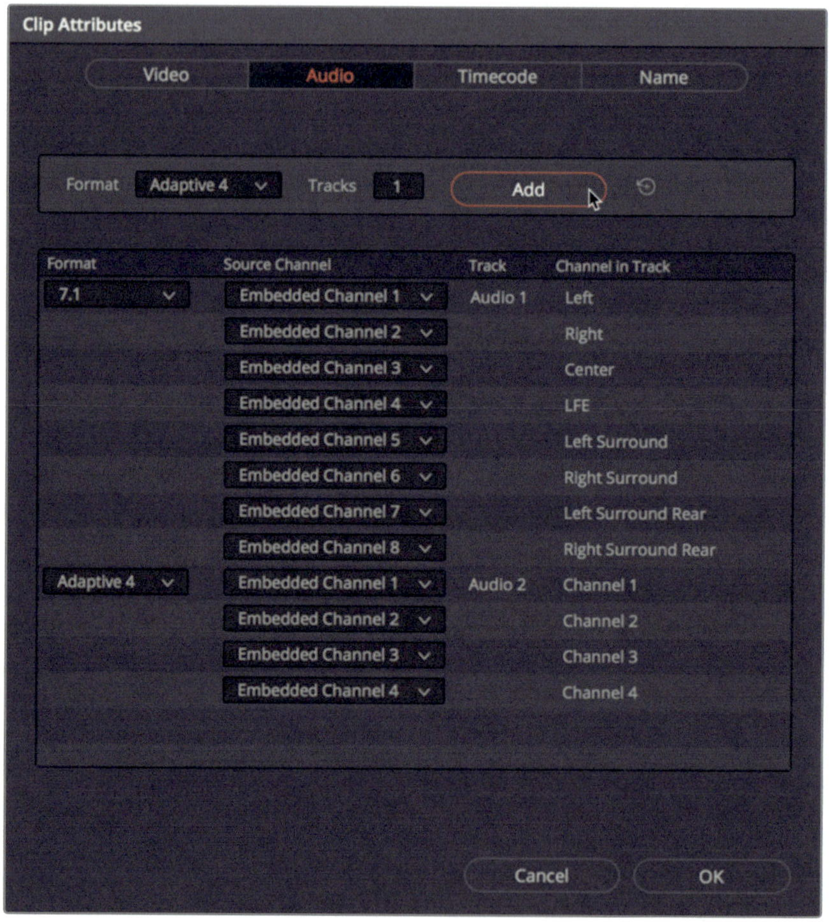

The Adaptive 4 audio track is added to the clip's attributes.

4 Change the Embedded Channel settings in the adaptive track to Embedded Channel 1, Embedded Channel 4, Embedded Channel 5, and Embedded Channel 6.

Now let's add another channel mapping configuration to the same clip.

5 In the Clip Attributes settings, set Format to Mono. In the Tracks field, type **4**. Click Add. Assign each of the Embedded Mono Channels to a different channel: 1, 4, 5, and 6. Click OK.

You now see all the added audio channels in the preview players. Let's add the revised clip to the timeline. Remember, in DaVinci Resolve 16, the Fairlight page automatically creates tracks to match the source clip's channel mapping.

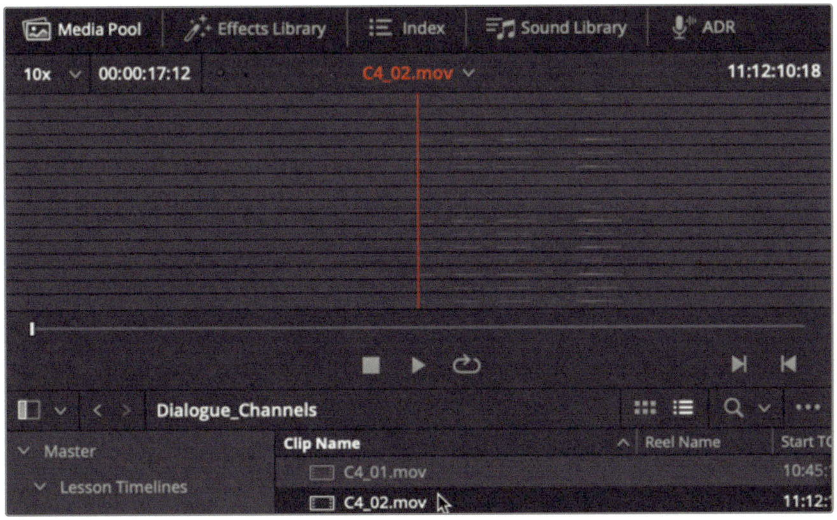

6 Drag the Vertical Slider to the left to reduce the height of the existing timeline track. Move the playhead to the beginning of the timeline.

7 Drag the C4_02.mov clip from the Media Pool to the empty space in the timeline.

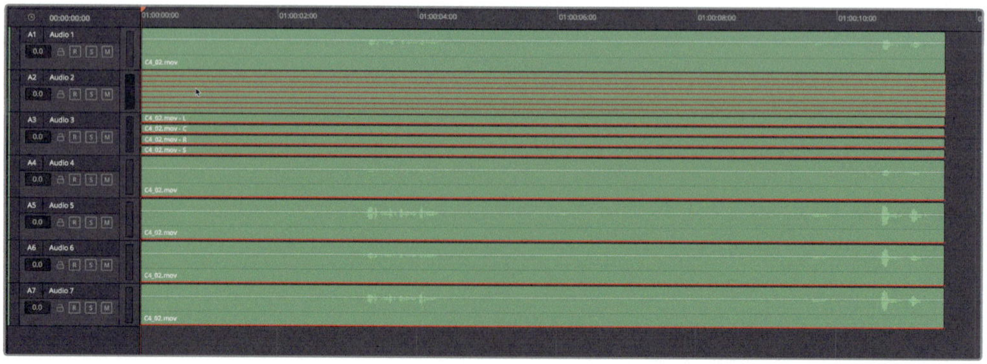

Exactly as expected, the new audio clip includes a 7.1 track, an Adaptive 4 channel track, and four independent mono tracks, each with a different audio channel. Chances are, you'll never actually need to make an audio clip with this exact configuration, but now that you have, you can confidently create any audio channel mapping you need for your projects.

Let's narrow down the clip to just the last four mono tracks.

8 Right-click the A1 track header, and choose Delete track.

9 Repeat step 8 to delete the 7.1 track and the Adaptive 4 channel tracks.

Finally, you can listen to each mono audio channel to evaluate the various microphone sources and select your favorite source to use for this scene. Along the way, you can use some of the skills that you've learned in the previous lessons to simplify the process.

10 Zoom in on the tracks vertically to clearly see the contents of each mono track.

Next, you'll mark a playback region around Emiliana's line, "Plot a course to put me next to him." Once marked, you can loop playback around that line while you listen to the separate tracks.

11 Mark an in point (press I) at 01:00:07:00 and an out point (press O) at 01:00:08:18. Turn on Looped Playback, and press Option-/ (slash) in macOS or Alt-/ (slash) in Windows to start looped playback. Swipe the Mute buttons on for all four tracks, and then systematically un-mute one track at a time. Or, you can click the Solo buttons, as necessary. Feel free to drag the volume overlay on any of the mono clips to increase or decrease the volume level.

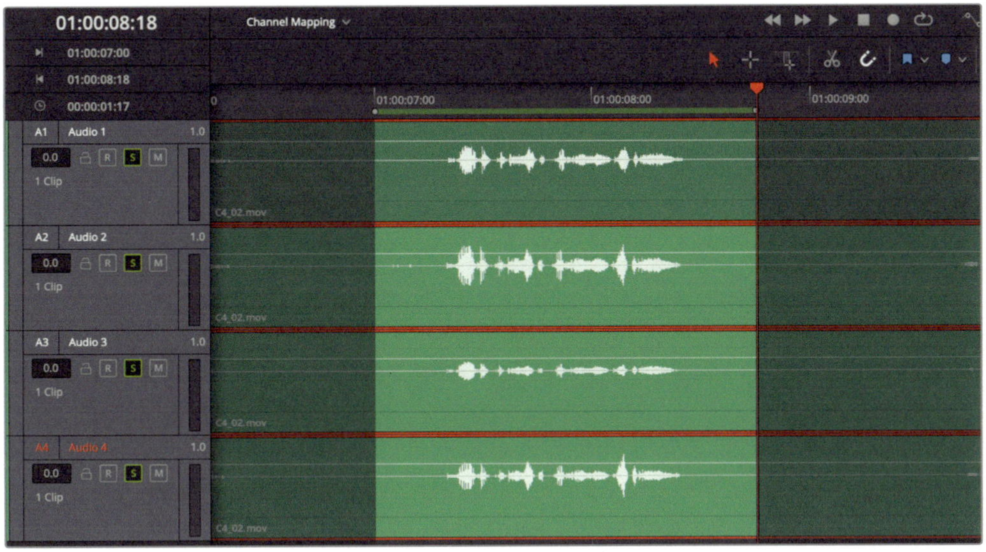

12 Delete all but your favorite mono track. If you aren't sure which one to choose, keep the A1 track.

The audio in track A1, formerly channel 1 in the 7.1 clip, sounds like a mix of the channels on A2 and A4 that was created on the set by the production sound mixer.

NOTE It's okay if you can't hear much difference between the channels. Training your ears to hear subtle audio differences is similar to colorists training their eyes to recognize subtle differences in shadows and flesh tones. You probably were able to eliminate the sound on A3 because it sounds more muffled than the others. That also happens to be the audio channel in the edited scene.

Deleting Unwanted Source Audio Tracks

In this exercise, you'll delete the 7.1 and Adaptive 4 audio tracks from the source clip.

1 In the Media Pool, right-click the C4_02.mov clip, and choose Clip Attributes.

2 In the Clip Attributes window, hover the pointer on the far-right side of the 7.1 track area until a trash can icon appears. Click the trash can icon to delete that channel configuration from the source clip.

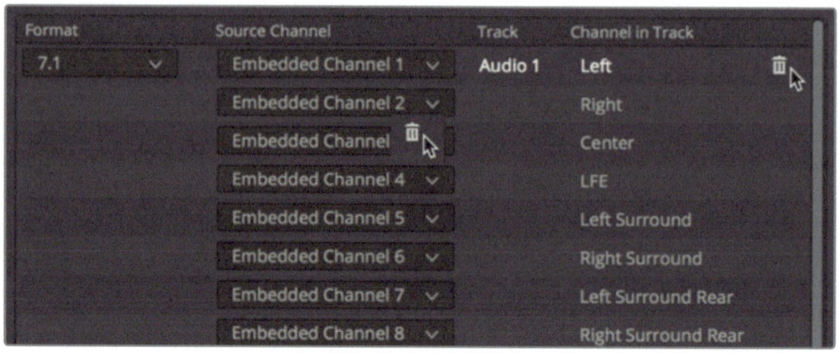

3 Delete the Adaptive 4 channel track. Click OK to close the Clip Attributes.

The clip now comprises four mono tracks, each containing a different embedded audio channel. The original eight embedded channels are still part of the clip's media, and can be accessed through the source Clip Attributes window at any time.

Changing a Clip's Source Channel in the Timeline

Now you can bring this example full circle by opening the original edited timeline and changing the mono channel exposed in the lime timeline clip to a different mono channel.

1 Open the 4 Multi-Channel Dialogue timeline.

2 Hide the Media Pool.

3 Right-click the second to last clip (lime), and choose Clip Attributes.

4 In the Source Channel pop-up menu, choose Embedded Channel 1 or your favorite source channel. Click OK.

You not only learned the secrets of Resolve's flexible audio channel settings but also evaluated the channels in a timeline clip and changed the clip's mono channel accordingly.

Performing Checkerboard Editing on Dialogue Tracks

Once a scene is edited and the best audio channels are chosen in the timeline, it's time to split and move all the dialogue clips to separate tracks for each character. This process is referred to as *checkerboard editing* because of the way the tracks look when the process is completed.

The primary reason for separating out the dialogue for each character is so each character has an individual track when you're adjusting volume, equalization, and dynamics processing; and creating the final mix. This editing technique is essential for all projects whether they are scripted narrative, unscripted, dramatic, corporate, or documentary.

In the next series of exercises, you'll follow a checkerboard editing workflow that starts with watching the scene, and creating a track for each character.

1 Open the 4a Dialogue Editing Start timeline.

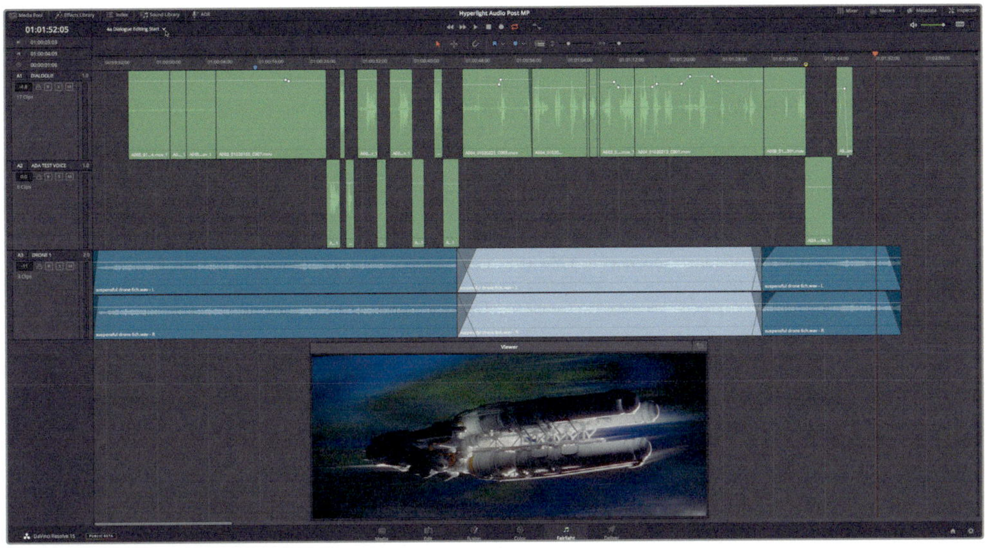

For this scene, all the dialogue clips are properly mapped mono clips with the correct source channel.

2 Play the scene and listen to how many characters have speaking parts. (Hint: computer voices count as characters.)

Three characters have speaking parts: Emiliana, Philip, and ADA. So, each will need a separate dialogue track. ADA's voice is already on a separate track, whereas Emiliana and Philip still need separate tracks. So, let's leave Emiliana's clips in the upper track, and move all of Philip's dialogue to a new track.

3 Right-click the A1 Track header, and choose Add Tracks. In the Add Tracks dialog, set "Number of tracks" to 1, Insert Position to Below DIALOGUE, and Audio Track Type to Mono. Click Add Tracks.

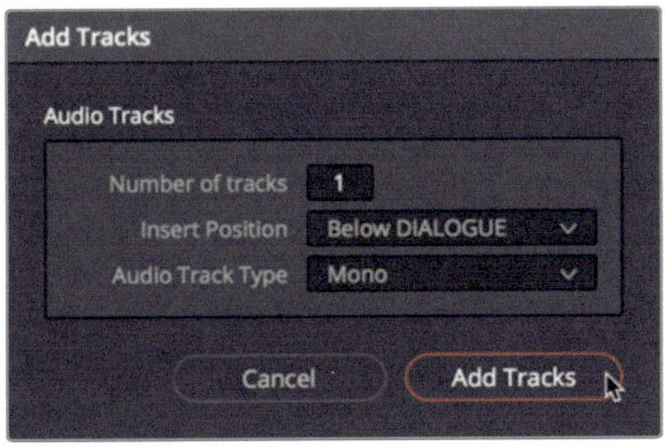

4 Change the names of the first two tracks to: A1, **EMILIANA**, and A2, **PHILIP**.

Let's also change the A2 and A3 track colors to help differentiate the clips within the checkerboard edit.

5 Right-click the A2 PHILIP track header, and choose Change Track Color > Yellow. Change the color of the A3 ADA TEST VOICE track to Tan.

The tracks are set and ready for separate clips...almost. Many dialogue editors will do one more thing before splitting up the dialogue tracks....

Resetting Clip Levels

Because these exercises are based on real-world workflows, this is a good time to learn how to reset volume levels to their default settings. Why would you do this? Because, in the real world, most dialogue editors and audio professionals prefer to start with "clean" unaltered audio, so they can personally determine which tools and processing are necessary to prepare the best-sounding dialogue tracks for mixing—just as colorists clear all editor-applied grades and plug-ins before they begin grading, and makeup artists start with a clean makeup-free face instead of adding makeup on top of performer-applied makeup.

So, please keep in mind that if you are an assistant editor who spends tireless hours normalizing, keyframing, and tweaking dialogue levels on your scenes before passing them along to the audio department, your efforts are appreciated by the picture editor and anyone who screened the edit.

Meanwhile, seasoned dialogue editors will likely begin each new scene by resetting all the levels. This is quite easy to accomplish in the Inspector. First, you'll select all the dialogue clips in the A1 track.

1 In the A1 track, drag the Selection tool around all of the clips in the track to select them.

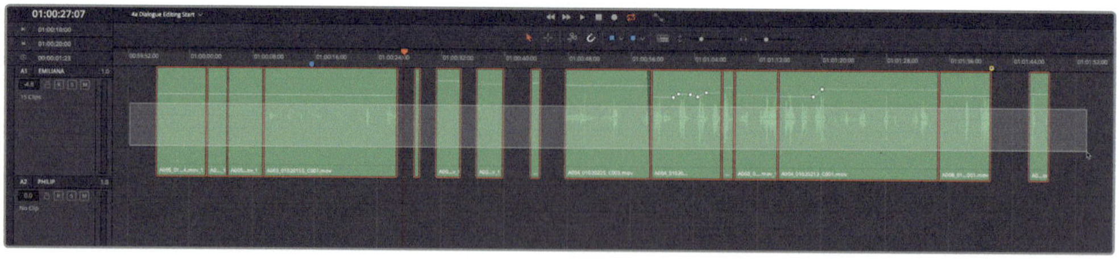

2 Show the Inspector.

The Inspector header indicates that Multiple Clips are selected and can be modified in the Inspector.

3 Click the reset arrow for the Clip Volume control to reset the levels for all of the selected clips.

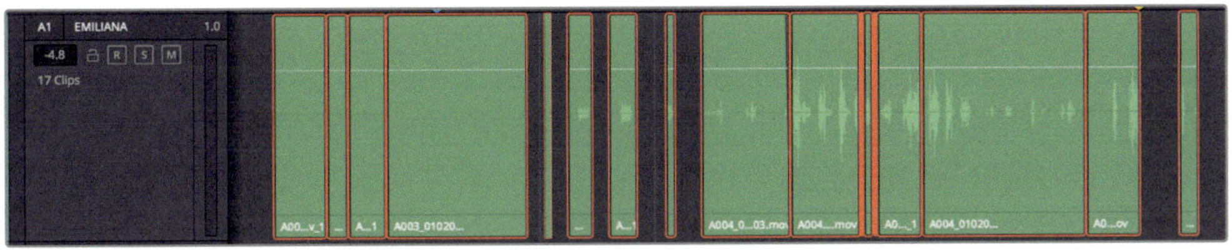

The volume overlay for each clip returns to the 0.0 dB setting to indicate that no change is applied to the original sources' sound levels. Also, there are no keyframes on any of the clips.

4 Select the clips in the A3 ADA TEST VOICE track, and in the Inspector, reset the clip volume level.

5 Hide the Inspector.

With the tracks created and the levels reset, you are ready to move the dialogue clips to the correct character tracks.

> **NOTE** To start with clean dialogue, you would also check to see if any plug-ins were applied to the clips. Such clips are easy to spot because they have an fx icon next to the clip name in the timeline. You'll work more with plug-ins in another lesson. To clear all the plug-ins, clip panning, or EQ applied to the timeline clips, you can right-click and choose Remove Attributes. Then check the boxes for all of the above to remove them at once for all selected clips. This technique can also be used to reset volume levels.

Using Shortcuts to Move Clips Between Tracks

Sure, you could manually drag clips between tracks using your mouse. But professional audio editing is all about precision editing and shortcut-driven, efficient workflows. By design, the Fairlight page is a fully-loaded digital audio workstation packed with easy shortcuts that enable lightning-fast audio editing. For example, moving clips between tracks using keyboard shortcuts is similar to moving text between lines in a word-processing program. In fact, the shortcuts to cut, copy, and paste are identical to the shortcuts in most word processors.

In this exercise, you'll use shortcuts for everything from selecting tracks, to moving the playhead, and editing the selected clip to a different track.

The keyboard shortcuts for clip editing are identical in macOS, and Windows. Their only differences are the modifier keys. In macOS, you'll use the modifier keys Command and Option, whereas on a Windows system, you'll use the modifier keys Ctrl and Alt.

For the first part of this exercise, you'll first use the menu to apply a command. Then, you'll use the shortcut the next time you use that command. After that, you'll be ready to fly through your workflow using only shortcuts to finish the job. And don't worry about memorizing shortcuts; chances are you already know them, and if not, you'll pick them up after a single use.

1 Press Home to move the playhead to the beginning of the timeline. Mute the A4 DRONE 1 track.

2 Choose Timeline > Timeline Destination Selection > Audio Destination A1 to select the A1 track. While the menu is visible, look at the other Track Destination Selection options and shortcuts.

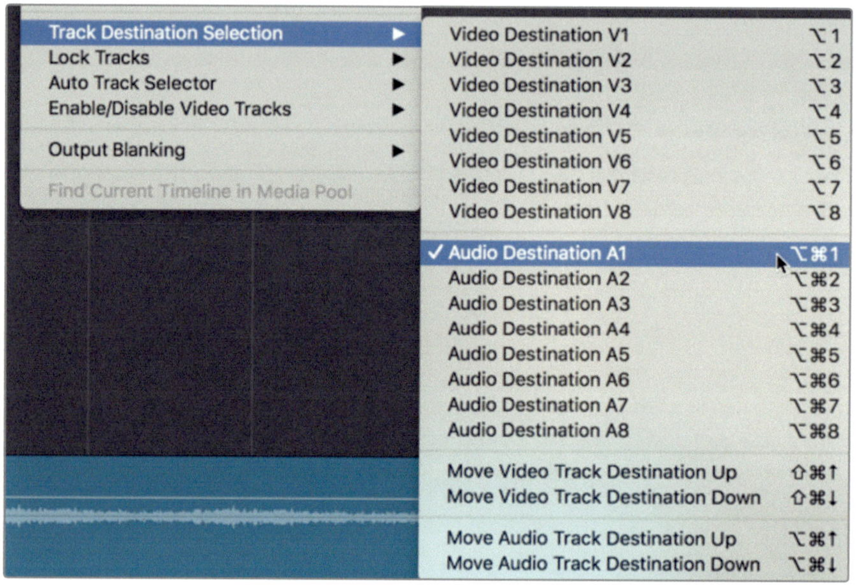

Notice that the keyboard shortcut to select the A1 track is Option-Command-1 (macOS) or Alt-Ctrl-1 (Windows). Also, in the Track Destination Selection options, you'll see that the shortcut to Move Audio Track Destination Up or Down uses the same macOS/ Windows modifier keys along with the Up Arrow or Down Arrow. If you look closely at the list of shortcuts from Audio Destination downward, you'll see that the Audio shortcuts all use the same two modifier keys. So, the good news is that the only thing you need to remember is the two modifier keys (the two keys closest to the Spacebar) and track number or the Up and Down Arrows. Easy.

You already use the JKL shortcuts for navigation; so, in this exercise, you'll use the Down Arrow or Up Arrow to move between clips, because when a track is selected, the arrow keys move the playhead to the next clip in the track and select that clip automatically. Let's try it.

3 With the A1 track already selected, press the Down Arrow to select the first clip in the selected track.

This clip doesn't contain dialogue, so for now, let's leave it and move on. Pressing the Down Arrow key moves down the timeline, stopping at the head and tail (first and last frame) of each clip as it goes.

4 Press the Down Arrow as many times as necessary to move down the timeline to the head of the fourth clip.

5 Press the Spacebar to play the clip. When playback is finished, press the Up Arrow, if necessary, to move the playhead to the tail of the clip and select it.

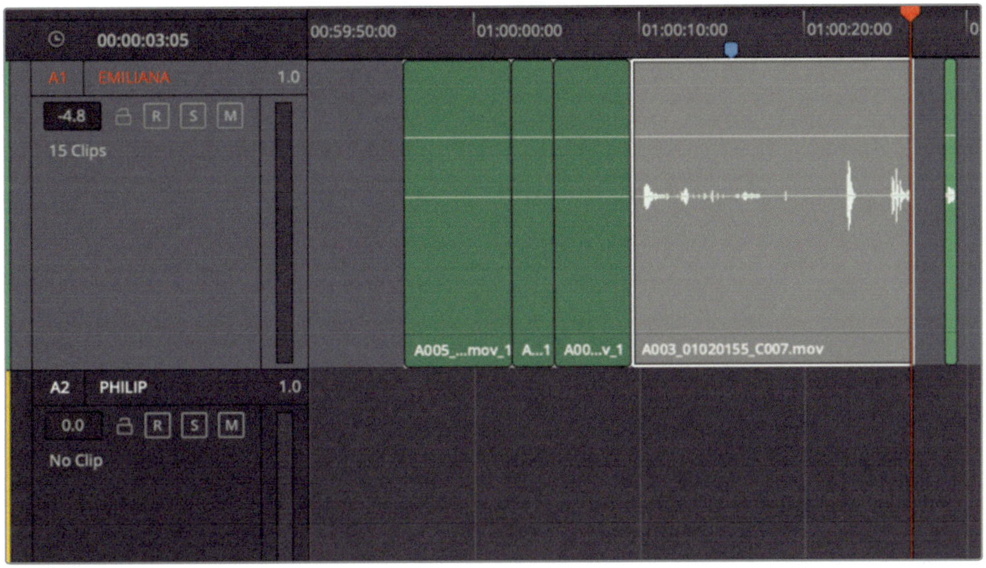

Clearly, this is Philip's dialogue clip. So, you'll now use a command to cut the clip. Once cut, you can select Philip's track as the destination and paste the clip in the new track. Let's look at the Edit menu to verify which shortcuts you'll need.

6 Click the Edit menu, and hold down the mouse button for a moment to look at the common edit commands and shortcuts.

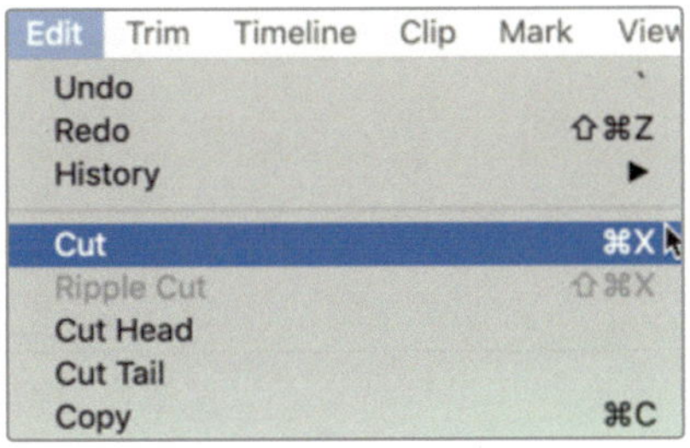

The shortcut for Cut is Command-X (macOS) or Ctrl-X (Windows).

The shortcut for Copy is Command-C (macOS) or Ctrl-C (Windows).

The shortcut for Paste is Command-V (macOS) or Ctrl-V (Windows).

Most likely, these are exactly the same shortcuts that you use for text editing as well.

7 Choose Edit > Cut.

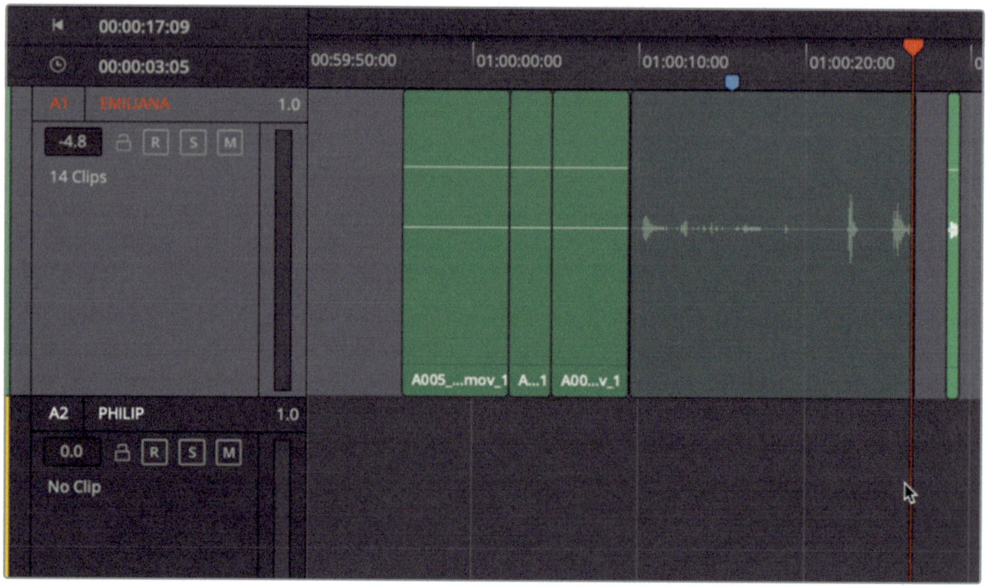

The Selected clip turns semi-transparent to indicate that it has been cut from the current position and is ready to move and paste elsewhere.

8 Press Command-Option-Down Arrow (macOS) or Ctrl-Alt-Down Arrow (Windows) to select the A2 track.

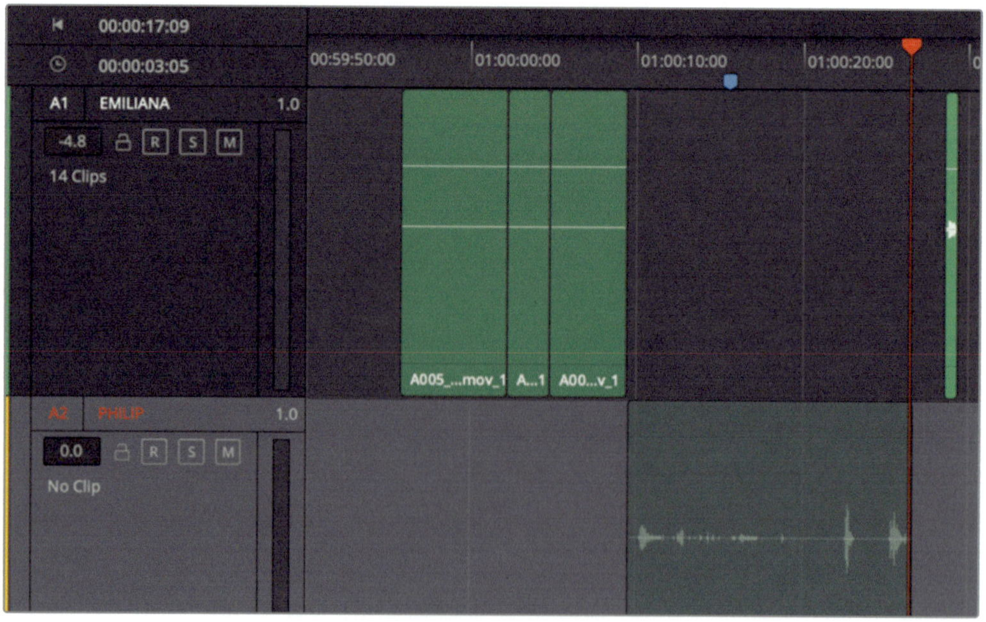

The semi-transparent clip appears in the exact same position on the A2 track.

9 Press Command-V (macOS) or Ctrl-V (Windows) to paste the clip into the new track.

The clip is moved. Keep in mind that cutting-and-pasting destinations are actually based on the playhead position. So be careful not to move the playhead after you cut a clip, if you are simply moving it to a different track.

10 Press Command-Option-Up Arrow (macOS) or Ctrl-Alt-Up Arrow (Windows) to select the A1 track.

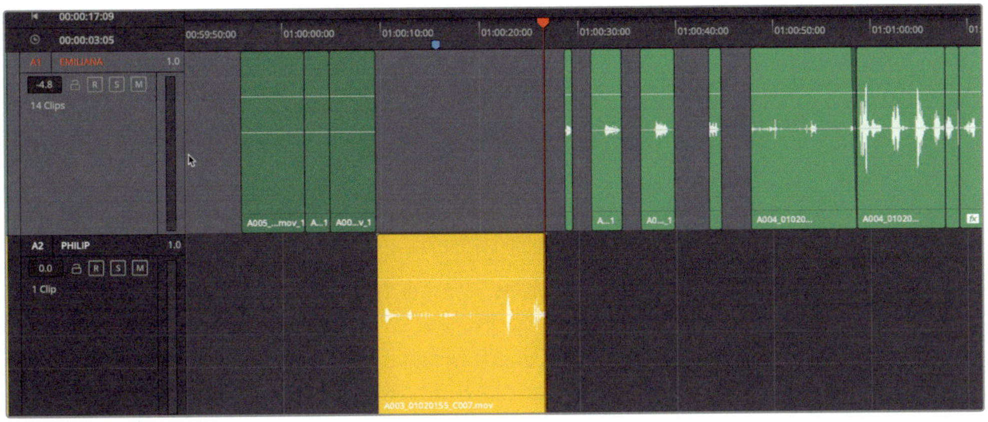

The A1 track is selected and ready for you to locate and move Philip's other clips. Notice that the clip that you pasted into Philip's track is yellow because you previously changed the track color. All the clips you move to Philip's track will also turn yellow while they are on that track.

Finishing the Job

You've done excellent work setting up the dialogue tracks and moving the first clip to Philip's track. You're hired! As an intern. No money, of course, but you'll gain lots of experience and infrequent compliments.

To finish separating out the dialogue tracks, you're now on your own. It's up to you to use the same shortcuts to locate and move Philip's remaining dialogue clips to the A2 track. If you make mistakes, no worries, just choose Edit > Undo and try again.

When you are finished, change the A1 track color to Orange.

A few things to consider before you get started. Some of the audio clips include production sounds such as breathing and rustling clothes. Breathing, sighs, snorts, giggles, and so on are considered part of the dialogue performance and should be moved to the associated character's track. If the scene were ever dubbed in another language, the actor performing the new lines would also need to perform the breaths, signs, and other vocalized sounds.

Clothing movement, footsteps, and other naturally occurring production sound *(nat sound)* picked up by the microphones are often replaced with Foley sound recordings. You'll work more with Foley sound effects and other sound design elements in Lesson 6. In the real world, dialogue editors are also responsible for splitting and moving all non-dialogue production sound to a separate production effects (PFX) track.

For now, just move clips into the track that belongs to the character responsible for the sound.

Don't worry if the heads or tails of some of the clips are a little rough. You'll clean them up after they are moved to their own tracks.

Have fun! When you are finished, you can move on to the next section. If you don't finish, you're fired—as an unpaid intern—but please still move on to the next section.

Cleaning Up a Checkerboard Edit

Now that the checkerboard edit is complete, you can clearly see how this technique got its nickname. In this exercise, you'll zoom in on the tracks to watch and listen to the timeline. As you go, you'll pause playback and clean up any sloppy heads or tails that stand out during playback. You'll employ two of DaVinci Resolve 16's new features: the Fixed Playhead mode and loop jog scrubbing.

1 If you completed all the previous exercises, you can continue to work with the 4a Dialogue Editing Start timeline. If not, open the timeline 4b Dialogue Tracks Split to continue.

2 Move the playhead to the first clip in the A3 track. This is ADA's first line of dialogue in the scene (01:00:26:11).

3 Zoom in horizontally until you can see only two orange clips on A1, two yellow clips on A2, and three or four tan clips on A3.

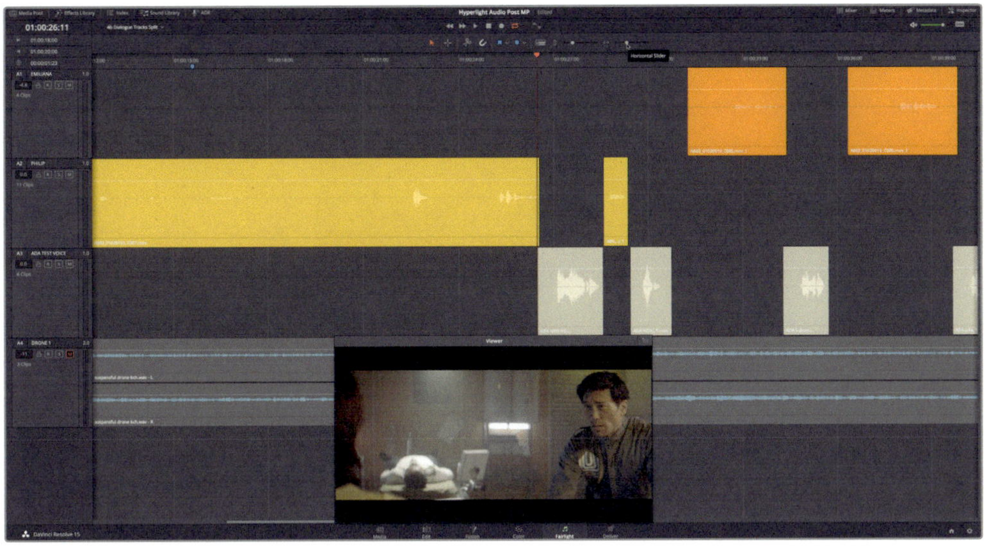

Adjusting the zoom level is great for focusing on specific clips during playback, but what happens when the playhead reaches the end of the visible area of the timeline?

4 Start playback from the current position, and stop playback after the playhead reaches the end of visible section of the timeline.

The playhead moves from left to right, and when it reaches the right edge of the timeline, it continues again from the left as it plays the next section of the timeline. This is just like turning the pages of a book. This common playback mode works for most activities, but when you are trying to focus on the sound of the clips, it can become a little distracting. Let's change it.

Using Fixed Playhead Mode for Scrolling Playback

Normal playback mode is highly functional for both picture editing and audio editing. However, at times, you need a seamless playback experience so that you can focus all your attention on the sounds and waveforms of the clips as they scroll beneath the playhead. In that case, you can turn on Fixed Playhead mode. This mode is terrific for listening to and watching clips, in part because you can always see ahead to the clips that are coming up.

1 Press Shift-Up Arrow to move the playhead to the blue timeline marker.

The playhead returns to the blue marker which is now conveniently centered in the middle of the timeline ruler. This marker, placed right before Philip's first line of dialogue, is a great place to start this exercise.

2 Open the Timeline View Options menu, and click the Fixed Playhead button.

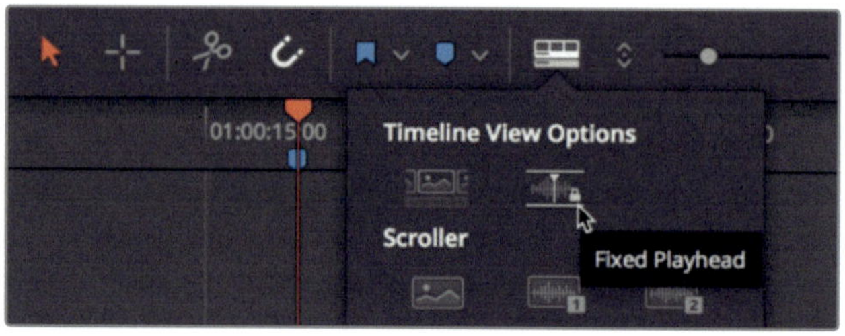

3 Start playback. As the clips scroll beneath the fixed playhead, focus on what you hear, and which clip is coming up next. When Philip says, "Boom. There was this explosion and it surrounded the whole ship," stop playback.

While admiring the Fixed Playhead mode's scrolling playback, you probably heard a few rough patches in Philip's dialogue.

4 Use the JKL keys to scroll to around 01:01:04:01, right before Philip says, "We've been trying to fix it."

> **NOTE** In Fixed Playhead mode, you can press the JKL keys or drag the timeline ruler left or right to scroll the timeline forward or back. To move the fixed playhead to a different position, you need to turn off Fixed Playhead mode in the Timeline View Options menu.

5 Start playback again, and focus on the end of the current clip and the head and tail of the next clip. When Philip says, "Two hours ago," stop playback.

Clearly, Philip's dialogue clip that includes the words, "went space walking," needs some work at both its head and tail. At this point, you'll go back to normal playback mode to clean up the clip. However, the next time you need to focus on audio playback, you can simply turn on Fixed Playhead mode and let the timeline scroll!

Trimming Synced Audio in the Timeline

For this exercise, you'll zoom in on the errant clip in the A2 track and trim the head and tail to clean it up. Keep in mind that this audio is synced, so you can't physically move the clip without also forcing it out of sync with the picture. Also, because the clip is between two other clips on the same track, you'll need to make sure that as you trim one clip you don't accidentally create a problem in another clip.

1 Open the Timeline View Options menu, and click the Fixed Playhead button to deselect that option.

2 Move the playhead to 01:01:08:08, which is directly over the clip that you'll be working on. Zoom in horizontally until you can clearly see the clip beneath the playhead, as well as most of the neighboring clips on either side.

3 Right-click the clip at 01:01:08:08, and choose Clip Color > Lime.

At its current size, parts of the waveform are cut off at both the head and tail of the clip. Also, the bright green color makes it easy to identify the clip as a work in progress. Let's mark in and out points around that section of the dialogue to include the word or phrase just before and after the current clip. Then, you can determine exactly what is wrong and how to fix it without compromising the other clips.

4 Press R to enter Range Selection mode. Drag a selection rectangle around the clips on the A2 track from around 01:01:06:00 to 01:01:11:00.

> **TIP** You can see the timecode for the in and out points in the timecode display in the upper-left of the timeline, and use it as a guide while you drag your selection.

5 Press A to go back to standard Selection mode.

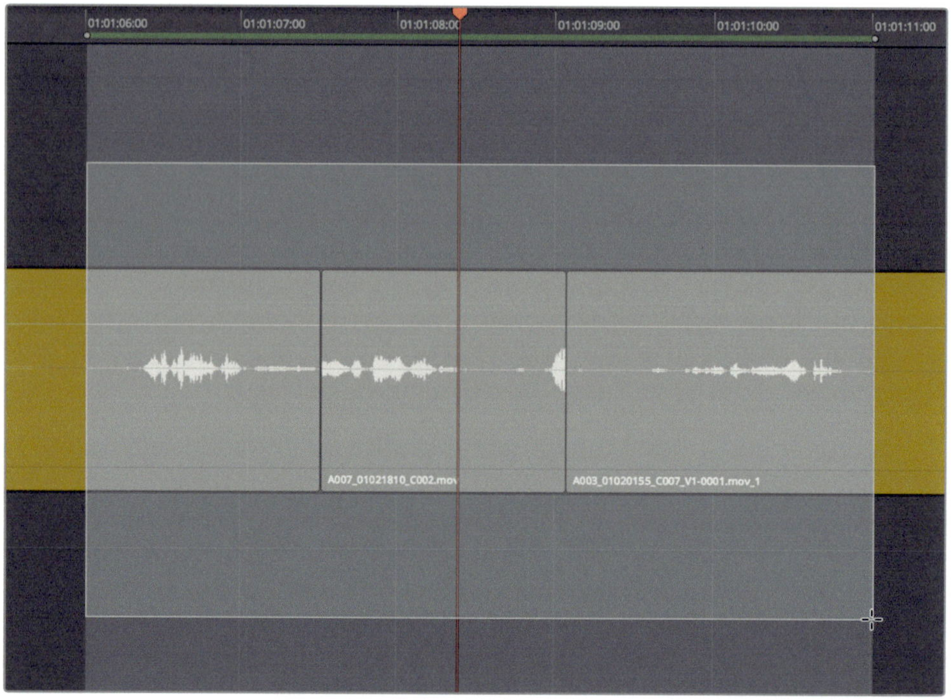

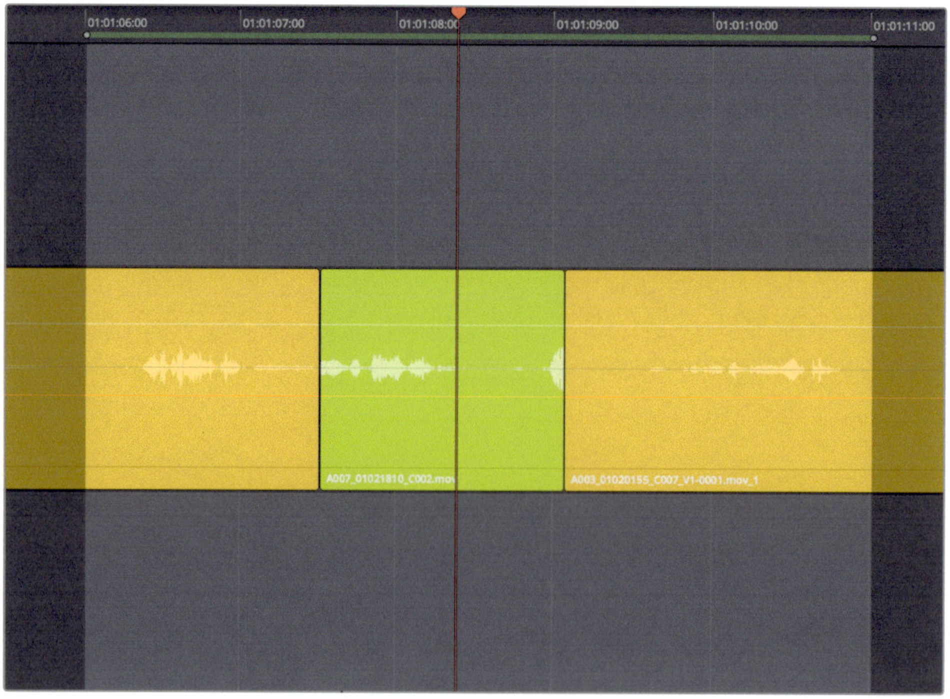

6 Turn on Looped Playback, if necessary. Press Option-/ (slash) in macOS or Alt-/ (slash) in Windows to start looped playback. After you listen to the section a few times, stop playback.

The first clip ends with Philip saying, "Since it happened," followed by a large intake of breath. The middle clip starts in the middle of the word "you" as Philip says, "You went space walking." To fix the first edit point between dialogue clips, you can extend the head of the lime clip to include the entire word "you." Don't worry about cutting some of Philip's breath sound from the tail of the outgoing yellow clip. Dialogue editing is about making the dialogue clear and intelligible. The breath sounds are important but also incidental to the dialogue.

Remember that when you trim an audio clip in the Fairlight timeline, you see the entire waveform of the clip, which makes it easy to see where words and phrases start and end.

7 Drag the head of the lime green clip to the left to include the entire word "you." Use the waveform as a guide as you extend the clip.

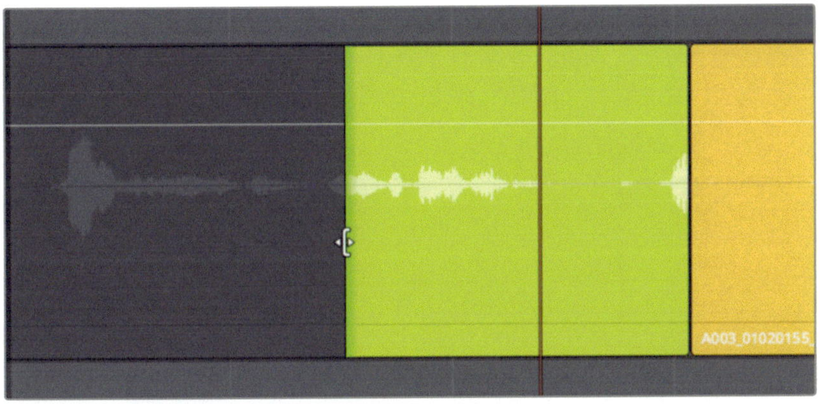

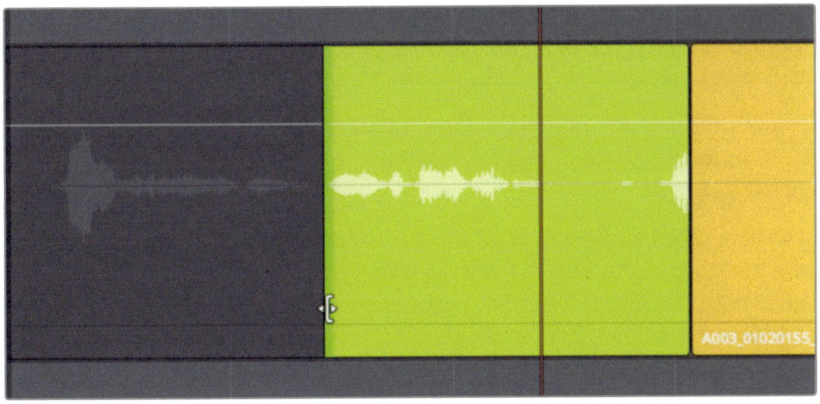

8 Drag the tail of the lime green clip to the left until you remove all of the waveform (the partial word) at the end of the clip.

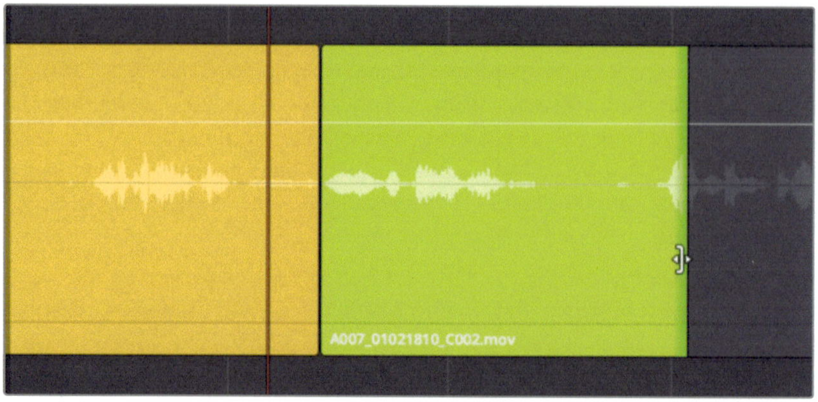

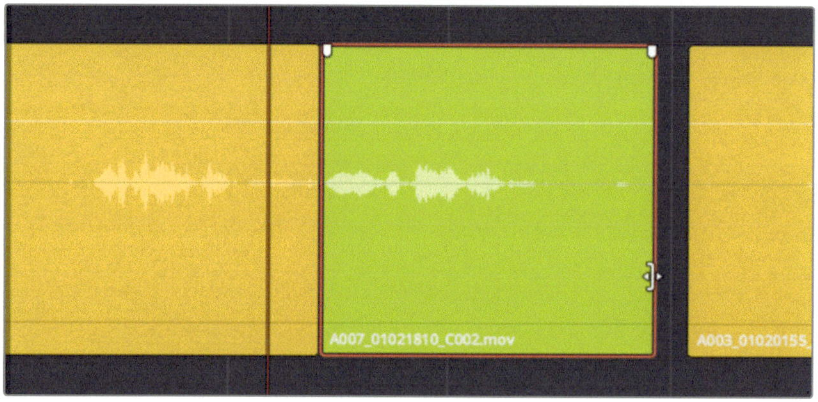

9 Play from the in point to the out point to hear your edits in the timeline. During playback, listen to the performance, and don't focus on the gap between the clips. You'll fill that in with tone in the next lesson.

10 When you are finished evaluating that section of dialogue, stop playback.

What are your thoughts? Does the dialogue sound natural? Does anything distract from the words or performance?

Did you notice that Philip takes two breaths before his line, "Then just two hours ago"? The first partial breath is at the end of the lime green clip while Philip is off-screen. The second breath is right before his line, "Then just two hours ago." The first partial breath is unnecessary and should be trimmed. The second breath is part of the performance.

To Breathe or Not to Breathe? The Secret to De-Breathing Dialogue

Breathing is part of dialogue. Inexperienced audio editors commonly remove purposeful (and sometime meaningful) breath sounds while trying remove unwanted noise from poorly recorded dialogue. In doing so, the dialogue loses its humanity and the performances suffer. If you see a character breathe, inhale, or exhale as part of the performance, it is important to include that breath with the dialogue.

If your talent takes excessively loud breaths, or they suffer from a cold or sinus condition that involves distracting whistling or wheezing, then by all means tone down the wheezing or replace it with less distracting breaths. But don't remove it entirely.

In only three circumstances should you completely remove all breathing from recorded voiceover/dialogue tracks:

1 In commercial advertising voiceovers in which the words spoken about a product or service are more important than the human speaking them. In this case, because of the types of microphones and intimacy of the recorded voice, breath sounds can and will distract from the message. So, for commercial advertising, the message is king, despite the presence of hot air.

2 When the character isn't a living, breathing creature, such as a talking computer like ADA. In this case, the computer doesn't breathe, so it is natural and even expected that the voiceover will be devoid of breath.

3 The character is alive but based on the science-fiction/fantasy concept that it doesn't breathe air to speak. (Or perhaps a character telepathically communicates with another character so all of his dialogue is conveyed non-verbally.) Whatever the situation, just keep enough of the sounds generated to vocalize words consistent with the character. For example, if your mermaid character breathes water through gills, and is talking underwater, you might want to hear water sloshing through the gills, but with no audible air movement through the mouth.

Increasing Size and Waveform for Easier Trimming

One advantage of working in the Fairlight audio page is that you can zoom in as much as necessary to clean up your dialogue tracks. In this exercise, you'll increase the height and width of Philip's dialogue track to better see the breath waveforms. Then, you'll apply Track Waveform Zoom, which is a new feature that increases the height and visibility of the waveform without increasing volume. The enlarged waveform can be used as a guide for easier triming.

1 Select the A2 PHILIP track, if necessary.

2 Move the playhead over the middle of the lime green clip.

3 Drag the Vertical Zoom slider all the way to the right to dramatically increase the height of the selected track.

4 Press Command-= (equals sign) in macOS or Ctrl-= (equals sign) in Windows to zoom in horizontally.

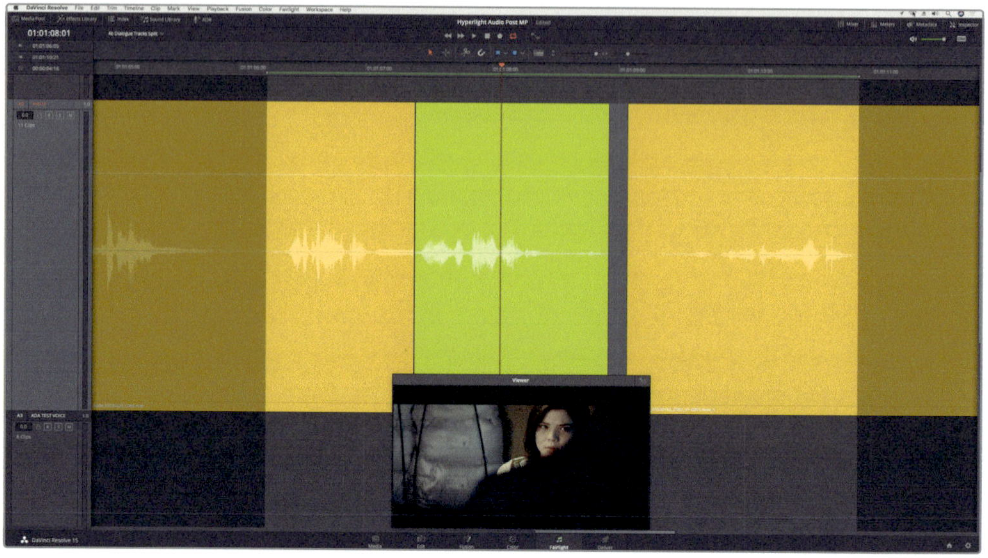

The playback range from 01:01:06:00 to 01:01:11:00 should now dominate the timeline.

5 Right-click the lime green clip, and choose Track Waveform Zoom > Zoom In All Tracks. Feel free to repeat this step several times to increase the size of the zoomed waveform.

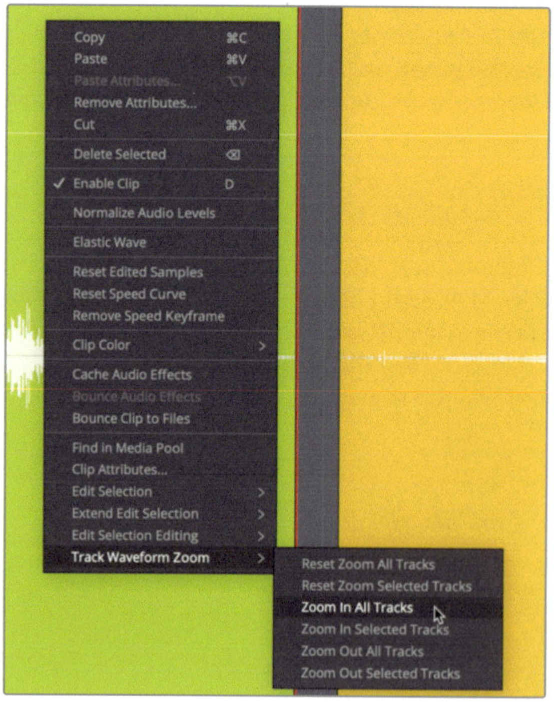

In the larger waveform, you can clearly see the waveforms for the separate breaths and the ambient room sound that was picked up by the microphones. In the next lesson, you'll use room sound (room tone) to fill the gaps between the dialogue clips. For now, let's focus on fixing those breath sounds.

6 Mark a new playback range from 01:01:08:00 to 01:01:10:00. This range cuts off the dialogue but makes it easier to focus on the breaths between words.

   The breath sound at the end of the lime-colored clip represents a fairly large increase in the waveform, so you should be able to trim it just by using the waveform as a visual guide. Keep in mind that Philip is off-screen at this point, so there's no reason to include this breath, just the sound of him moving across the room.

7 Move the playhead just before Philip's breath sound at 01:01:08:15.

8 Trim the tail of the lime clip to the playhead.

9 Listen to the play range.

Right-click the lime-colored clip and choose Track Waveform Zoom > Reset Zoom All Tracks to reset it.

10 Press R to enter Range Selection mode. Click any empty space in the timeline to clear the timeline marks. Then, click the lime clip to mark that clip for playback. Press A to enter standard Selection mode.

11 Listen to the green clip. When you are finished, press Option-X (macOS) or Alt-X (Windows) to clear the in and out points.

The end of the lime clip sounds much better with the removal of the partial breath.

Scrubbing Audio at the Sample Level Using Loop Jog

Scrubbing is the term used for listening to audio as you move the playhead back and forth over an audio clip in the timeline. When standard audio scrubbing is turned on, you can use it by pressing the JKL keys or by dragging the playhead through the timeline ruler. Loop jog scrubbing, on the other hand, provides a sample-level preview of the audio while scrubbing the playhead. Both types of scrubbing are available in the Timeline menu. Audio scrubbing is turned on by default, and you can toggle it on and off by pressing Shift-S. In this exercise, you'll use both standard audio scrubbing and loop jog scrubbing to preview and precisely trim the head of the next clip.

Why expend so much effort to trim the breath at the head of a clip? Because, in addition to editing spoken words and managing audible breaths, dialogue editors are also responsible for removing unwanted human noises such as lip smacks and tongue clicks. As you know, at the beginning of the next clip, Philip takes a breath onscreen as part of his performance. Right before that breath is a bit of mouth noise that should be eliminated. So, for this precision maneuver, you'll employ the power of loop jog scrubbing to determine how much to trim.

1 Move the playhead to the empty space between the lime green and yellow clips.

2 Drag the horizontal scroller at the bottom of the timeline to the right until you no longer see the lime green clip in the A2 track.

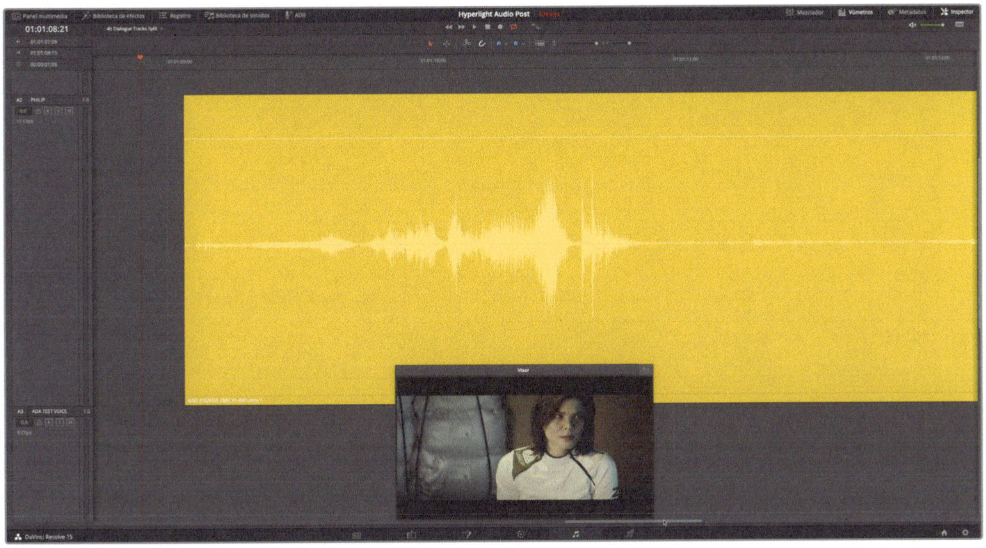

You should have a clear view of the beginning of the yellow clip.

3 Drag the playhead from the head of the clip to find the beginning of Philip's first words, "Then, just two hours ago."

With standard audio scrubbing turned on, as you scrubbed the playhead over the clip, you should have heard a bit of the tongue click and breath sounds.

4 Press Shift-S to toggle off audio scrubbing. Try scrubbing the beginning of the clip again.

This time, you can't hear any sound at all while dragging the playhead. Without audio scrubbing enabled, you are limited to using your eyes to judge the waveform while trimming, which is not recommended!

5 Press Shift-S again to turn on audio scrubbing.

6 In the Timeline menu, choose Loop Jog to enable loop jog mode.

7 Drag the playhead over the beginning of the clip once again. When you reach Philip's words, hold the playhead still.

This time, you could probably hear much more detail as you scrubbed the playhead. The fast-looping sound you hear is an 80-ms (millisecond) sample preview of the sound preceding the playhead.

8 Release the playhead. Press the Up Arrow to move the playhead to the beginning of the clip.

9 Press Command-, (comma) in macOS or Ctrl-, (comma) in Windows to open the Preferences panel.

10 Click the User button to show the User Preferences panel, and in the preferences sidebar, click the Fairlight tab.

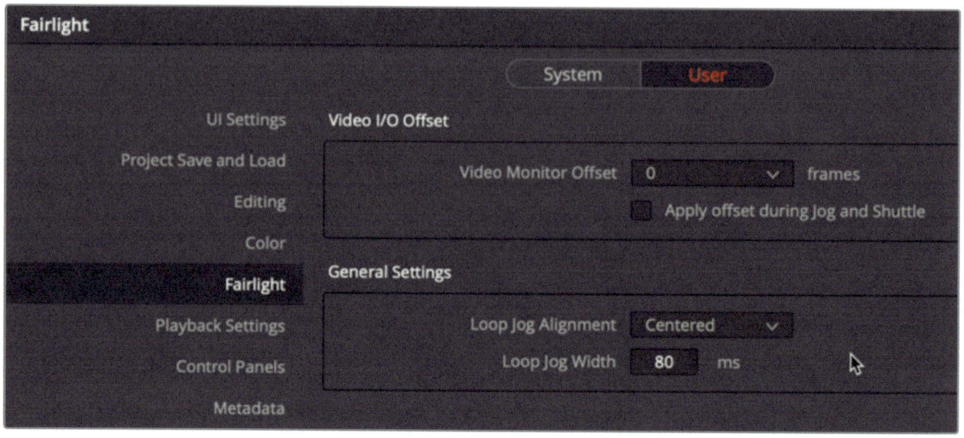

In the Fairlight General Settings, you'll see the Loop Jog Alignment pop-up menu and a field in which to set the Loop Jog Width. The default setting is 80 ms (milliseconds).

11 Change Loop Jog Alignment to Centered. Click the Loop Jog Width numeric field, and drag to the right to its highest setting, 2000 ms. Click Save.

Each second of video equals 1000 ms. So, by setting the loop jog amount to 2000 ms, and centering it around the playhead, you will hear one second before and after the playhead. Let's test this in the timeline.

12 Scrub the playhead over the beginning of the yellow clip and hold it in place to hear the loop jog cycle.

Well, those 2000 milliseconds loop exactly as predicted. Unfortunately, it isn't helpful at all in this situation. A 2000-millisecond playhead-centered loop jog cycle is handy for checking playback around edit points, but for cleaning up sound, you're better off using the default settings.

13 In the User Preferences of Fairlight General Settings, change the Loop Jog Alignment to Pre and the Loop Jog Width to 80 ms. Click Save.

With this pre-playhead setting, you can scrub forward slowly over the tongue-clicking sound and hear the repetitive looping sound of that clicking, which will sound like teeth chattering as you drag. When you no longer hear the chattering sound, you will know that you are clear of the tongue click.

14 From the beginning of the clip, slowly scrub the playhead forward while listening to the loop-jog click-chattering sound caused by the tongue. When you no longer hear the chatter, release the playhead.

15 Turn on snapping (click the magnet icon), if necessary. Trim the head of the clip to the playhead position.

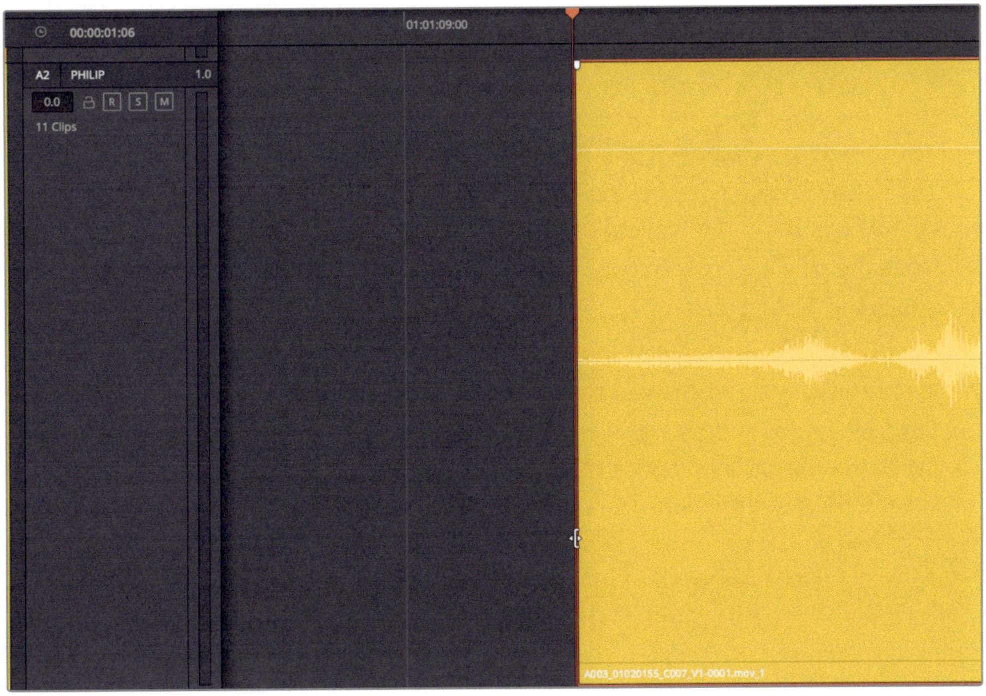

16 Right-click in the timeline and choose Track Waveform Zoom > Reset Zoom All Tracks to reset it. Play the clip from the beginning to hear the improvement to the head of the clip.

17 Choose Timeline > Loop Jog to deselect the scrubbing mode.

18 Zoom out of the timeline and tracks to fit all the clips in the timeline window.

Mission accomplished! Philip's dramatic performance at the head of the clip—including his intake of breath—sounds great. The dialogue track cleanup tasks are complete for now. And best of all, you've added a few new Fairlight tools and techniques to your pro-audio skillset.

Balancing Dialogue Clip Levels

Finally, you have reached the last major step of dialogue editing. You need to balance the dialogue levels for the entire scene so they are consistent throughout. To do so, you'll adjust the audio volume overlay on each clip while watching the track meters in the Mixer. During this process, you'll also apply keyframes to the volume overlay, when necessary, and apply clip normalization to automatically set the peak levels on some clips.

To further understand the task, let's take a closer look at the Fairlight Mixer and the target levels for dialogue.

> **NOTE** For the following exercises, you will continue working with the same timeline. If for any reason you didn't complete all the previous exercises, feel free to open and continue working with the **4c Dialogue Clip Levels** timeline.

1 Show the Mixer. Drag the left edge of the Mixer to the left to extend it until you see all four tracks (A1 through A4) plus the main output (M1)

   The Fairlight Mixer opens to the right of the timeline. You can easily identify each track by the color bar at the top and the track name in the middle.

2 Move the playhead to the beginning of the timeline.

3 Select the A1 EMILIANA track.

   As you can see, in the Mixer, the corresponding channel strip is also selected.

4 In the Mixer, in the A1 channel strip, click the Solo button to solo the A1 track.

   The Solo button in the A1 track header is also in the on position. Because the first clip in the A1 track is nat sound (natural production sound) without dialogue, let's move on to the second and third clips.

5 Create a playback range around the second and third clips in the A1 Track.

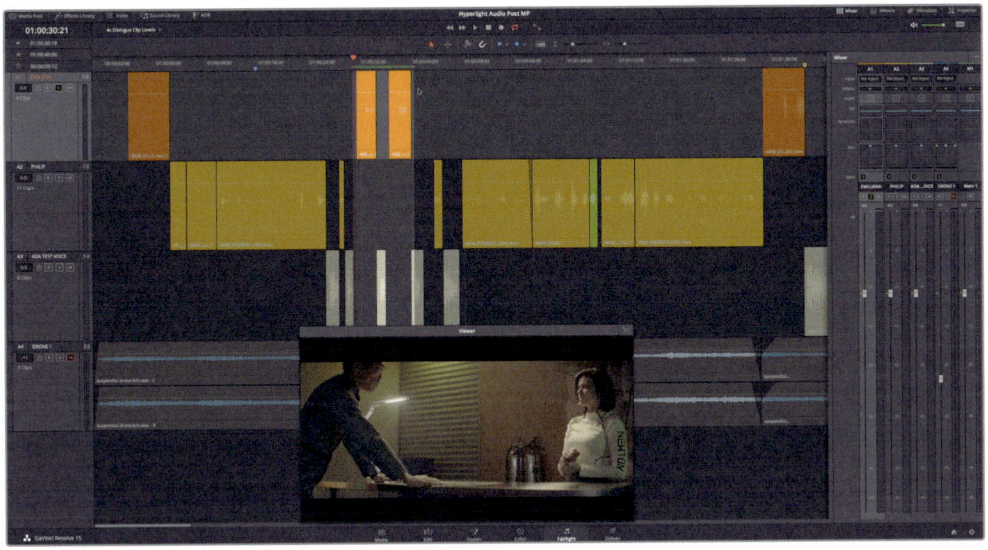

You'll work more extensively with the Mixer in Lesson 9. For now, focus on the meters at the bottom of each channel strip.

Reading the Meters for Dialogue

Not only are the meters in the Fairlight Mixer easy to read, they are also color coded so you can watch numerous tracks at once and quickly see if they are within the target range.

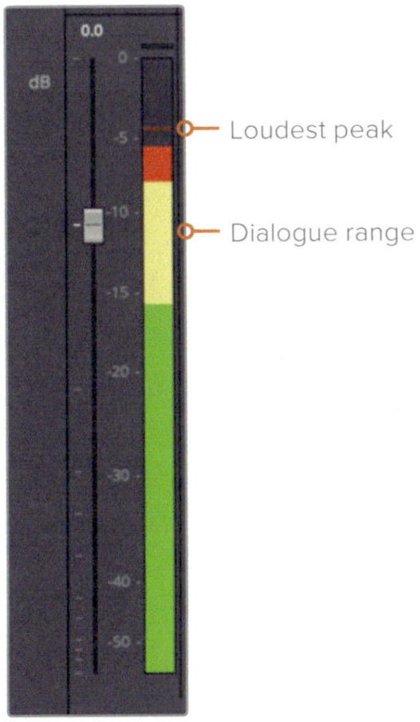

The solid color in the meter shows the current volume level on a decibel scale. These peak meters are "sticky," which means that they have a thin horizontal line that sticks to the loudest peak level, even when the current level is lower than the recent peak. Peaks within the green level area are safe and have no chance of being too loud. Yellow peak levels are good targets to aim for when working with dialogue. Red peak levels are getting close to being too loud and *clipping*, or going into distortion. When you get to the final mix, it is OK to let some of your peaks go into the red as long don't go too far into the red and don't remain there too long. However, at this point in the process, your dialogue levels should be well within the yellow range. The exceptions might be a loud scream or whisper, neither of which are in this scene.

Establishing Target Dialogue Levels

Whether your scene is narrative, voiceover, or a talking-head interview, you manage the dialogue level on each clip in the same way.

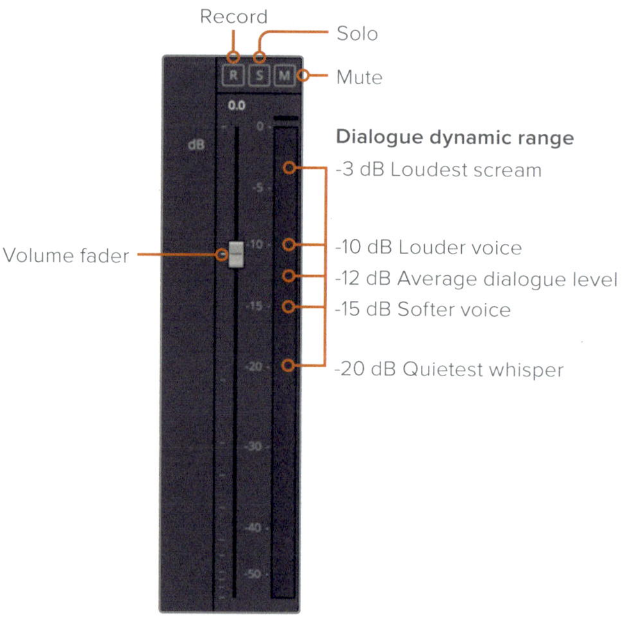

The average level should be around -12 decibels (dB) on this peak meter. However, context is also important. If you have characters who are bigger and louder, their levels may be closer to -10 dB. A quieter voice might sound better around -15 dB because it is a speaking voice. Blood-curdling screams could be as high as -3 dB, while hushed whispers may be as low as -20 dB.

Let's take a moment to see where Emiliana's dialogue falls on the meters in the second and third clips.

1 Turn on looped playback, if necessary.
2 Press Option-/ (slash) in macOS or Alt-/ (slash) in Windows to start looped playback.
3 Look at the meter in the A1 channel strip to determine the current dialogue level. During playback, watch the solid bar in the meter to see the average level. Also, look for the highest peak, based on the horizontal peak mark in the meter.

4 Stop playback.

The average in the first dialogue clip was around -27 dB while the average for the second clip was around -22 dB. In both cases, the clips are significantly lower than the target level of -12 dB.

Adjusting Clip Levels

Now that you know the levels of Emiliana's first two dialogue clips, and you know the level that you are aiming for in the meters, you can start adjusting each clip's volume overlay to raise that clip's level. Later, when you start pre-mixing tracks, you can use the volume fader in the Mixer to raise or lower the overall level of each track.

1 Zoom in horizontally as much as necessary to focus on the second and third orange clips.

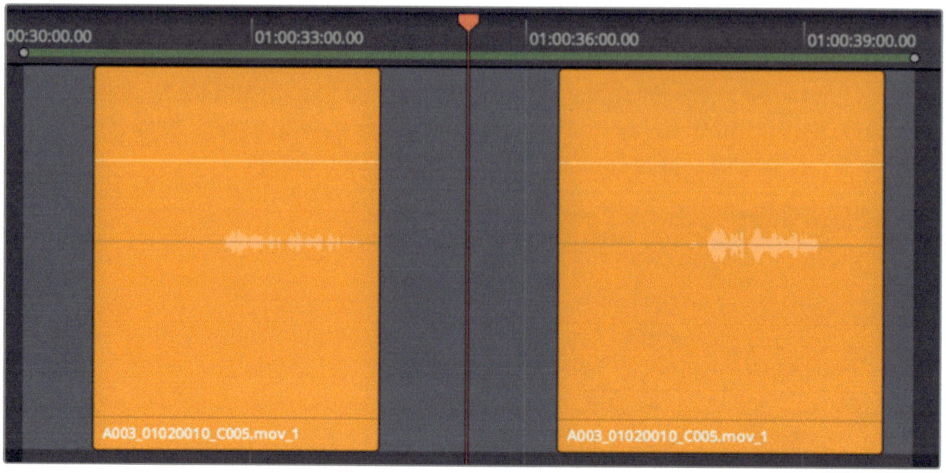

Remember, you want the adjusted level to average around -12 dB on the meter. Depending on the actor's voice and inflection, some fluctuation will occur between the peaks and the average. In this clip, the average (solid bar) is around -26 dB, while the peak is near -23 dB. So, the difference between the current average and the desired average (-12 dB) is approximately 14 dB.

2 Press R, and using the Range Selection tool, mark a play range around the first of the two orange dialogue clips. From this point forward, you will simply be asked to mark the clip or mark a play range.

3 Show the Inspector. Press A to go back to Selection mode, and select the clip.

4 In the Inspector, drag the Volume slider to the right until the field reads between 11.00 and 12.00.

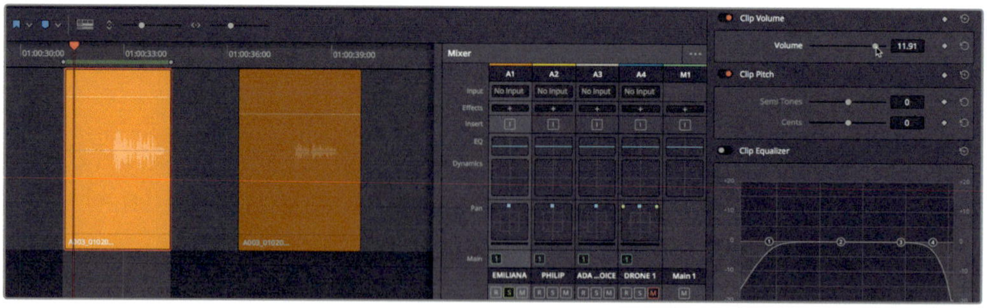

5 Play the marked clip to see if it is closer to the average target level of -12 dB.

The average is right on target. Now you can move on to the next clip. This time, you will adjust it on-the-fly during playback.

> **TIP** If you would like to display clip volume levels in the lower-left corner of a timeline clip, simply choose Fairlight > View Clip Info Display, and check the Clip Info Volume display option.

6 Mark the next dialogue clip in the A1 track. Then, select the clip with the Selection tool.

7 Start looped playback. During playback, drag upward on the clip's volume overlay. Continue adjusting the level using either the volume overlay or the volume slider in the Inspector until the average level is around -12 dB.

8 When you are finished, stop playback.

The Volume field in the Inspector should be between 8.00 and 9.00. Remember, the numbers in the Volume field represent the percentage of change applied to the original level of the source material. They do not specify the actual volume level of the output.

> **TIP** You can also drag left or right within the data field next to the clip Volume slider, or double-click within the data field and type a specific value.

Don't worry about finding the perfect number. As long as Emiliana's dialogue levels are consistent and those levels reflect the context of the scene, it should be fine. Later, you can finesse the levels of the entire track when you start mixing.

9 Scroll down to the last orange clip in the A1 track.

10 Mark the clip. Use either the Range Selection tool, or the Selection tool to adjust the volume overly.

11 Start playback and set the volume level for the clip.

This time, the clip level needs to be lowered by around -2.00 in the Inspector.

12 Choose Mark > Clear In and Out, or press Option-X (macOS) or Alt-X (Windows).

13 Press Shift-Z to fit all of the clips horizontally in the timeline. Press A for the Selection tool.

Excellent work. Hopefully, by the third clip you were more comfortable with the workflow and confident in your clip level settings. One of the wonderful things about audio post work is that it is pretty straightforward once you understand what you need to do and why.

Using Keyframes to Change Clip Levels

In Lesson 1, you used keyframes to lower the music levels during the voiceover sections of the soundtrack. In this exercise, you'll use keyframes to change the volume levels within a dialogue clip. When making these kinds of changes, it's a good idea to first find an average level that works for most of the clip's dialogue, and then apply keyframes to raise or lower only those sections that need further adjustment. For this exercise, let's move to the A2 track and work on two of Philip's dialogue clips.

1. Unsolo the A1 track, and select the A2 track.
2. In the A2 PHILIP track, zoom in on the two yellow clips right after the lime green clip.
3. Play the two clips to hear the dialogue and watch the levels to get a feel for the overall levels. Pay attention to the context and emotion in Philip's performance.

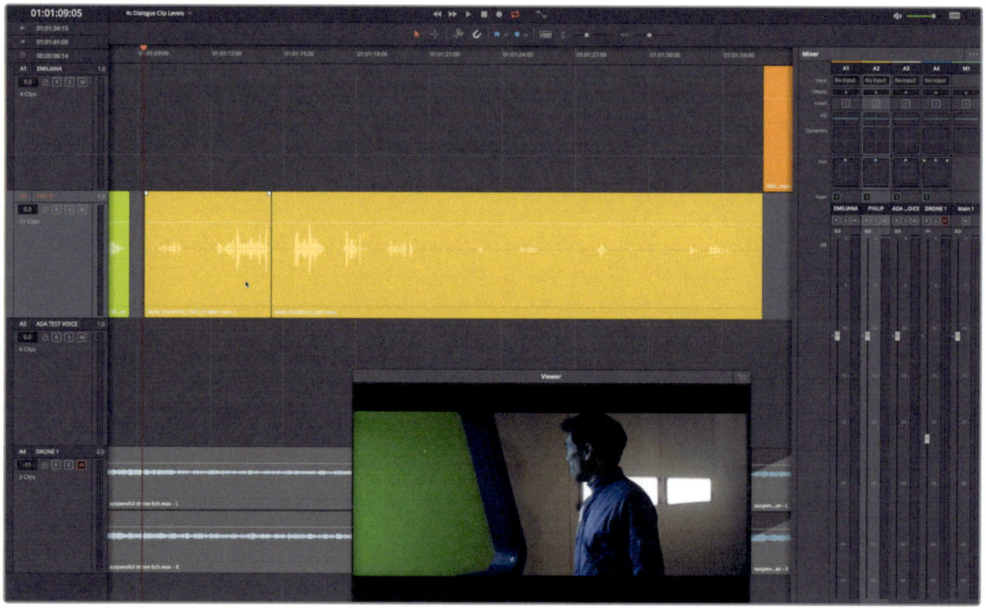

You probably noticed that his levels were right on target in some places and much quieter in others. However, dialogue isn't a one-size-fits-all proposition. When Philip gets excited about the explosion, his voice should be a little louder. And when he returns to Emiliana to tell her that she (the other Emiliana) died, he is a little choked up and his voice is understandably lower. To maintain the authenticity of the performance, you want to maintain some difference between his levels, but not too much. Let's add a few keyframes to bring Philip's levels closer to the target levels. Remember, the difference between a loud speaking voice (-10 dB) and an audible hushed whisper (-20 dB) is only 10 decibels.

NOTE In Lesson 9, you'll explore dynamics processing that can manage the dynamic range between the loudest and quietest speech levels.

Let's start with the first clip where Philip says, "Then just two hours ago...boom...there was this explosion, and it surrounded the whole ship."

4 Mark the clip. Zoom in as much as necessary to clearly see the entire waveform of the selected clip. Start looped playback, and watch the levels. When you are finished, stop playback.

Looking at the waveform in the selected clip, you can clearly see that the beginning of the clip is much quieter than the end. The tricky part with this clip is to raise the level of Philip's line, "Then just two hours ago," without cranking up the level of his breath sound.

5 Scrub the playhead over the clip, or press the JK or KL keys in combination to find where Philip's breath stops and his speech begins. Visually, this occurs when his head turns.

6 Move the playhead to 01:01:09:18.

You will set the first keyframe here. However, instead of clicking the volume overlay in the timeline to do so, you'll use the Clip Volume keyframe controls in the Inspector.

7 In the Inspector, on the Clip Volume controls, click the Keyframe button (which looks like a diamond).

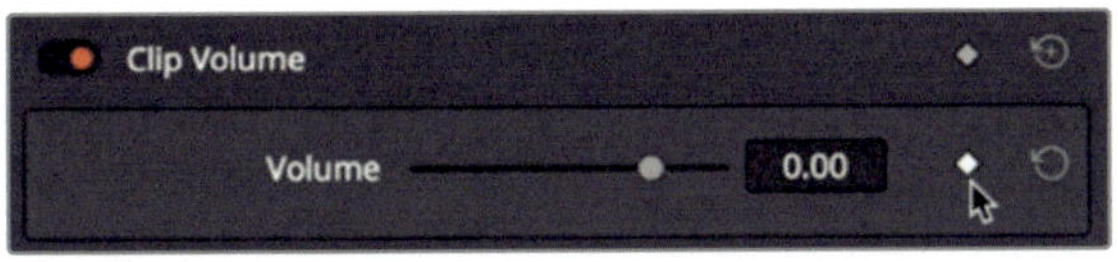

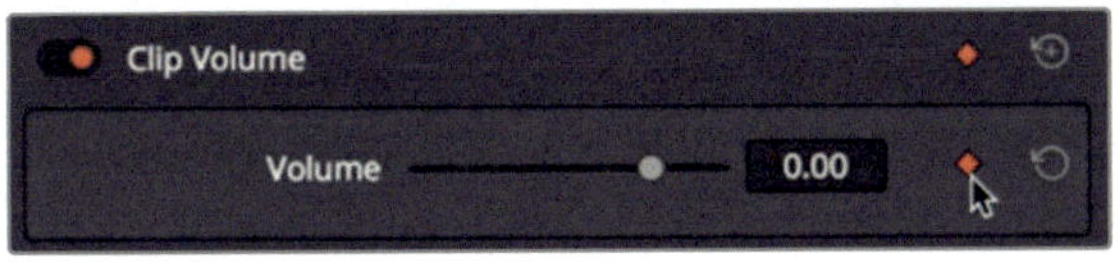

The Keyframe button turns red to indicate that a keyframe is set, and a corresponding keyframe is visible at the playhead position on the clip.

8 Press Left Arrow once to move the playhead one frame to the left.

9 In the Inspector, set another keyframe.

You'll add the next pair of keyframes by clicking the volume overlay. To set a keyframe on the volume overlay, Option-click (macOS) or Alt-click (Windows) the overlay.

TIP You can set keyframes using either the Selection or Range Selection tools in the Fairlight page. For this exercise, you'll use the standard Selection tool.

Here are a few guidelines for setting audio keyframes:

- As the name suggests, keyframes are used to set values on specific frames. So, you can create and move them based on full frames, regardless of your zoom level.
- Setting pairs of keyframes between sections of dialogue makes it easier to raise or lower each section as you go.
- After a keyframe is created, you can move it or select and delete it.
- To clear all the keyframes on a clip, in the Inspector, click the Clip Volume Reset button.

10 Scrub through the clip from the beginning, and stop after Philip says, "Then just two hours ago."

A fairly large gap exists between dialogue lines here, so you have plenty of wiggle room for setting and moving your keyframes.

11 Add a pair of keyframes to the volume overlay. These do not have to be right next to each other. Option-click (macOS) or Alt-click (Windows) to set each keyframe.

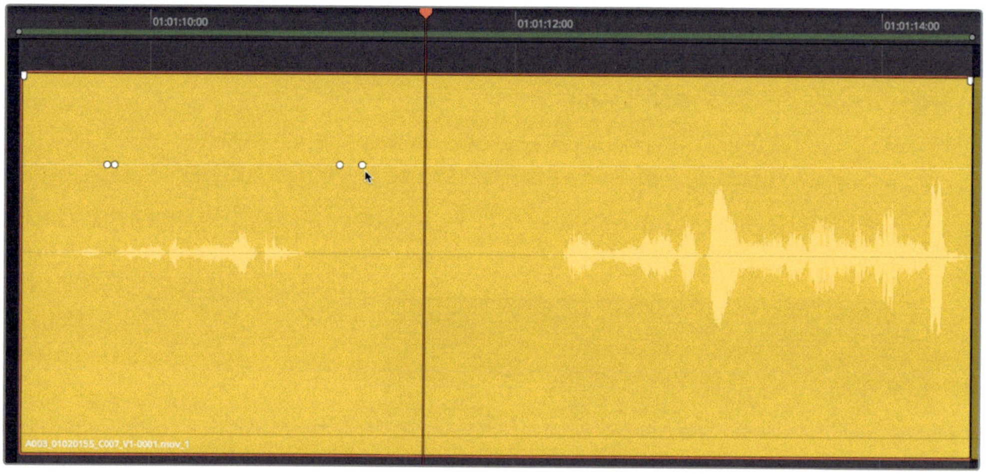

Now that you have more than one keyframe on the clip, the keyframe controls in the Inspector include arrows to move to the next or previous keyframe.

12 In the Inspector, click the arrows on either side of the Clip Volume Keyframe button to move the playhead to the previous or next keyframe.

You are ready to raise the volume level for the section of dialogue between the two pairs of keyframes. Keep in mind that the Volume field in the Inspector shows the level based on the current playhead position.

13 Between the pairs of keyframes, drag upward on the volume overlay until the waveform size looks slightly larger than the first part of the waveform in the next phrase of the clip—somewhere between 6.00 and 8.00 in the Inspector.

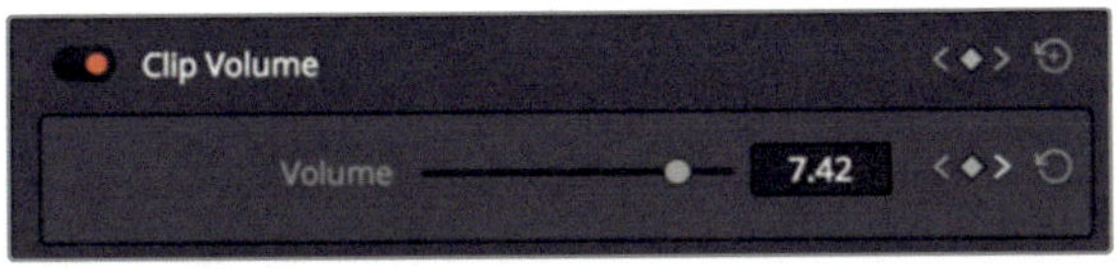

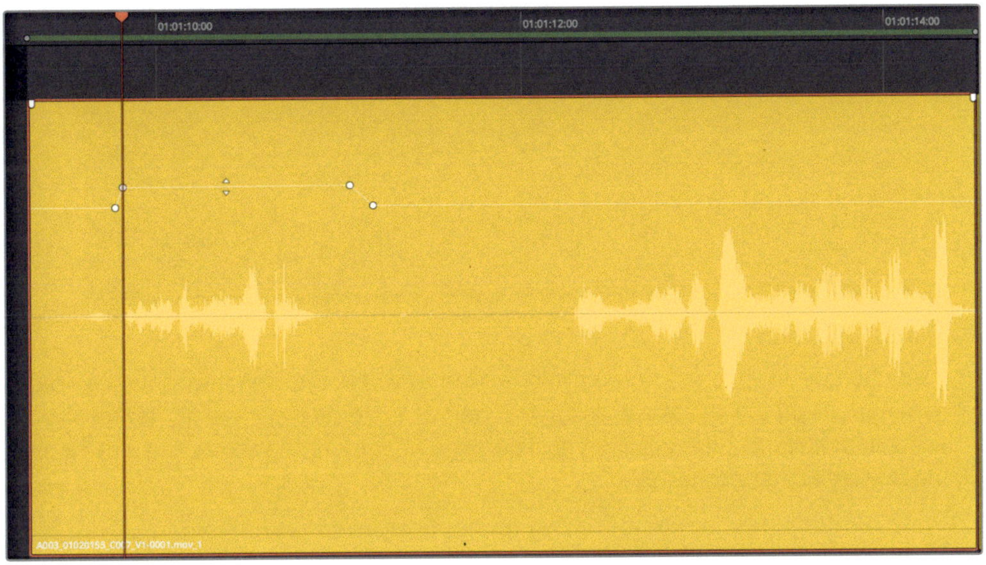

Your goal is to increase the volume of the next part of the dialogue just before the large spike in the waveform that represents the "plosion" part of the word "explosion". In fact, the waveform spike in "plosion" is a perfect representative of a *plosive*.

NOTE Plosives are power hits in the waveform usually caused by pronouncing syllables that include "B" or "P" sounds. In audio post-production, taming the plosives and sibilance "esses" are two of the fundamental tasks in preparing dialogue tracks for mixing. You'll work more with plosives and sibilance in the next lesson.

14 Set a pair of keyframes just before the powerful spike in the waveform between "ex" and "plosion." at around 01:01:13:00. Then, raise the volume level before the plosive to somewhere between 2:00 and 4:00 so that the average part of the waveform looks and sounds similar to the levels on the other side of the plosive.

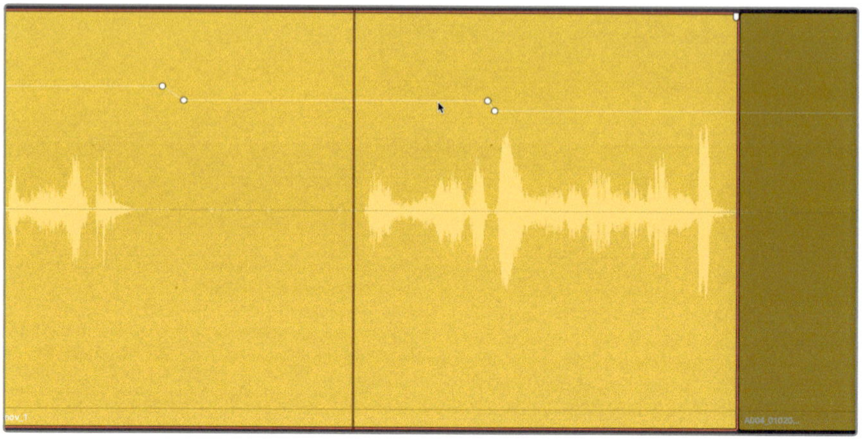

15 Play the finished clip.

The changes in volume are seamless, yet are all within the standard dialogue range. Did you notice that the levels are also in the yellow range on the meters? The more you work with dialogue, the quicker you'll become at recognizing levels and making the necessary adjustments.

More Keyframes, Please!

Now that you've set keyframes to balance Philip's dialogue levels within a clip, this is the perfect time to practice your skills. Take a few minutes to practice what you have learned and balance the levels in the next clip where Philip says, "We also sustained some hull damage but...its nothing critical...and...now you're here. Alive...and...you look just like her." This will be challenging but not unusually difficult to do. Just think it through as you go. A colorist must match visual levels with the previous and following clips, so you'll do the same here for dialogue. Because Philip is continuing a sentence from the previous clip, it's a good idea to match the first part of the clip to the previous clip. Then, add keyframes as necessary to raise other words or phrases. Trust your ears and the meters.

Philip isn't whispering; he is talking, albeit quietly in places, so don't leave his dialogue as low as a whisper (-20 dB). How much, and which phrases you adjust are up to you.

Also, remember that part of the waveform is only the sound of Philip crossing the room and leaning in. Just as you avoided raising the level of his initial breath, consider keeping those levels lower than the dialogue. Good luck.

When you are finished, if you have time, adjust the lime green clip and the clip before it so that the entire series of four clips sound good together. If you are feeling dialogue savvy, by all means finish all of Philip's track.

Applying Auto-Normalization to Balance Clips

If you think balancing dialogue levels is agonizing and tedious, DaVinci Resolve offers a "cheat." Resolve 16 includes automatic clip volume normalization in both the Edit page and Fairlight page, which is for quickly setting the peak levels in a clip without considering the context of the scene or the performance. Auto clip-normalization is like using the auto-white balance tool in the Color page. It's an easy way to set a starting point for an editor who needs to quickly set levels on audio clips.

Of course, now that you are a seasoned dialogue editor, you probably have no interest in using auto-normalization. However, it's still worth knowing about, and in fact, the one circumstance in which auto-normalization is most effective is when you are working with a voiceover track with little or no inflection or emotion to consider—a computer voice, for example. And it just so happens this timeline has such a track. So, you're in luck.

In this exercise, you'll apply normalization to the clips in the A3 ADA TEST VOICE track.

1. Select the A3 track.
2. Adjust the zoom levels in the track until you can clearly see the first five clips in the A3 track.
3. Right-click the first clip in the A3 track and choose Normalize Audio Levels to open the Normalize Audio Levels panel for that clip.

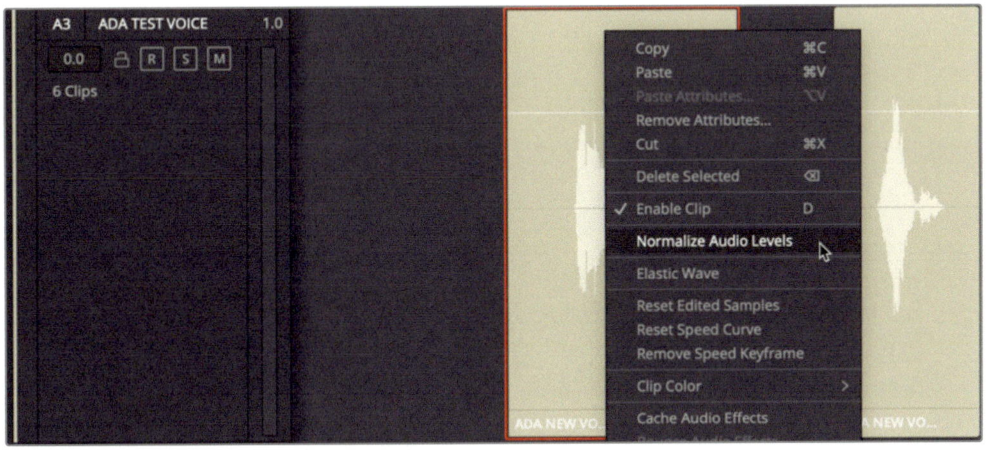

In the Normalize Audio Levels dialog box, you can select a peak Target Level in dBFS (decibels full scale). If the average level should be around -12, then the peaks may be closer to -9. Let's try it.

4 Set the Target Level to -9 dBFS, which is the default level, and click Normalize.

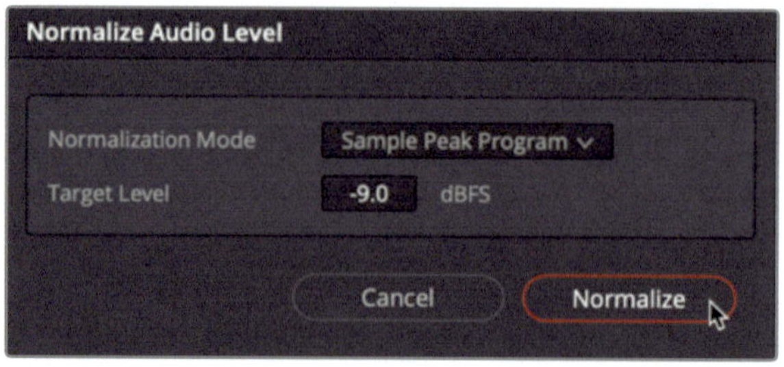

The volume overlay and waveform in the clip are reduced slightly. So, what actually happened? Normalization set the volume overlay so that the highest peak in the clip is at the volume level that you designated (-9 dBFS).

5 Play the clip, and look at the meters.

The clip's average level is around -11 dB. That's not bad for the level of a talking computer. Remember, balancing the dialogue clips starts with setting the levels within a clip, and then adjusting all the clips in the track to sound good together. If it turns out that all of Ada's dialogue seems a little too loud, you can always lower her track level, as necessary.

One added benefit of using Resolve's auto-normalization is that you can apply it to multiple clips at once.

6 Select the first five clips in the A3 track. Then, right-click one of the selected clips, and choose Normalize Audio Levels.

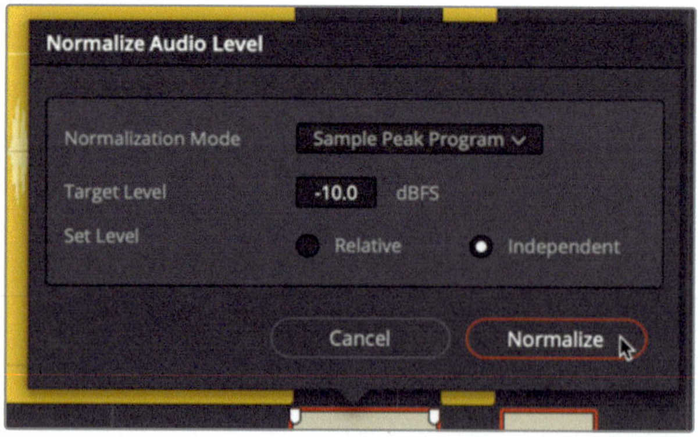

This time, the Normalize Audio Levels pane includes two Set Level options: Relative and Independent. The Relative choice treats the group of selected clips as if they were one clip, so the highest peak among the clips determines all of their relative levels. Independent, on the other hand, normalizes each clip separately based on each clip's peaks.

7 Set the Reference Level to -10 dBFS and the Set Level to Independent. Click Normalize.

8 Solo and Select the A3 track. Play the first five clips and watch the levels as you listen to the dialogue.

As you can see and hear, the levels don't exceed the target levels and are fairly well balanced. You may have noticed that some of the clips still seem louder than others. If your instinct while listening was to stop and investigate, you may want to manually adjust the levels to improve them, which means you have the makings of a dialogue editor!

You have completed the first level of dialogue editing! But cutting the tracks together and balancing the levels is only the first part of your job. Dialogue editing is complete when the tracks are strong enough to stand on their own. Don't worry, you'll get there in the next lesson.

Yes, mapping, splitting, trimming, and balancing dialogue levels takes time, patience, and skill. However, because the dialogue tracks are the most important audio element in your project, the end definitely justifies the means.

Lesson Review

1 True or False? Changing a track from two-channel stereo to single-channel mono automatically deletes the right channel from all the stereo clips in the timeline.

2 Which right-click menu item must you use in the timeline or the Media Pool to change a clip's channel mapping?

- a) Clip Channels
- b) Clip Attributes
- c) Clip to Track Ratio
- d) Dynamic Range

3 Where can you reset the volume levels of all of a track's clips at once?

- a) Track header
- b) Mixer
- c) Media Pool
- d) Timeline menu
- e) Inspector

4 True or False? You can use common text editing keyboard shortcuts to cut, copy, and paste audio clips in the timeline.

5 What color is visible in a tracks meters when the levels are optimal for dialogue?

- a) Yellow
- b) Green
- c) Red
- d) Blue

Answers

1 False. Changing a stereo track to mono will play the first channel of audio in the timeline clips; however, it does not affect the channel mapping of the timeline clips.

2 b

3 e

4 True. You can use common text-editing shortcuts to cut, copy, and paste clips in the timeline: Command-X (macOS) or Ctrl-X (Windows) to cut, Command-C (macOS) or Ctrl-C (Windows) to copy, and Command-V (macOS) or Ctrl-V (Windows) to paste.

5 a. Optimal dialogue levels (between -15 and -10 dB) appear in the yellow color range on the track meters and Mixer.

Lesson 5

Repairing and Replacing Unwanted Sounds

Have you ever heard the saying, "We'll fix it in post"? Well, now it's time to make good on that promise. Many extraneous sounds are recorded with the dialogue during production. Some are unavoidable like squeaky dolly wheels, floor creeks when an actor crosses the room, and the crackling of clothes near the talent's wireless microphone.

When is the best time to fix problems in the dialogue tracks? It's always better to fix them sooner rather than later. As the dialogue editor, or all-purpose sound editor, it's up to you to find and fix the issues in your tracks before you pass the project on for mixing. Whether you need to remove noise, replace words, tame sibilance, or patch gaping holes, sooner or later your dialogue tracks will need one or all of these common repair techniques.

Time

This lesson takes approximately 50 minutes to complete.

Goals

In this lesson and Lesson 6, "Advanced Dialogue Repair," you'll work with many of Fairlight's dialogue triage tools to find and fix a wide range of problems. As you will soon discover, sound issues are cumulative, and what may seem insignificant now will only amplify and haunt you later when the dialogue tracks are processed and enhanced in the mix.

Chances are, if you're new to dialogue editing, problems lurk in those tracks that you never thought to look for or deal with until now. Rest assured, by the time you finish with this set of dialogue repair lessons, you'll be armed with a solid understanding of dialogue track trouble-shooting and which Fairlight tools you'll need to fix issues that arise in your own projects.

NOTE The exercises in this lesson build on the tools and skills that you learned in previous lessons. If you skipped ahead to this lesson, you may need to review the previous two lessons as a guide to skills and tools already covered.

Preparing the Project

So far, you've learned how to build soundtracks, record voiceover, and edit dialogue using the basic tools and shortcuts necessary for most of your audio editing needs.

In this lesson, you'll use basic and advanced editing tools to step through common real world dialogue issues and learn how to fix them. Let's start by opening the first timeline and looking at the to-do list in the Markers index.

1. Open the timeline, "5 Dialogue Repair and Replace Start."

   This is a completed version of the timeline you worked on in the previous lesson.

2. Show the Markers index in list view.

   You can see the spotting list of elements to repair, fix, or replace in this timeline.

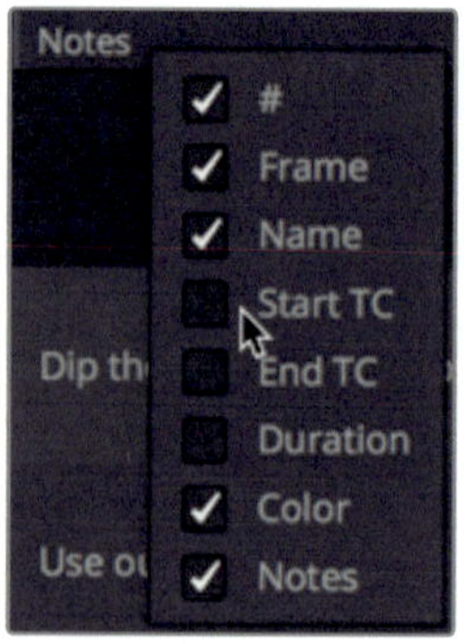

3 Right-click one of the column-headers and deselect the checkboxes for Start TC, End TC, and Duration to hide those columns. Drag the column headers left or right to reposition them as shown in the following figure.

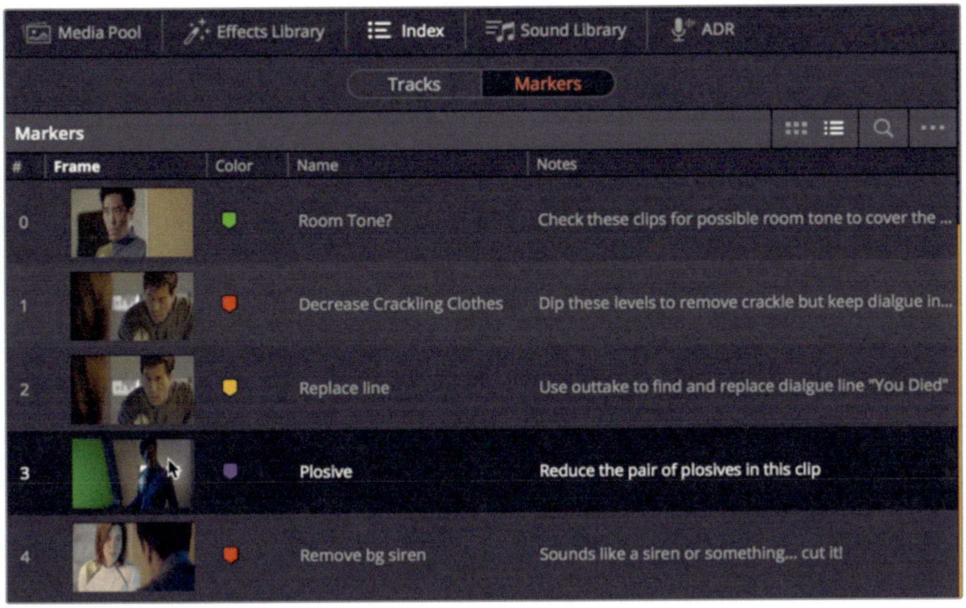

4 Double-click the icon for the purple marker, titled "Plosive," to move the playhead to that marker.

5 Hide the Index.

The timeline is ready for action, and you have a road map in the marker index to help you navigate from one issue to the next along the way.

Exploring the Edit Selection Tool

When it comes to editing audio quickly, there are two primary options: physical hardware or mouse and keyboard tools. For editors who prefer physical controls with lightning fast results, there is the Fairlight Audio Editor, which comes in both desktop and console models and includes dozens of specialized picture keys, touch-sensitive knobs, and a clutch-action shuttle wheel to drive nearly every function on the Fairlight Page interface. For mouse and keyboard editors who work fast and fluid with their mouse-du-jour, DaVinci Resolve 16.2 has a brand new Edit Selection Mode with a powerful multifunctional editing tool that changes function based on where and how you click and offers a robust array of pointer-based advanced editing options using simple mouse clicks and the occasional modifier key or user-defined keyboard shortcuts.

The secret to mastering the Edit Selection tool is to first learn how it changes based on where and how you click. Another consideration is whether the playhead is moving. When stationary, playback will always start at the new edit selection whenever you press the Spacebar. So, you can click anywhere in the timeline and instantly preview the audio from that point without having to move the playhead. During playback, the Edit Selection tool is completely uncoupled from the playhead so you can continue editing while listening to playback or even while recording.

In this exercise, you'll explore the nuances of the Edit Selection tool, so you'll be ready to use it in the next section.

6 Click the Edit Selection Mode button to change the pointer to the Edit Selection tool.

7 Click the pointer in any empty space in the timeline to set an edit selection point.

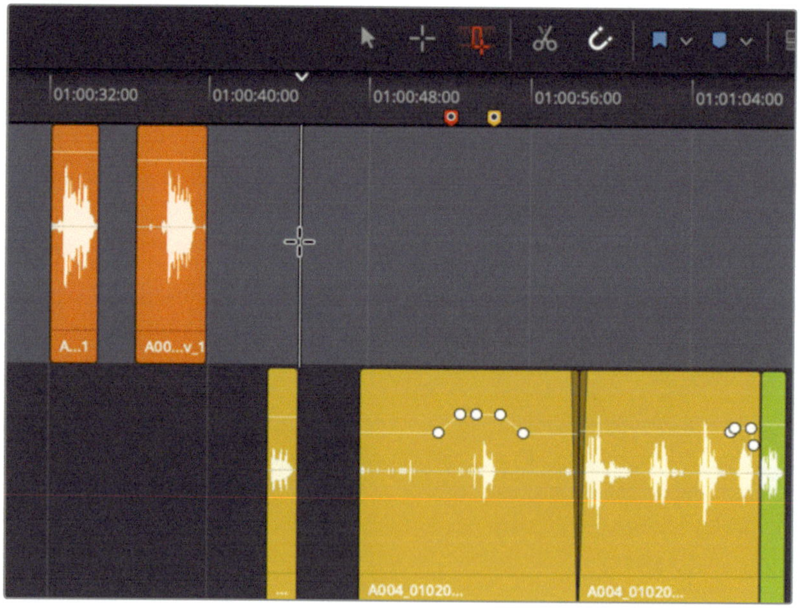

The edit selection point is indicated by a vertical white line. Notice that the playhead only shows in the timeline ruler directly above the edit selection point.

8 Press the Spacebar to start playback from the edit selection point. Stop playback and start again to see that it restarts from the edit selection point just like the stop and play again feature that you used in Lesson 1.

9 Continue playback and click another empty space to set a new edit selection point. Press the Spacebar twice to stop the current playback and start from the new position. Stop playback.

10 Move the pointer over the upper half of the first clip in the A1 track and click to set an edit selection point. Then zoom in horizontally to the timeline so you can clearly see the clip.

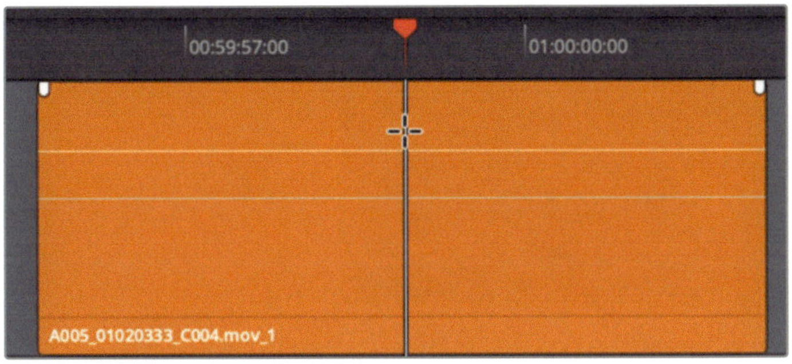

When the Edit Selection tool is over the upper half of a clip, it looks like a crosshair, similar to the Range Selection tool, and can be used to set a selection point or drag a selection range.

11 Drag the pointer over the upper half of the clip to drag a range.

12 Press option-/ to play the selected range.

13 Click anywhere in the upper half of the clip to clear the selection range.

14 Double-click the upper half to mark the entire clip. Click again to deselect the clip.

15 Move the pointer over the lower half of the clip to change the cursor to the Grabber tool (hand).

16 Click the lower half once with the Grabber tool to select and mark the entire clip.

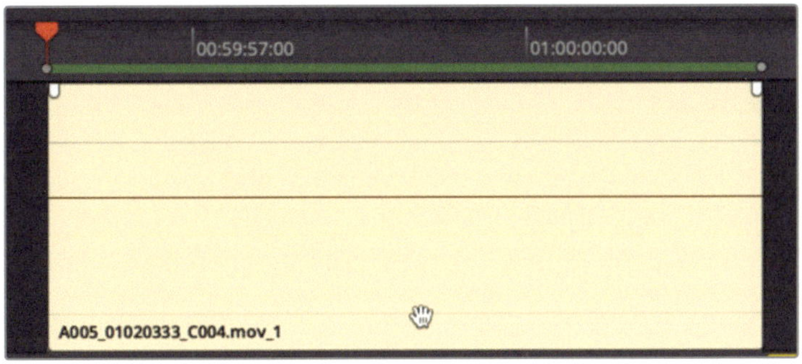

The Grabber tool is similar to the Edit Selection tool and can be used for selecting and moving clips in the timeline. Additionally, in Edit Selection mode you can trim the head and tail of clips by dragging the edges, and adjust the volume overlay just as you can in standard Selection mode or Range Selection mode.

17 Press Shift-Z to fit the timeline clips horizontally to the timeline.

18 Click the empty space below the purple marker to set a new Edit point and playback position at the marker.

Now that you are familiar with the Edit Selection tool, let's put it to work in the next series of exercises.

NOTE The Edit Selection tool combines many of the selection, editing, and playback features that you're already familiar with so that you can continue working as you have, without the need for switching tools or selection modes.

Reducing Plosives with Keyframes

One of the most common and easy-to-fix problems in your dialogue are the eruptions of *plosives*. As mentioned in the previous lesson, plosives are "power hits" in dialogue recordings that are usually caused by the energetic pronunciation of syllables that start with the letters "P", "B," and sometimes "K" or "Ch." Often, *pop-screens* are placed in front of the microphones in studio recordings to tame those popping "Ps". Nonetheless, they still occur, and will need to be dealt with. If not, when you mix dialogue, you'll end up over-processing your dialogue tracks to tighten the excessive dynamic range caused by those plosives. Instead, by simply taming them using keyframes, the levels will fit the rest of the track, leaving it natural sounding and needing less compression in the mix. Most plosives can be tamed using two pairs of keyframes. Go to the purple "Plosive" marker, if necessary. It marks the "plo-sion" portion of the word "explosion" that was called out as a plosive in the previous lesson.

1 In the A2 PHILIP track, zoom in to the yellow clip beneath the purple marker.

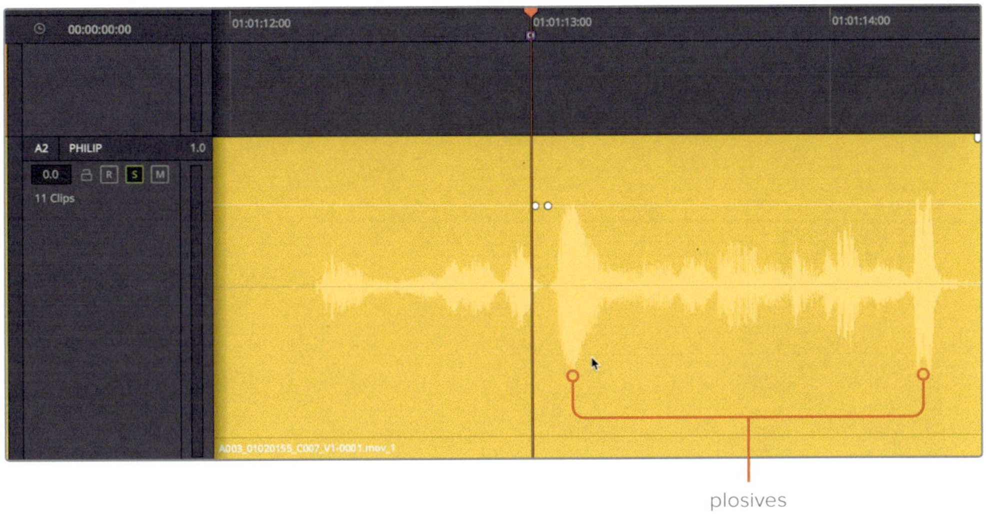

For this example, the first set of keyframes is already in front of the plosive. All you need to do is add another pair.

2 Place a pair of keyframes just to the right of the plosive.

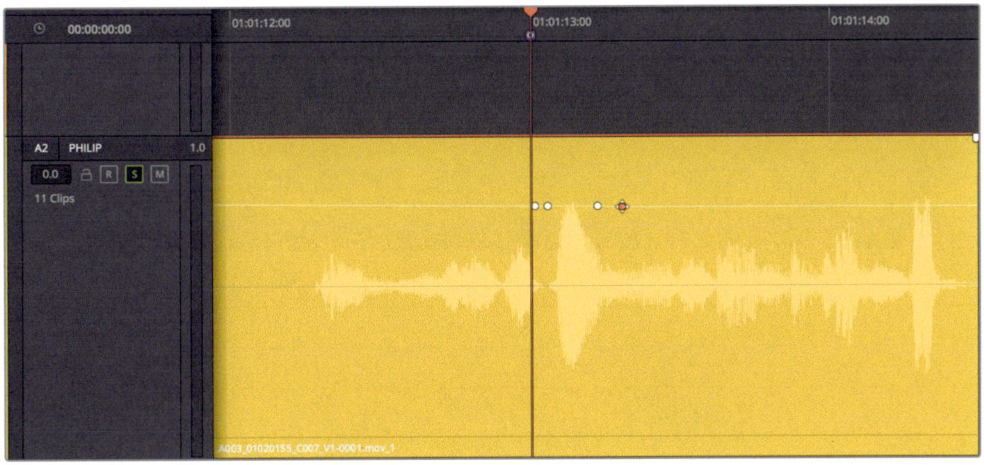

Your goal is to reduce the height of the plosive by at least a third without compromising the inflection and natural sound of the speech.

3 Drag down on the overlay between the pairs of keyframes until its height is about one-third lower than the original peak. You can use the next highest peak as a guide. Mark the clip and listen to it once while watching the meter in the track header.

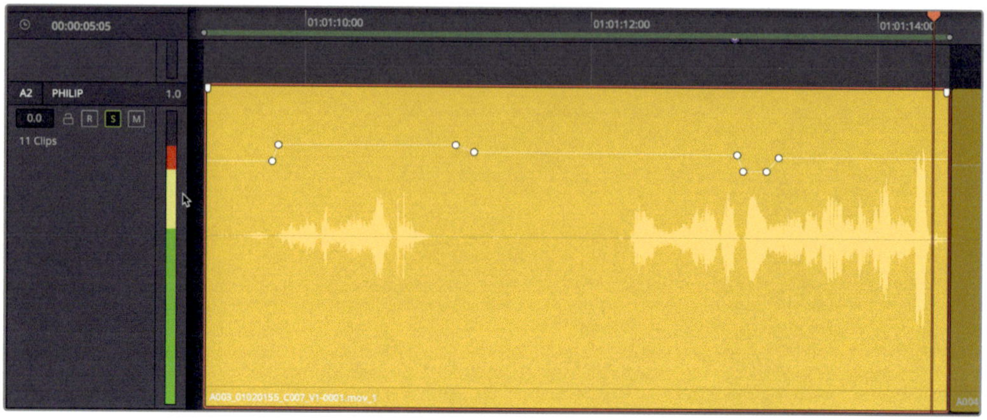

The volume levels for most of the clip are in the yellow range, as they should be for a balanced dialogue clip. However, one syllable at the end peaks in the red. That red in the meter is a signal that you probably have another plosive. Let's reduce it.

4 Add a pair of keyframes on either side of the plosive at the end of the clip. Then, lower the volume level of the plosive until it matches the height of the other two loudest waveform peaks in the clip.

5 Start looped playback and watch the track meter. Lower the plosives until they no longer push the meter into red.

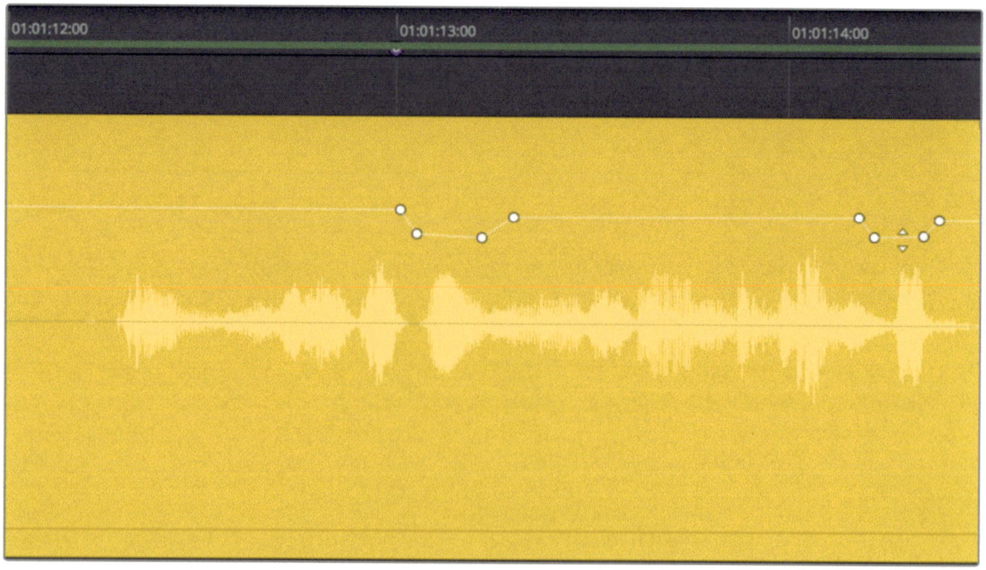

6 Stop looped playback, and clear the play range.

That's all there is to plosives. Find them, fix them...all good. And as you can see and hear, by carefully lowering the plosives, the clip still retains plenty of dynamic range between its highest and lowest levels.

As an efficient dialogue editor, you'll get in the habit of tackling plosives as you balance your dialogue tracks. If a vocal scream or another sound needs to jump out, let it! Otherwise, you can retain the impact of the speech while lowering just the levels of individual, over-enthusiastic syllables.

Eliminating Distracting Sounds Between Dialogue Clips

One of the most straightforward issues that you'll find in dialogue tracks is the abundance of distracting sounds. But what qualifies as a distracting sound? It is any unnecessary sound that may distract from the dialogue. It could be human noise, or background noise picked up through the talent's microphone. However, it is in spaces between words that you'll usually find them.

In this exercise, you'll find and remove three noises using three different methods of extraction: splitting and trimming, using the Range Selection tool, and placing keyframes. You have already worked with these tools in previous lessons, so the steps you'll follow here will utilize many of your new skills to perform the tasks at hand.

1. Select the A1 track. Move to the last marker in the timeline (red). Use the zoom sliders or keyboard shortcuts to zoom in to the last clip in the A1 track.

   At a glance, the clip doesn't reveal any obvious issues.

2. Mark the clip, and play the clip once. If you don't hear anything out of place, increase the volume level of the clip, and listen to it again.

   Even on spaceship sets, the sounds of helicopters, motorcycles, airplanes and sirens can contaminate a recording. In this case, a distant siren sound can be heard between Emiliana's lines. You may be thinking that it is barely perceivable and will eventually be covered up by sound effects and music. Maybe...But because those other audio choices are up to the director, it's always best to hand over clean dialogue tracks that play seamlessly. Plus, this happens to occur in a very intimate scene with just two people talking in a room. No conveniently distracting sound effects sources would be logical in that context. In fact, the director may choose to let the scene play out with sparse, if any, musical underscoring. When you aren't sure if you can safely leave a sound in your dialogue tracks, here is a rule to live by: "When in doubt, cut it out!"

   Keep in mind that anytime you remove the audible room ambiance between words, you'll later have to replace it with room sound in a separate track. So, let's remove the siren. You could trim off the beginning of the clip, but the room sounds good until the siren. Let's just trim the part that requires removal. This is also an opportunity to use Loop Jog scrubbing to determine where the siren sound starts.

3 Choose Timeline Loop Jog. Raise the clip's volume curve as much as necessary to clearly hear the room and siren sound between Emiliana's words.

4 Using the waveform as a guide, move the playhead just before Emiliana's first spoken word in the last clip.

With the Edit Selection tool, you can scrub the playhead over any clip by holding shift-command while you drag the pointer over the clip. As a bonus, shift-command scrubbing the Edit Selection tool temporarily Solo's the clip's track so you will hear only the clip that you are scrubbing. Let's try it on Emiliana's clip.

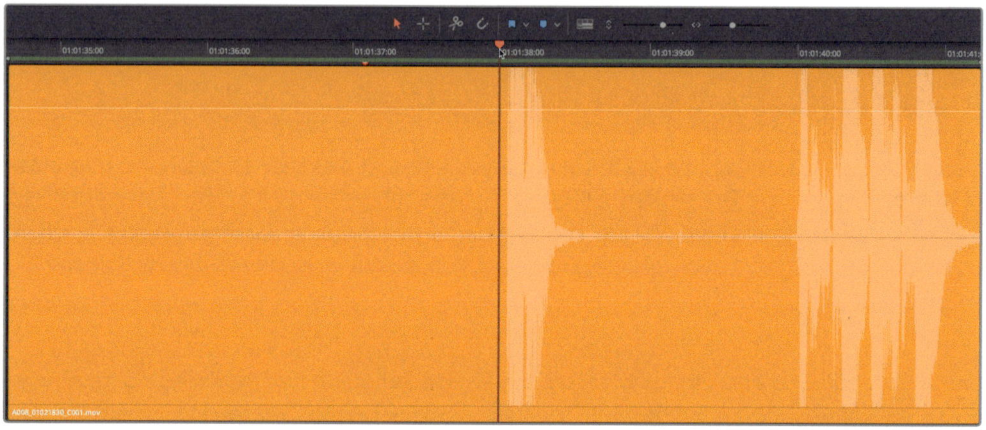

5 Shift-command scrub the playhead back and forth over the beginning of the clip to hear the siren just before her words. Scrub toward the beginning of the clip until you no longer hear the siren. That is where you will start an edit selection range (around 01:01:36:14).

6 In the upper half of the clip, drag an edit selection range from the playhead to the right just before the first word. Then press delete to remove the current edit selection range.

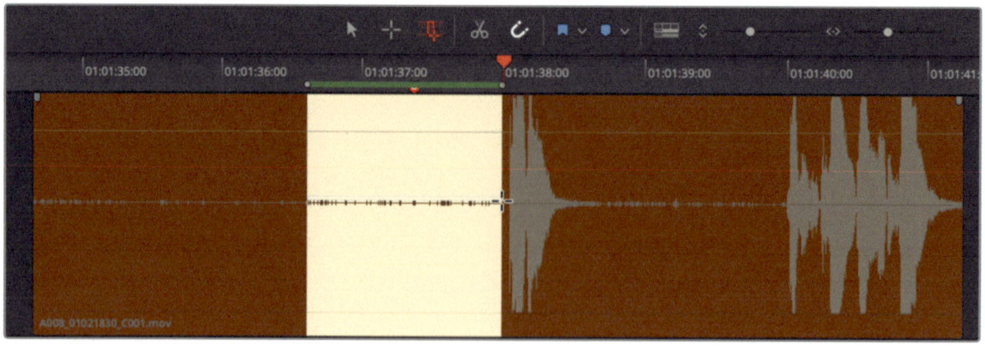

Just like that, the first siren is gone. Next, you'll use the Range Selection tool to remove the siren before Emiliana's second phrases in this clip.

7 Play the section between her phrases.

Once again, the beginning of the room sound is okay. However, a noise in the middle (visible as small spike in the waveform) is followed by the slow return of the distant siren. Let's create a selection that includes both the noise and the siren. The trick to this method is not to worry about getting it perfect. Just select as best you can to delete the unwanted sound. You can then trim the heads and tails surrounding the gap as necessary.

8 Drag a selection range from just before the noise to just before she speaks. Press Delete.

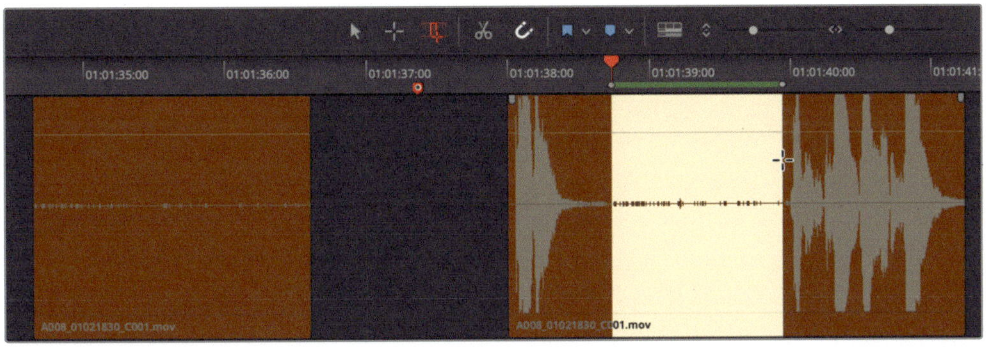

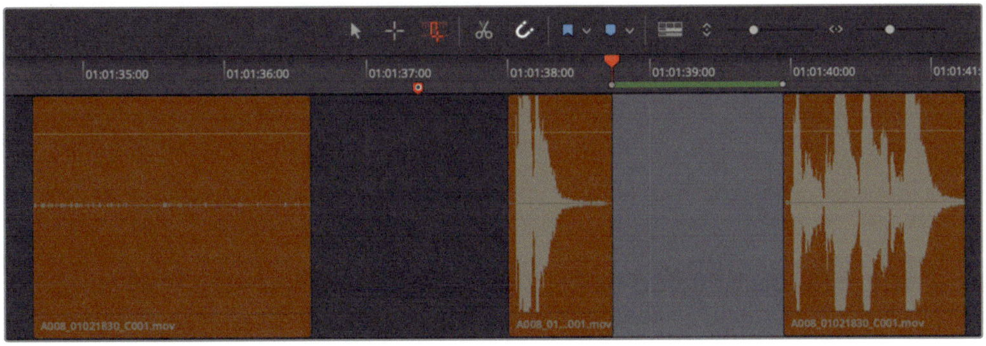

Now that both sirens are removed, you can return the clips to their previous volume levels. This is a good opportunity to apply clip normalization.

9 Select all three of the remaining orange clips. Right-click one of the selected clips, and choose Normalize Audio Levels.

10 Set the Reference Level to -8, and the Set Level to Relative. Click Normalize.

11 Play the last three clips in the A1 track to hear how your edits sound. If you accidentally clipped off any of Emiliana's words, trim the heads or tails of the clips to recover the full words.

Emiliana's words in her last section of dialogue should now be free of distraction.

What Is Foley Sound?

Foley sound consists of any character-driven sound effects caused by characters interacting with their onscreen environments. Foley sounds are named after Jack Foley, a legendary sound editor at Universal Studios, who originally developed the technique of recording reenactments on a stage. Foley sound replaces the original production audio for everything from fist fights to footsteps and clothing movement.

Dipping Annoying Sounds with Keyframes

Now, you'll move to the next red marker where you will reduce a few clothing crackles in one of Philip's dialogue clips. Ideally, all the non-speaking production sounds will be replaced with Foley sound recordings. Yes, on feature films, even the clothing sounds are reenacted, rerecorded, and added back into the soundtrack for sonic authenticity. Why go through all of that trouble? Because during production, the microphone priority is on capturing dialogue and, therefore, the slightest movement of clothing can often sound like a thundering herd of wrinkles every time a character moves. Because this scene is from a short film, a clothing pass on the Foley stage may not be in the budget. In that case, you must make the most of what you have. For this project, you will salvage what you can in the production sound, and reduce or remove the distractions along the way.

In this exercise, you'll identify the clothing crackles in a clip. Then, you'll use keyframes to carefully dip the levels directly over those crackles to reduce them.

1 Go to the red marker titled "Decrease crackling clothes." Show the viewer, and place it in the lower-third of the screen. Zoom in on the clip in the A2 track.

2 Mark and play the clip.

Can you hear the crackles? If you can hear it, chances are the audience will too, especially in a quiet theater with powerful speakers.

As you can see, the clip already includes some keyframes applied while balancing the clip levels. Unfortunately, while turning up Philip's dialogue, the clothing crackles were also intensified, so now his jumpsuit sounds like it's made of potato chip bags.

Let's start with the first crackle, which looks like a spike in the waveform at around 01:00:50:03.

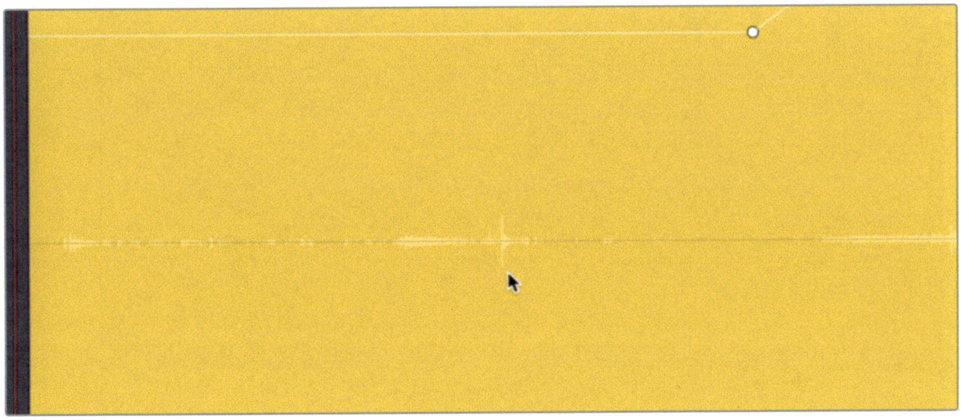

3 Option-click (macOS) or Alt-click (Windows) to add a keyframe on the volume overlay directly above the spike in the waveform. Add keyframes on either side of this first one so you have a total of three keyframes.

NOTE Feel free to zoom in as much as necessary to see the clip's waveform as you manipulate the keyframes.

4 Drag the center keyframe down, watching the spike in the waveform diminish as you lower the volume above it. Continue dragging down to the lower-third of the clip and release the keyframe.

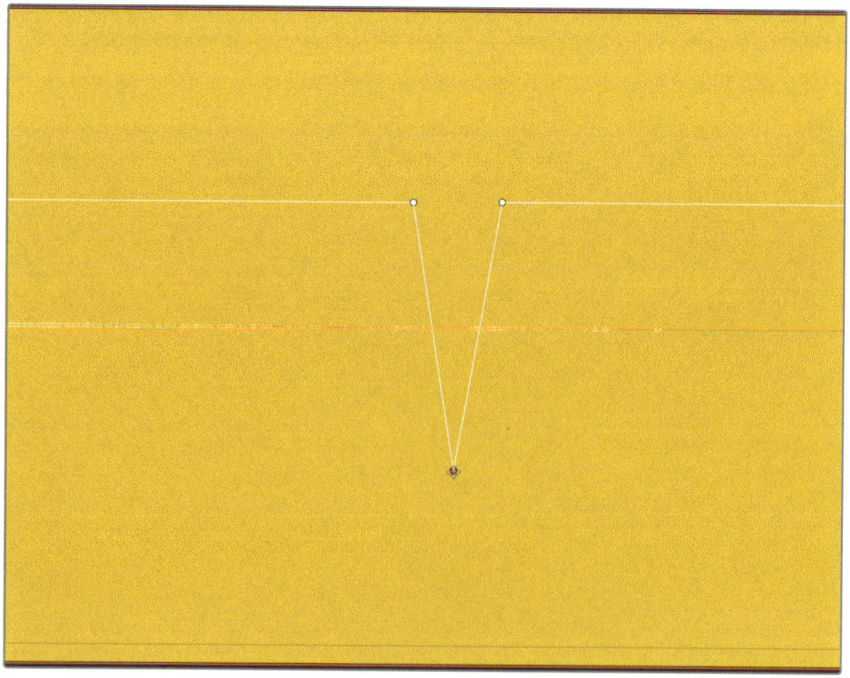

5 Create a wide playback range around the keyframes, and adjust the left and right keyframes during playback to smooth the transition between keyframes. Also, drag the center keyframe down as much as needed until you no longer hear the specific crackle. When you are finished, the playback should sound seamless with no audible remains of the original annoying crackle.

6 When you are finished, clear the playback range. Zoom out, if necessary, and play the clip from the beginning to hear your keyframes in context.

Placing three keyframes worked great for specific spikes in the sound. You can also use this technique with four or more keyframes to carefully reduce or remove a sound in the waveform.

Let's try doing so on the next section of the clip to reduce the crackling sound right before Philip's line, "You died." This sound is critical to the storyline and shouldn't have any distractions that take away from the impact of the words and performance.

A keyframe is already above the crackle. For this example, you'll zoom in to the clip, as necessary. Then, you'll add several keyframes, and just before he speaks, lower all the sound to match the levels earlier in the clip. Finally. you'll add additional keyframes to reduce the specific crackle (spike) without altering Philip's dialogue level. You have already learned the skills necessary for this exercise, so the steps will be general guidelines.

7 Zoom in on the clip, as necessary. Scrub the playhead to the beginning of Philips' first word. Leave the playhead there as a guide. Set a keyframe at the playhead position.

8 Adjust the volume overlay before the playhead to match the height of the earlier clip section. Add as many keyframes as you need to dip the volume level of the crackle and smoothly return to the original dialogue level for Philips' dialogue line. Play that section and make changes until you are satisfied with the results.

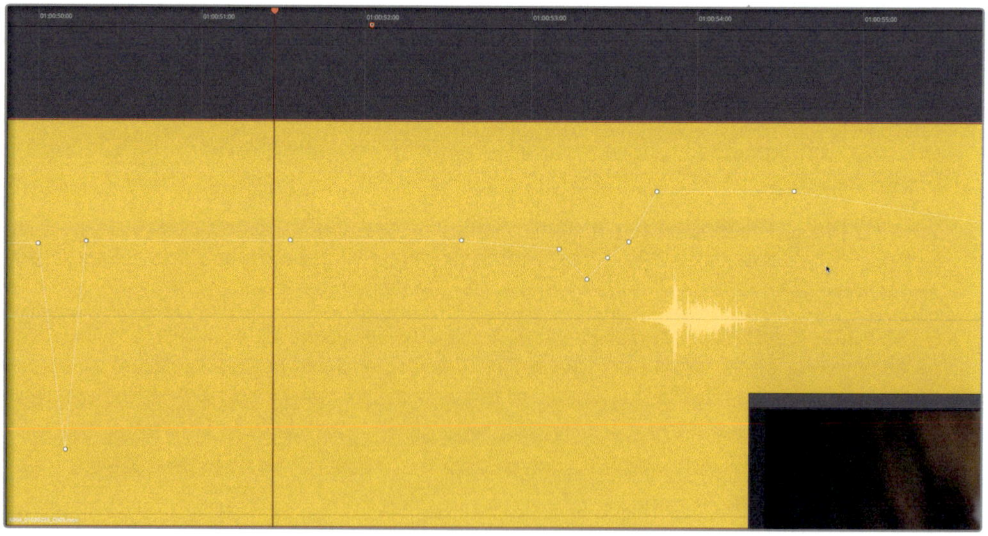

But wait, there's more! The crackle is gone, but you also removed Philips' breath prior to the second crackle. You've already discovered that audible breathing is important to retain for human characters, so let's recover that part of the clip's waveform.

9 Just after the first omitted crackle (3-keyframe dip), raise the overlay until you can see the ambient waveform and hear some of Philip's movement and breathing. Play the clip, and adjust the breathing level until it works with the rest of the clip.

10 Zoom out of the clip until you can see the clips on either side, as well as the A1 and A3 tracks.

Whew! That's much better. The distracting crackles are eliminated and Philip is breathing normally in his performance. But you may have already discovered just one major issue that remains. A big, fat crackle occurs right at the end of Philip's crucial line. If you already noticed the aforementioned dialogue defect and were wondering how you might tackle it, you have the aural instincts of a dialogue editor!

> **NOTE** If you didn't finish all the previous steps, feel free to open the timeline "5a Dialogue Decrackled" now to catch up and continue.

Replacing Words from Outtakes

When a piece of dialogue is unusable, you have two choices. Re-record it with ADR, or replace the line with sound from one of the scene's *outtakes*. To support this process, dialogue editors compile a list and cues for all unsalvageable dialogue. The first strategy is to use the same line from another camera angle or take, so you're in luck. Another full take of this scene is waiting in the Media Pool.

In this exercise, you'll employ new Fairlight audio editing features to drag, drop, and finesse the replacement dialogue in the track.

1 Show the Media Pool bin list. Click the Icon View button to view the bin contents in icon view.

2 In the Hyperlight Audio bin, select the "Files for Repair and Replace" bin.

3 Select the A003_01020143_C006.mov clip to load it into the preview player.

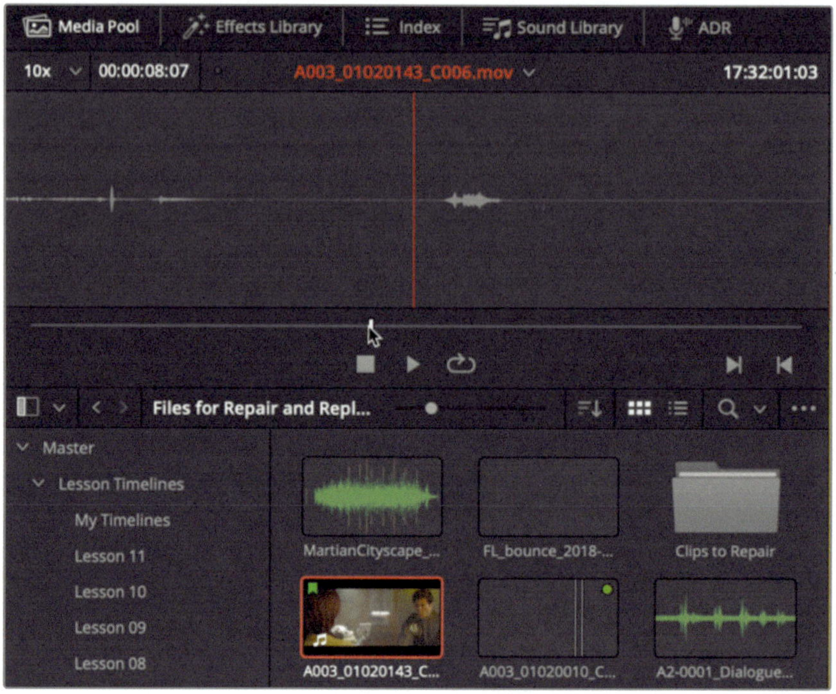

4 In the preview player, change the Zoom pop-up menu to 20x.

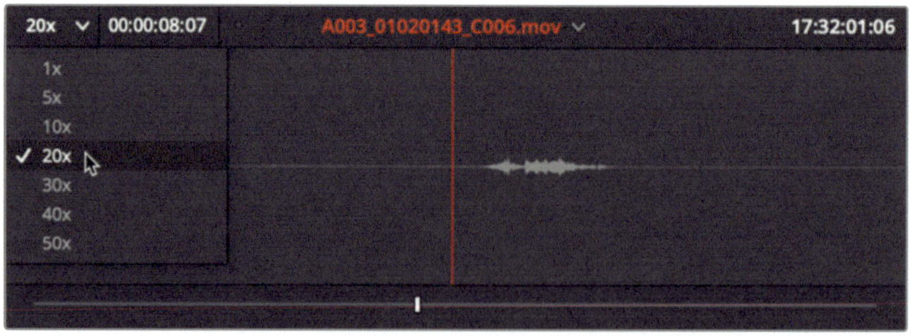

You can now see more of the clip's waveform in the preview player.

5 Press Home to move the playhead to the beginning of the clip. Use the JKL keys to play through the clip and find the portion where Philip says, "You died." Remember, Shift-L is fast forward, and Shift-J is fast rewind. (Hint: You'll find the line at around 17:32:01:08 in the source timecode field in the upper-right corner of the preview player.)

6 Change the Zoom pop-up menu to 30x to get a better view of the dialogue line you want to mark.

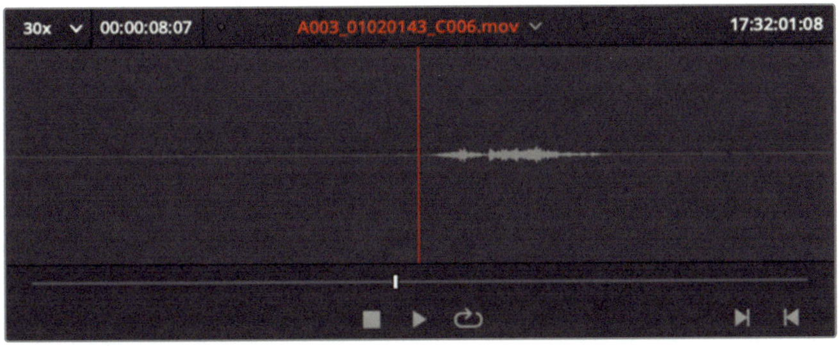

7 Use the JKL keys to move the playhead just before the line, "You died," and press I to mark an in point. Move the playhead to the end of the line, and press O to mark an out point.

> **TIP** When marking words or phrases, mark as close to the words as possible for easier syncing in the timeline. You can always extend the head or tail of the clip, if necessary, once it is in the timeline.

Did you notice that the outtakes version of the line is clean and clear, without any annoying clothing crackles? All you need to do is prepare the timeline for the new clip.

Working in Audio Layer Editing Mode

Resolve has two audio editing modes. The default mode is Overwrite Editing mode. In this mode, when two clips overlap in the timeline, whether by trimming or dragging clips, the material in the lower level of that overlap is removed from the timeline. Alternatively, in Layered Audio Editing mode when two clips overlap, they are automatically placed on separate audio track layers. This relocation is all done invisibly within the track until you show audio track layers.

To try using this feature, let's first change the timeline editing mode.

1 Choose Timeline > Layered Audio Editing to change the timeline mode.

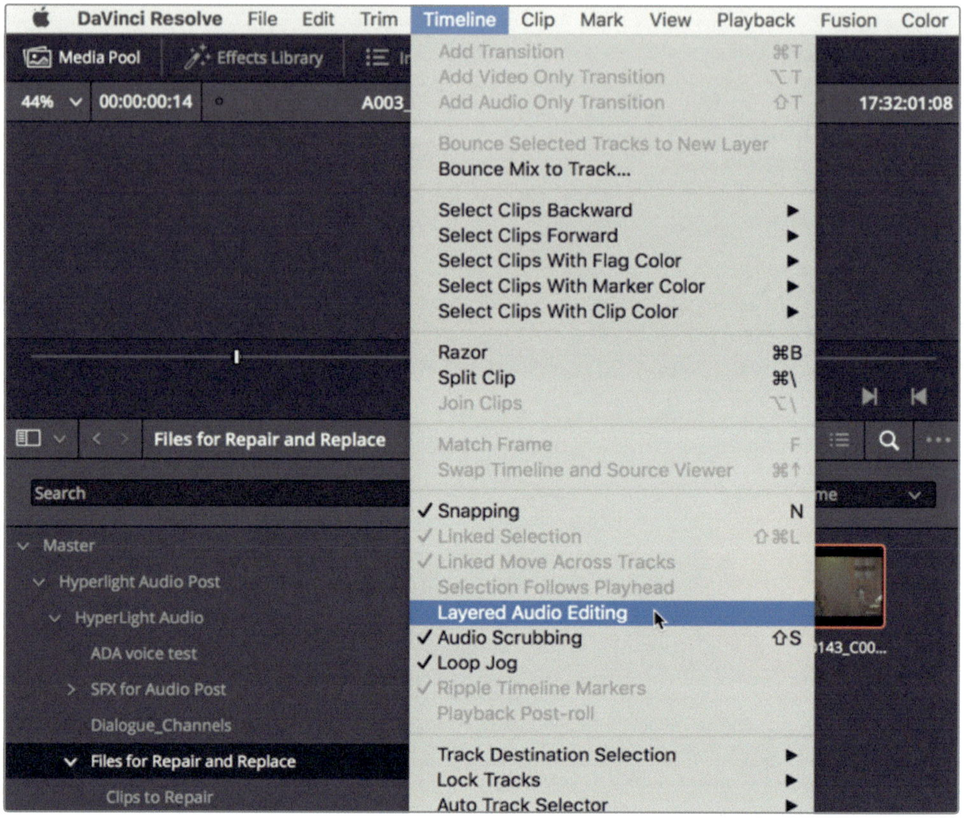

Next, to simplify this editing process, let's edit the new clip to the A1 track, so you can move it down to overlap the original clip in the A2 track. For this exercise, you can use either the Range Selection tool or the Edit Selection tool to perform this editing maneuver. If you are using DaVinci Resolve 16.2 or higher, use the Edit Selection tool.

2 In the timeline, click the Edit Selection Mode button, if necessary.

3 Scrub the playhead to the end of the phrase, "You died," and and click an edit selection point to use as a guide.

4 Select the A1 track, if necessary. Notice that the edit selection point moves up to the selected track. Turn off Snapping (magnet icon) so you can freely align the new clip in the timeline.

5 Drag the marked clip from the Media Pool preview player to the A1 track, but don't release the mouse button.

6 Drag the new clip toward the right until the tail of the clip aligns with the end of the waveform (just before the edit point).

The clip is in place, and because it is on a separate track, you can hear both versions of Philip's line at the same time during playback. Also, with the play range already marked, you can initiate looped playback. Finally, the clip you added is already selected, so you can nudge it to the left or right to sync the clips.

7 Select the clip in the A1 track.

8 Start looped playback and listen to Philip's line. The sync should be pretty close. If the clips are noticeably out of sync, press the . (period) and , (comma) keys to nudge the clip left or right one frame at a time until the clips sound in sync.

9 In the new clip, drag up on the volume overlay until the waveform amplitude (height) is about the same as in the original clip.

With the clips in sync, you can now move the clip to the proper track.

Zoom in on the clips until you can clearly see the waveforms of both clips in the timeline.

10 Right-click the new clip, and choose Clip Color > Green so you can more easily differentiate the new clip from the previous clip.

11 In the A1 track, select the lower half of the new green clip. Drag the clip down to the A2 track. Do not release the clip yet.

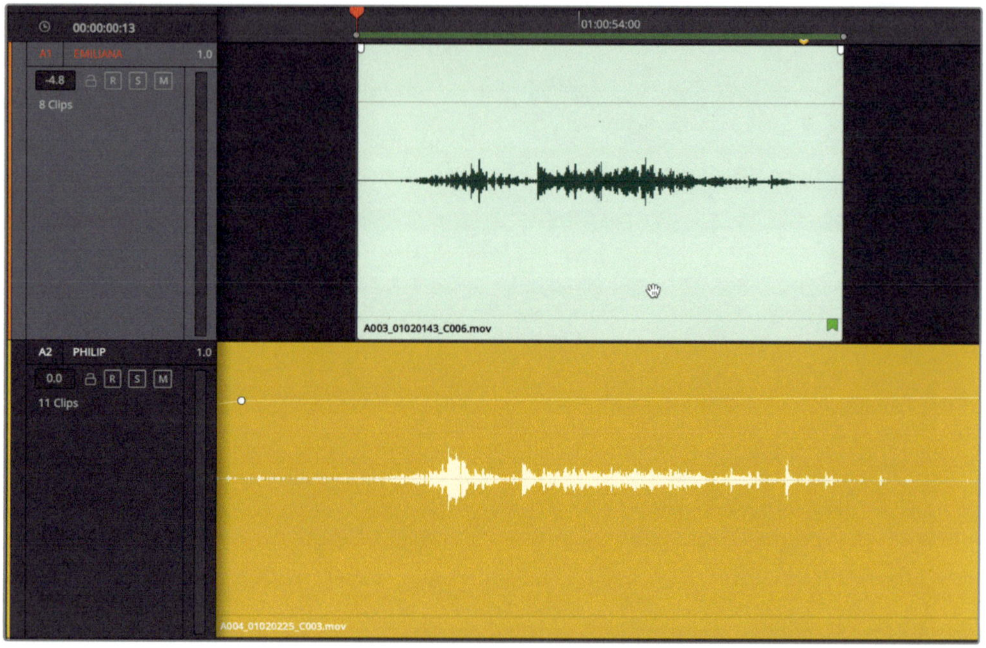

Before you release the clip, you can see the semi-transparent waveform of the new clip as it overlaps the original clip's waveform. This is why the clips become semi-transparent when you edit them! And this is also the best time to use the Edit Selection's Grabber tool to align the waveforms into perfect sync.

NOTE If you are using Range Selection mode to perform this edit, you can simply use the Cut shortcut to cut the clip, and then select the A2 track and JKL keys to align the semi-transparent clip with the playhead before pasting.

12 Drag the clip until you are happy with the alignment of the waveforms. Focus on aligning the visible peaked sections of the waveform that represent words or syllables. When you're done, release the clip. Since the clip is in an upper layer in the track, you can select it anytime and nudge or drag it further to refine the position.

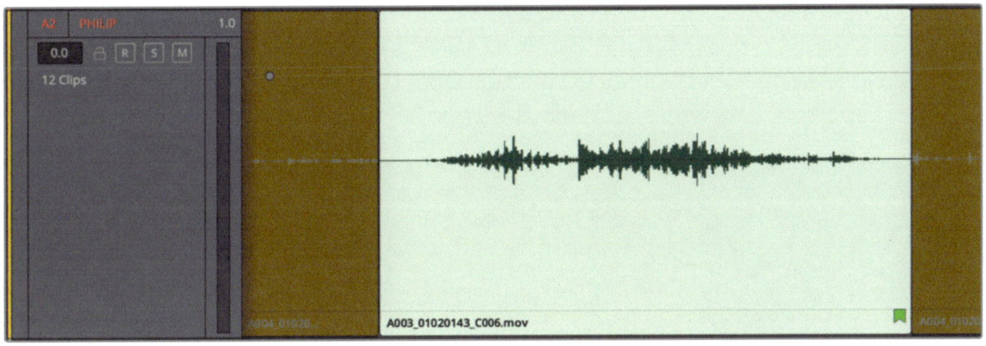

13 Clear the play range.

14 Choose View > Show Audio Track Layers, and play the newly added clip.

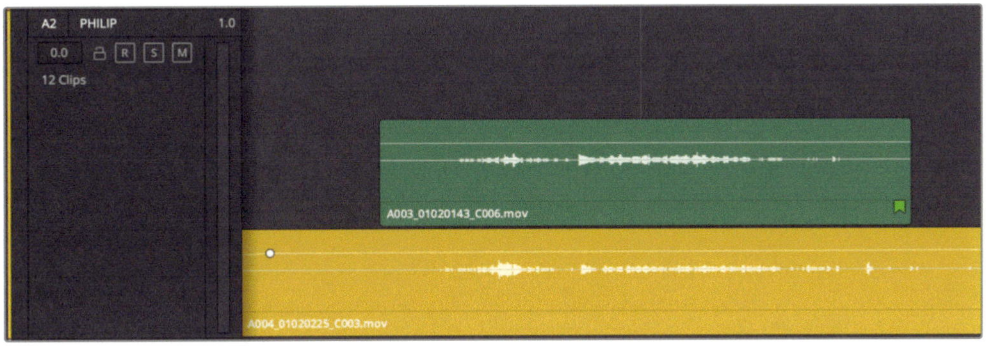

The new clip sounds great until it ends when you hear a click from the clip that was previously in the timeline. No problem; you'll simply extend the new clip to cover the click and add a fade while you're at it.

15 Drag a wide playback range around the stacked clips. Then, while listening to playback, extend the head and tail of the new green clip to cover up any jarring background noise just before or after the line. Finally, add short fades at the head and tail of the clip.

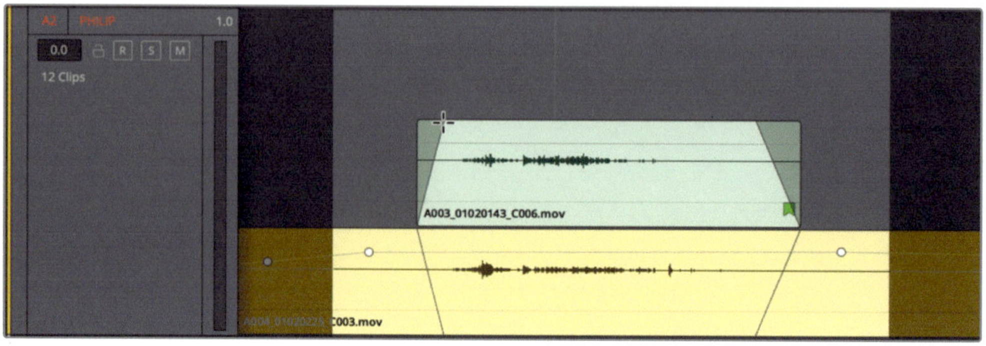

16 When you are finished, clear the play range, and turn off the Show Audio Track Layers view option.

17 Zoom out until you can see the yellow and green clips in their entirety.

18 Play the timeline from 01:00:48:00 through 01:01:01:15. Stop playback after Philip says, "The hyperlight core came out a few days ago."

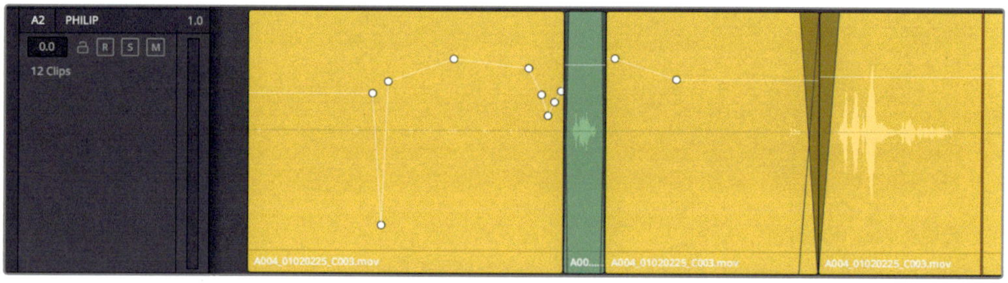

You've successfully replaced Philip's dialogue line from an outtake. However, you may still find it difficult to focus on the dialogue without being distracted by the fluctuations in the background sound from clip to clip. Also, a few stray distracting sounds remain to be fixed.

NOTE If you didn't finish all the previous steps, feel free to open the timeline "5b Dialogue Repair Finished" now to catch up and continue.

Filling in Gaps with Room Tone

Nothing screams amateur video more than hearing background sound change between every cut, or worse, plunging into a black hole of silence between dialogue lines. In dialogue tracks, silence is deadly! Unless your scene takes place in the vacuum of space, actual silence should never be present in your soundtrack.

You may think, why not just cover up the silence with tracks of sound effects and ambience? But that's like throwing an artificial shadow across a character's face to hide the fact that you didn't color correct the flesh tones. Remember, the dialogue tracks are the most important elements in your soundtrack, so you should treat them as such.

The acoustic solution to filling the gaps and making dialogue sound seamless is room tone, which is the "silence" recorded on the set at the end of a scene to match the "silence" between spoken words. As those quotation marks suggest, room tone silence isn't really silent. Even soundstages designed for optimal sound recording have some ambient sound.

The room tone is usually recorded at the end of each scene by the location sound mixer who recorded the original dialogue. However, because you lack a recorded room tone for your current scene, you'll need to find some room tone in the existing dialogue clips. Once you find a usable piece of room tone, you can either copy and paste it beneath each gap to patch the holes, or stitch it together to make a long bed of room tone that you can apply consistently throughout the track. For this lesson, you'll use the latter method, which is the preferred method used by most dialogue mixers for a more consistent and seamless dialogue scene.

1. Right-click any track header, and choose Add Tracks.
2. In the Add Tracks dialog, set the "Number of tracks" to 1, Insert Position to Below ADA TEST VOICE, and Audio Track Type to Mono. Click Add Tracks.

   An empty A4 track appears in the timeline between the A3 and A5 tracks.
3. Change the name of the new track to **ROOM**.

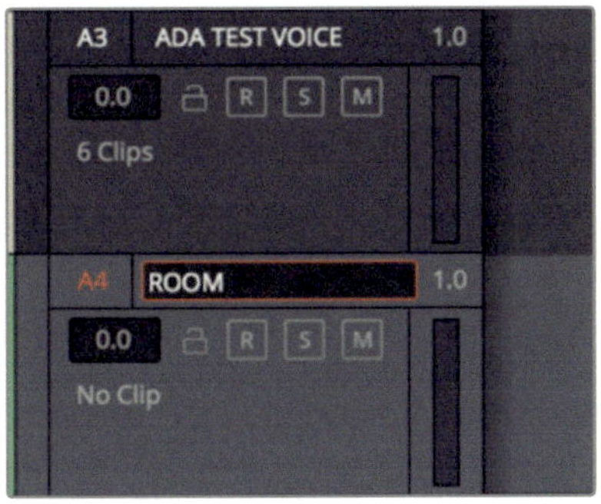

With the room tone track created, you need to find a usable piece of room tone to place in it.

Finding Usable Room Tone in the Timeline

When you need to create your own room tone from material already in a track, here are some basic guidelines to follow:

- Find the longest section of existing room tone in the scene.
- Copy and paste the content you want to test into a separate track.
- Raise the volume during the test to more easily detect any subtle noises within the clip.
- Narrow down your choice to the cleanest portion, or as in this case, the room tone section with the fewest additional sounds.
- Test the tone by listening to it looped. Remove any audible elements at the head or tail that could give away the repetition.

With those guidelines in mind, let's find some usable room tone. Luckily, in the beginning of the timeline, you have three possible candidates for room tone.

1. Zoom in on the first three clips in the timeline. These include the first orange clip in the A1 track, and the first two yellow clips on the A2 track.
2. Raise the volume overlay on all three clips until you can fully see the waveforms.
3. Drag a play range around all three clips. Listen to the playback. Listen again while watching the viewer to identify the source of some of the noises you hear in the background.

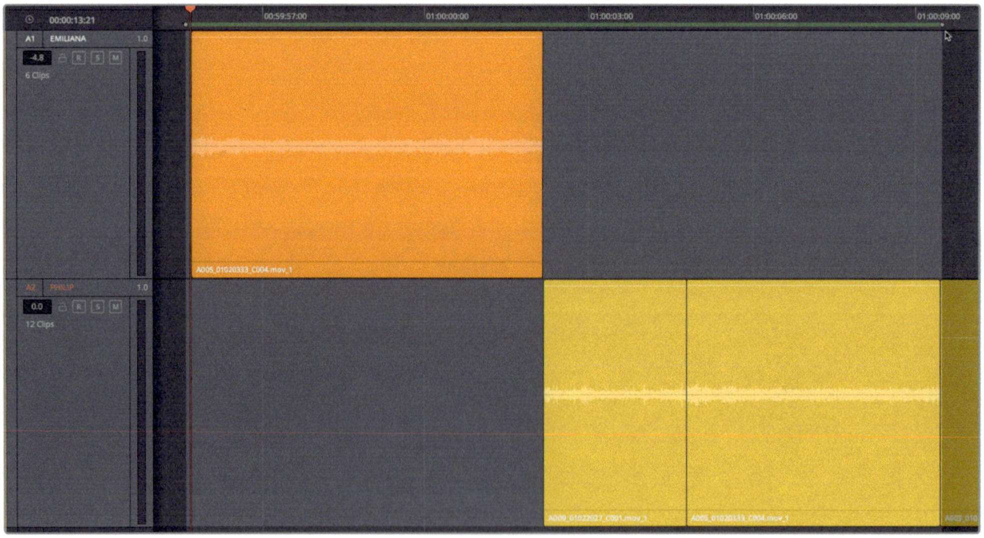

What are your thoughts? Most likely, you discovered that a clip that looks like a good candidate for "silent" room tone may not be useful at all. In fact, all these clips are plagued with just enough distracting sounds that none of them can be used as-is. A possible candidate is the middle section of the third clip; unfortunately, that usable section is only about one-and-a-half seconds long. It's time to "go fish" and look for additional room tone elsewhere. Perhaps you can find an additional room tone candidate at the end of the first clip in the A1 track.

4 Clear the play range.

5 Click the tail edge of the clip in the A1 track to see the unused waveform to the right of the current clip.

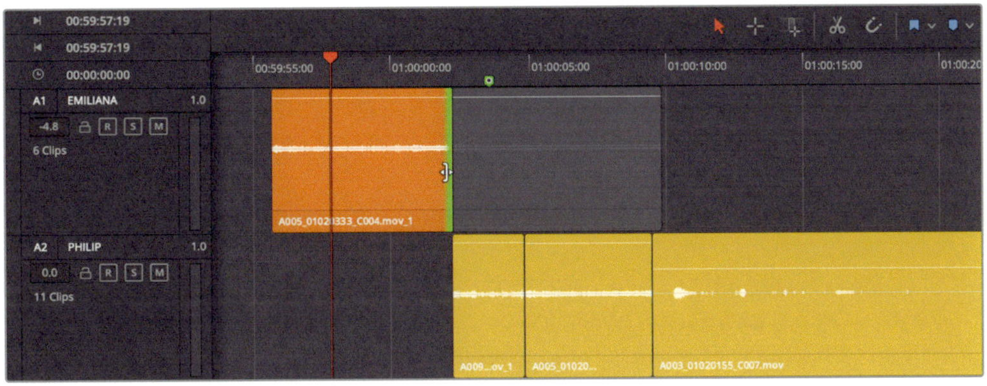

It looks as if at least ten seconds of room tone precedes a pronounced spike in the waveform.

Let's paste a copy of the clip into the A4 track to test it out.

6 Select the first clip in the A1 track. Copy and paste the selected clip into the A4 track.

7 Double-click the volume overlay on the first three clips in the A1 and A2 tracks to reset their volume levels.

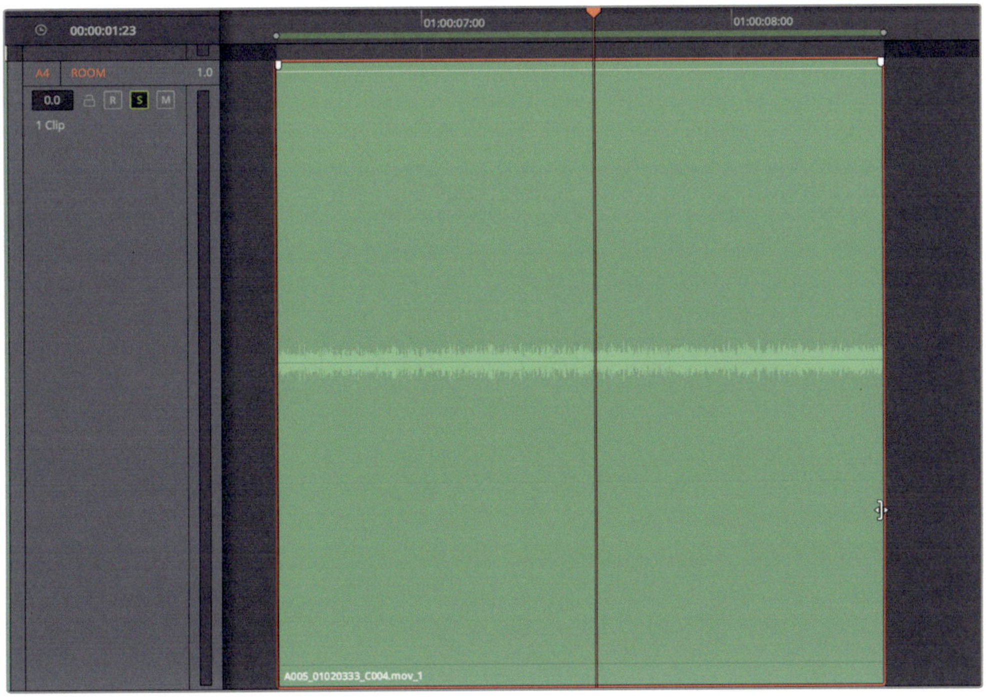

8 Drag the tail of the A4 clip to the right until you see a pronounced spike in the waveform. Then, trim the tail back to the left to end the clip it just before the spike.

9 Solo the A4 track, and mark the clip. Turn on looped playback, if necessary, and listen to the clip.

   You will undoubtedly hear noises within the clip. Some are more pronounced than others. Now let's narrow the play range to a cleaner section at the beginning of the second half of the clip.

10 With the Edit Selection Tool, drag a selection range between 01:00:03:09 and 02:00:07:17.

11 Start looped playback and listen to the new range. When you are finished, clear the play range.

   This slice of room tone is pretty good. One minor sound is audible near the middle that may or may not be an issue because you can cut it out and stitch it together with the other two room tone pieces in audio track layers.

Stitching Room Tone in Audio Track Layers

This method of *stitching* room tone is exactly what it sounds like. You take short pieces of room tone and stitch them together with crossfades to form a seamless "fabric" of audio. Audio track layers in the Fairlight page make it easier than ever to stitch together your room tones.

In this exercise, you'll copy, paste, overlap, and crossfade the best sections of room tone in track layers to create a room tone for the scene. Along the way, you'll implement many of the skills that you've already learned. For this exercise, you'll use the standard Selection tool (arrow).

1 Right-click the selection range and choose Edit Selection Editing > Trim to Selection to trim the clip in the A4 track to fit the current selection range. Clear the range.

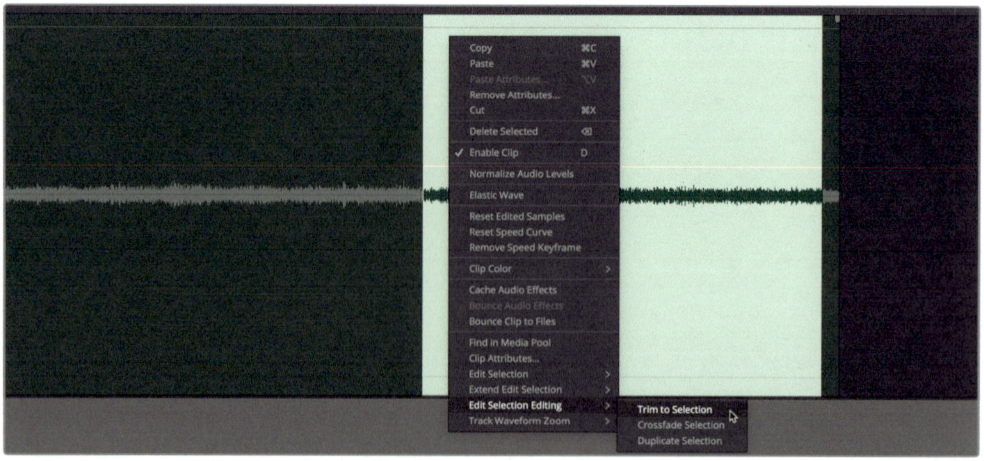

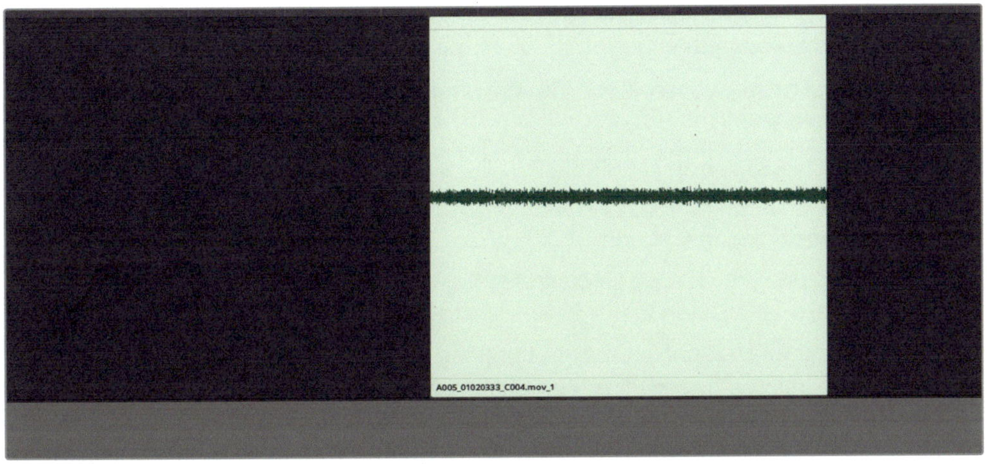

2 Press A for standard selection mode. Show audio track layers.

3 Select the clip in the A4 track. Copy the clip, and without moving the playhead, paste the clip twice.

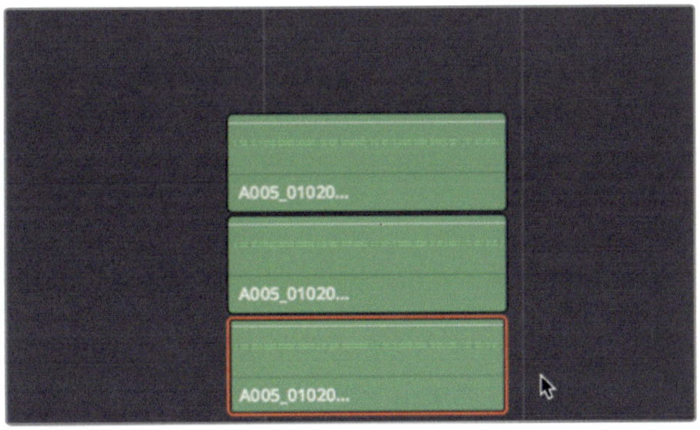

Two copies of the clip appear in layers above the original selected clip.

4 Drag the upper two clips down to the lowest layer, and place them end to end until you have three consecutive clips in the same layer.

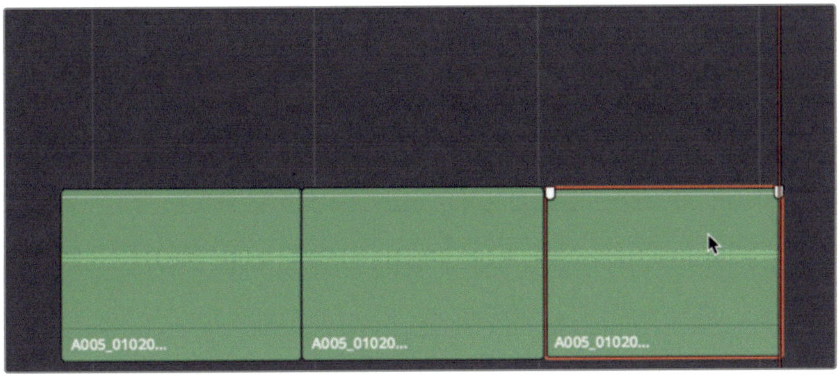

5 Play the three sequential clips in the A4 track, and listen for any noises that draw attention to the repetition.

A distant sound about two-thirds into the original clip breaks the illusion of continuous ambient room sound.

6 In the A4 track, delete the second and third clips.

7 Scrub the last third of the clip to find the distracting sound. Then, trim away the last third of the clip including that sound.

8 Copy and paste the clip two times and string the copies next to the original, just as you did previously. Listen to the three clips together.

You can hear a few slight sounds, but nothing too distracting. Also, keep in mind that the volume levels of these clips are much higher than they will be in the final mix. Let's proceed with the next step, which is stitching them together. Although they sound okay back to back, you can hear a slight change at each cut. To remedy this, you'll stack the clips on alternate tracks and add crossfades.

9 Zoom in on the three clips in the A4 track.

10 Select the second clip, and drag it up one layer and to the left to overlap the first clip by about one-tenth of the previous clip. (Provide enough overlap to create crossfades).

11 Drag the third clip beneath the second so that it overlaps by the same amount as the previous clips.

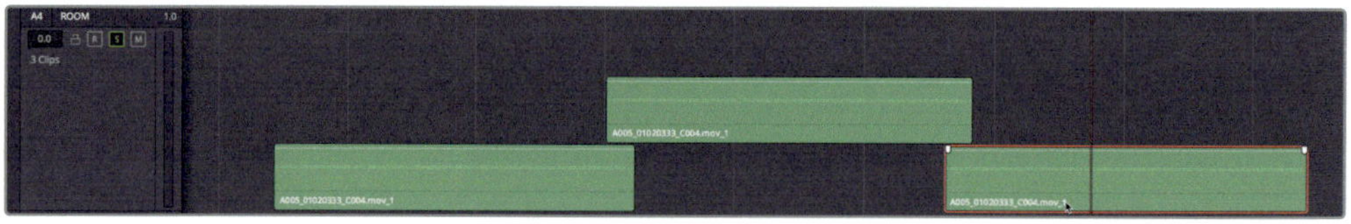

12 Copy and paste the second clip, drag it to the right, and place it on the second layer so that it overlaps the third clip by the same amount.

So far, so good. Hopefully, you can see the checkerboard, quilt-like appearance of your room tone clips. Rather than continue working with one clip at a time, let's now select and paste all four at once.

13 Zoom out horizontally so that you have room to copy and paste all four clips next to each other in the track.

14 Select and copy all four clips. Move the playhead to the right to align the semi-transparent copies in position. Then, paste the copies.

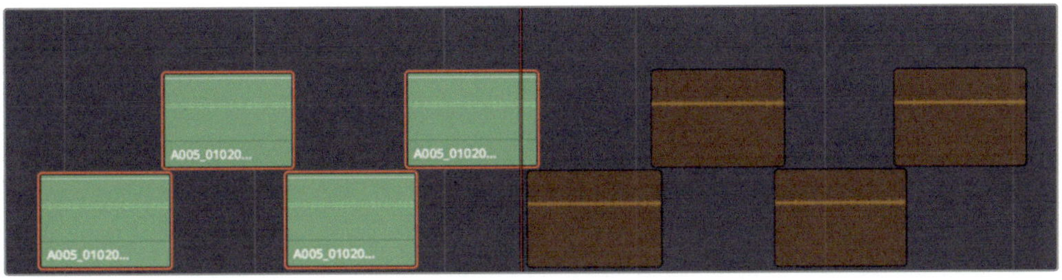

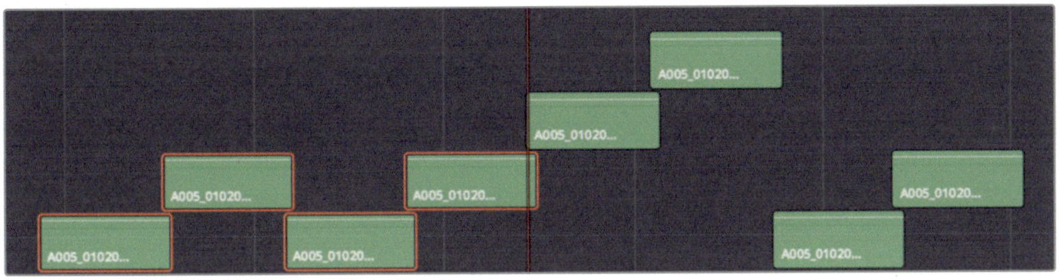

If your pasted clips don't quite look like the screen shot, no worries. They will sound the same regardless of their position as long as they overlap in layers. If your clips do look like the screen shot, you can easily fix the pattern by lowering the first clip on the third layer to the lowest layer. Once the pattern is fixed, you'll extend the chain of clips one more time.

15 Drag the fifth clip down to the lowest layer to fix the two-layer pattern.

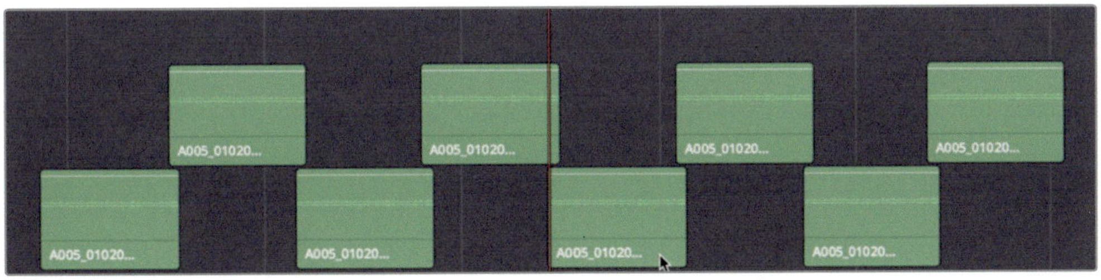

16 Select and copy all eight clips, and paste them above the first set of eight.

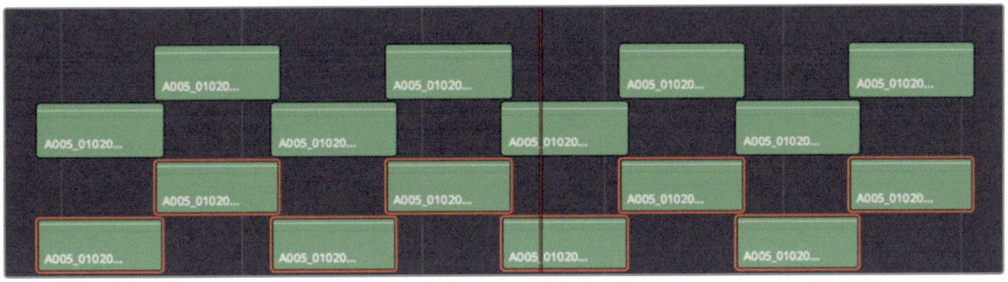

17 Select all eight clips in the upper two layers and drag them into position to continue the original pattern. If necessary, move any clips on a third layer down into the first two layers.

18 Zoom out so you can see the entire timeline.

You could continue to paste all sixteen clips to finish the track, but that won't be necessary. In a few minutes, you'll bounce these clips into a single rendered clip which you'll use instead. First, you'll need to return the volume to the default level, and add cross fades.

Copying and Pasting Audio Attributes

Just as you copied and pasted clips, you can also copy and paste specific clip audio attributes including volume, EQ, and plug-ins. So, all you need to do is set the volume and fade handles on one clip and paste those attributes to all the other clips.

1 Move the playhead over the second clip in the A4 track. Select the A4 track, and zoom in to the selected clip.

> **NOTE** You selected the track merely to make it easier to zoom in on it. Doing so is not necessary for changing clip volume or copying and pasting clip attributes. When a track is selected, clips beneath the playhead in selection mode appear gray.

2 Double-click the selected clip's volume overlay to reset the level. Drag fade handles on either side of the selected clip to cover the overlap with the lower clips. Right-click the clip and choose Copy, or press Command-C (macOS) or Ctrl-C (Windows) to copy the selected clip.

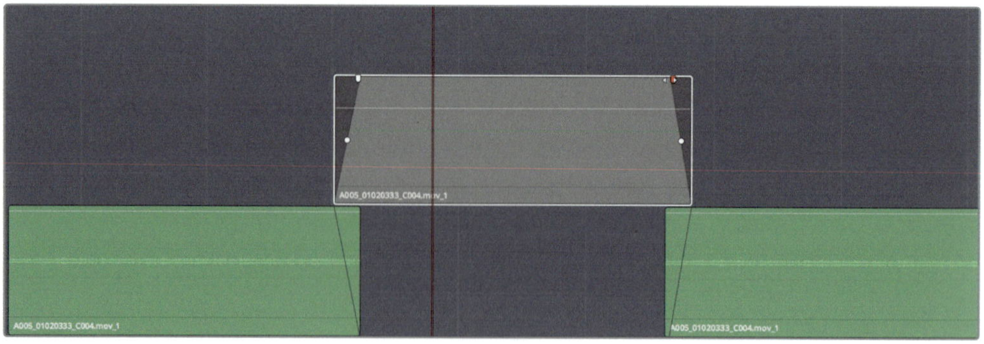

3 Click any empty space in the timeline to deselect the track and the selected clip.

The semi-transparent copy disappears, as well. You need to apply fades only to the clips in the upper layer, so let's paste attributes to those.

4 Zoom out as needed to see all of the clips in the A4 track.

5 In the upper layer of the A4 track, select all eight clips. Right-click any of the selected clips, and choose Paste Attributes.

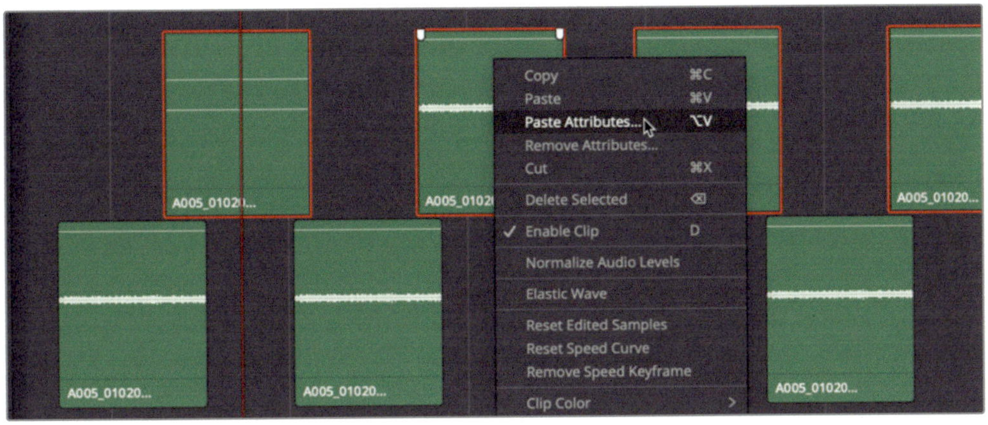

The Paste Attributes dialog box appears and indicates that attributes from the original clip will be pasted to eight clips.

6 In the Paste Attributes dialog box, select the Audio Attributes checkbox, and deselect Plugins and Equalizer. By doing so, you will paste only those attributes that involve the clip volume. Click Apply.

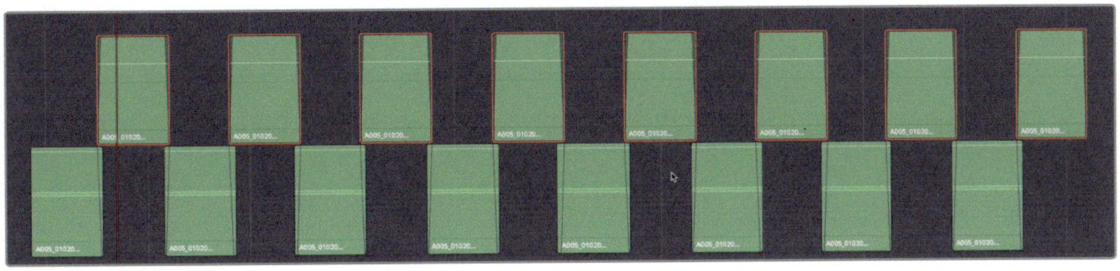

Wow! Just like that, all eight clips now have the same fades and volume levels. Let's do it one more time for the bottom layer.

7 Double-click the volume overlay on the first clip in the bottom layer to reset the volume level.

8 Right-click the clip, and choose Copy. Select all of the clips in the bottom layer. Right-click one of the selected clips, and choose Paste Attributes.

The Paste Attributes dialog box opens with the previous settings already selected.

9 In the Paste Attributes dialog box, click Apply.

10 Solo the A4 track, and play through the stitched clips. Feel free to adjust the fades to suit yourself, although they should sound pretty good as they are.

You've successfully copied the clips and stitched them together in a matter of minutes. Now, let's bounce (flatten) them into a single rendered clip that you can use throughout this scene.

Bouncing a Selection to a New Layer

The room tone clips that you stitched together work fine; however, they will become a bit cumbersome to move around and work with later in the scene. Fortunately, you can simplify that process by bouncing the clips to a new layer.

Bouncing is the audio equivalent to flattening a layered image or rendering a composite video. In this case, bouncing creates a new audio file in your capture location and Media Pool. In the Fairlight page, you can bounce a selected play range to a new layer, or bounce your output to a new track in the timeline. For this exercise, you'll set a play range around the clips in the A4 track, and bounce them to a new layer.

1 In the A4 track, mark an in point at the head of the first clip. Then, mark an out point at the tail of the last clip.

The timecode display indicates that the duration of the play range is around 41 seconds. That is a respectable length of room tone to work with. The trouble is, it's still in little pieces. Your goal is to create one long and seamless piece of room tone based on the clips in the A4 track. Because you may need to copy and paste this content later, it's better that it doesn't start and end with fades baked into the rendered file.

2 Zoom in as necessary. Then, drag the Fade handle at the tail of the last clip to the right to remove it.

3 Select the A4 track. This is a necessary step for bouncing to a new layer in the same track.

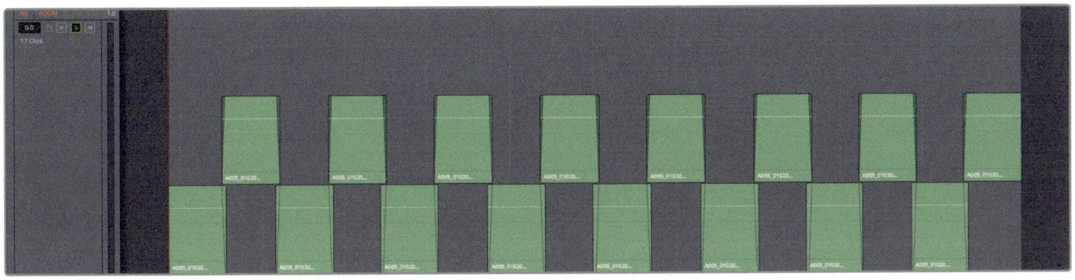

4 In the Media Pool, select the “Files for Repair and Replace” bin.

Just as with audio recordings, your bounced files will appear in the selected bin.

5 Choose Timeline > Bounce Selected Tracks to New Layer.

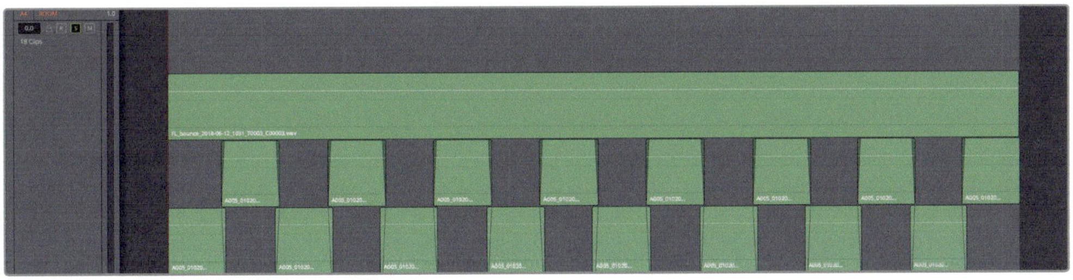

Voila! A new bounced clip the length of the play range appears above the other clips in the selected track. The source clip is also visible in the selected bin in the Media Pool.

The bounced clip's name starts with "FL" (for Fairlight), continues with "bounce" to indicate that it is a bounced file, followed by the current date.

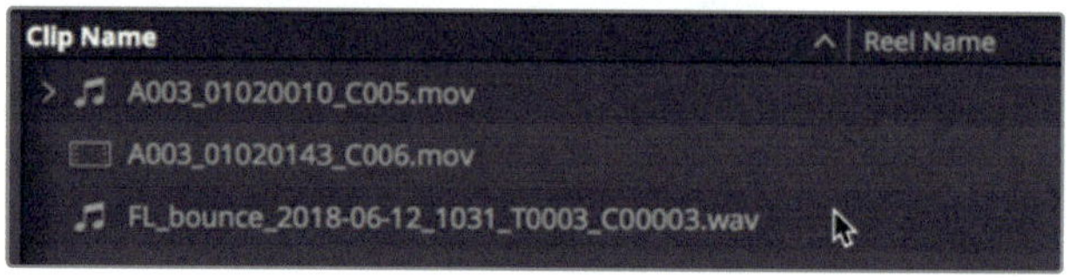

At this point, you can keep the original stitched clips in the timeline beneath the bounced layer, disable them, or delete them from the track. For educational purposes, and to preserve any sentimental attachment you may have to your first stitched patchwork room tone track, let's keep it and make a new version of the timeline with the stitched clips deleted to finish the job.

6 In the Media Pool Timelines bin, right-click your current timeline, and choose Timelines > Duplicate Timeline.

7 Name the duplicate timeline **5 Dialogue with Room Tone**, and open that version of the timeline, if necessary.

8 In the A4 ROOM track, select and delete all of the short room tone clips in the lower two layers.

Now that you have one solid piece of room tone to work with, you can direct your collective skills toward finishing the ROOM track.

NOTE The Fairlight Audio Editor includes two options for automatically generating room tone or room ambience to fill a selected range. These options are Fill and Back/Front Fill and generate a new bounced clip for the duration of a marked range, based on the audio currently in the clipboard. You can learn more about working with the Fairlight Audio Editor to control the Fairlight page in the DaVinci Resolve Reference Manual available under the Help menu.

More Room Tone Please!

In this exercise, you'll work on your own to complete the room tone track. The big-picture goal is "wall-to-wall tone for the entire scene." So, you will need to copy and paste the bounced clip as often as needed to extend from the first to the last frame of the scene. Use the track layers to overlap the copies and fade the top layer, as necessary.

Once you have the clips in position, you'll need to set the volume level based on the room sound already in the scene. Using the Mixer and Philip's first dialogue clip as guides, determine an appropriate volume level for the room tone before his first line. (Hint: Solo Philip's track, set a play range for the room sound in Philip's clip, and in the Mixer, watch the meter for the A2 track to determine the average level.)

When you know the target volume level for the room tone, set all the copied clips to match that level. (If you aren't sure what level to aim for, try using -45 dB). You'll need to manually adjust each clip, or set one clip and copy and paste its attributes to the others. Clip Normalization will not work here because it is based on peaks and goes only as low as -30.

Once you finish building out the room tone track, hide the audio track layers and listen to the timeline to hear how it sounds overall.

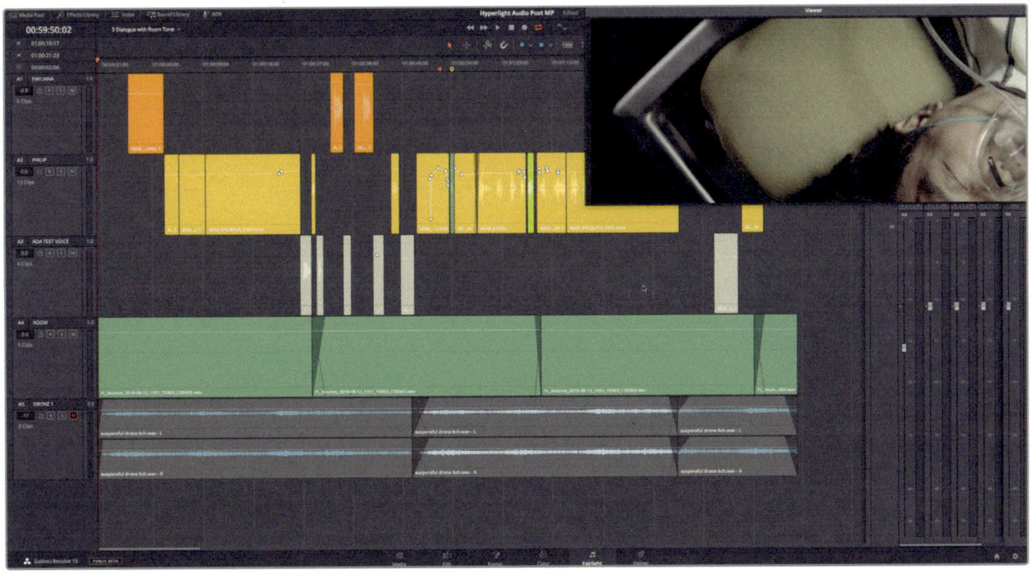

The scene has come a long way. It isn't perfect, but it's darn close. Did anything jump out at you and make you want to stop payback and fix it? Specifically, did you notice the creak (either the floor or the table) when Philip shifted his weight before he says, "The hyperlight core"?

At this point, any time you hear something distracting that needs fixing in the dialogue tracks, fix it! Pay special attention to sounds that could be cut out or trimmed away.

Move the playhead to 01:00:58:11. Zoom in to the playhead and Philips' track. Trim the tail of the clip on the left of the playhead just enough to remove the creaking sound. Play the section to see how it sounds without the distraction. And, as always, save your work.

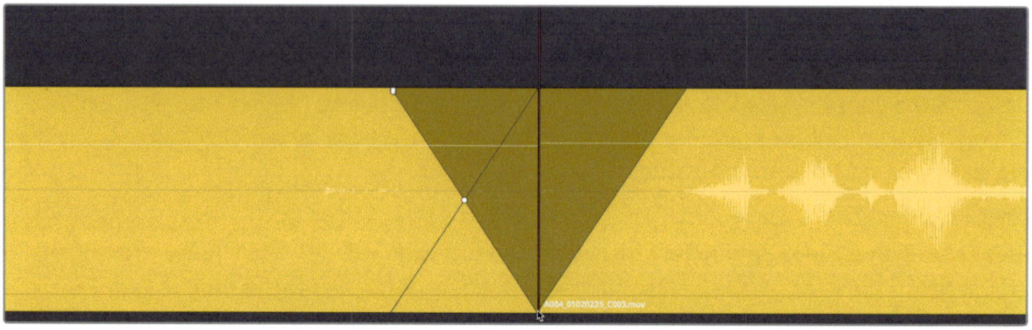

You've completed the dialogue editing and clean-up for this specific scene. Whether you found the experience fun, challenging, or tedious, it's all a necessary part of creating professional-sounding dialogue tracks. Coming up next, you'll explore additional tools, plug-ins, and techniques for troubleshooting and enhancing dialogue tracks.

NOTE If you didn't complete all the steps and would like to hear an example of the finished scene, open and play the timeline, "5d Dialogue Tracks Finished."

Lesson Review

1 Where can you see all your timeline markers and use them as a spotting list? (Choose all that apply.)

- a) Metadata
- b) Media Pool
- c) Timeline pop-up menu
- d) Index
- e) Sound Library Marker list

2 Which tool allows you to add in and out points to your timeline?

- a) Selection tool
- b) Range Selection tool
- c) Razor tool
- d) Mark Selection tool

3 Which tool can you use to reduce plosives in dialogue tracks?

- a) De-ploser plug-in
- b) Ex-plosive plug-in
- c) De-Esser plug-in
- d) Gate
- e) Keyframes

4 What tool is useful for selecting and deleting specific portions of a clip?

- a) Edit Selection tool
- b) Overwrite tool
- c) Lift tool
- d) Insert Gap tool

5 True or False? To create a seamless track of room tone, you can record tone from the built-in Oscillator.

Answers

1 d

2 b

3 e

4 a

5 False

Lesson 6

Advanced Dialogue Repair

In the previous lesson, you focused on basic dialogue track repairs, word replacement, and room tone. In this lesson, you'll tackle other common dialogue issues including powerline hum, noisy clips, and sibilance. These types of issues require additional tools such as FairlightFX repair plug-ins, clip equalization, and dynamics processing.

Time

This lesson takes approximately 40 minutes to complete.

Goals

Finding the Problem

First, let's open and examine the timeline that was created for these examples.

1 Open the timeline, "6 Repairs and FairlightFX."

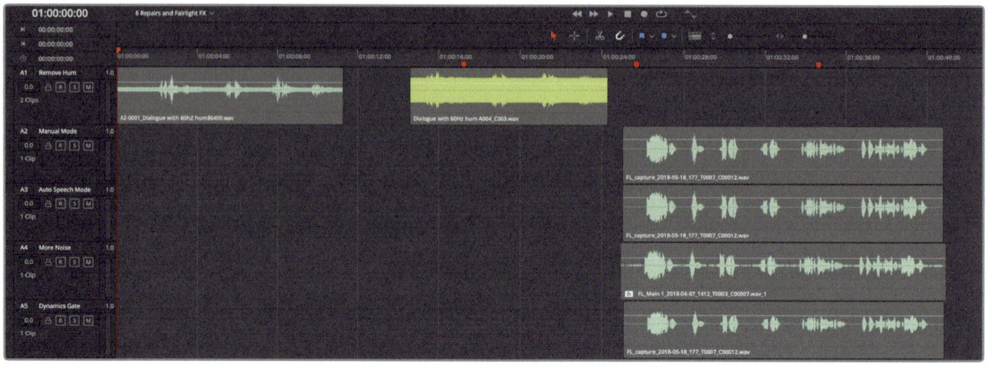

This timeline includes six clips in five separate tracks. Each clip has temporarily been disabled. Also, tracks A2 through A5 contain versions of the same clip.

2 Select the clip in the A1 track, and press D.

The clip is enabled and ready for playback.

3 Play the first few seconds of the clip.

A clear background hum contaminates the entire clip. This problem is fairly common and can have a number of causes including poor audio cable shielding, incorrect power grounding, or when haphazardly overlapping audio cables and power cables on the set. This hum can't be removed with keyframes or editing. Instead, you'll employ one of Resolve's powerful new FairlightFX repair plug-ins.

NOTE Power cables and audio cables can coexist when recording as long as they run parallel to each other and cross at perpendicular angles.

Repairing Dialogue with FairlightFX Plug-Ins

DaVinci Resolve 16 includes a full set of built-in FairlightFX plug-ins, along with support for third-party macOS Audio Unit plug-ins. DaVinci Resolve Studio extends support to third-party VST plug-ins.

In Lesson 1, you applied an Echo plug-in to a sound effect clip. In the next series of exercises, you'll work with three powerful FairlightFX repair plug-ins including the De-Hummer, Noise Reduction, and the De-Esser. Plugins can be applied to clips, tracks, or buses. You'll work more with track- and bus-level FX and processing in Lesson 7, "Mixing and Sweetening the Soundtrack." For now, you'll focus on applying plug-ins to specific clips.

1 Show the Effects Library.

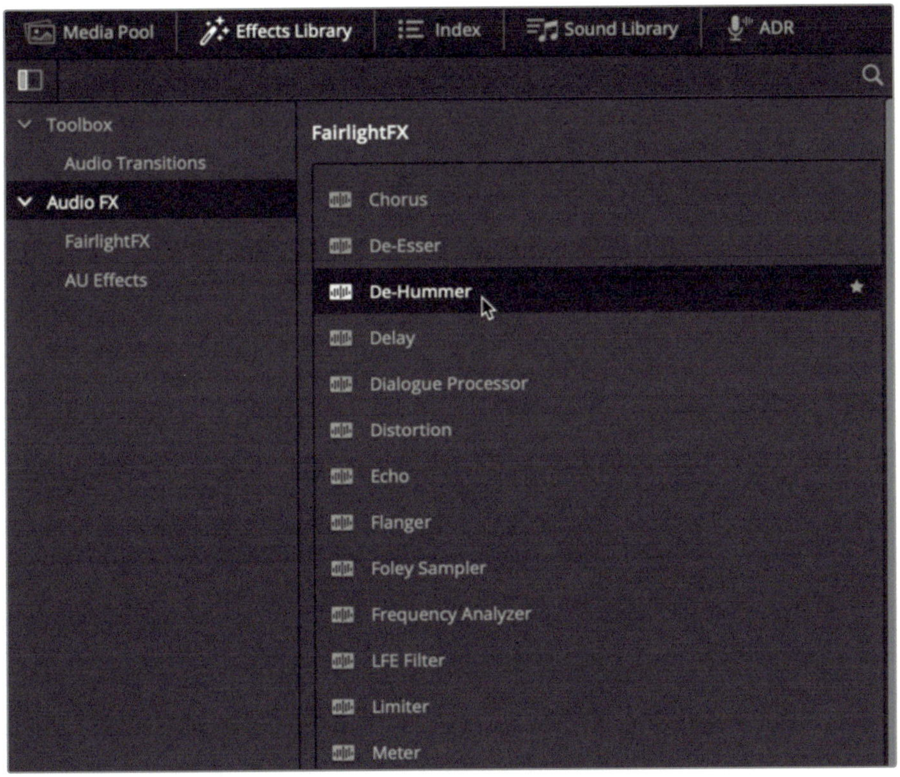

The FairlightFX plug-ins appear at the top of the Audio FX list.

2 In the FairlightFX list, drag the De-Hummer plug-in onto the first clip in the A1 track.

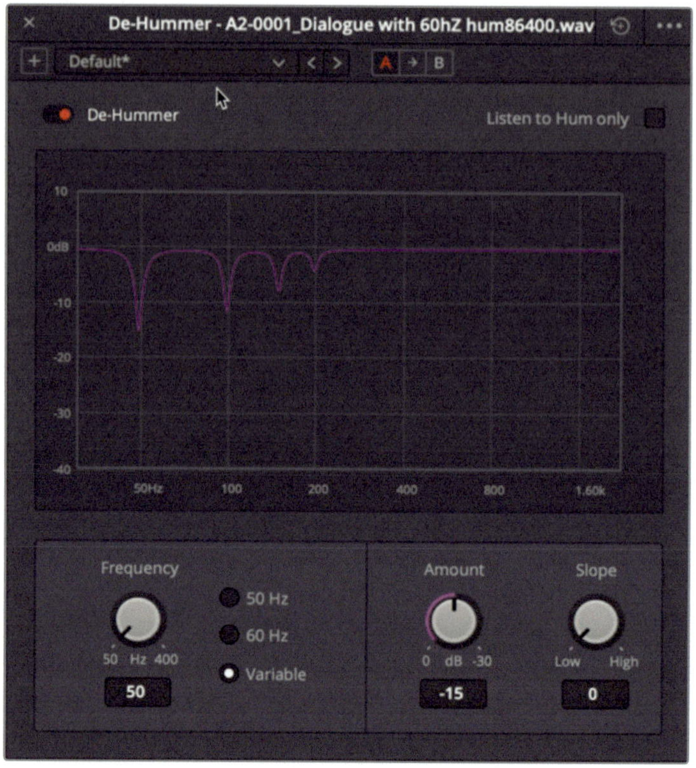

The De-Hummer dialog box opens with controls to specify the type of hum in the clip. You don't have to be a hum expert to determine the frequency. Chances are, it's a powerline hum. If so, it will have a 60-Hz frequency in America or 50 Hz in other parts of the world.

The graph display across the bottom displays low frequencies to the left and high frequencies to the right. The purple notches in the graph indicate which frequencies are currently reduced. The default De-Hummer setting is for a fundamental frequency of 50 Hz followed by harmonic frequencies. Harmonics are naturally occurring multiples of the fundamental frequency, often called *overtones*. These harmonic frequencies give sound it's added color and character. However, when it comes to annoying hum contaminating your dialogue, you'll probably want to eliminate the fundamental and the harmonic frequencies. Luckily, the FairlightFX De-Hummer does most of the work for you.

Notice that the controls in the lower left of the De-Hummer include presets for 50 Hz and 60 Hz.

3 In the Frequency section, click the 60 Hz button to set the De-Hummer to notch out the 60 Hz frequency.

The notch filters (purple notches) shift toward the right so that the deepest notch (fundamental) is now 60 Hz, and the ensuing harmonics have shifted accordingly.

NOTE When you're working with a hum that is not at 50 Hz or 60 Hz, you can click the Variable setting and adjust the Frequency dial until you reduce the correct frequency.

4 Mark the clip, and start looped playback.

5 While listening to playback, drag the Amount dial to increase the amount of the frequency that is removed. Drag until you no longer hear the hum.

You can control the slope of the harmonics, also referred to as the *bias*, with the Slope dial. Depending on the dominant overtones, you may need to use the Slope dial to shift the slope toward higher frequencies. You can also tweak the slope to remove the harmonics while retaining the fundamental frequency.

6 Drag the Slope dial slightly toward the right to strengthen the different harmonic notch filters, and until you hear the least amount of hum during playback.

Each FairlightFX plug-in includes a Bypass switch to turn a plug-in off or on. Bypass switches are a quick way to hear a clip with and without the plug-in applied. Let's use the Bypass switch to toggle off and on the plug-in to hear the before and after versions of the clip. The Bypass switch is red when the filter is turned on.

Plug-in on

Plug-in off

7 In the upper-left corner of the De-Hummer dialog, click the red Bypass switch to turn the plug-in off. Continue playback, and then turn the plug-in on.

8 When you are finished removing the hum, stop playback. Close the De-Hummer dialog.

Now that you've seen and heard the power of a single FairlightFX De-Hummer plug-in, let's try a more challenging clip.

Applying Multiple Plug-Ins for Cumulative Results

So far, the De-Hummer plug-in has worked great for eliminating a moderate hum in the dialogue clip. However, sometimes you'll encounter much more prominent issues that require multiple plug-ins to repair. In this exercise, you'll work with the same clip, but this time the hum is louder.

1 Select the second clip in the A1 track, and press D to enable the clip.

At a glance, you can clearly see that something is wrong with this waveform. Any time you see a really thick center in a waveform, it indicates a constant and fairly loud sound.

> **TIP** If you ever encounter a noisy-looking clip like this in your own projects, protect your ears before listening. In other words, DIM playback or move your headphones from directly over your ears before you listen.

2 Listen to a few seconds of the clip to hear the excessive hum.

3 In the FairlightFX list, drag the De-Hummer plug-in onto the second clip in the A1 track.

4 In the Frequency section, click the 60 Hz button to set the De-Hummer to notch out the 60 Hz frequency.

5 Mark the clip and start looped playback. While listening to playback, drag the Amount dial to increase the amount of the frequency that is removed. Stop playback.

Unfortunately, increasing the Amount dial to the full -30 is not enough.

6 Drag another De-Hummer plug-in onto the clip. The dialog for the second plug-in opens above the first dialog.

If you look closely at the top of the De-Hummer dialog, you'll see (2 of 2) after the clip name in the dialog header, indicating that the dialog is for the second of two plug-ins in order of processing.

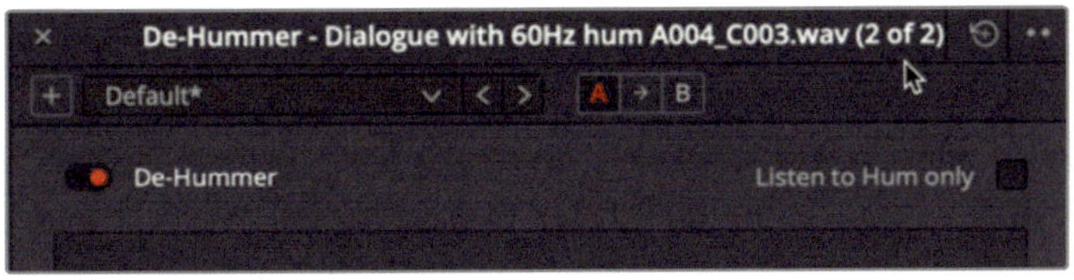

7 Drag the header at the top of the (2 of 2) De-Hummer plug-in toward the right to move it away from the first De-Hummer dialog. Position the plug-in dialogs so that you can easily see them both during playback.

8 On the second De-Hummer dialog, click the 60 Hz button.

9 Start looped playback. Lower the amount on the first De-Hummer to -20.

10 Slowly increase the Amount on the second dialog until the hum is completely gone. Around -20 should do the trick.

11 When you are satisfied, close both dialogs and stop playback.

12 Mute the A1 track.

Now you have first-hand experience working with Resolve's powerful De-Hummer plug-in. Keep in mind this was an extreme example created for this exercise. In most cases, a hum will be less pronounced and easier to eliminate. The takeaway from this exercise is that this plug-in is highly effective and easy to apply to your timeline clips as needed.

Reducing Noise

Noise comes in many shapes, sizes, and intensities. Whether it's music, background conversation, the roar of a sports car or purr of a kitten, anything may be considered noise when it contaminates the clarity of your dialogue tracks.

When applying noise reduction, you can sometimes reduce the noise easily, and sometimes you need to reduce it incrementally with multiple plug-ins and processes. In this exercise, you'll encounter a more tenacious noise than the simple hum you removed in the previous exercise. This time, you'll use a new plug-in with the goal of achieving noise *reduction*. Why not noise *removal*? Because in many cases, noise covers a lot of territory and frequencies and attempting to remove it entirely may strip your dialogue of its tone and texture.

Let's start by listening to the clip in the A2 track to determine the type and intensity of the noise.

1 Select the clip in the A2 track, and press D to enable the clip.

2 Mark the clip, and start looped playback.

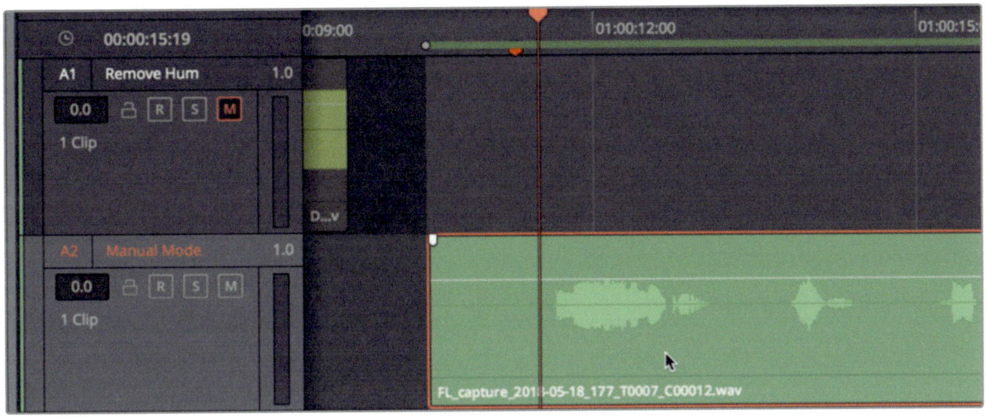

This clip is one of the outtakes from the voiceover recordings in Lesson 2. The noise you hear is very common and can be caused by poor connections, damaged cables, or weak microphone batteries.

Regardless of how the noise "invaded" this recording, it's there. So, you'll deal with it using the FairlightFX Noise Reduction plug-in in the Effects Library.

3 Stop playback.

4 In the FairlightFX list, drag the Noise Reduction plug-in onto the clip in the A2 track.

The Noise Reduction dialog opens with a variety of controls including a graph that shows the spectral analysis of the frequency (Hz) and the strength (dB) of the signal during playback. Detection, Smoothing, and Output controls are also present across the bottom of the dialog.

The FairlightFX Noise Reduction plug-in is based on *spectral subtraction* and can be set to automatically detect noise in selected sections of dialogue or to manually learn a selected section of noise and then extract that noise print from the signal.

The two mode buttons below the graph are for selecting Manual mode or Auto Speech mode. Manual mode learns a "noise print" and extracts it from the overall signal. Auto Speech mode uses powerful algorithms and spectral analysis to detect and extract speech from the signal, thereby reducing the noise.

Let's try both methods, starting with the default Manual mode. First, you need to set a play range to isolate a section of the noise.

5 Mark a play range from the beginning of the clip to just before the first speech waveform. Also, since the noise level is pretty low in this clip, raise the volume level of the clip until you can clearly see the noise waveform on the clip.

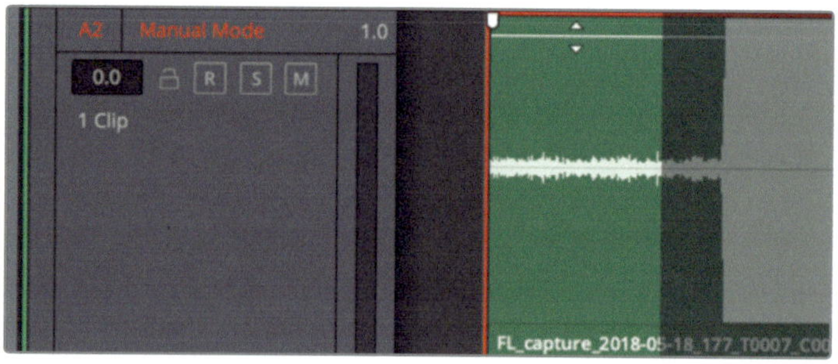

6 Click the Learn button to create a noise profile during playback. Start playback of the noise-only portion of the clip.

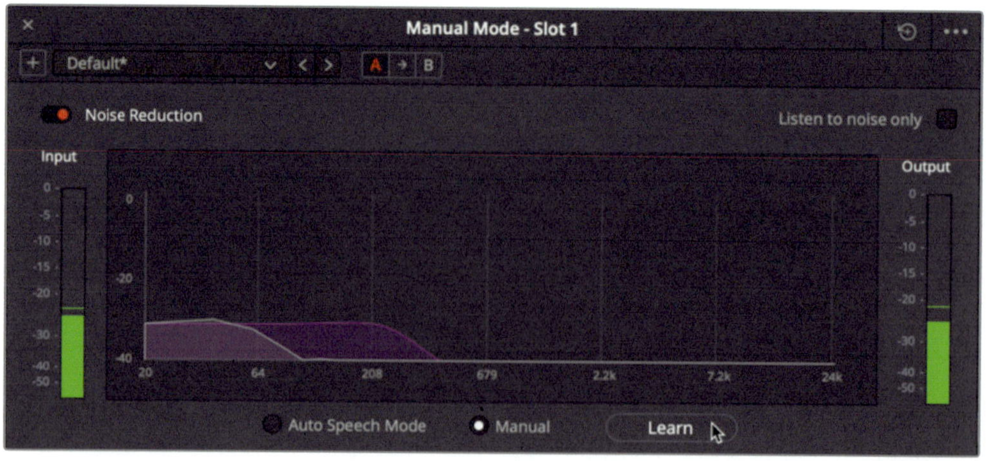

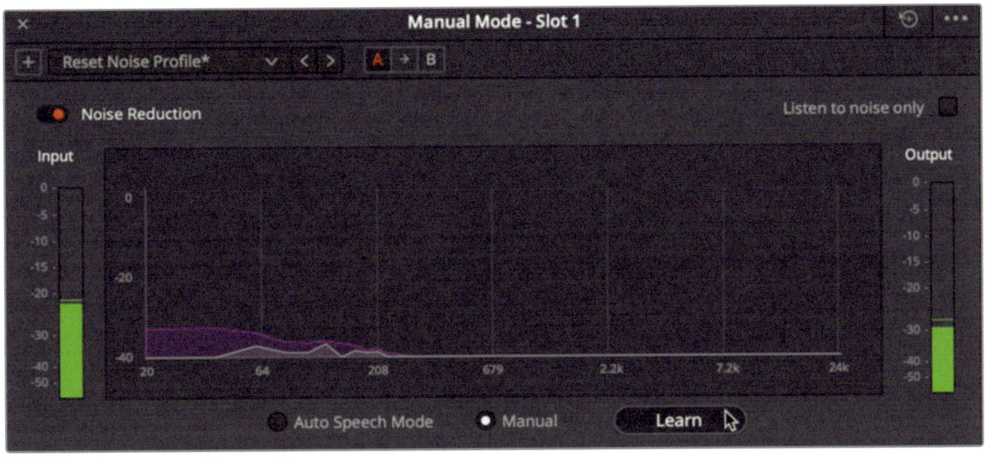

In Manual mode, the purple overlay in the spectral analysis shows the noise profile that is targeted for subtraction from the signal. The noise profile continues to update during playback until you click the Learn button again.

7 Click the Learn button again to save the current noise profile.

The learned noise profile is subtracted from the rest of the clip's signal during playback.

8 Stop looped playback. Lower the volume level on the clip until the waveform height matches the clips below. Extend the play range to include the entire clip.

9 Start looped playback again to hear the clip with the noise print subtracted. Click the Bypass button off and on to hear the clip without and with the Noise Reduction plug-in applied. When you are finished, stop playback, and close the Noise Reduction dialog.

Just like that. The noise is gone. Let's try it again in Auto Speech mode on the clip in track A3.

NOTE You can click the Reset button in the upper-right corner of the plug-in dialog to return a plug-in to its default settings. You can also reset just the noise profile in the Noise Reduction preset menu. For this example, you'll use a separate clip for comparison.

10 Mute the A2 track, and select the clip in the A3 Auto Speech Mode track. Press D.

11 In the Effects Library, drag the Noise Reduction plug-in onto the clip in the A3 track.

12 In the Noise Reduction dialog, click the Auto Speech mode control. Start looped playback.

Once again, the noise disappears while the spoken words remain. This time, in the fifth dialogue phrase, you'll notice a slight audible dip in the word "discrepancy." You can use the Detection controls to fine-tune the noise reduction. In this case, let's increase the Attack time from +1.4 milliseconds to +170.7 milliseconds.

The Attack control is used primarily in Auto Speech mode and controls the duration of the detection time of the noise profile. A slower attack time (lower value) updates the noise profile more quickly and is best applied to noise that varies quickly. A faster attack time (higher value) updates the noise profile more slowly and is better for a consistent noise, as in this clip.

13 Set a play range around the fifth phrase in the clip. Start playback of the phrase.

14 Drag the Attack dial to the right to increase the attack speed to around +170.7 milliseconds.

Changing the attack value helped, but the level still dips a little at the end of the word "discrepancy" because the volume level of the end of that word is much lower than the other words in the clip. You can always balance the levels of the individual words using keyframes.

15 Stop playback. Close the Noise Reduction dialog. Clear the play range. Mute the A3 track.

You've witnessed the power of Resolve's Noise Reduction plug-in. In the next exercise, you'll apply multiple plug-ins to the same clip to reduce a more prominent noise.

MORE INFO You can find detailed information on each FairlightFX plug-in and its specific controls in the DaVinci Resolve 16 manual available in the application's Help menu.

Combining Manual and Auto Speech Modes with Multiple Plug-Ins

In the previous examples, the Noise Reduction plug-in has worked well for eliminating a slight noise in the voiceover clip. However, sometimes you'll encounter much more prominent noise issues such as the din of crowd noise or the constant cacophony of traffic. When that happens, rather than tackle all the noise at once at the potential expense of your dialogue, you may need to use multiple plug-ins to reduce noise incrementally. In this exercise, you'll work with the same clip, but this time the noise is much louder.

1 Select and enable the clip in the A4 track.

At a glance, you can see a fairly strong signal between the dialogue phrases. This usually suggests a steady background sound and is rarely a good thing in a dialogue clip.

2 Mark the clip, and start looped playback.

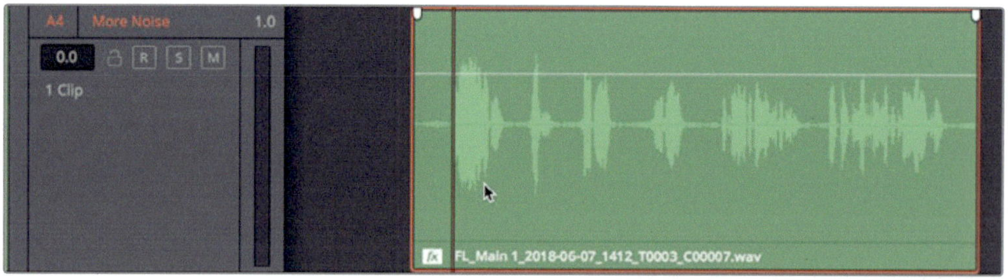

Wait, where is the noise? If you look at the clip in the A4 More Noise track more closely, you'll see an *fx* icon before the clip's name, which indicates that fx plug-ins have already been applied to it.

3 Show the Inspector.

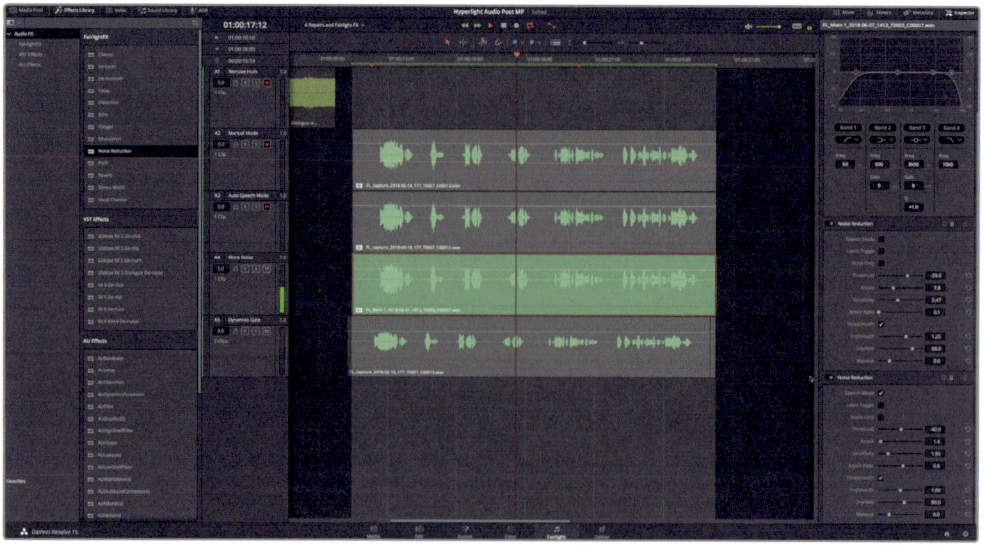

Here, you'll see two Noise Reduction plug-ins applied to the clip. Effects processing works from the top down, so whichever plug-in is highest in the list is applied first. The second plug-in is applied to the results of the first plug-in, and so on.

4 In the Inspector, click the Bypass buttons on both Noise Reduction plug-ins to turn them off. Continue playback.

Now you can hear the noise.

5 Click the controls button for the first Noise Reduction plug-in to show the plug-in's control dialog.

Based on the settings, you can see that this Noise Reduction plug-in is using a noise print in Manual mode to reduce the noise. Also, the Threshold setting is set higher (-29.6) to accommodate a louder noise floor.

6 Turn on the plug-in, and listen to the difference it makes in the noise level.

7 Show the controls dialog for the second plug-in, and turn it on.

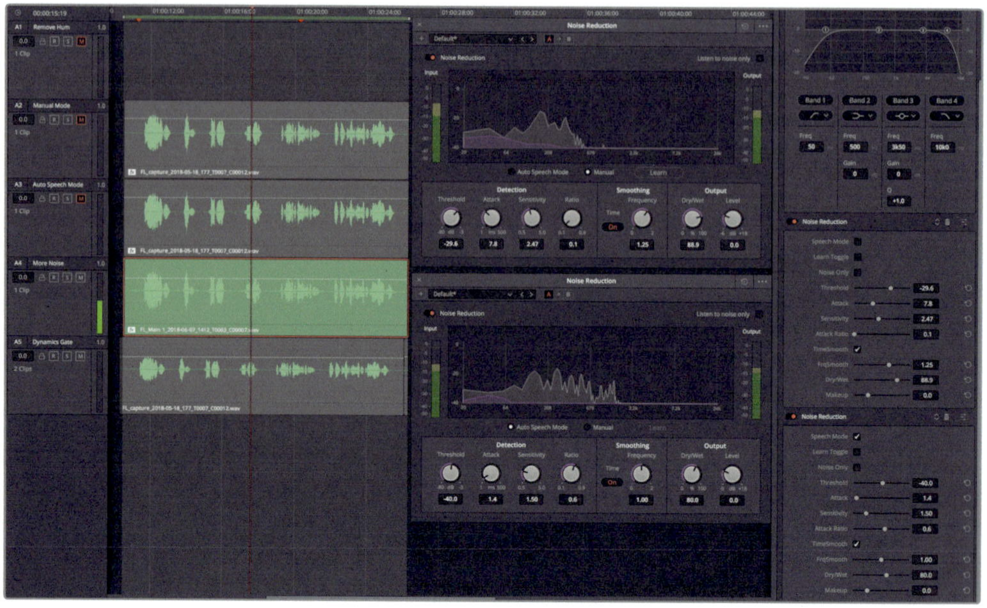

The second plug-in is using Auto Speech mode to clean up the remaining noise left by the first plug-in. Looking at the plug-in parameters in the Inspector, you'll see a Noise Only checkbox that corresponds to the "Listen to noise only" checkbox in the upper-right corner of the Noise Reduction window. This mode is like inverting a key when you are compositing an image, but applied to noise reduction. It lets you monitor the part of the signal that is being removed. Let's listen to the first Noise Reduction plug-in in Noise Only mode.

8 In the Inspector, turn off the second Noise Reduction plug-in.

9 Start playback. In the Inspector, select the Noise Only checkbox for the first Noise Reduction plug-in.

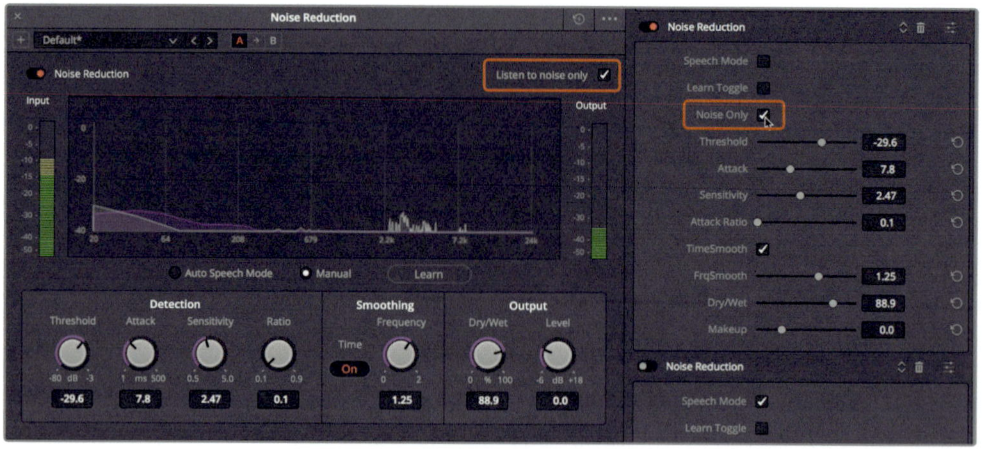

The corresponding "Listen to noise only" checkbox on the Noise Reduction window is automatically selected.

You now hear only the part of the signal that is being removed, which includes some of the dialogue frequencies. If you can clearly make out vocal patterns or words in Noise Only mode, those same frequencies and tonal elements are being stripped from the dialogue. When finessing noise reduction controls, it's always a good idea to listen to noise reduction in Noise Only mode to make sure you aren't excessively diminishing the vocals. Remember, it may be better to stack multiple noise reduction filters instead of being too heavy-handed with just one.

10 Deselect the "Listen to noise only" checkbox on the Noise Reduction window. Turn on the second noise reduction plug-in.

11 Close both plug-in dialogs, and stop playback.

12 Mute the A4 track.

The more you work with noise reduction, the easier it gets. The goal in this lesson is to understand which repair plug-ins are available in DaVinci Resolve 16, whether they work alone or in combination in Manual or Auto Speech modes.

Next, you'll work with the last repair plug-in to deal with another very common dialogue problem.

What Does a Dialogue Mixer Do?

In audio post-production, a dialogue mixer focuses on delivering great-sounding dialogue tracks at optimal levels to add to the rest of the mix. In doing so, dialogue mixers take the edited dialogue tracks to the next level by introducing compression to narrow the dynamic range, as well as attenuate (reduce) detracting frequencies while boosting frequencies that enhance the clarity of each voice.

Identifying Sibilance

Two of the most common issues that occur in dialogue recordings are plosives and excessive sibilance. You already know how to spot and tame plosives using keyframes. Vocal sibilance, on the other hand, is generated by breathy, consonant sounds such as S, T, Z, Sh, and Ch, and results in an unpleasant tonal harshness in the upper midrange frequencies between 4 kHz and 8 kHz.

The type of microphone used for recording and the proximity of the microphone to the subject contribute to sibilance, as well as the amount of air movement and power required to produce sibilant sounds. Phrases like "excessive esses" and "she sells sea shells by the sea shore" are sure to increase the vocal sibilance levels in nearly any recording.

Some spirited debate is ongoing among dialogue mixers regarding when the best time is to de-ess dialogue. Some believe it should be after compression and during the mixing stage. Others believe all equalization ought to be done before compression, and so the debate continues. While dialogue mixers may disagree on the order and methods of processing and equalizing tracks, they have the same singular goal in mind: to deliver the best-sounding dialogue tracks.

In the next few exercises, you'll cross the boundary between dialogue editing and dialogue mixing to explore sibilance and the ever-present esses. Let's start by listening to some sibilant-rich dialogue.

NOTE If you didn't complete all the previous noise reduction exercises, open the timeline, "6a Reduced Noise," to catch up and continue.

1 Unmute the A2 track. Make sure that all of the other tracks are muted.
2 Mark a play range around the fifth phrase in the A2 track, "I cannot explain this discrepancy."
3 Play the phrase several times to determine if you can hear the sibilance.

If you are new to dialogue editing and mixing, the first thing you need to do is recognize excessive sibilance. Once you know what you're listening for, you'll notice it everywhere, and more importantly, you'll know how to reduce it in your own projects.

The best way to pinpoint irritating frequencies is to boost and sweep a bell curve to find them. Let's use the 4-band parametric equalizer in the Inspector to find and increase the sibilant range to more easily identify it.

4 Select the clip in the A2 track, if necessary, to show it in the Inspector.
5 In the Inspector, turn on the Clip Equalizer.

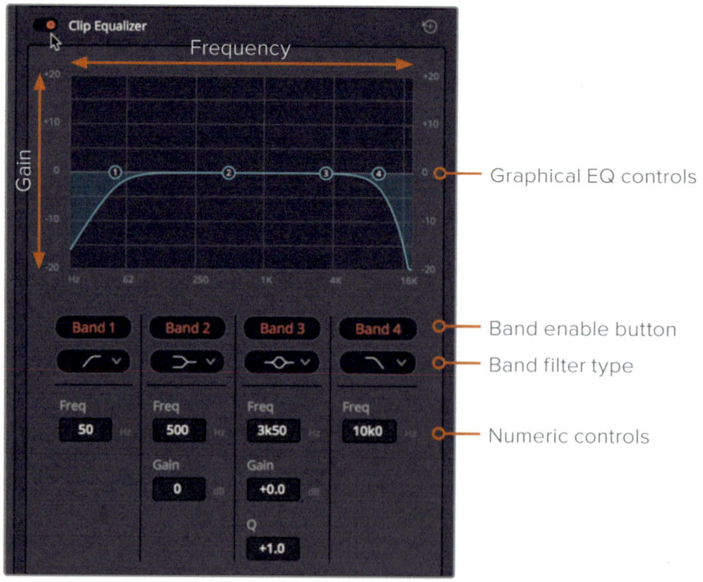

In the graphical EQ controls area, you can see and manipulate the frequency graph. The numerical controls at the bottom of the equalizer give you more precise control over the frequencies and gain of each band.

If you look more closely at the blue graph, you can see that the default setting resembles a clip with fade handles applied to the head and tail. The steep slope at the left edge of the curve is actually a high-pass filter that allows high frequencies to be heard while eliminating the lowest frequencies. The slope on the right side of the curve is a low-pass filter that allows lower frequencies to be heard while eliminating the highest frequencies. (In the next exercise, you'll apply both the high- and low-pass filters to alter the recorded voice in a clip.)

For now, let's use the Band 3 control, which is in the same frequency range as the sibilance.

NOTE You'll work more extensively with equalization in Lesson 8.

6 Click the band filter type pop-up menu below the Band 3 button to see the types of frequency filters available for bands 2 and 3.

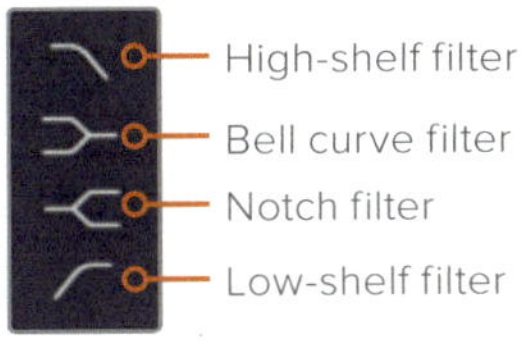

High- and low-shelf filters are similar to high- and low-pass filters, but the shelf filters attenuate (lower) unwanted frequencies, whereas the pass filters completely block unwanted frequencies. You can apply bell-curve filters to boost or attenuate frequencies anywhere on the graph and narrow or widen their range of influence using the Q control. A notch filter is used for completely removing, or cutting, specific frequencies.

For this exercise, you'll sweep a midrange frequency bell curve in Band 3 to boost or attenuate frequencies during playback and isolate the talent's voice. Let's start by turning off all the bands except for Band 3. Then, you'll sweep the bell curve to evaluate the radio dispatcher's recorded voice.

NOTE During playback, dragging the graphical controls left or right through the EQ graph is referred to as sweeping, and is the most effective way to hear changes and identify areas that need work.

7 In the Clip Equalizer controls, click the red band enable buttons for Bands 1, 2, and 4 to turn them off.

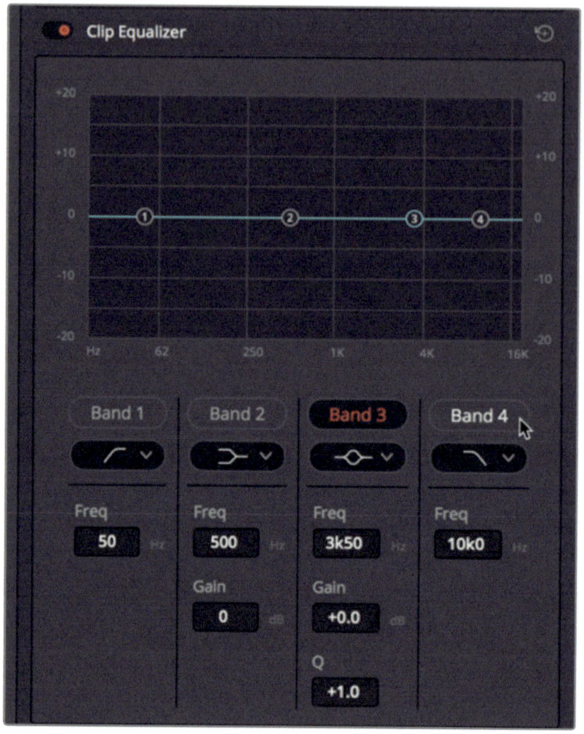

NOTE Clip equalization in the Fairlight page adjusts the tonal quality of individual clips, whereas track equalization in the Mixer adjusts the entire track.

8 Start looped playback. During looped playback, drag the Band 3 bell curve handle up, and sweep left and right between 4K and 8K. Listen for the unpleasant range where the esses are most irritating (between 7K and 8K). Continue playback, and drag the bell curve down in the same position to reduce that frequency range.

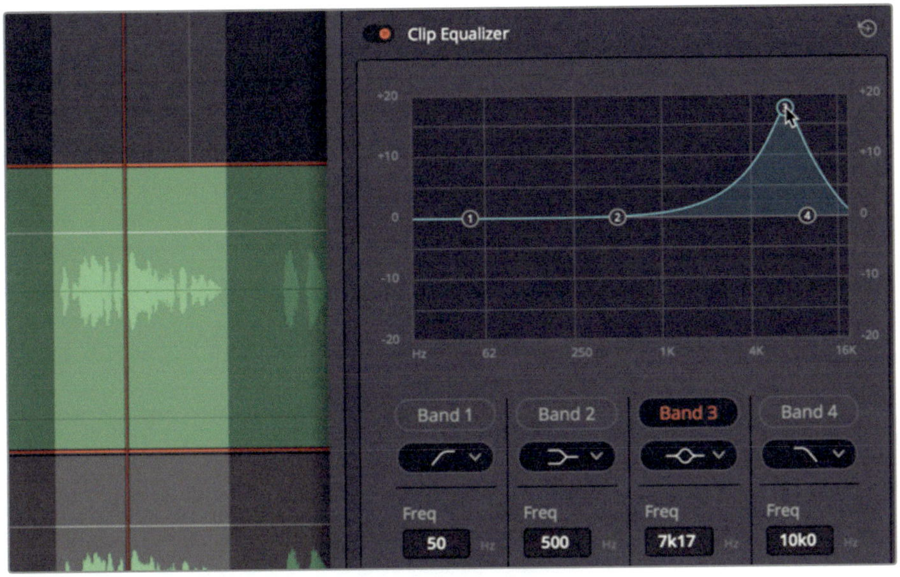

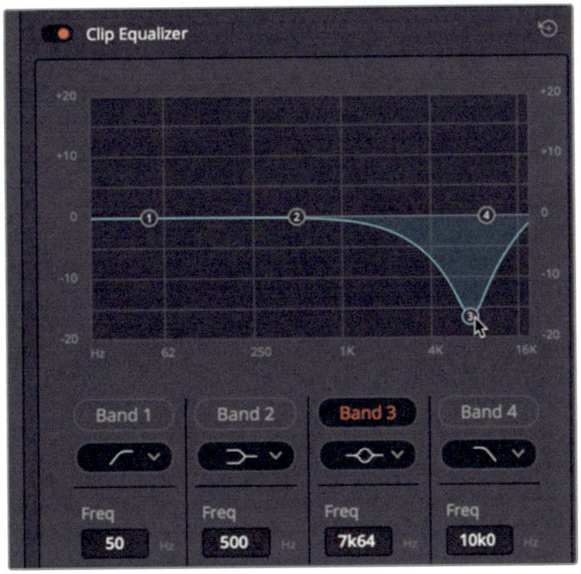

The esses are greatly diminished, but so is the quality and tone of the voice. One of the secrets to working with equalization is to make changes in small increments and use multiple processing effects in combination.

9 When you are finished, stop playback, and turn off the Clip Equalizer.

Now that you are familiar with the sound of excessive sibilance and where to find it in this section of the dialogue, let's use the De-Esser FairlightFX plug-in to reduce it.

Reducing Sibilance with the De-Esser

The FairlightFX De-Esser is a specialized EQ and compressor plug-in that uses a narrow frequency range to control vocal sibilance.

1 In the Effects Library, drag the De-Esser plug-in onto the clip in the A2 track.

The De-Esser dialog includes a graph showing which part of the signal the controls are set up to adjust. A set of meters show reduction and output, so you can see a side-by-side comparison that indicates which parts of the signal are affected, how much the targeted signal is reduced, and what level is being output.

2 Start looped playback. Drag the handle down as far as it goes, and then sweep the notch filter back and forth between 6K and 8K.

3 Under the Frequency Range controls, click the bell-curve filter button to reduce a wider range of frequencies.

Once again, as with the 4-Band EQ, too much of the vocal tone and character is diminished along with the esses. Remember, less is more. Only reduce what is necessary. Later, you can increase the amount of filtering or add another De-Esser in the effects chain to improve the results. You'll know your De-Esser plug-in is set too high if you hear a dialogue lisp in which the T and S sounds are no longer intelligible.

4 Under the range controls, click the Notch button to return to the notched frequency range. Slowly raise the handle toward the 0 dB line to decrease the amount of reduction to around -10 dB.

Ideally, back off the filtering as much as you can to reduce the sibilance while retaining the vocal quality.

Notice that no matter how much you reduce the targeted frequencies, the output level remains the same. Remember, de-essers are actually specialized compressors—and in this case, the built-in compression under the hood ensures that your output volume remains intact.

5 Click the Bypass button to toggle the plug-in off and on and audition the before and after versions of the dialogue phrase with the De-Esser plug-in applied.

You may think this is such a minor improvement, why bother? It's for the same reason that you fix the flesh tones during color correction. The results may seem minor in this example, but sibilance is cumulative and becomes more pronounced the farther along you go in dialogue mixing.

This clip has two working plug-ins: the Noise Reduction and De-Esser. They should work well for the entire clip.

6 Close the De-Esser dialog. Set a play range for the entire clip, and start playback to hear the clip with both the Noise Reduction and De-Esser plug-ins applied. When you are finished, stop playback.

7 Mute the A2 track. Hide the Inspector and the Effects Library.

As you can see and hear, the FairlightFX repair plug-ins work well independently or in combination to improve the sound of your dialogue.

What Is Dynamic Range in Audio Post-Production?

When you are working with audio, the dynamic range is the difference between the loudest and quietest levels in a signal. A track's dynamic range is very similar to visual contrast within a shot. A dialogue track with a high dynamic range has very loud and very quiet elements within the track, such as a character whispering and then screaming in the same scene. A dialogue track with a low dynamic range is fairly flat, such as in a commercial voiceover in which the volume level of the talent is even from start to finish.

Dynamic range also comes into consideration when performing noise reduction because the range between the noise level and the dialogue levels can affect how much noise reduction you can apply. The signal-to-noise ratio (SNR) is a mathematical calculation based on the dynamic range between the noise level and the dialogue level, and is expressed in decibels (dB). For example, if your voiceover track has an SNR value of 50 dB, it is fifty times louder than the noise in the same track. The higher the SNR value, the better sounding the signal and less prominent the noise.

Gating Low-Level Noise

Now that you have crossed into the realm of using mixing tools to improve dialogue, let's look at another common mixing tool that you can use to reduce or eliminate low-level background noise.

1. Select the A5 Dynamics Gate track, and enable the clip in the A5 Dynamics Gate track.
2. Mark the clip .
3. Show the Mixer.
4. In the A5 channel strip, double-click the Dynamics controls to open the Dynamics controls panel.

The Fairlight Mixer includes the four most common dynamics processors in one easy-to-use panel. The compressor narrows the dynamic range by lowering the loudest peaks and bringing them closer to the lowest peaks. The expander, in contrast, enlarges the dynamic range to increase the difference between the loudest and quietest peaks. The limiter and gate work on opposite ends of the signal to avoid exceeding a target level (limiter) and to prevent sounds lower than a set threshold from being heard (gate).

Let's focus on using the gate to reduce or silence the part of the signal that is below a given threshold.

Let's turn on the gate while listening to the clip.

5 Start looped playback to hear the clip with low-level noise.

6 Click the Gate button to turn on the gate dynamics processing.

The noise is no longer audible, and the voiceover sounds pretty good. The dialogue sounds processed in a few places, but it's a start. So, what happened?

In the controls beneath the Gate button, you'll see that the Threshold is set to the default level of -35 dB, which means the gain (volume) of any signal that is below -35 dB is reduced.

7 Watch the Gain Reduction meter during playback. You'll see that the Gain Reduction is active between dialogue phrases whenever there is no talking and the signal falls below the threshold.

The Range control sets the maximum amount of gain reduction that is applied to the signal when it falls below the gate threshold.

8 Drag the Range control to the right to increase the amount of reduction by the highest amount, 60 dB.

The graph and Gain Reduction meter show that the signal below the Threshold (-35 dB) is receiving the highest amount of gain reduction (-60 dB). That's overkill for this signal.

9 Double-click the Range control to reset it to the default level.

Gating your low-level signal is very effective. However, it's best to set the Threshold as close to the noise level as possible. Otherwise, you will also remove all the breath sounds and syllables that taper off below the Threshold.

TIP When adjusting audio controls, it's often easier to start with the controls set to their extreme levels, either the high or low, and then slowly back them off or add them during playback until you achieve your goal.

10 During looped playback, adjust the Threshold level to lower it all the way to the left (-50.0 dB). You'll hear the background noise again. Then, slowly raise the Threshold until you no longer hear the noise.

11 Drag the Range control all the way to the left, and slowly increase the Range value until you no longer hear the background noise during playback.

12 When you're finished, stop playback.

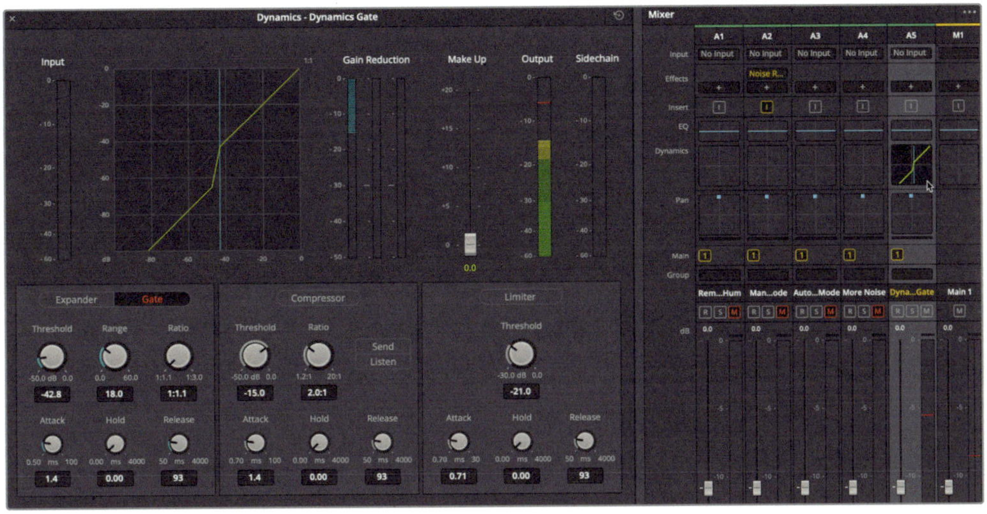

Notice that the graph in the channel strip matches the graph on the Dynamics panel.

13 Close the Dynamics Panel, and hide the Mixer.

All these methods for removing background noise are useful and get easier to use the more you work with them. Keep in mind, that you can apply the FairlightFX plug-ins individually to clips or combine up to six per track. While gate processing is available on the track via the Mixer, sometimes you need to combine the noise reduction and gate for the best results.

TIP When a dialogue clip is contaminated with both hum and noise, remove the hum first so the noise reduction plug-in can home in on the noise, which is likely a different frequency range than the hum.

Testing Your New Dialogue Repair Skills in a Real-World Clip

Before you move on to editing waveforms in the next section, let's take five minutes to put your new audio repair skills to the test.

Your goal in this exercise is to use your ears and knowledge to quickly diagnose the issues in a clip's audio tracks. Then determine the most effective and efficient way to repair the sound using tools that you've already learned. For this exercise, you'll step away from the Hyperlight project to work with a clip from a real-world cooking show.

To keep things moving, and keep you thinking, there will be some loose guideline steps. Remember, this is also a timed exercise that should be completed in less than five minutes. Let's get started.

1 In the Media Pool, select the Additional Footage bin at the bottom of the Master Bin list. This bin contains only one clip, which is the one you will be repairing shortly.

2 Right-click the A001_07081653_C002_v2.mov clip inside the Additional Footage bin and choose > Create New Timeline Using Selected Clips.

3 In the New Timeline dialog, Name the timeline Cooking Clip Audio, and then click Create,

Since this clip would likely start on the Edit page in editorial, let's do that as well.

4 Go to the Edit page and play the beginning of the clip and use your ears to troubleshoot.

What do you think? The timeline only shows one channel in a stereo track, so you can't use your eyes to troubleshoot. Hopefully, your ears recognize the common audio issues. Chances are you immediately noticed the heavy noise, refrigerator hum, and the fact that Jamie's voice is coming from the right speaker only. Although the repair plug-ins are also available in the Edit page, let's go back to the Fairlight page where you'll have full access to all your new repair tools and workflows.

5 Go to the Fairlight page and expand the track height for a larger view of the audio track.

As expected, it is a stereo clip with only one usable channel of dialogue, and the usable dialogue channel is full of noise and hum. With that information...fix it! The clock is ticking.

If you need a little guidance, feel free to use the bulleted hints below.

- Change the clip's attributes from Stereo to Mono with only the good dialog channel exposed. Then change the track to mono.
- Apply the De-Hummer plug-in to the clip. Drag the Amount dial to -30 and sweep the Frequency dial until the high-pitched hum disappears. If necessary, try clicking the Listen to Hum Only checkbox to hear only the hum until you find the target frequency (approximately 101 Hz). Then back off on the Amount as much as you can while still eliminating the sound (approximatley -20 dB).
- Apply the Noise Reduction plug-in and try either the Auto Speech mode or Manual mode using the Learn button and a range to make a noise print.
- Experiment with different settings in the plug-ins to refine the sound, such as adjusting the Slope in the De-Hummer while listening to Jamie's voice to find a position where the hum is removed and his voice is less impacted by the notch filters (about 0.9)

6 When you are finished, stop playback.

Congratulations! You just completed a real-world dialogue repair job in mere minutes. If you'd like to see an example of the finished clip, open the timeline, "6 Real World Audio Repairs Finished."

Dealing with Processor Intensive Plug-Ins

In DaVinci Resolve, there is no limit to the number of plug-ins that you can add to a clip. However, as your tracks get more complex and plug-ins become more processer intensive you may discover that your computer lacks the processing power to play all your audio tracks and effects in real time. When that happens, there are two ways to alleviate the burden that the processor-intensive audio effects have on your workstation.

The first option is to cache your audio clips with plug-ins, and the second option is to bounce audio clips with plug-ins. The first method "bakes" the plug-in effects in a cache file and updates the waveform accordingly. This method does not create new media, can easily be turned on or off while you work, and the cache file automatically updates every time you change any of the plug-in settings. Bouncing audio clips renders a new piece of media with the plug-in effects baked in as part of the waveform. Let's go to the previous timeline and take a look at both methods.

1 Open the timeline, "6 Repairs and FairlightFX."
2 Unmute the A1 track if necessary.
3 Right-click the first clip in the A1 track and choose Cache Audio Effect.

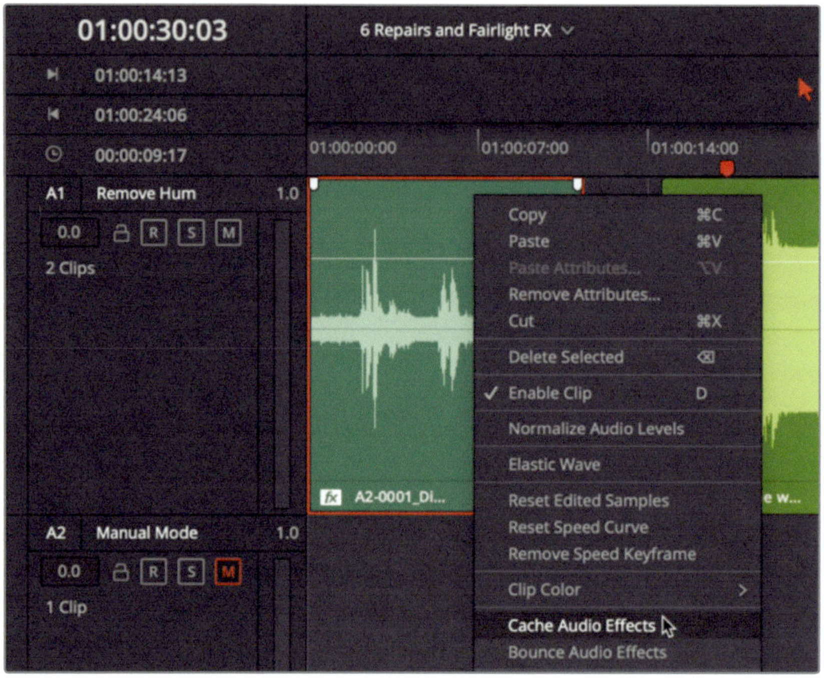

The cached clip's waveform updates, and a cache badge in the lower-left corner lets you know that this audio clip has cached effects.

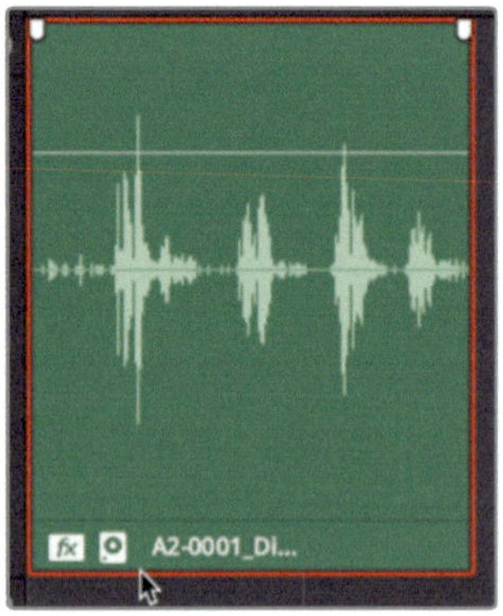

4 Select the cached clip and open the Inspector. Here you will also see the cache badge next to the plug-in for the cached clip.

You can still use the bypass switch and open the plug-in settings from the Inspector to change any of the parameters in the plug-in's settings window. Now let's cache and then bounce the second clip.

5 Right-click the second clip in the A1 track and choose Cache Audio Effect.

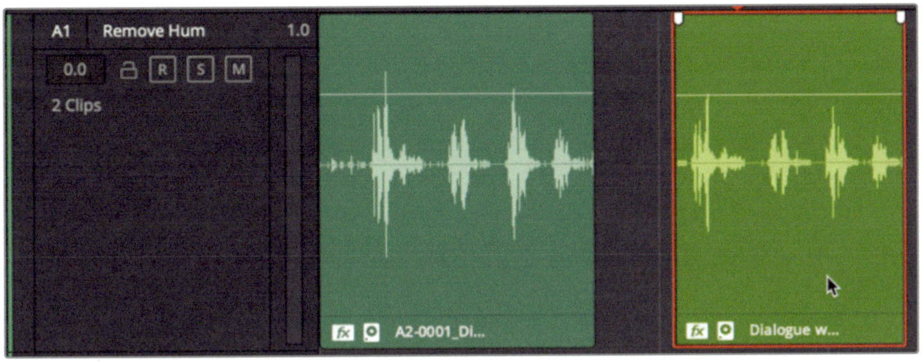

As you can see by the cached waveform, the results are nearly identical to the previous clip.

6 Right-click the second clip in the A1 track and uncheck Cache Audio Effect.

7 Right-click the second clip and choose Bounce Audio Effects. A new rendered clip appears in the timeline. The original clip is still there, just beneath the bounced clip.

8 Choose View > Show Audio Track Layers to see the rendered clip on the top layer and the original clip disabled underneath.

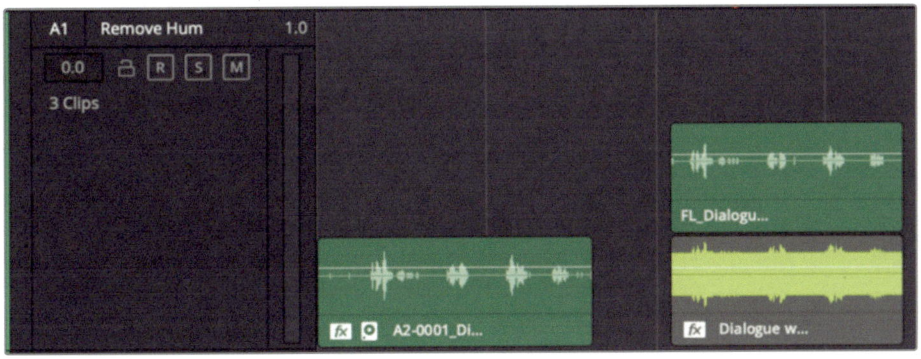

9 Hide the audio track layers.

Now that you've seen how to cache and bounce clips with audio effects, you can apply these processor-saving techniques to your own projects.

Removing Clicks at the Sample Level

So far, you've decreased your biggest noise issues. Now let's explore a common problem that is much smaller in scale but equally significant. Previously, all your work has been performed at the frame level. But you're ready to dive much deeper into the waveform to perform microsurgery at the sample level. In this set of exercises, you'll take a close look at a single frame of audio and then find and mark a digital click that is embedded in a waveform. And to make things even more interesting, this common irritant is located right in the middle of an instrumental track, so no noise reduction or replacement outtakes are available to fix this issue.

Sample-level editing requires a new set of skills and patience. For starters, you need to abandon most of the shortcuts you've been using for navigation. You'll learn the rest as you go along.

Let's start by opening a different timeline containing a single music clip.

1 Open the timeline, "6b Fix Music Cue."

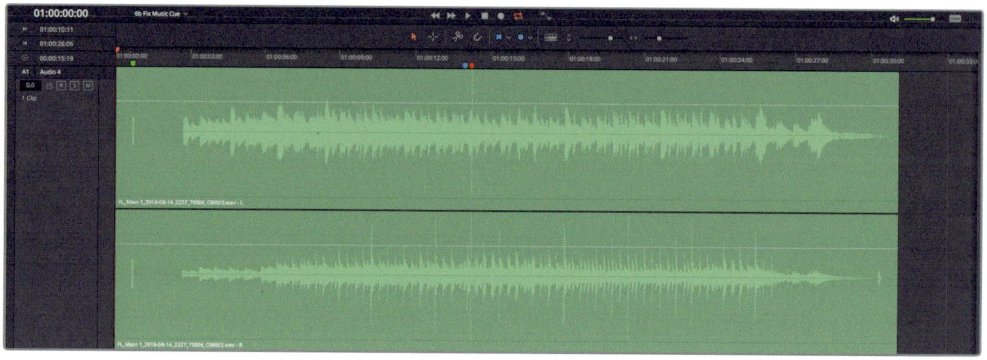

At a glance, the waveform appears to be full of diminutive spikes that may or may not be issues. The even spacing between the spikes in the waveform usually indicate a percussion beat. However, the only way to be sure is to listen to the track.

2 Play the clip. During playback, listen to the music while you watch the playhead move along the waveform.

Did you notice the digital click (glitch) near the middle of the track?

3 Press Shift-Down-Arrow to move the playhead to the blue timeline marker. Play the clip from the blue marker. When you hear the distracting click, stop playback.

Digital clicks and pops can be as aurally irritating as gnats at a picnic. However, before exterminating this digital annoyance, you should look at some of the navigation shortcuts you'll need.

4 Move the playhead to the green marker near the head of the clip.

Below the green marker you'll see a slightly thicker and less prominent spike in the waveform. This is actually the *2-pop*, which is a single frame of 1-kHz tone placed two seconds before the start of picture or sound as an audible sync mark.

5 Zoom in horizontally until you can clearly see the 2-pop's waveform.

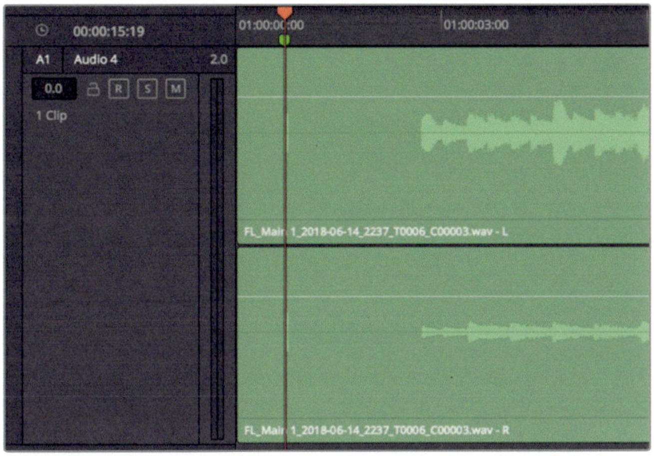

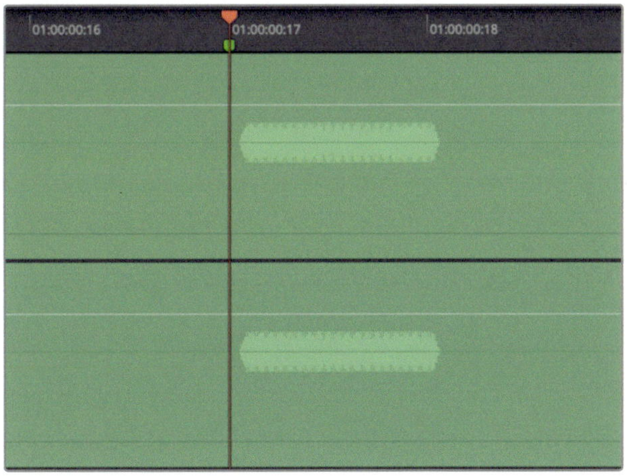

6 Press the Right Arrow once to move one frame forward. Press the Left Arrow to move one frame back.

Even though the 2-pop is only one frame in duration, the playhead still scrubs the clip so you can hear the pop in either direction.

7 Drag the playhead over the middle of the 2-pop, and zoom in closer until the waveform spreads from edge to edge across the timeline.

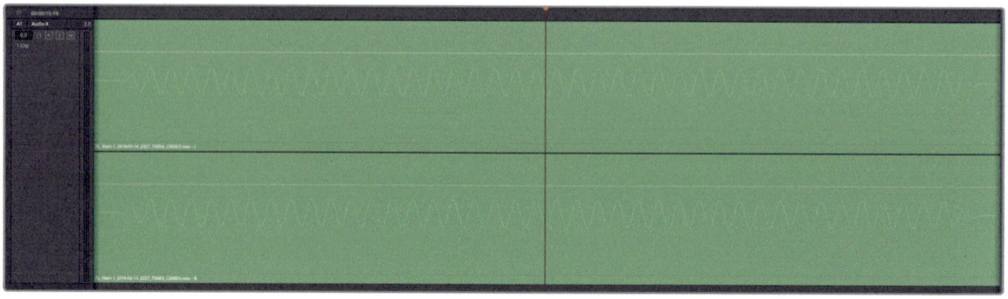

Notice the waveform's pattern within that single frame.

8 Press the Right Arrow and Left Arrow to scrub the single frame again, both forward and backward.

Single-frame navigation is the smallest available navigational method when you're using either the JKL or Left and Right Arrow keys.

9 Drag the playhead over one peak in the waveform. Then, zoom in further until you see individual samples (dots) that make up the waveform. Zoom in horizontally as far as you can to see the individual samples.

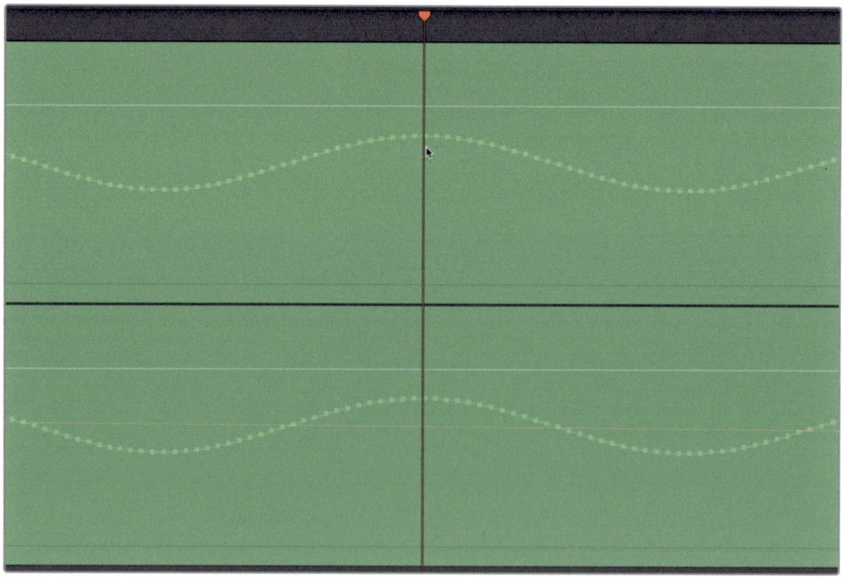

Welcome to the sample level of the waveform. Just as molecules are the building blocks of matter, samples are their equivalent in digital audio.

10 Press Shift-Z to fit the clip to the timeline.

Now that you've seen how to get to the sample level of your clips, it's time to move on to the digital click in the music track.

Isolating the Audible Glitch as a New Clip

Because you are probably new to editing samples in the Fairlight page, it's a good idea to first create a play range and split the clip around the digital issue. Splitting the clip won't actually help you with sample editing, but it makes navigating around your issue much easier.

1 Move the playhead to the red timeline marker.

2 Zoom in so that both the red marker and the click's waveform spike are visible in the timeline.

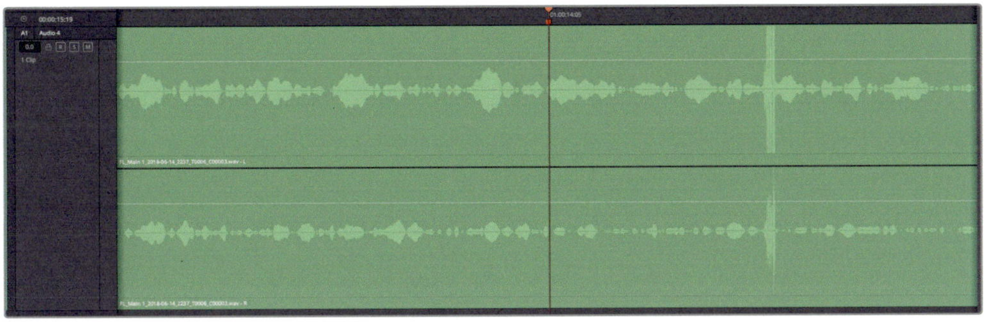

3 Press the Right Arrow key to scrub forward one frame.

The click's waveform is nowhere to be seen when you move one frame to the right.

4 Press the Left Arrow to move the playhead back one frame to the red marker.

5 Drag a play range around the click waveform. Then drag down slightly on the selection to split it at the play range.

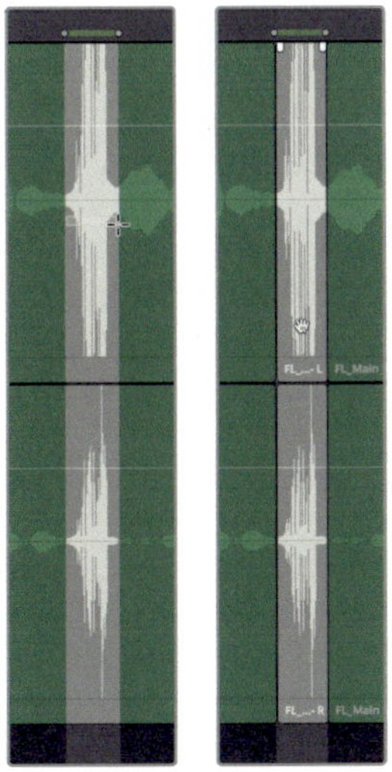

Why create a separate clip for the click? Because doing so enables you to navigate to the head and tail of that separate, diminutive clip using the Up Arrow and Down Arrow keys, just as with any other clip. So, you'll never get lost, even at the sample level.

6. Press A to switch to Selection mode.
7. Press the Up and Down Arrow keys to practice navigating between the head and tail of the new clip.
8. Drag the playhead over the middle of the newly created clip, and zoom in until you clearly see the clip's waveform and samples.

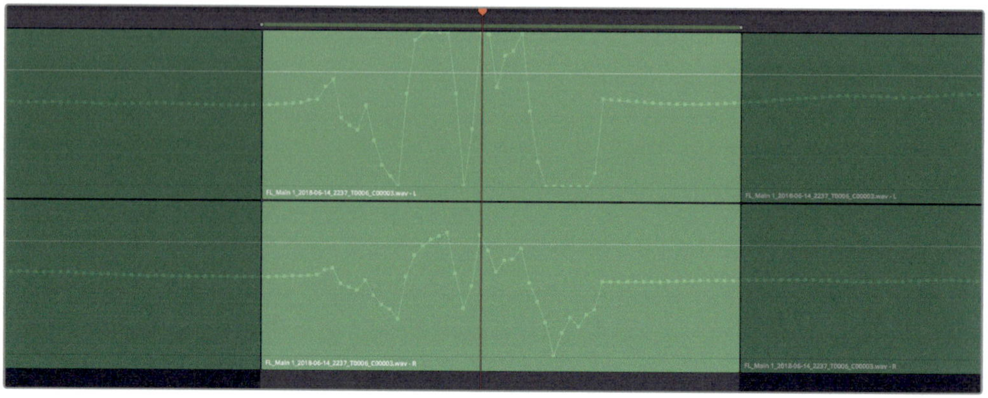

9. Press A to switch to standard selection mode.

   You've isolated the issue and have a clear view of the audible glitch. Next, you'll redraw the waveform to eliminate the errant sound altogether.

Editing Samples to Correct the Waveform

Your goal in this exercise is to redraw the waveform to eliminate the drastic peaks that make up the glitch. You don't want to silence the section by dragging the waveform all the way down because that, too, will be audible. Instead, you'll draw a smooth horizontal line to connect the samples at the left with the samples at the right and continue the existing waveform. Because this music clip is in stereo, you'll need to redraw the waveform on both the left (upper) and right (lower) channels.

1. In the upper channel, click one of the samples prior to the drastic change in the waveform. The sample turns red to indicate that it is selected. You can now drag up, down, right, or left to redraw the waveform.

2 Drag the sample to the right to create a smooth horizontal waveform that connects with the existing samples at the tail of the clip.

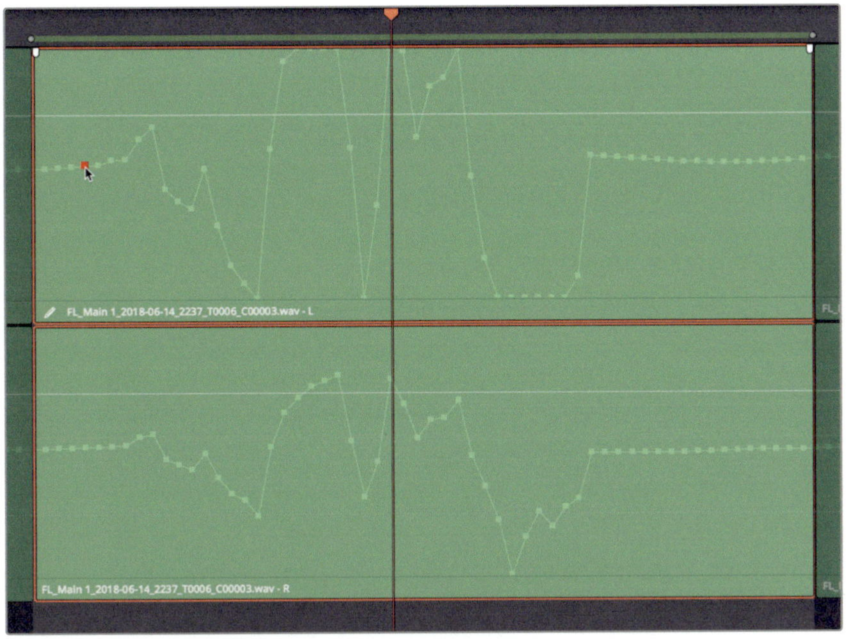

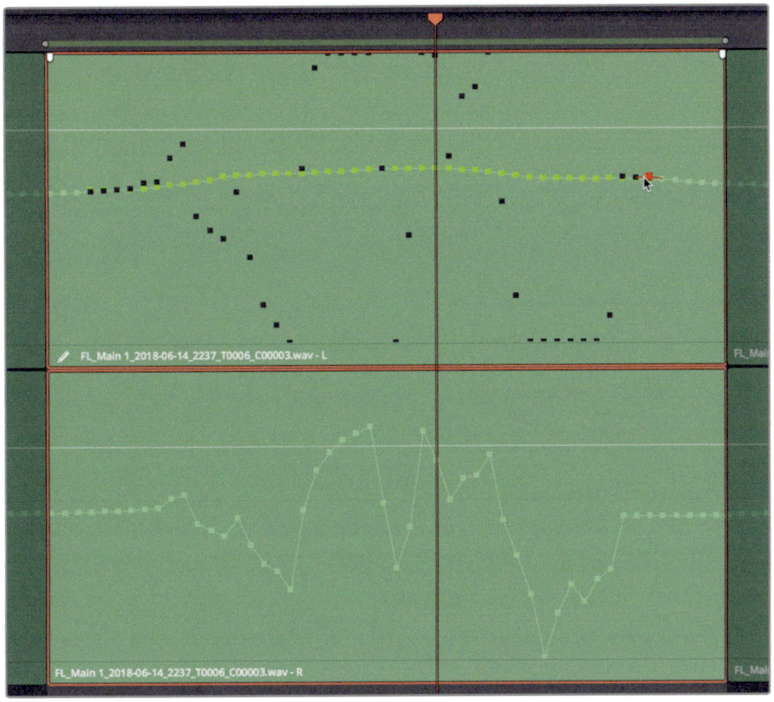

Don't worry if your sample drawing isn't perfect. You can drag back and forth as much as needed to smooth it out. The secret to editing at the sample level is to draw smooth lines (but not necessarily straight lines) that reconnect with the samples on either side of the change.

NOTE To reset a clip's samples, right-click the clip, and choose Reset Edited Samples.

3 Repeat steps 1 and 2 in the lower channel.

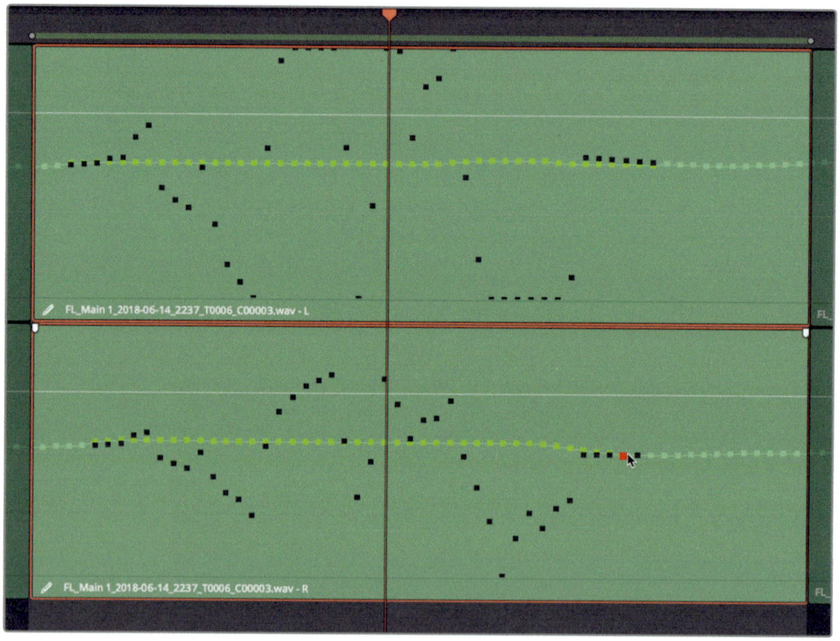

4 When you are finished, clear the play range.

5 Press Shift-Z.

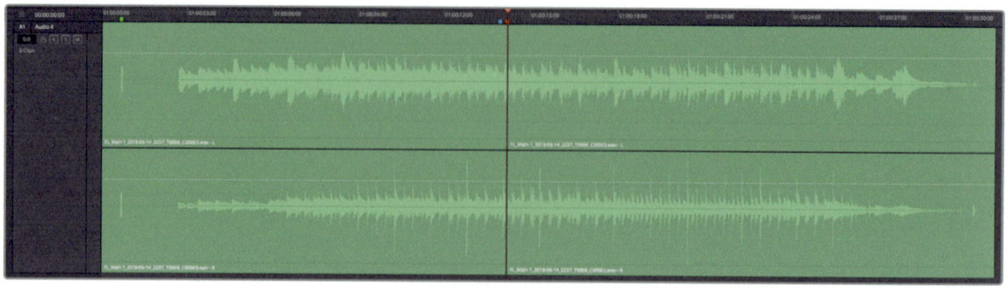

6 Play the clip from the blue marker to hear the repaired glitch. If you hear the edit, go back in and try to smooth it out. You'll know it's right when you can't hear any indication that the waveform was ever edited.

You just performed a seamless "glitch-ectomy" in the middle of a busy instrumental music clip. You can apply this same sample-level technique on dialogue clips to remove noises, or even shave the "s" off of a plural word. When you get used to sample-level editing, you can do it without first splitting the clip. For now, splitting the clip is like using training wheels for sample-editing newbies and can be a time saver for seasoned editors, as well.

Retiming Audio with Elastic Wave Time Controls

The last audio repair tool that you'll work with is Elastic Wave audio retiming, which is a fast and easy keyframe-based way of dynamically retiming audio by condensing or stretching different parts of a waveform without changing pitch. This is commonly used for retiming sound effects, commercial voiceover cues, and matching ADR recordings to the lip movement onscreen.

In this exercise, you'll use Elastic Wave keyframes to retime an ADR recording to match the original dialogue in the timeline.

1 Open the timeline, "6 Elastic Wave Start."

This clip is a bounced composition of the best recorded takes.

2 Choose View > Show Audio Track Layers.

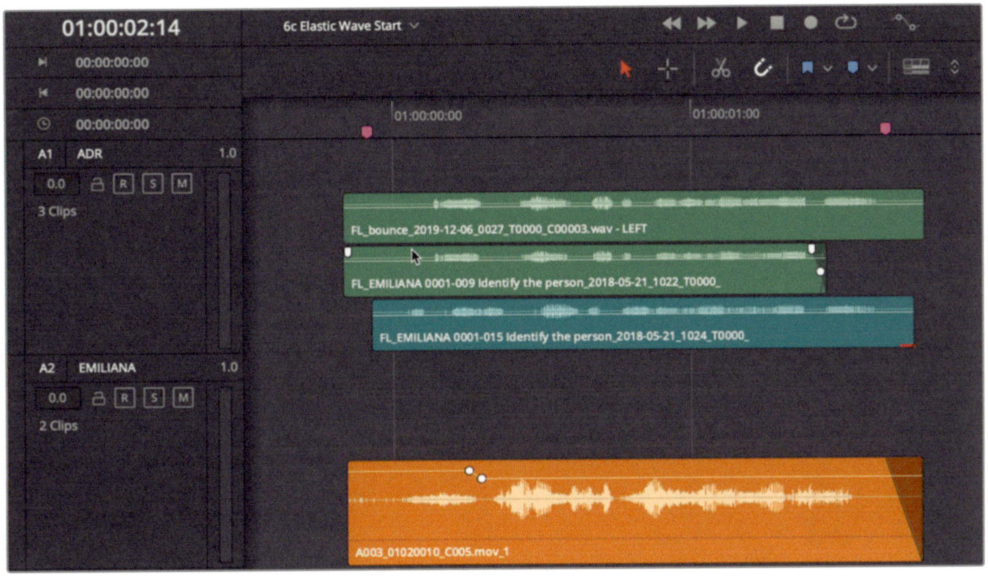

Now you can see two ADR recordings in the lower layers, and the bounced clip in the upper layer.

3 Select both clips in the lower layers and press D to disable them. Then Hide Audio Track Layers.

4 Mark the clip in the A1 track to set a playback range.

5 Solo tracks A1 and A2 and start looped playback. Listen to the clips together to determine how well they match. Stop playback.

Clearly, they don't match very well. Let's use the Elastic Wave keyframes to retime the clip in the A1 track to match the clip in the A2 track.

6 Right-click the clip in the A1 track, and choose Elastic Wave to reveal the Elastic Wave retiming controls.

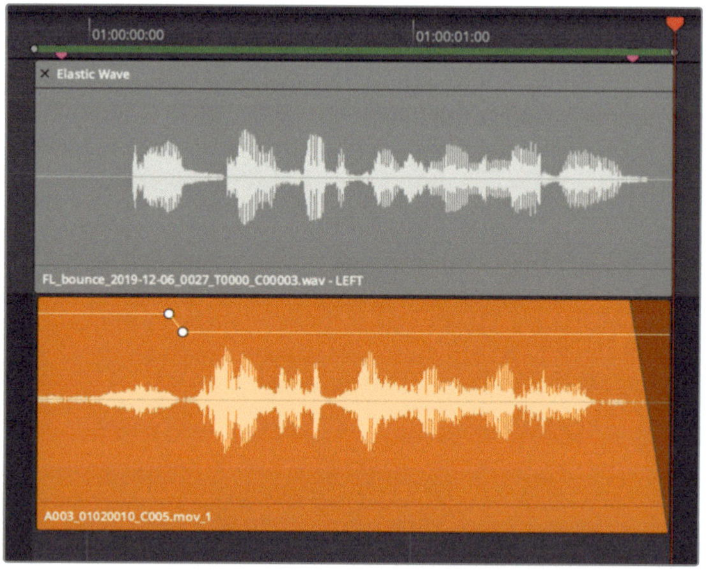

Now you can retime the entire clip or add time keyframes to stretch or condense specific areas of the waveform using keyframe markers.

7 Move your pointer over the right edge of the clip to see the Trim tool. Move the pointer toward the left over the clip until it changes to the Retiming tool.

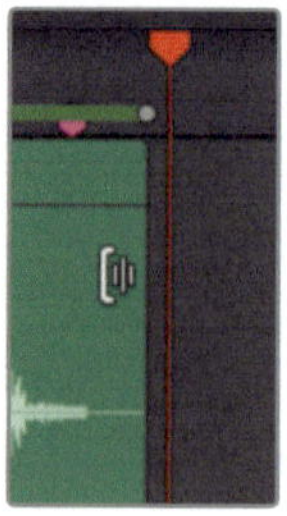

8 With the Retiming tool showing, drag the right side of the clip toward the right to stretch, and then drag toward the left to condense the entire clip.

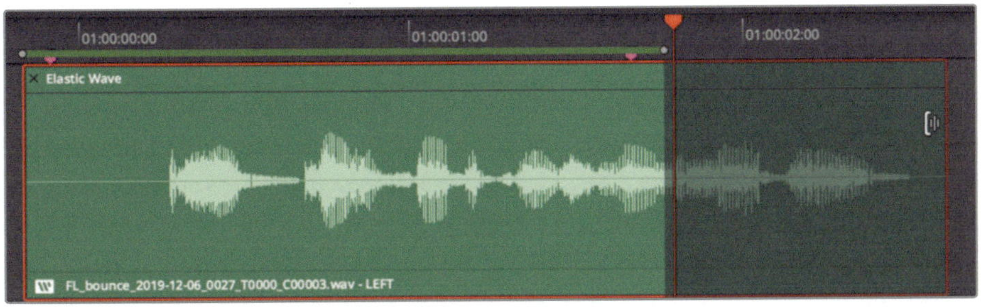

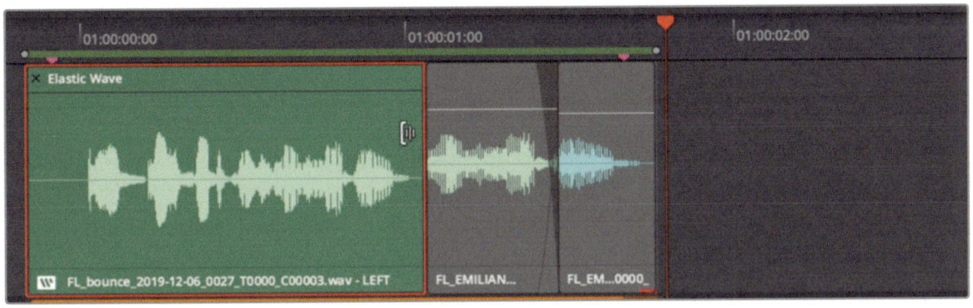

9 Stretch the clip back toward the right until the waveform for the last word in each clip aligns. You can do this visually, or by playing the end of the clips as needed until the timing of the word "Lab" is the same on both clips.

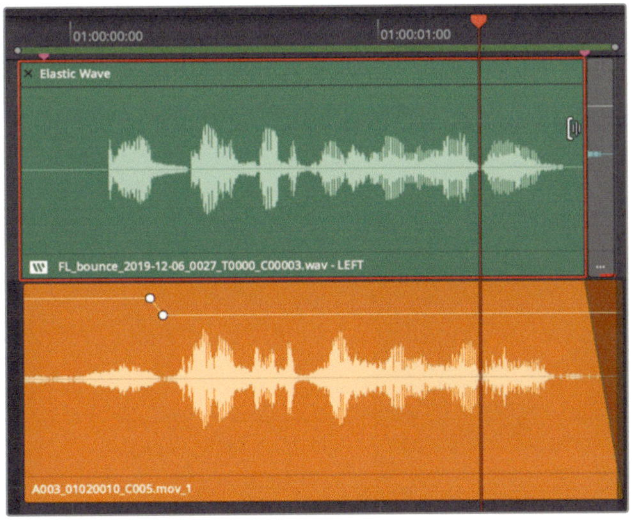

10 Start looped playback and listen to the timing of the clips together.

It's amazing how just retiming the clip slightly and aligning the last word really helps match the entire clip. However, there is still room for improvement.

Using Elastic Wave Speed Keyframes to Retime the Waveform

To finesse the timing of specific areas of the waveform within the clip, you simply add speed keyframes. Once you add a time marker, you can drag it left or right to stretch the wave accordingly. Speed changes based on keyframes only affect the waveform from that keyframe to the neighboring keyframes applied to that clip. Using an audio clip's waveform as your guide, you can use multiple speed keyframes to match the waveform of one performance to the waveform of another, in order to make the timing match.

In this exercise, you'll use Elastic Wave speed keyframes to further improve the timing between the two clips. Let's start with a keyframe near the middle of the clip after the word "identify." To set a time keyframe, you simply command-click the pointer over the waveform.

1 Move the pointer over the middle of the clip, just after she says, "identify."

2 Command-click the clip near the lowest point in the waveform (between words) to add a time keyframe.

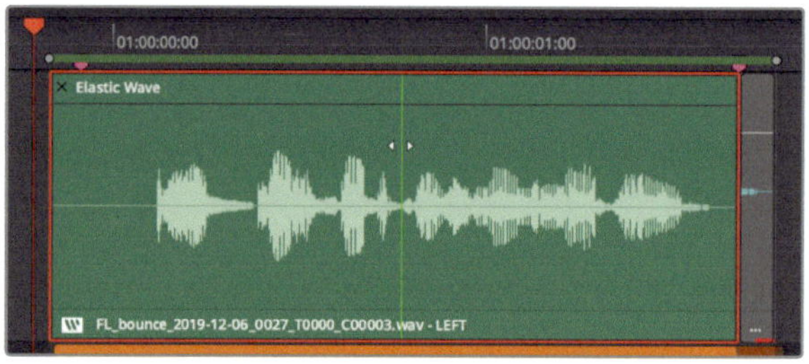

Since this part of the waveform lines up perfectly in both clips, you'll use this time keyframe to hold this position and prevent the waveform changes from rippling beyond this point to the end of the clip.

3 Add another time marker near the beginning of the waveform, at the gap after the "I" syllable in the word "identify."

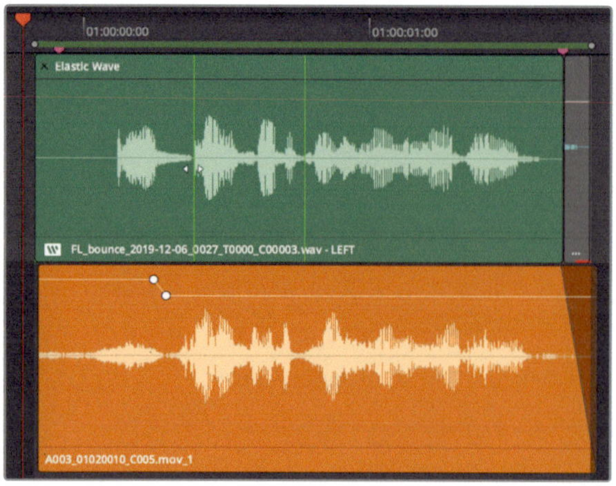

If you wish to move the position of a keyframe relative to the waveform, you can hold command while dragging the keyframe left or right. To delete a keyframe, just hover over the keyframe, and then right-click and choose Remove Speed Keyframe. You can also reset the entire speed curve by choosing Reset Speed Curve from the same right-click contextual menu.

4 Drag the first time keyframe slightly to the left to improve the timing of the waveform to the right of the keyframe. Drag the second time keyframe slightly to the right to align the waveform peaks in the last part of the clip. Feel free to add more keyframes and experiment with adjusting the clip's timing.

5 Once you are finished adjusting the waveform, play the two clips together to hear the results. The timing should be very close, if not identical.

For the real test, the ADR clip with the retimed audio should stand alone in sync with the video clip and mouth movement.

6 Unsolo the A2 track. Keep the A1 track soloed. Start looped playback and listen to the modified clip while watching the video. If it looks like Emiliana is speaking the words, based on her lip movement, your retiming is a success!

7 When you are finished with the Elastic Wave retiming tools, click the close button (x) in the upper-left corner of the clip. An Elastic Wave badge in the lower-left corner of the clip indicates that the waveform has been retimed. You can double-click the badge to show or hide the retiming controls.

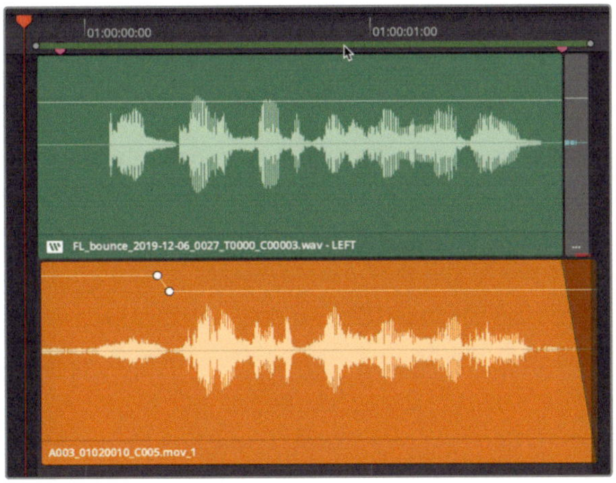

NOTE All Elastic Wave retiming adjustments you make in the Fairlight page appear in the Edit page as variable speed effects, accessible using the Retime controls. However, not all retiming effects created on the Edit page can appear as Elastic Wave retime effects on the Fairlight page.

More Audio Repairs Please!

You covered quite a few new tools and techniques in this pair of dialogue repair lessons including keyframing plosives; removing annoying noises between words; gating low-level noise; repairing plug-ins including the de-hummer, noise reduction, and de-esser plug-ins; and, finally, performing sample-level repairs and waveform retiming. Whew! As a dialogue editing and mixing trainee, you rock!

Now, you can put some of those skills to the test, and choose your own audio adventure from the following list. You may need to go back to the corresponding section in this lesson or Lesson 5 to refresh your memory.

In the "5 Dialogue Work Finished timeline:"

- Select a word or phrase in one of the dialogue tracks and replace it with the same word or phrase from the outtake clip in the Files for Repair and Replace bin.
- Create a 30-second room tone clip from a selection in the existing timeline dialogue clips. Create a new room track. Copy and paste the selected clip. Stitch the clip together in audio track layers. Bounce the stitched room tone clips to a new layer.
- Select the group of clips in the A2 track that contained the clothing crackles. Right-click and remove audio attributes to clear all the volume changes. Set keyframes to balance the clip and reduce the crackles.
- Choose a clip with sibilance and plosives. Use the De-Esser plug-in to reduce the sibilance and keyframes to reduce plosives.

In the "5f Reduce Noise" timeline:

- Delete the De-Hummer plug-in from the clip in the A1 track. Apply a new De-Hummer plug-in and adjust it to remove the hum.
- Remove the plug-ins from the clip on the A3 track, and try reducing the noise using the Noise Reduction plug-in. Try using Manual and Auto-Speech modes for the best results.
- If you're up for a challenge, delete the plug-ins from the clip in A4, and try reducing the noise. Use as many de-noise plug-ins as you need. Consider gating the low-level noise once it has been reduced.

In the "5 Fix Music Cue" timeline:

- Right-click the first part of the music clip, and choose Show in Media Pool. Drag the music clip from the Media Pool to the empty space below the A1 track to create a new track. Mute the A1 track. Use sample-level editing to remove the glitch from the music clip.

Coming up in the next lesson, you'll step out of the dialogue editor's seat and into the creative role of the sound designer.

Lesson Review

1. True or False? You can apply up to six plug-ins to a clip and apply unlimited plug-ins to a track.
2. Which FairlightFX plug-in is designed to remove powerline hum from an audio clip?
   a) Noise Reduction
   b) Hum Remover
   c) De-Hummer
   d) Un-Hummer
3. True or False? When using the FairlightFX Noise Reduction plug-in, the Manual mode utilizes the Learn button to create a noise profile.
4. Which FairlightFX repair plug-in reduces excessive sibilance?
   a) Shusher
   b) Ess-Remover
   c) De-Essinator
   d) De-Esser
5. True or False? The FairlightFX Sample Editor plug-in lets you see and edit the audio waveform at the sample level.
6. True or False? Loop Jog uses in and out points in the timeline to loop playback.
7. How do you enable the Elastic Wave retiming controls?
   a) Use the Elastic FX plug-in
   b) Choose Elastic Wave from the right-click menu
   c) Enable Elastic Wave in the clip Inspector
   d) Select the clip and click the Elastic Wave button in the toolbar
8. 8. Which modifier key is used to add and adjust time keyframes in the Elastic Wave retiming controls?
   a) Command
   b) Shift
   c) Spacebar
   d) Option

Answers

1 False. You can apply unlimited plug-ins to a clip and up to six plug-ins to a track.

2 c

3 True

4 d

5 False

6 False

7 b

8 a

Lesson 7

Enhancing the Soundtrack with Sound Design

Creating the aural illusion of a scene involves both the imagination and creative problem-solving skills of the sound designer. Outstanding motion picture sound design has helped film and television audiences travel through deep space, thwart alien invaders, evade dinosaurs, escape unnatural disasters, and survive the zombie apocalypse.

Once an audience has "been-there-done-that" an onscreen *soundscape* (audible landscape) becomes part of the accepted "norm" for similar onscreen experiences. A sound designer's goal is to create a rich, realistic, and emotionally compelling soundtrack that exceeds the audience's expectations without drawing attention away from the story.

Time

This lesson takes approximately 60 minutes to complete.

Goals

In this lesson, you will become a sound designer and be responsible for inserting sound effects; manipulating sounds to fit the scene; and adding height, width, and depth to the soundtrack. Along the way, you'll also learn some powerful new DaVinci Resolve 16 tools, plug-ins, and shortcuts.

What Does a Sound Designer Do?

The sound designer is responsible for creating sound effects and subtle sound layers to enhance the feeling and mood of a scene, while also managing the post-production sound process. In many cases, the sound designer is also the supervising sound editor. Think of the sound designer as an acoustic head chef responsible for the overall sound experience. And like a chef, the sound designer oversees every element on every track as the sounds are cumulatively blended, balanced, and mixed into the final soundtrack.

Preparing the Project

Throughout this lesson, you'll explore and work with a variety of different timelines. Let's start with a timeline containing two finished scenes from Hyperlight that are excellent examples of creative sound design. In this exercise, you'll open the first timeline and set up the marker index.

1. Hide all open panels, if necessary.
2. Open the "7 Sample Sound Design Scenes" timeline.

   The timeline opens with four stereo audio tracks; tracks A2 through A4 are muted.
3. Show the viewer, and then resize and move it so it fits into the lower-right half of the screen.
4. In the Index, show the Markers panel. Change the Markers index to list view, if necessary. Show only the columns Number, Frame, Color, Name, and Notes in that order from left to right.

5 Drag the vertical boundary between the Markers index and the timeline to the right and enlarge the Markers index until you can read the entire Name and Notes columns.

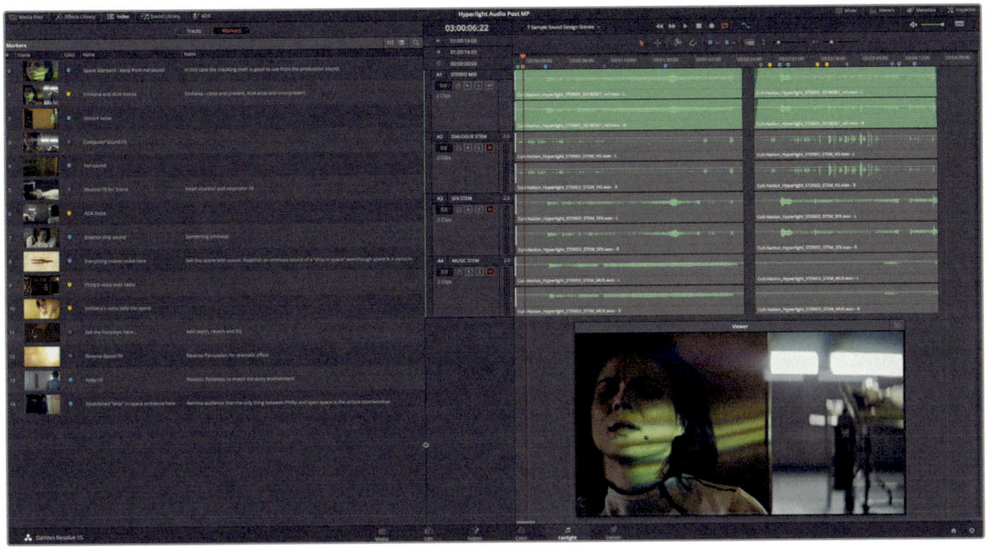

6 Hide the Index.

7 Press Shift-Z to fit the clips into the visible timeline area. Do not play the timeline yet.

The marker index is ready when you need it, and the timeline is ready to go.

NOTE The exercises in this lesson build from the tools and skills that you learned in the previous lesson. If you skipped ahead to this lesson, you may need to review previous lessons.

Understanding Diegetic versus Non-Diagetic Sound

All soundtrack elements fall into one of two categories: sound occurring within the scene that the characters can hear, or *diegetic sound*, and *non-diegetic sound* that is added in post for the benefit of the audience. Spoken and unspoken dialogue that the characters think or say is diegetic because it comes from the characters within the context of their onscreen world. Narrative voiceover that the characters can't hear or respond to is non-diegetic.

Non-diegetic music, such as the score, can be a major contributor to a scene's emotional impact. Other non-diegetic sounds include drones, pulses, percussive beats, whines, warbles, stingers, alarms, ticking, heartbeats, tinkling bells, and cymbal swells that are added to the soundtrack specifically to manipulate the audience's emotions. Often, these non-diegetic sounds reflect the innermost thoughts and emotions of the characters within the scene.

When you start a new scene as a sound designer, you focus on the diegetic sounds that must be included with the dialogue to reinforce the characters' actions and location. These include Foley effects, such as footsteps, source sound effects generated by objects onscreen, and any other background or ambient sound effects unique to the location.

Through television and films, audiences have been well-trained to recognize the sounds of common locations. For example, a busy hospital scene needs diegetic sounds such as beeping monitors, respirators, alarms, public address announcements paging doctors and staff, and the distant sounds of ambulance sirens to meet the audience's audio expectations for that location.

A professionally designed soundtrack is a powerful and cost-effective way to set a mood and maximize the emotional impact of a scene. For example, when the soundtrack includes gentle uplifting music, birds chirping outside, and laughing children, the mood will be positive and hopeful. In contrast, ominous music and cliché sounds such as squeaky carts, dripping faucets, and buzzing flies can completely alter the mood as effectively as a blood curdling scream.

Once the diegetic sounds have been established, the sound designer often adds layers of atmospheric background and mood-enhancing sounds to manipulate the audience and evoke tension, excitement, or elation, depending on the context of the scene.

Analyzing the Soundtrack Stems

A seasoned sound designer reads the script and meets with the director, editor, and composer to discuss the overall style and feel of the project. They watch the film one scene at a time and make a *spotting list* of elements that need to be added, enhanced, or replaced.

To better understand the power of sound design, let's play the first scene and listen to the finished mix. Then, you'll watch the same scene again and listen to each of the three fundamental components (*stems*) that make up the soundtrack: dialogue, sound effects, and music. With each listening pass, you'll gain a better understanding of the cumulative effect of these elements.

1 Play the first clip in the timeline, and try to identify which sounds (other than dialogue) are diegetic sounds that the character Emiliana can hear, and which sounds are purely for the audience.

   What did you think of the soundtrack? As promised, this soundtrack demonstrates exceptional sound design and music composition based on director Nguyen-Anh Nguyen's artistic vision.

Clearly, the musical score was designed to evoke tension and fear to help the audience feel the character's internal emotions. Let's play the scene again and listen to only the dialogue stem in the A2 track. In this case, the dialogue stem is also a great example of diegetic sound.

Soundtracks comprise three primary components: the dialogue, the sound fx (SFX), and the music. Many tracks are mixed down to create each of these fundamental stems in the final mix.

2 Mute the A1 STEREO MIX track. Unmute the A2 DIALOGUE STEM track. Play the clip again to hear only the finished dialogue stem with the dialogue tracks and production sound.

It takes only a few seconds to realize just how empty and unrealistic the location and set seems without the embellished diegetic sound effects and the non-diegetic accents and music. Suddenly, the scene has lost credibility and your focus may drift toward defects in the set rather than the emotional story onscreen. If you imagined specific sound effects that you would like to hear, you are thinking like a sound effects editor and sound designer.

Did you notice the excessive crackling of the space blanket when Emiliana yanked it off her back? This is one of the rare exceptions when crackling fabric movement picked up on the talent's microphone during production actually works in the final soundtrack.

3 Unmute the A3 SFX STEM track. Play the clip again, this time listening for both diegetic sound effects that work with the dialogue to complete the scene and non-diegetic sounds used to enhance the audience's experience.

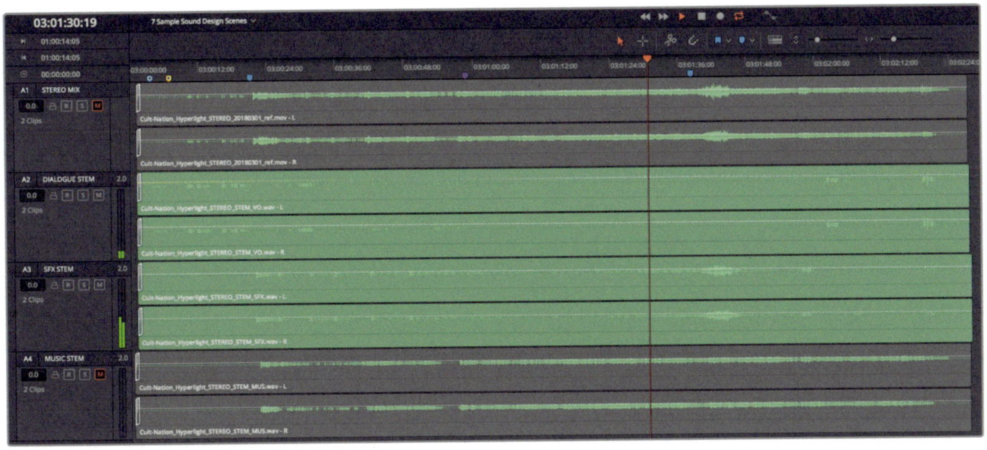

Did you notice all of the layers of sound used to increase tension? Let's listen one more time to only the dialogue and music.

4 Mute the A3 track, and unmute the A4 track. Play the clip once more.

The score is powerful enough to carry the drama, but it doesn't convincingly establish the location or enhance the tension without the sound effects track.

Now, let's listen to the final stereo mix for the scene that comes right after the primary dialogue scene in which Philip tells Emiliana that her counterpart from his reality is dead.

5 Unmute all the tracks, and solo the A1 track. Play the second clip in the timeline. During playback, listen for sounds that help establish the location. See if you can identify sounds added only for dramatic effect. Listen for changes in the dialogue or footstep sounds that suggest the physical space and surfaces.

What did you think of this scene? Once again, the excellent sound design really adds authenticity to the story and the location. Hopefully, listening to these soundtrack stems has given you a greater appreciation for and understanding of sound design and the power of both diegetic and non-diegetic sounds.

NOTE Feel free to listen to each of the separate stems in tracks A2, A3, and A4 for second clip in the timeline, before moving on to the next section.

Creating a Reversed Percussion Transition Effect

Throughout this lesson, you'll work with a variety of sound manipulation tools available only in the DaVinci Resolve Fairlight page. However, to apply speed changes to a clip, you need to do so on the Edit page. Speed changes allow a sound designer to change the duration of a clip or reverse a clip to create a stylized transitional sound between scenes. In this exercise, you'll use both speed techniques to create a unique transitional sound effect. Let's use the Marker index to locate a finished example of a transitional sound effect already placed in the mix.

1 Show the Index. In the list, double-click the thumbnail for marker #13 to move the playhead to that position.

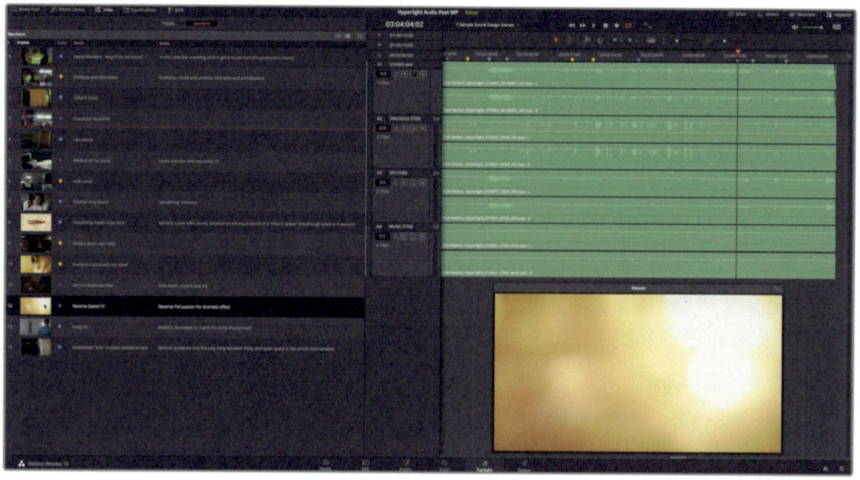

2 Solo the A1 track, if necessary. Play the clip from the marker's position to see the visual transition and hear the reversed audio effect.

Reversing a percussion crash or other sound effect is a fun way to *button* a scene (decisively end the scene). The first step toward creating that effect is to determine its duration.

3 Mark an in point at the purple Reverse Speed FX marker. Press the JKL keys to play forward in slow motion and stop playback at the last frame of the transition (03:04:07:13). Mark an out point.

4 In the upper-left corner of the timeline, look at the timecode display area to see the duration of the marked range.

The play range is 3 seconds and 11 frames (03:11). So, that should also be the approximate duration of the sound effect you use for the transition.

5 Clear the play range.

6 Hide the Index, and show the Media Pool.

7 In the Media Pool, choose Hyperlight Audio > SFX for Audio Post > Sound Design FX > **Sound Design Layers Pile Driver.wav**. Then, change the preview player zoom level to 1x.

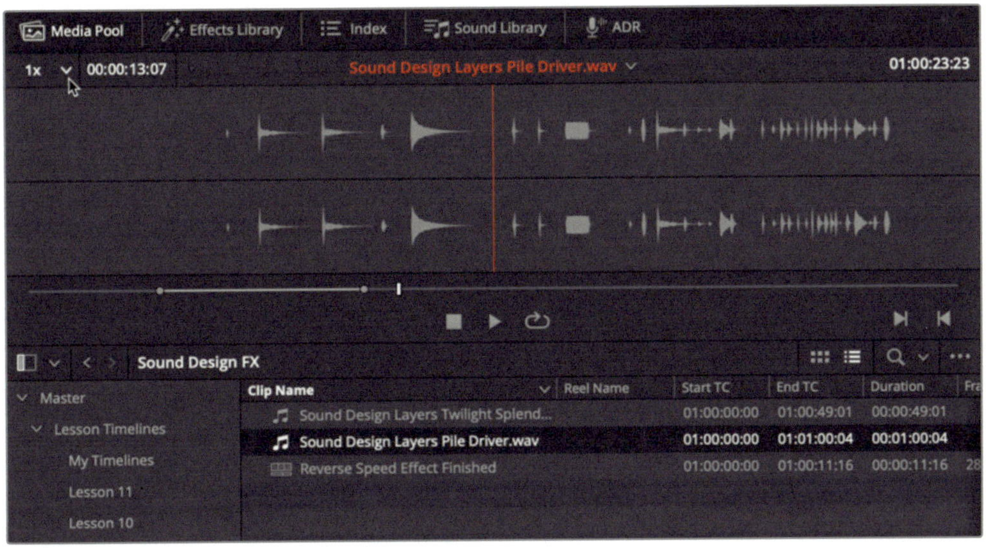

This clip contains a series of percussion sounds recorded using the Pile Driver MIDI drumkit sound from Apple Logic Pro X.

8 Change the preview player's zoom level to 5x.

Now you can see a closer view of the waveforms for each recorded sound. Waveforms with a defined triangular shape that start loud and taper off are a good candidate for a reverse speed effect.

NOTE To save time, the clip has already been marked with in and out points containing three different percussion hits.

Previewing a Clip in Reverse Speed

The preview player in the Fairlight page Media Pool lets you preview your source media in both forward and reverse playback. So, you can "test drive" an effect's reverse sound before editing it into the timeline. In this exercise, you'll listen to the three marked clips in forward and reverse playback, and mark one to edit into the timeline.

1 In the preview player, drag the scrubber bar to the marked in point.

2 Press L to play the three marked percussion hits. After the third percussion hit, press K.

3 Press J to preview the percussion hits in reverse. After the third reversed percussion hit, press K.

Which reversed percussion hit do you think would sound best under the visual transition? Let's go with the third one because it has the longest tail and the most defined slope from the original hit to the end. Instead of dragging the clip to the existing timeline, let's experiment by applying speed changes in a new timeline.

Creating a New Timeline for SFX Experimentation

Like most art forms, sound design can be fun, yet messy. So, it's always a good idea to experiment with sound effect creation and manipulation in an isolated timeline. By doing so, you can unleash your creative ideas without worrying about adversely affecting other audio clips in the timeline. Plus, when the effect is a success, you can use that timeline as a template for creating similar effects.

1 Move the playhead just before the third hit, and press I to mark a new in point.

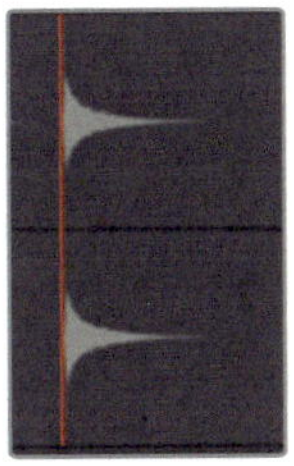

2 In the Sound Design FX bin, right-click the **Sound Design Layers Pile Driver.wav** clip, and choose Create New Timeline Using Selected Clips.

The New Timeline dialog box opens with the Timeline Name field selected.

3 In the Timeline Name field, type **Reverse Speed Effect**, and press Return.

The new named Reverse Speed Effect timeline opens.

4 Hide the Media Pool.

5 Zoom in on the timeline clip until you can clearly see its waveform.

6 Mark the clip, and check its duration in the play range duration field.

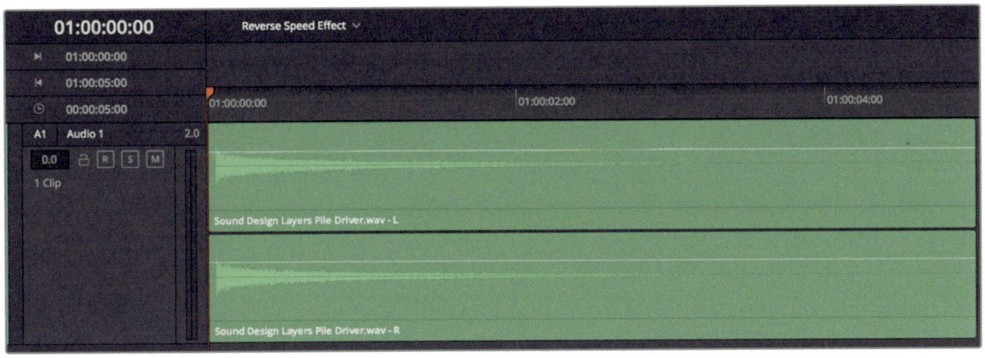

The clip's duration is around five seconds (05:00). If you recall, the desired duration for the transition was 03:11. In the next exercise, you'll change the clip's duration and thus change the speed in the process.

NOTE Because this percussion sound includes a long tail, you could simply trim several seconds from the end. In this case, you'll leave the clip intact and simply change the speed/duration as necessary.

Applying Speed Changes to a Clip

To change this clip's duration using a speed change, you first need to go to the Edit page. There, the speed changes are available for each clip in the contextual (right-click) menu. In this exercise, you'll first duplicate the clip so the original is retained as a reference. Then, you'll open the timeline in the Edit page, and resize the audio track. Finally, you'll change the clip's duration and audition it at the new speed.

1 Clear the play range.

2 Right-click the clip, and choose Copy. Move the playhead down the timeline to around eight seconds (08:00), and paste the clip.

3 At the bottom of the page, click the Edit button, or press Shift-4, to show the Edit page. Then press Shift-Z to fit the clips to the timeline.

4 Drag the divider bar upward between the video and audio tracks to raise the audio track to the upper-half of the timeline.

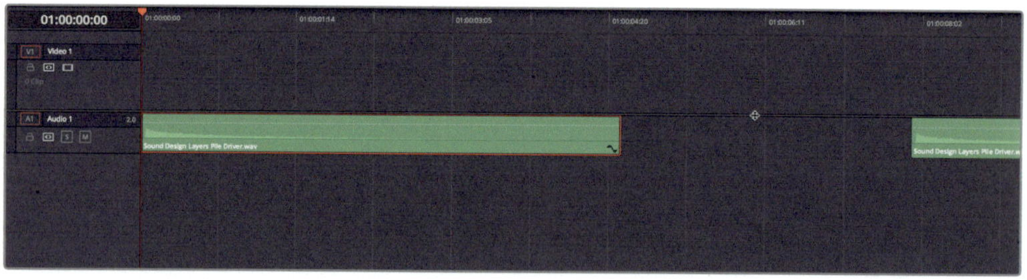

5 Drag the A1 track header downward to increase the height of the audio track.

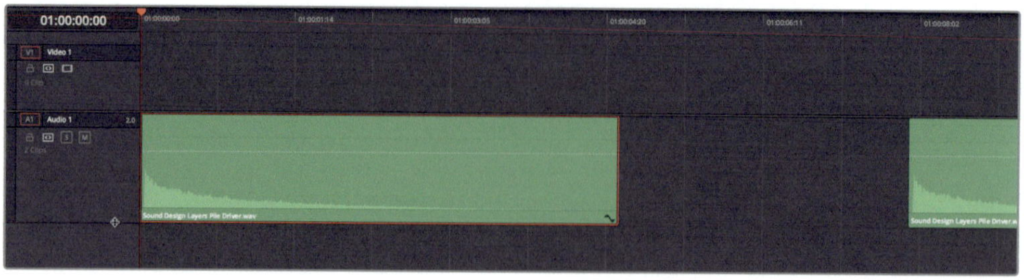

Now you can clearly see the audio waveform of the clip in the Edit page timeline.

6 Move the playhead to the end of the first clip (01:00:05:00). Then, right-click the first clip and choose Change Clip Speed.

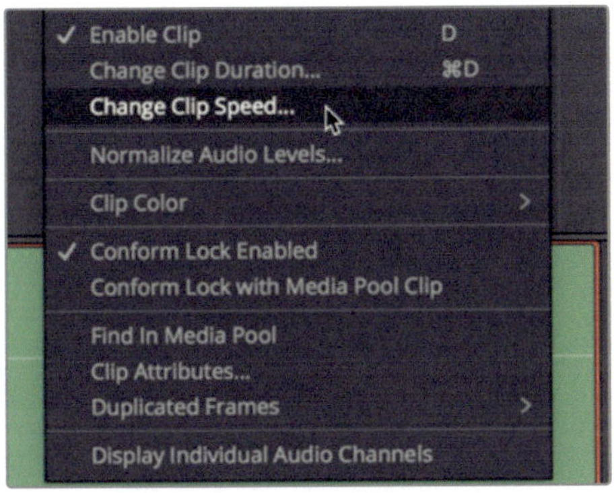

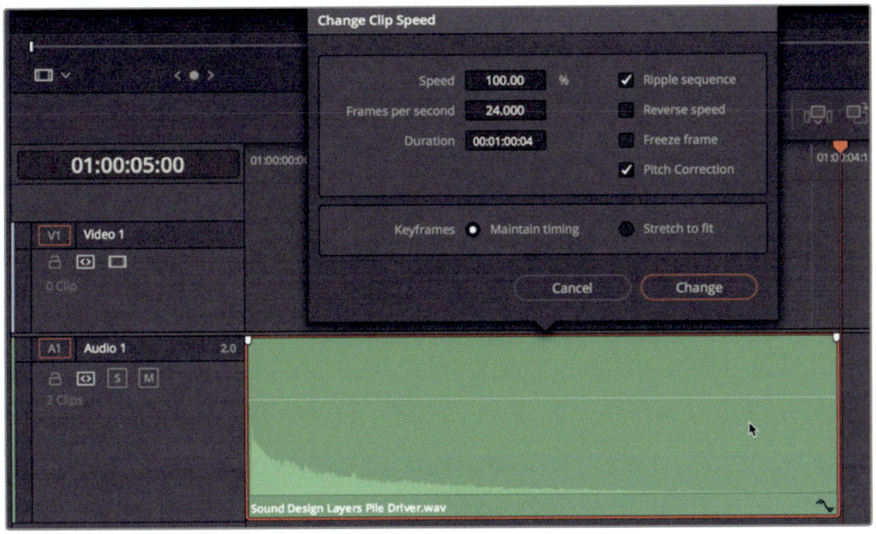

The Change Clip Speed dialog box includes fields that show the Speed, Frames Per second, and Duration. Any changes applied to the clip's speed will be reflected in the Frames Per second, and Duration fields. On the right side, you'll see three additional options that are important for sound design: Ripple Sequence, Reverse Speed, and Pitch Correction.

Let's start by increasing the clip's speed by around 35%, which will in turn decrease the duration by the same amount. The Duration field shows the length of the original media clip without in and out points.

7 Select the Ripple Sequence option.

Doing so will change the duration of the clip in the timeline. If you leave the Ripple Sequence option deselected, the clip remains at its current length, but the waveforms reflect the speed change and additional material is included in the visible portion of the clip.

8 Drag the Speed field value to the right to increase the speed to 135%. Click Change, or press Return.

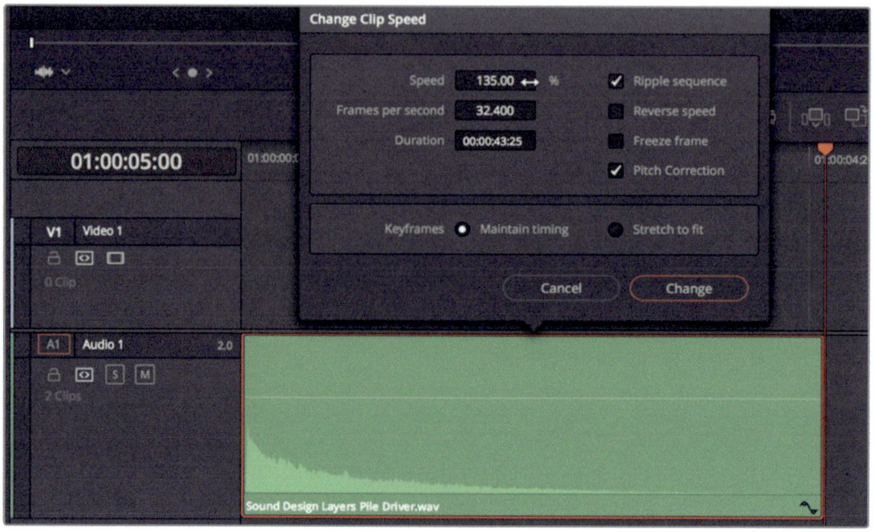

9 Move the playhead to the end of the first clip.

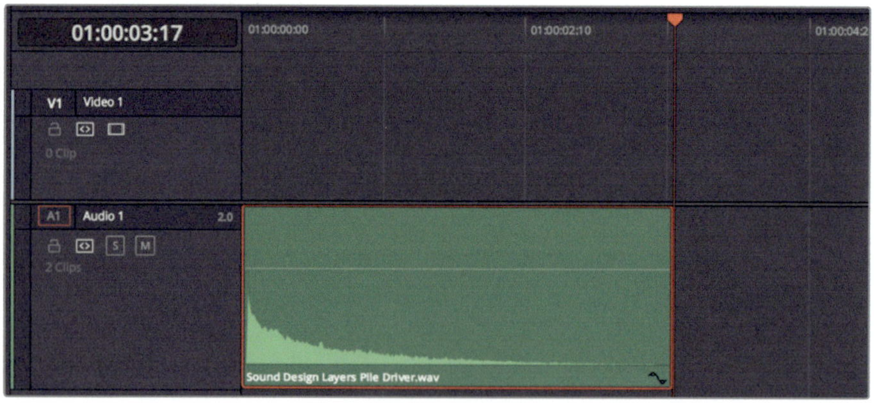

The clip now ends at 03:17, which is really close to the 03:11 duration of the transition.

10 In the timeline, scrub the playhead over both clips to hear and compare the sounds of each clip.

They sound the same, except the second one is longer. But how is that possible? Their pitch is the same because the default setting in Resolve is to employ pitch correction during speed changes. Depending on the type of sound that you are manipulating, extreme speed changes will sound processed, with or without pitch changes.

11 Right-click the first clip, and choose Change Clip Speed. In the Change Clip Speed dialog box, select the Reverse Speed checkbox.

The negative Speed and Frames Per Second values indicate the reversal of the clip. Press Return to apply the change.

12 Press Shift-7 to go to the Fairlight page. Play the reversed clip.

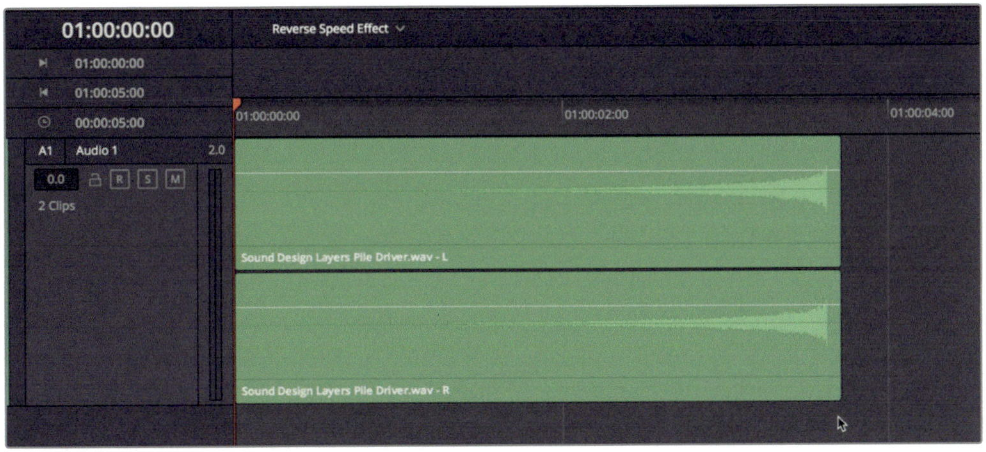

As expected, the reversed clip sounds like the opposite of the original clip. Now, instead of a loud percussion hit that slowly decades (fades out) over time, the reversed clip slowly fades in, builds to an exciting crescendo, and stops abruptly.

You marked a source clip in the Media Pool, made a new timeline, changed the clip's speed, and reversed it. Awesome! This trick is used all the time in soundtrack and music production. Chances are, now that you are aware of the reverse-clip effect, you'll frequently notice it more in other projects.

NOTE If you didn't complete the previous exercise, open the 7 "Reverse Speed Effect" timeline to catch up.

Syncing Sound Effects to Picture

One of the daily challenges for sound designers and the sound effects editors on their team is to precisely sync audio clips to video cues. In this exercise, you'll copy the sound effect that you just created and paste it into a new track in a different timeline. Then, you'll use a new visual feature to sync the sound effect to the picture.

1. Move the playhead to the beginning of the reversed clip. Right-click the reversed clip, and choose Copy.
2. In the timeline pop-up menu, open the "7 Sample Sound Design Scenes" timeline.
3. Right-click any track header, and choose Add Track > Stereo.
4. Move the playhead to the purple marker at 03:04:04:02.
5. Select the A5 track, and press Command-V (macOS) or Ctrl-V (Windows) to paste the clip in the selected track.
6. Zoom in to the timeline until you can see the clip's waveform.

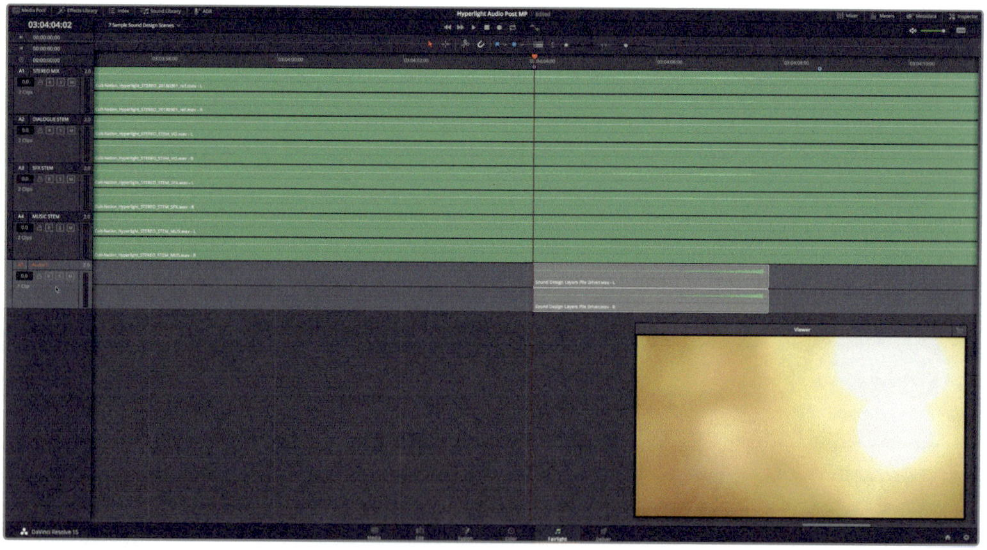

7. Solo the A4 and A5 tracks. Play the clip in the timeline and watch the video to see how it works with the picture and music.

   It's close. But in professional audio, close doesn't cut it. Luckily, the DaVinci Resolve Fairlight page includes an awesome set of visual aids to assist with clip timing.

Using Video and Audio Scrollers for Precise Synchronization

The Fairlight audio timeline includes a set of scrollers that let you compare a frame-accurate filmstrip of the video track with any track's audio waveform. You'll find the controls to show and hide these scrollers in the timeline options menu.

1. Drag the viewer to the upper-left corner of the screen to make room for the scrollers at the bottom of the timeline.
2. Click the timeline options pop-up menu to show all the timeline options. Under the Scroller header, click the Video and Audio 1 icons.

3. Click any empty space in the timeline to hide the pop-up menu.

   At the bottom of the timeline, you'll see the video scroller and audio scroller 1, which shows the selected track A5 - Audio 5.

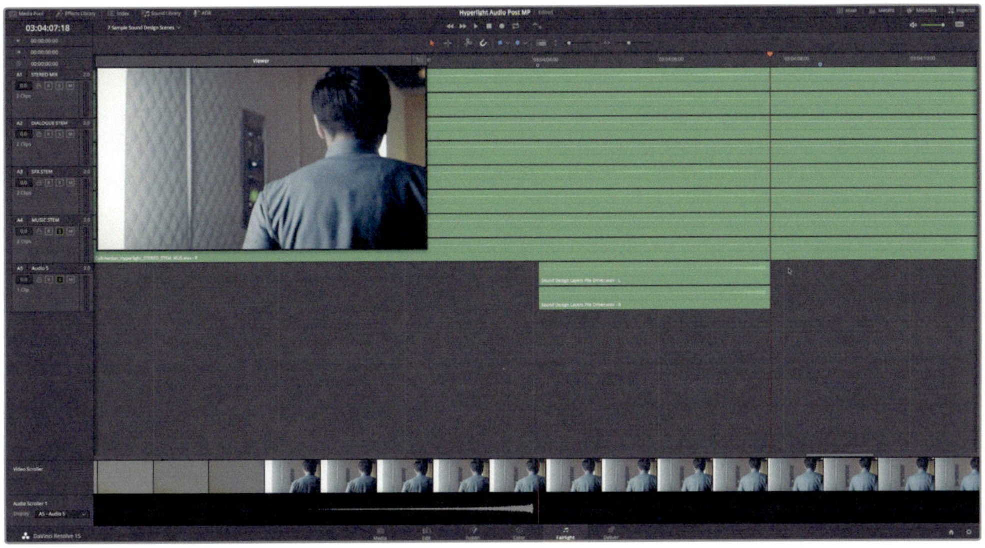

4 If another audio track is showing, click the Audio Scroller 1 Display pop-up, and choose A5 - Audio 5.

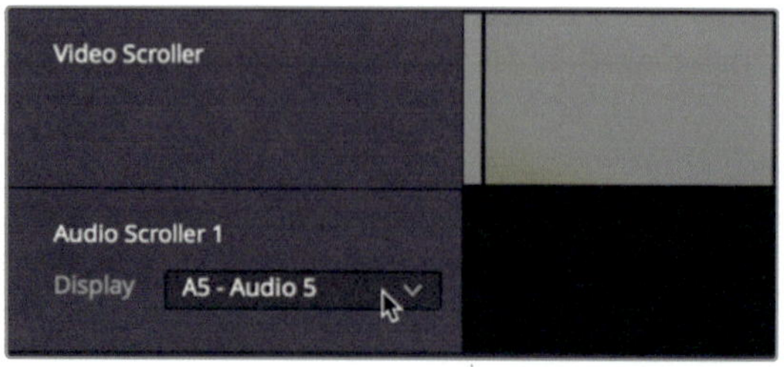

In the Video Scroller, each filmstrip frame corresponds to a single frame of video. Let's change the zoom level of the filmstrip to include the neighboring clips.

5 In the Video Scroller, right-click the filmstrip, and choose Zoom Level > High.

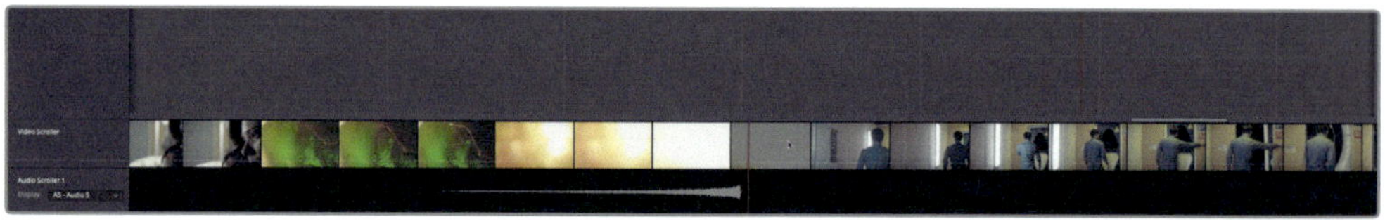

Instead of seeing every frame, the filmstrip now includes a higher range of frames so that you can see frames from shots on either side of the video transition. The red vertical lines on the audio and video scrollers represent the timeline playhead.

6 Right-click the filmstrip, and choose Zoom Level > Low.

7 Move the playhead back to the purple marker.

As you can see by looking at the video scroller, this is not actually the first frame of the transition. However, you can drag the scroller left or right to move the playhead to the desired position.

8 Drag the video scroller to the right until you see the first frame of the transition. It is the first frame after the green-lit close-up shot of Emiliana.

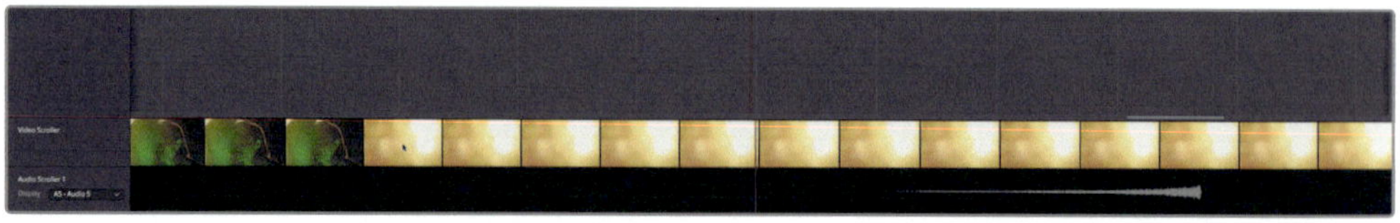

9 Play the timeline while watching the video scroller, and stop playback when you see the end of the visual transition.

The timeline playhead and corresponding playhead on the scrollers clearly show that the waveform finale and end of the transition are close. You could nudge the clip into position; but instead, let's use the scrollers and some common keyboard shortcuts to fix the timing. When working with video scrollers, you can double-click any frame to move the playhead to that frame. So, if you align the playhead to the end of the waveform on the clip, you can use editing shortcuts to paste it precisely at the frame you select on the scroller.

10 Using the reversed clip's waveform as a guide, move the playhead to the last frame of the waveform (not the clip).

11 Select the reversed clip, and press Command-X (macOS) or Ctrl-X (Windows) to cut the clip.

A semi-transparent version of the clip remains synced to the playhead until you paste it.

12 In the video scroller filmstrip, double-click the first frame after the video transition: the medium shot of Philip walking down the corridor.

The playhead and the semi-transparent version of the clip move to the selected frame.

13 Press Command-V (macOS) or Ctrl-V (Windows) to paste the clip

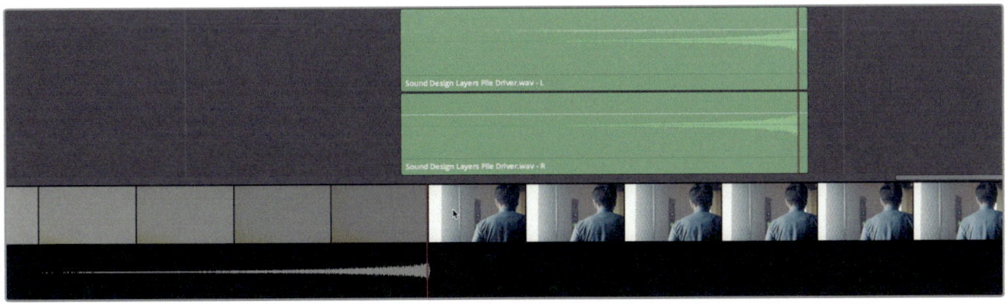

The audio transition ends just as the new shot begins.

14 Play the transition from the purple marker to audition it in sync with the video.

This time it sounds perfect! Let's hide the scrollers before moving on to the next section.

15 Unsolo the A4 and A5 tracks.

16 In the Timeline options menu, deselect the Video and Audio 1 scroller icons.

As you just discovered, reverse audio effects are easy to create and sync to picture. Best of all, you now have the skills to add reverse transitional audio effects to your own soundtracks.

NOTE When using keyboard shortcuts to cut, copy, and paste, you can also press the JKL keys to reposition the transparent clip before pasting it.

Doubling a Track to Thicken the Sound

Another common sound design and mixing technique is to double a track to "thicken" or "fatten" the sound. Doubling a track involves duplicating the contents of one track into another track, and adding effects to one or both tracks. The combination of the two tracks sounds thicker and richer than the original. This technique is commonly applied to instrument tracks and background vocals in music production, and to ethereal voices and sound effects in audio post. In this exercise, you'll double ADA's voiceover track and apply pitch changes and other effects to thicken the sound. First, let's listen to a finished example to hear the differences between Emiliana's and ADA's voices.

1. In the "7 Sample Sound Design Scenes" timeline, solo the A1 track.
2. Play the beginning of the first clip from the yellow marker to the second blue marker and listen to the differences between the two voices. Stop playback after ADA says, "I cannot explain this discrepancy."

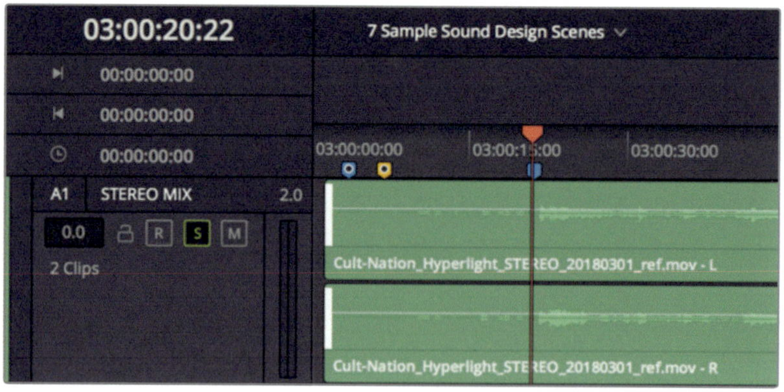

Both voices are clear and easy to understand. Emiliana's voice sounds real and present, as if she really were only a few inches away, whereas ADA's voice sounds more processed and omnipresent (from all directions), as though it's coming from multiple nearby speakers. You'll learn more about dialogue mixing, dynamics, and equalization in the next lesson, but for now, you'll focus on the sound design for ADA's voice to transform it from a clean human voice to one that sounds as if it is computer generated.

NOTE A "clean" or "dry" sound is one that has no effects or processing applied.

Let's open a different timeline in which you can double ADA's VO track and experiment with her voice.

3 Open the "7 ADA VO test" timeline. Press Shift-Z to fit the clips horizontally into the timeline.

This is the dialogue scene that you worked with in the previous lessons. Track A3 contains the test VO recorded using an actual computer-generated voice. Track A4 is the human voice recording that you edited in Lesson 3. This track is muted. Tracks A5 and A6 are empty tracks for doubling and experimenting with ADA's voice. First, let's listen to the computer voice, and then play the clean voiceover recording in the A4 track.

4 Show the viewer, if necessary, and move it to the lower-right corner of the screen.

5 Move the playhead to the first blue timeline marker, and press I to mark an in point. Move the playhead to the second blue marker, and press O to mark an out point.

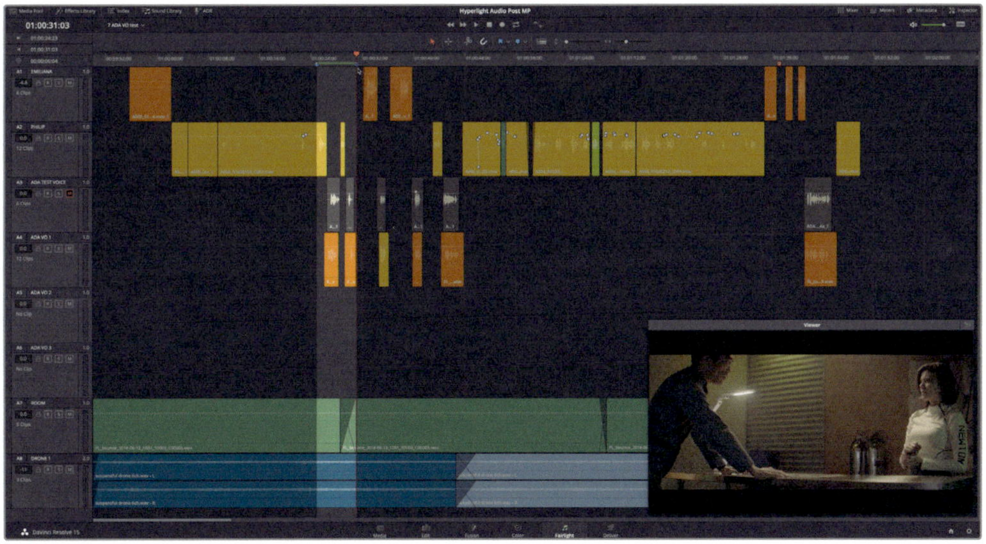

6 Press Option-/ (slash) in macOS or Alt-/ (slash) in Windows to play from the in to out points and hear ADA's first two phrases with the computer voice.

7 Mute track A3 and unmute track A4. Play from the in to out points again to hear the human version of the same phrases. When you are finished, press Option-X (macOS) or Alt-X (Windows) to clear the in and out points.

Let's thicken the sound of ADA's voice by copying the clips from track A4 to tracks A5 and A6.

NOTE The term "doubling a track" used to refer to creating only two versions of a track in tape-based music recording. However, when using digital audio workstations, you can make as many tracks as you need to thicken and enhance your sound.

Copying and Pasting Multiple Clips from One Track to Another

The secret to copying and pasting multiple clips from one track to another is to start with the playhead on the first frame of the track. Technically, you can place the playhead anywhere when you copy and paste. However, by always starting with your playhead in the same position, you can guarantee that the clips you paste will be in the exact same positions as the clips you copied.

1. Press Home to move the playhead to the first frame of the track.
2. Drag the Selection tool across all the clips on the A4 track to select them.

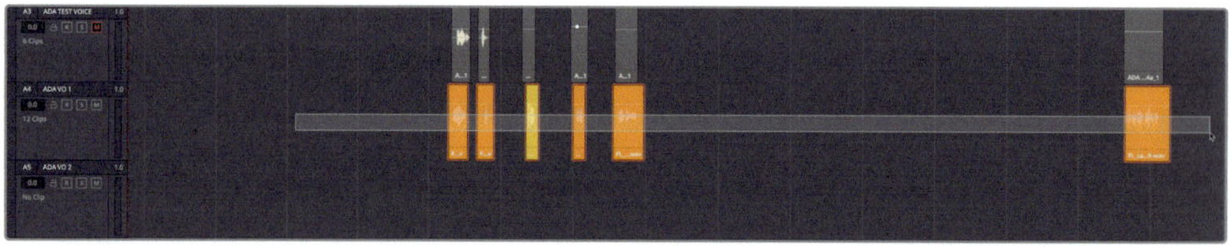

3. Copy the selected clips. Select the A5 track, and paste the semi-transparent clips into the A5 track.

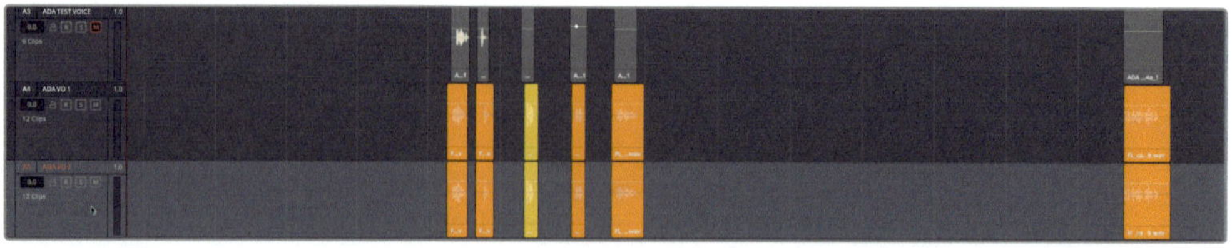

4. Select the A6 track, and paste the clips again.

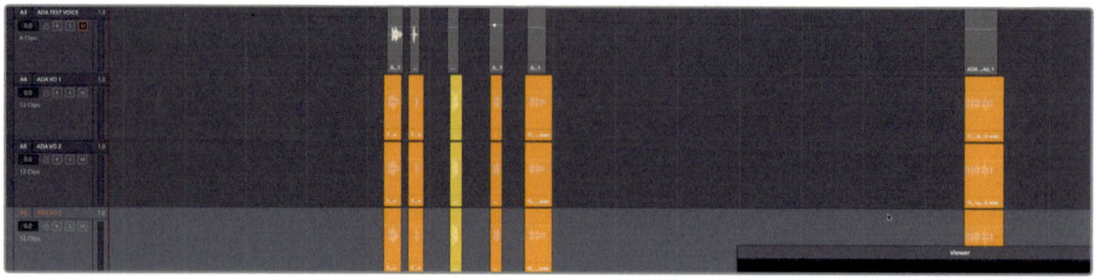

Now you have three tracks containing identical clean versions of the ADA VO clips.

Applying Pitch Changes to Clips in the Inspector

One easy way to vary the sound between two identical tracks is to make changes to the pitch. If you recall, the Inspector includes pitch controls for every clip. In this exercise, you'll experiment with pitch changes on several clips to hear the result mixed with the original sound. Then you'll apply the pitch changes to all the clips on the track. Let's keep the A4 track unchanged, and experiment with the pitch on the clips in the A5 ADA VO 2 track.

1 Select and solo the A5 track. Create a play range for the first two clips in the A5 track. Show the Inspector.

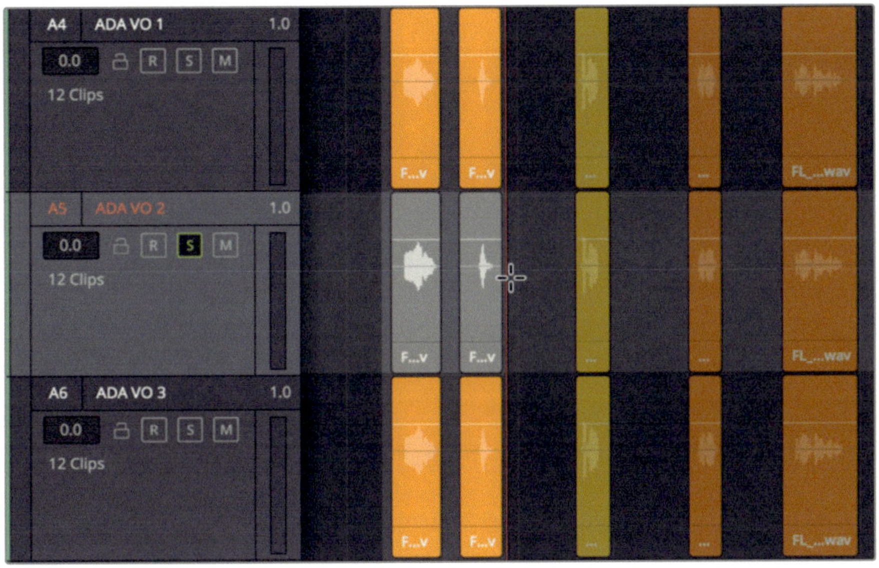

The Clip Pitch controls let you lower or raise a clip's pitch by up to 24 semitones. With 12 semitones in an octave, you can pitch a clip up or down by two full octaves. The Cents controls manipulate the clip by hundredths of a semitone and are used for fine tuning a clip's pitch.

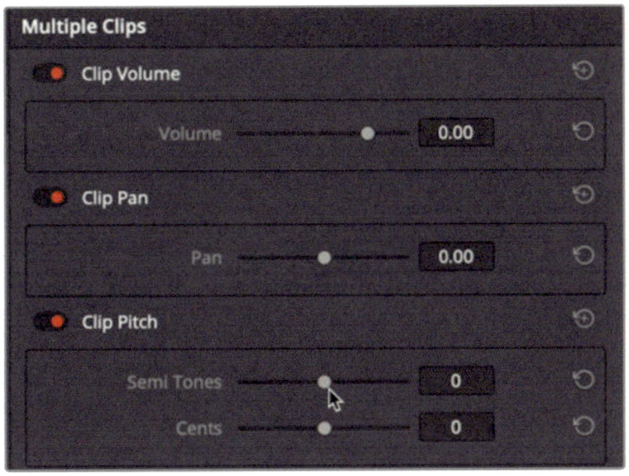

Dragging the pitch controls to the left lowers the pitch and is designated by a negative number. Conversely, dragging to the right raises the pitch. The best way to understand the power of pitch control is to actually hear it in action.

2 Start looped playback. During looped playback, drag the Clip Pitch slider to each of the following settings: -1, -5, -10, -20, 0,1, 5, 10, and 20. When you are finished, stop looped playback.

As you just heard, subtle pitch changes can sound natural, while more radical changes can completely transform a voice from human to that of demonic creature (-20) or a talking mouse (+20).

3 At the far right of the controls, click the Clip Pitch Reset button.

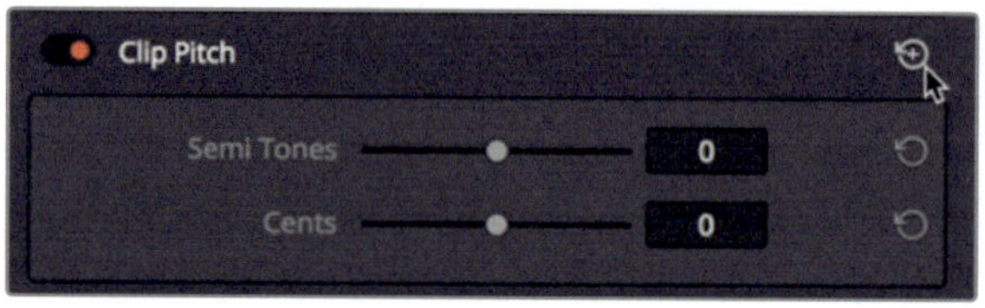

Next, you can try dragging the Cents slider to introduce smaller incremental pitch changes. Let's do so while also listening to the A4 track to hear the "doubled" effect.

4 Solo the A4 track so that you can hear both the A4 and A5 tracks during playback.

5 Start looped playback. During looped playback, drag the Cents slider slowly to the left, and then to the right to hear the slight changes in pitch.

6 When you are finished, stop looped playback and reset the pitch controls on the A5 track.

The pitch controls in the Inspector are perfect for making both large and small pitch changes to individual clips. However, if you need to change the pitch for an entire track, it's better to use the FairlightFX pitch plug-in.

Thickening Cumulative Sound with Track Level Pitch and Volume Changes

In this exercise, you'll work with the Pitch FairlightFX plug-in to change the pitch on both the A5 and A6 tracks. You'll introduce a slight pitch change on the A5 track to produce a subtle thickening. Then, you'll pitch change the A6 track by a full octave to add a deeper, low-end quality to the voice. After the effects are applied, you'll adjust the volume levels of the pitched tracks to create a more blended, "composite" sound.

1 In the Effects Library, drag the Pitch plug-in onto the A5 track header.

The Pitch dialog includes Semitones and Cents controls, as well as a percentage control for the Dry/Wet output.

2 Start looped playback. Drag the Cents control left to around -30.

3 Solo the A6 ADA VO 3 track to hear it with the other two ADA VO tracks.

4 Drag the Pitch plug-in onto the A6 track header. Lower the Pitch on the A6 ADA VO 3 track by -12 semitones.

As you can see, each plug-in panel includes the track's name, so you can differentiate between each dialog. This is also convenient when making changes to more than one plug-in at a time.

> **NOTE** Mixing multiple versions of the same audio clip is very similar to the process of image compositing in which you apply blend modes to multiple versions of the same video clip. When compositing images, you often lower the opacity of one or more of the blended clips for different effects. Similarly, in audio track doubling, you adjust the volume levels on different tracks to blend the various sounds without overpowering the original track.

Once you have added pitch changes to your tracks, you can adjust the volume levels right in the Inspector. Let's start by lowering the level of the A6 track.

5 Click the A6 ADA VO 3 track header to select that track. In the Inspector, drag the Track Level Volume slider to the left until the heavily pitched voice is barely audible. Your goal is to include it as an accent to enhance and thicken the sound of the original track. If you aren't sure what level to use, try a volume level around -20.

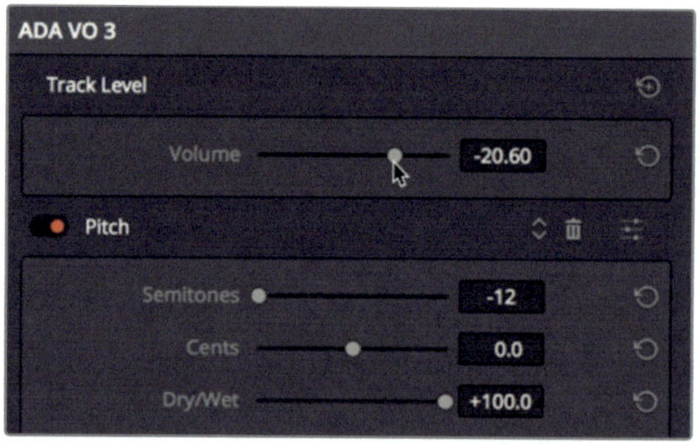

6 Stop looped playback. Close the Pitch controls.

Perhaps you're thinking, why bother adding the -12 semitone track if you are turning it down so much? It's all about the finesse. A track pitched an octave lower can really beef-up the low end and thicken a sound. To prove that this track is worth the trouble, let's toggle the track's Solo button off and on during playback.

7 Start looped playback, if necessary. Click the Solo button on the A6 track off and on several times to hear the combined tracks with and without the A6 track. When you are finished, leave the Solo button turned on.

   Hearing is believing.

8 Continue with looped playback. Select the A5 ADA VO 2 track, and lower the track volume to around -7.20 to better blend it into the mix.

9 Unsolo the A4, A5, and A6 tracks. Set a play range between the two blue markers. Start looped playback of the new range.

10 During looped playback, mute the A5 and A6 tracks to hear the clean voiceover. Unmute the A5 and A6 tracks to hear the doubled results.

11 Hide the Inspector.

Excellent. Not only did you successfully thicken ADA's voice, but you also learned how to apply subtle and dramatic pitch changes to clips and tracks. However, ADA still sounds human. In the next exercise, you'll apply another plug-in to increase the number of voices from the A4 track.

> **NOTE** Sound designers and music mixers sometimes use 12 or more tracks of the same audio clip with slight pitch changes applied to make the cumulative sound more powerful and help it stand out in the finished mix.

Multiplying Voices Using the Chorus Plug-In

Another sound design technique to thicken sound is to us a Chorus plug-in. As the name suggests, this plug-in emulates the sound of multiple simultaneous voices to suggest a vocal chorus. Like a real chorus, slight timing and pitch differences between the voices can thicken the overall sound. The FairlightFX Chorus plug-in samples the input and slightly detunes it (using minor pitch changes) and mixes it with the original signal. The differences between voices can be slight or dramatic depending on the settings.

1 Solo the A4 ADA VO 1 track.

2 In the A4 track, set a play range around the first two clips.

3 In the Effects Library, drag the Chorus plug-in to the A4 track header.

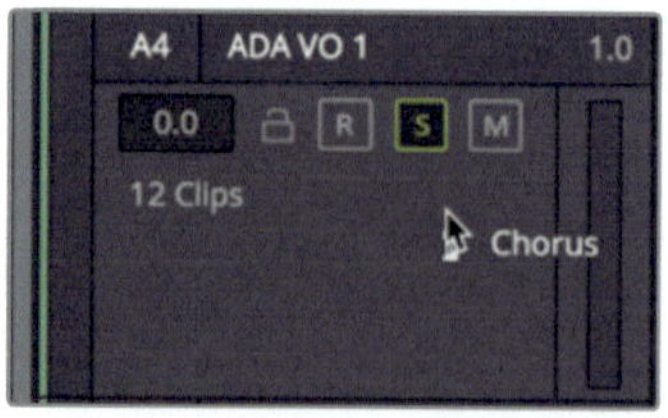

The Chorus dialog box opens, displaying a set of graphs: on the left is the graph of the time difference between voices, and to the right is an animated graph of the signal after the altered parameters are applied.

Let's listen to the first two clips with the default Chorus preset applied, and then listen to the same clip with the Dramatic preset.

4 Start looped playback.

5 In the Chorus preset pop-up menu, choose Dramatic to apply and audition that preset during playback.

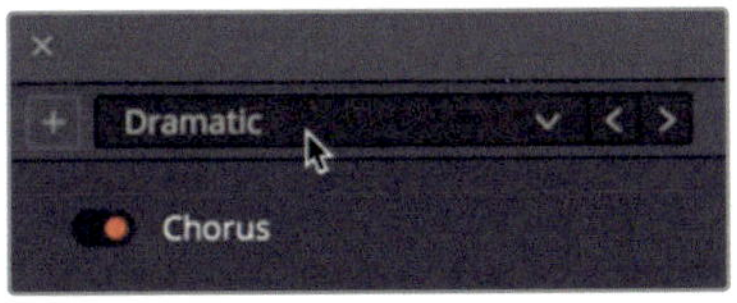

Wow. The clip now sounds fuller and more dramatic.

6 Continue looped playback, and unsolo the track to hear it played along with the other ADA VO tracks.

7 Toggle the Chorus Bypass button (red switch) off and on during playback to hear the cumulative effect of the Chorus plug-in on the original sound, and the combination of all three tracks. When you are finished, leave the Bypass button off (red).

The Dramatic preset is perfect for ADA's voiceover; however, let's take a moment to try the manual settings to create your own chorus sound.

8 Solo the A4 track. Start looped playback, if necessary.

9 In the upper-right corner of the Chorus dialog, click the Reset button to return to the default settings.

Let's start by adjusting the Delay settings to the far left of the controls area. The Delay settings control the length of delay between the original sound and the Chorus effect.

10 Drag the Delay knob to the right to increase the length of the chorus, as indicated by the delay lines displayed in dark blue in the graph.

11 Under the Voice controls, drag the Separation control to increase the time separation of the delay voices.

12 Under the Width controls, drag the Expansion control to set the left/right length differences (phase offset) for stereo output.

The Chorus plug-in includes a low-frequency oscillator (LFO) that is mixed with the signal to create modulation effects. The green bar in the Modulation controls area shows the oscillating pattern.

13 In the Modulation controls area of the Chorus dialog, watch the bright green dot as it oscillates back and forth on the green Modulation graph. Drag the Frequency dial to the right to increase the oscillator's frequency and, therefore, the speed of the modulating pattern.

14 Drag the Pitch control to the right to increase the amount of pitch change for a heavier *vibrato* (fluctuations in pitch) effect.

15 Drag the Level control to the right to increase the *tremolo* (fluctuations in volume level) effect.

Below the Modulation graph are six waveform shape buttons: (from left to right) sine, triangle, saw 1, saw 2, square, and random.

16 Click the random shape button on the far right.

The random oscillator shape adds a synthetic feel to the chorus effect that will help dehumanize ADA's voice. The overall effect will be much more pronounced when you increase the Dry/Wet Output control.

The last two controls to the right are Feedback and Output. The Feedback controls include an Amount control to adjust the percentage of the signal fed back to the Chorus Delay line, and a Blend control that lets you change the amount of signal that bleeds into the opposite channel for stereo signals. Increasing the Feedback Amount value adds more of the Chorus effect to the signal, while lowering this value adds more of the inverted Chorus effect to the signal. A value of 0 has no effect, whereas the more you increase or decrease the Amount value, the more gradually pronounced the Chorus sound becomes.

17 Drag the Amount control to the right to 99.

Increasing the Dry/Wet Output control will increase the percentage of the chorus effect that is audible.

18 Drag the Dry/Wet control to the right to 100% to hear only the chorus effect.

19 Drag the Dry/Wet control to the left to slowly lower the wet percentage during playback.

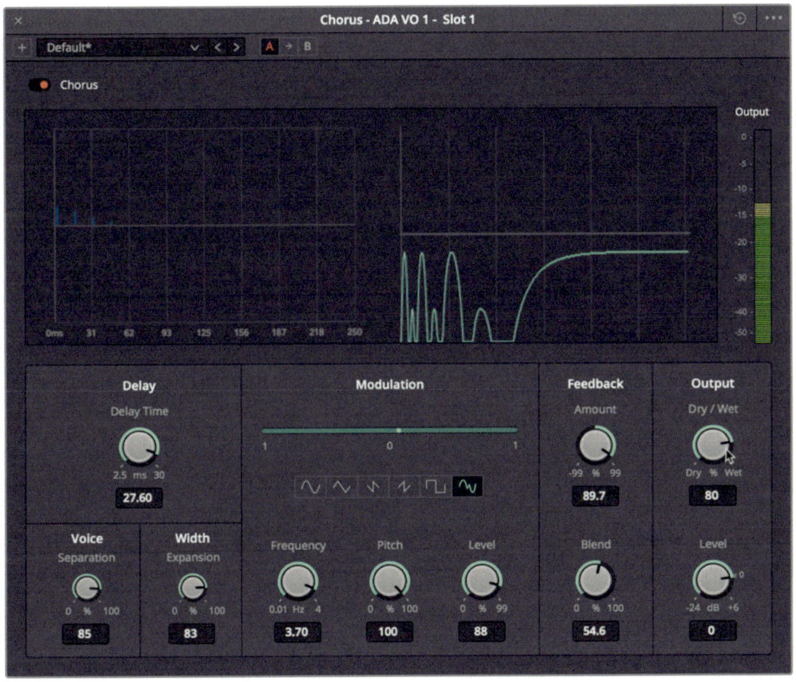

20 Experiment with the different settings. When you are finished, reset the plug-in controls, and choose the Dramatic preset. Close the Chorus panel.

21 Unsolo the A4 track, and play the dialogue scene from the first blue marker to hear the new and improved (dehumanized) voice with the other characters.

ADA's voice no longer sounds like this author's feeble attempt at recording voiceover. You'll later get an opportunity to further alter ADA's voice. For now, let's move on to the next sound design technique.

Adding Depth with Time-Based Plug-Ins

When it comes to motion picture storytelling, both the director of photography and the sound designer must consider height, width, and depth when composing a scene. Just as the director of photography uses light, shadows, reflections, and focus to add scenic depth between the foreground and background elements, the sound designer uses reverb and time-based plug-ins to add audible delay, reflections, and reverberation in a scene to depict depth in the soundtrack.

Evaluating Foley Sound Effects

You have two ways to add Foley sound effects to your soundtrack. The first is to record each cue as a live performance during playback, which requires a recording studio and Foley artist to perform the various movements in sync with the picture. You can apply the Fairlight ADR tools to Foley recordings the same way you use them for dialogue replacement. The second method for adding Foley sound effects is to acquire them from a sound library or service. This method may be less expensive but requires more editing because purchased effects won't be perfectly timed to your onscreen action. Whichever solution you use, you'll still need to enhance the sound effects to fit the spatial elements of the scene.

Let's start by opening a different timeline and listening to a variety of Foley footsteps. As you work with the different examples, you'll see that many of the techniques that you used to edit and enhance dialogue have been applied to the Foley footsteps.

1 Open the 7 "Footsteps" timeline.

  This timeline includes examples of several Foley footstep recordings and sound design techniques.

2 Hide the viewer. Show the Inspector, Effects Library, and the Markers index in list view.

3 Modify the Markers list columns to clearly see the Name column.

4 Mark the first clip in the A1 track, and select the clip to see it in the Inspector.

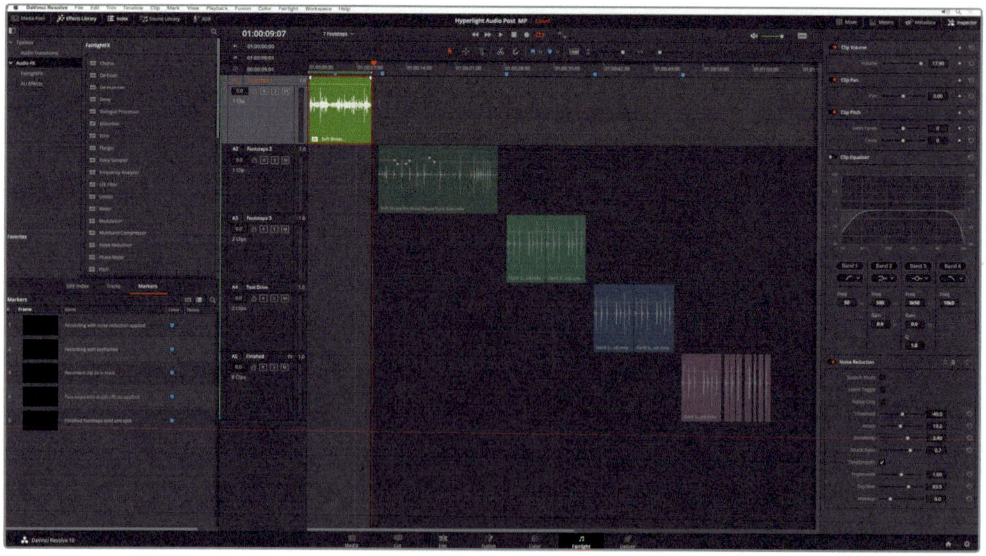

In the Inspector, you can see that this clip has a Noise Reduction plug-in.

5 Start looped playback. In the Inspector, click the Noise Reduction Bypass button to toggle the plug-in off and on during playback.

Did you notice that the soft shoes in this recording were a bit squeaky when they bent and flexed? If so, you have the discerning ears of a sound designer! Of course, these squeaking shoes may offer more realism than most scenes call for—but if the shoe fits....

Still, the Noise Reduction plug-in significantly reduced the background noise, demonstrating that effects designed to improve dialogue clarity can also be applied to sound effects.

Let's move on to track A2 Footsteps 2 to hear the next example.

6 Mark and select the clip in the A2 track.

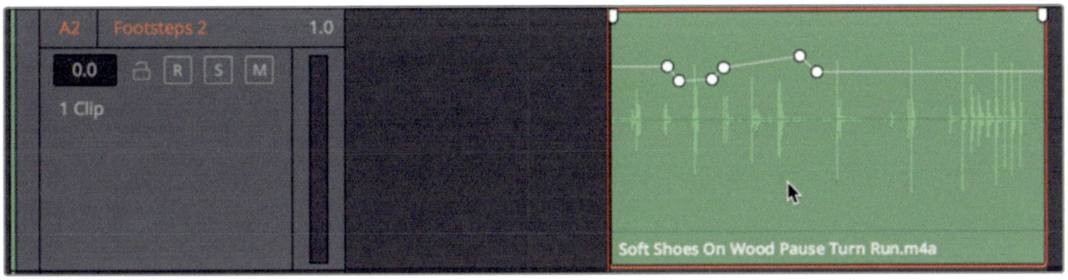

Foley sound effects acquired from a sound library or service usually include clip names that describe the types of shoes, surface, and action used in their recording. For example, the clip on the A2 track is named **Soft Shoes On Wood Pause Turn Run.m4a**.

7 Play the clip. Listen to hear if the recording lives up to its name.

That's about as generic a sound as you can get for soft shoes on wood. In this example, no plug-ins were applied to either the clip or the track. However, the volume overlay includes keyframes to balance the levels of the footsteps. One thing that is missing from this clip is the sense of space. Are these footsteps in a hallway? A bedroom? A confined bathroom or a resonant cathedral?

8 Set a play range around the two clips in the A3 Footsteps 3 track.

This track includes two identical clips named **Hard Shoes On Wood.m4a**.

9 Play the clips and listen to the sound of the hard shoes on wood.

Did you notice that this clip has a bit of unpleasant low-level noise between steps? Let's apply a Gate to clean up the track.

10 Select the A3 track. Show the Mixer. On the A3 channel strip, double-click the Dynamics control.

11 In the Dynamics controls panel, click the Gate button to turn on the default Gate dynamics processing.

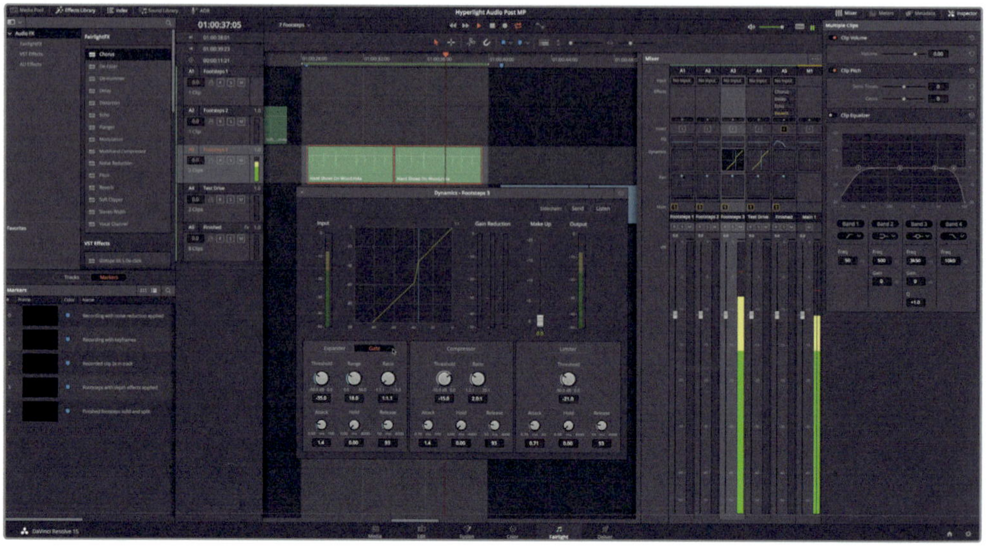

The Gate processing instantly cleaned up the sound of the individual footsteps. Notice that the gate was also applied to the A4 track, which includes the same two footstep clips.

12 Stop playback. Hide the Dynamics window and the Mixer.

Once again, the footsteps live up to their name, but nothing else. It's time to give these footsteps some depth.

Applying Multiple Time-Based Plug-Ins

Time-based effects use varying amounts of delay to manipulate the perception of time, and therefore, the aural illusion of space. An obvious example is echoing footsteps in a cave. The more delay between those echoing footsteps, the larger the perceived size of the cave.

One of the best ways to explore different plug-ins is to add them all to a track and "test drive" each plug-in during playback. In this exercise, you'll audition different plug-ins using bypass switches and presets to add realistic depth to the Hard Shoes on Wood Foley clips. Along the way, you'll also explore different types of FairlightFX time-based plug-ins.

1 Set a play range around the two clips in the A4 Test Drive track.

The A4 track already has Gate processing applied to it.

2 Select the A4 Test Drive track header to show the track parameters in the Inspector.

3 In the Effects Library, drag the following plug-ins to the A4 Test Drive track header: Chorus, Delay, Echo, and Reverb.

4 Close all the plug-in windows. Hide the Effects Library and the Index.

5 In the Inspector, Option-double-click any of the plug-in effect headers to compress all of them at once. Then, in the Inspector, click the bypass switch on each plug-in to turn them all off.

The effects are applied, and now you can test each one during looped playback. In the previous exercise, you used the Chorus plug-in to thicken a voiceover track. Now, you'll use the default Chorus setting to add a bit of perceived depth to an otherwise "flat" Foley recording.

6 Start looped playback. In the Inspector, click the Chorus bypass switch to turn on the effect. Toggle the bypass switch off and on to evaluate the effect in the track.

The effect is subtle, but it definitely adds a hint of depth to the track, as well as weight and grit to footsteps. Let's try the next example.

7 Continue looped playback. Turn off the Chorus effect, and turn on the Delay.

Clearly, this effect's default settings are more pronounced and distracting.

8 In the Inspector, to the far right of the Delay controls, click the Customize button to open the Delay panel.

The FairlightFX Delay is a standard stereo delay effect that you can use for track doubling, early reflection generation, and harmonic enhancement.

9 Drag the Delay Time control all the way to the left to remove the delay, and then slowly drag it to the right to add a slight delay to the footsteps. Try a value between 30 and 35 milliseconds (ms). Set the Dry/Wet amount to around 15 so that only 15% of the output includes the delayed effect.

10 Toggle the Delay bypass off and on to compare the track without and with the delay. With the delay turned on, you should hear a slight delay that creates the illusion of sound reflecting off the ceiling and walls.

It's amazing that a small amount of delay can make a sound seem as if it is coming from an actual space. How is that possible? Because our hearing is based on binaural perception (two ears) and the differences in loudness, frequency, reflections, and reverberations in those ears. Your job as the sound designer is to apply the amount of delay necessary for the space depicted on the screen.

11 Close the Delay panel. In the Inspector, turn off the Delay plug-in.

12 Continue looped playback. Turn on the Echo plug-in.

This Echo default sounds a bit like a tap-dancing lesson.

This classic Echo effect includes a graph showing the timing and intensity of the generated echoes on each channel, as well as an Output meter that displays the level of the resulting signal.

Let's apply some of the Echo presets.

13 Show the Echo panel. Try each of the presets starting with Fast Ping Pong to hear four variations of Echo applied to the footsteps.

14 When you are finished, close the Echo panel and bypass the effect.

If you recall, you used the Echo Fast Ping Pong preset in Lesson 1 when you were building that exciting trailer soundtrack. It worked great for that effect but isn't as effective here. Echo is a powerful tool, but a little bit goes a long way. When you need it, you'll know where it is and how to apply it.

> **NOTE** You'll find detailed information about all the FairlightFX plug-ins and their controls in the DaVinci Resolve 16 User Manual, available in the application's Help menu.

Last, but not least, is the all-powerful Reverb plug-in, which is one of the most useful effects in audio post-production.

15 Start playback, if necessary, and turn on the Reverb plug-in.

You are probably surprised at how bad the Reverb effect sounds. That isn't the plug-in's fault; it's actually due to the Gate that is applied to the track. In this example, the gate removes some of the low-level sounds necessary for simulating spatial sound reflections and reverberation.

16 Show the Mixer. In the A4 Test Drive track, open the Dynamics panel and turn off the Gate processing. Close the Dynamics panel.

Suddenly, the footsteps sound as if they are in an actual space. Unfortunately, now the low-level noise is also audible, which shows that presets and default settings will only get you so far in sound design.

How Loud Should Sound Effects Be?

The first rule of sound effects levels is that there are no rules, other than not making sound effects too loud or too quiet. Remember, dialogue is king, so all the sound effects are there to support story, characters, mood, and audience perception. How loud should sound effects such as Foley footsteps be? That depends on the scene, the environment, and the artistic instincts of the sound designer. If the soundtrack is supposed to sound "realistic," for example, footsteps in a snow-covered field will probably sound much quieter than tap shoes in a high-rise stairwell. The audience's expectation for sound levels also changes with the type of shot in view. A close-up shot of work boots crunching through the snow should be mixed louder and with more "presence" than a wide shot of the same boots on snow. Keep in mind that sound design is a creative art form, and therefore a sound designer and director can choose to use completely unrealistic sound levels just for dramatic or comedic effect. For example, a child walking into her sleeping parent's room could have footsteps that sound like a charging elephant, or footsteps so light she seems to glide across the room.

Here are a few guidelines for starting levels:

Dialogue: Average speech -12 dB (-10 dB loud voice to -15 dB soft voice, -6 dB scream, -20 dB whisper)

Music: Underscoring during dialogue -18 to -20 dB, music without dialogue -12 to -15 dB

Sound Effects: -10 to -20dB

Loud SFX: Explosions and other highly impactful SFX peak at -6 to -8 dB

Balancing Clip Levels Before Adding Effects

Just as with dialogue tracks, you should first balance your effects tracks levels in the context of the scene. Also similar to working with dialogue tracks, balance the effects clips within the track. Try not to apply the track fader until you are ready for mixing.

1 In the Inspector, turn off the Reverb plug-in.

2 Using the Selection tool, select both clips in the A4 track so you can adjust their clip levels simultaneously in the Inspector.

3 Start looped playback. Watch the A4 track's levels in the Mixer.

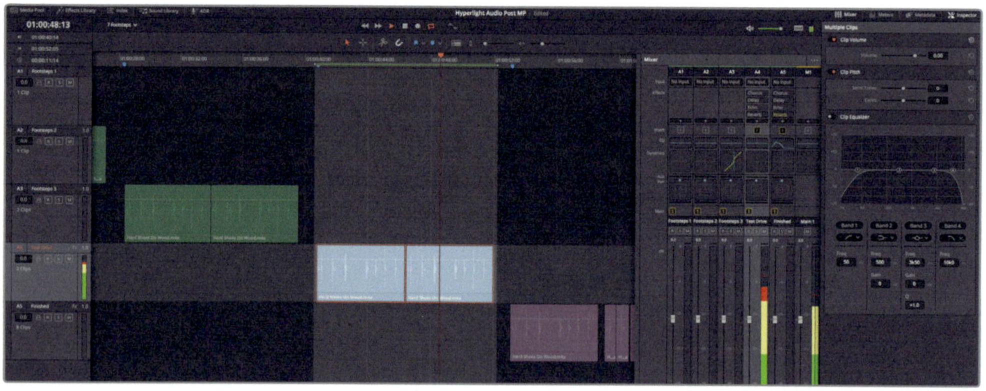

The levels are peaking well into the red, so they definitely must be lowered. Because these clips don't have to balance with any dialogue, you can set them so their peaks are within the yellow range of the meter, around -18 to -12 dB.

4 In the Inspector, lower the Multiple Clips Volume level to around -7.30.

5 Hide the Mixer.

At this level, the slight background noise is no longer noticeable, so you can proceed with the Reverb without gate processing.

How Do We Interpret Spaces Through Sounds?

Chances are that you can instantly recognize the sound of a dog barking and even guess the general size and breed of the dog based on its pitch and power. Even in the dark, with no visual information, we can guess our proximity to the dog and assess whether it is nearby, in the house next door, in a concrete garage, or at the bottom of a deep well.

How do we determine all that information from sound alone? The human brain can instantly process a sound based on recognition. If we don't recognize a sound, we guess at it's nature and origin based on similarities with other familiar sounds. We also estimate the size and distance of the source based on its volume level, and estimate its location based on triangulating the signal with our built-in stereo sensors (ears).

To identify the general size and surface of a space, we subconsciously use three acoustical elements: the original sound, the early reflections on surfaces, and the late reverberations of the sound bouncing off the surfaces in the area. Bright early reflections indicate hard smooth surfaces such as tile, marble, or glass; whereas dull, muffled, or muted early reflections suggest grass, books, carpet, or padded walls.

Sound designers and sound mixers apply these acoustic principles to emulate realistic spaces and events when building and mixing soundtracks.

To ensure clean recordings, sound stages and recording studios are acoustically treated to absorb and minimize early and late reverberations. Then, spatial simulators, such as the FairlightFX Reverb, are applied to those non-reverberant recordings to simulate the acoustic space appropriate to the scene.

Simulating Spaces with Reverb

Resolve's powerful FairlightFX Reverb plug-in can instantly add a sense of scale and space to your tracks. This spatial simulation plug-in displays a graphical 3D cube interface with which you can control corresponding reflections and reverberations based on the size and shape of the "room." Additionally, various controls let you take a "dry" recording and make it sound as through it's in a live, reverberant space.

1 Select the A4 track.

2 In the Inspector, turn on the Reverb plug-in. Double-click the Reverb plug-in header to expand it. Click the Customize button to show the Reverb dialog.

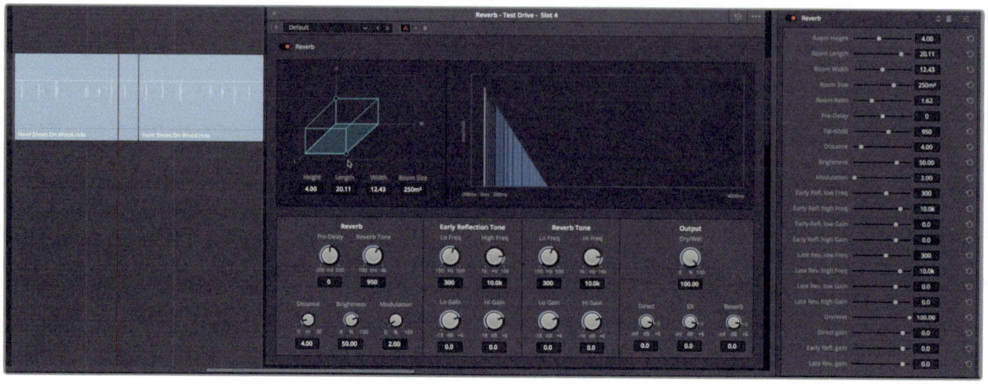

The Reverb dialog controls are fairly self-explanatory because they are based on the size of a simulated 3D room in which you can adjust the combination of the original signal (Direct), early reflections (ER), and late reverberations (Reverb).

In the upper left of the dialog, the teal-colored 3D rectangle represents the room size. The graph to the right roughly visualizes (from left to right) the reverb's effect on the audio signal. The controls at the bottom let you fine tune the Reverb timing, as well as Early Reflection Tone and Reverb tone. The controls in the lower right adjust the levels of the Direct signal (white vertical line), Early reflections, and Reverb levels as depicted on the graph in blue.

Let's start by listening to the first two presets, as they are a great way see and hear the extreme differences between small and large spaces.

3 Start looped playback. In the Reverb dialog, choose the Bathroom preset. Toggle the Bypass button off and on to hear the footsteps without and with the Bathroom preset.

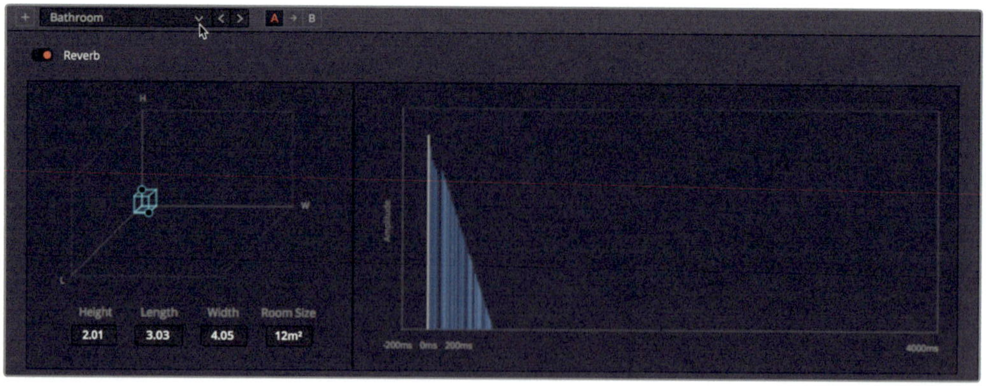

Notice that the quick, reflective sounds from the Bathroom preset give the aural impression of a small space with highly reflective surfaces.

4 Choose the Cathedral preset to hear the enormous-sounding space created by increasing the Reverb Time, Early Reflection Tone, and Reverb Tone.

Notice that the size of the room displayed to the left correlates to the length of the reverb tail to the right.

As mentioned earlier, textile surfaces such as marble and wood emit varying degrees reflective qualities. In the Reverb dialog, the Early Reflection Tone and Reverb Tone areas offer equalization controls to independently modify the tone of the reflections and reverb to suit the textile characteristics of the room. Let's compare the sound of three different small spaces, each with different Tone settings. Listen for the differences in reflective sound between the Studio presets and the Bathroom preset.

5 Continue looped playback. Choose the Studio2 preset, and listen to the dull, reflection-absorbent sound. Then, choose the Bathroom preset to hear a reflective, "tiled" room sound. Next, try the Studio1 preset. Switch between these three presets to compare their Early Reflection Tone and Reverb tone settings.

Could you hear the difference in the tone, early reflections, and reverberations of the presets? Both of the Studio presets have duller, less reflective sounds, which suggest sound-absorbing materials on the walls and ceiling. The Bathroom preset, on the other hand, has heavy early reflection that indicates more reflective surfaces such as tile, glass, or polished metal.

Let's manually change the size and sound of the room to a long corridor.

6 Stop playback. Reset the Reverb plug-in.

7 Drag the control handle (dot) in the front-right corner of the 3D room to create a rectangular room with the following dimensions: Height 4.00, Length 30.00, Width 5.00.

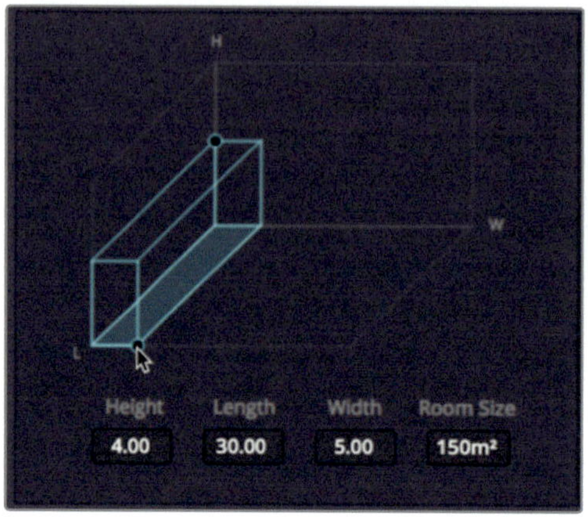

8 Start looped playback. In the Reverb controls, drag the Reverb Time to the right and left to extend and shorten the reverb length. Try setting it at around 1400 ms.

This 3D shape shows a long corridor, so let's adjust the sound to match the space. To do so, you'll use the distance control to increase the simulated distance between the virtual source and the virtual listener.

9 Drag the Distance slider to around 20.00 m (meters).

Next, you'll adjust the Brightness control to change the shape of the decay time for the higher frequencies. At maximum brightness levels, the high frequency decay time is identical to any other frequency. Lower brightness levels result in shorter decay times and a duller sound.

10 Drag the Brightness control to the right and left to hear the difference, and set a final level around 25.00%.

The Output controls allow you to combine all audio processing results into a single output signal. These controls include a Dry/Wet control to determine the percentage of the original to processed signal, along with independently adjustable Direct, ER (early reflection), and Reverb (late reverberation) controls.

Let's tweak the Output controls.

11 While looking at the blue graph, drag the ER control to around -30 dB to see and hear the impact of a less reflective surface. Click the Reverb's bypass switch on and off to hear the clip without and with the reverb.

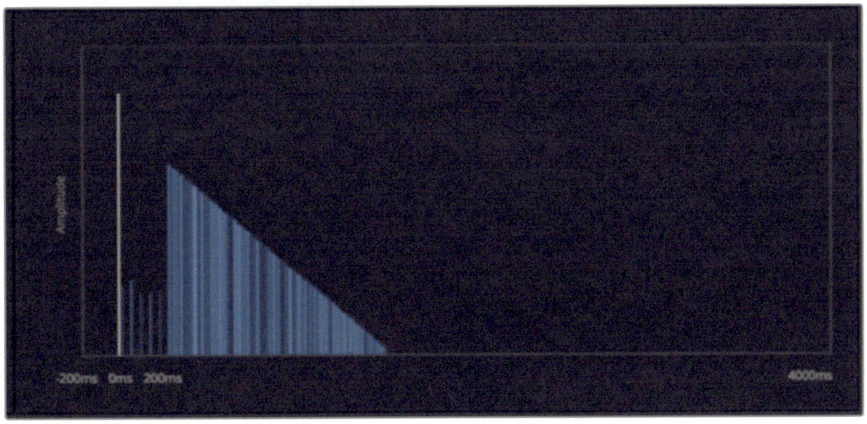

You can see that the early reflection is depicted by the loosely spaced vertical blue lines between the direct signal (white line) and the blue slope of the late reverberations.

12 Drag the ER slider to the right to around -4.5 dB to increase the early reflections.

13 While watching the graph, drag the Reverb control to the left to lower the level of the late reverberations to around -35.5 dB.

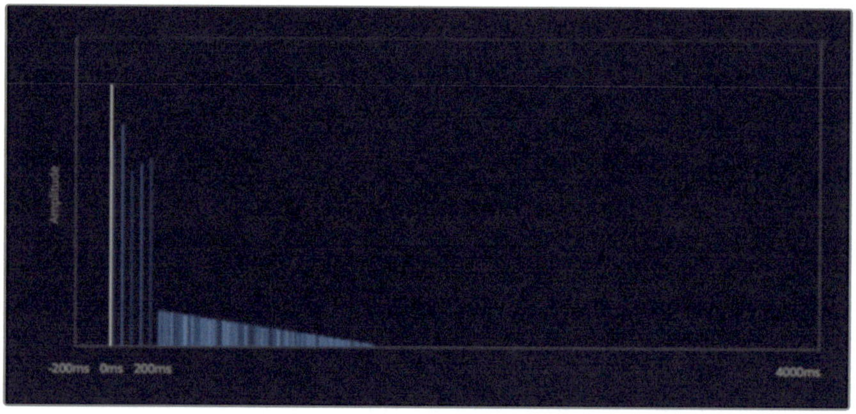

You can hear and see the minimal reverberation in the graph.

14 Set the Reverb to around -12.0 dB. Stop playback.

Fabulous. You've created an awesome-sounding corridor. Let's save it.

Saving a Customized Preset

After you have customized a preset, you can save it for use in another scene in the same project, or in another project altogether. In this exercise, you'll save the settings that you just created for the Corridor reverb.

1 Click the + (Add Preset) button.

2 In the Reverb preset dialog, type **Corridor**. Press Return, or click OK.

TIP When naming presets and sound effects, it's always a good idea to give them clear, self-explanatory names. For example, a corridor is always a corridor, while a hall could be a hallway or a concert hall.

The Preset pop-up updates to show the current preset, Corridor. You can select this preset from the list at any time, and even update it and resave as a different preset.

3 With the Corridor preset selected, increase the Reverb Time to around 2922 ms, and in the Output controls, increase the Reverb level to around -5.1 dB.

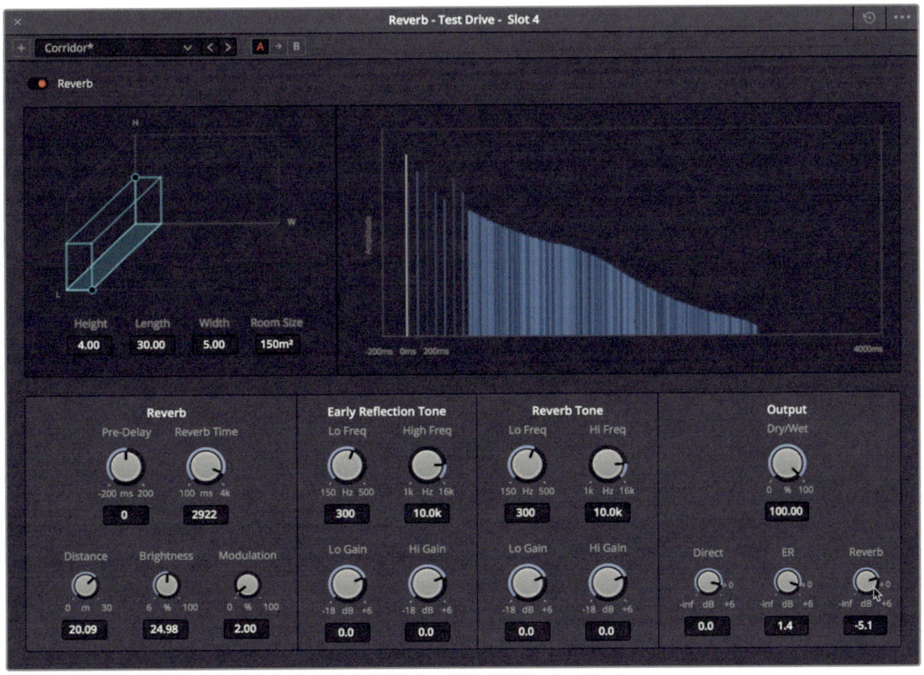

An asterisk after the preset's name indicates that changes were applied to the current preset. You can save over the original preset, or save a new version. Let's save these changes as a different preset.

4 Click the Add Preset button.

A Reverb Preset dialog opens with the option to update the current preset or create a new one.

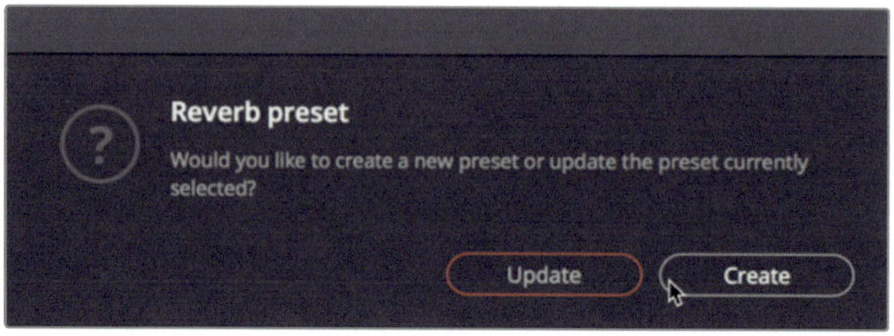

5 Click Create. Name the new preset **Corridor 2 wet**, and press Return.

6 Click the preset menu to see the two new plug-ins in the list.

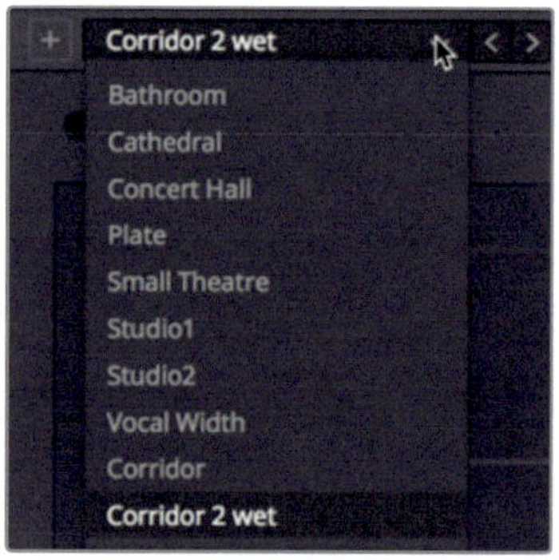

7 Start looped playback. Switch between the Corridor and Corridor 2 wet presets to compare them. When you are finished, stop playback.

The best part of saving presets is that you can use them in any Resolve project on your system.

Mixing and Matching Time-Based Plug-Ins

Because you already have four plug-ins applied to the A4 Test Drive track, this is a great time to experiment with using the plug-ins separately and in combination. Before moving on, take a moment to test these plug-ins, presets, and controls alone and in combination.

For example, try combining the Chorus and the Reverb plug-ins for a more realistic sound. Keep in mind that you can use the bypass switches to toggle the plug-ins off and on while you work. Also, feel free to try the parameter controls in the Inspector or on the effect's panel to experiment with different settings. Best of all, you can open as many effect dialog panels as needed while you work. Have fun. When you are finished, stop playback and close all of the open plug-in dialog windows.

Creating Physical Sound Barriers with EQ

Another element of professional sound design is limiting high frequencies to create realistic aural obstructions to match the source sound to the listener's perspective.

You can limit frequencies of an individual clip in the Inspector or EQ an entire track in the Mixer.

In this example, you'll further transform the footsteps so they sound as though they are coming from a corridor on the other side of a sealed door in a space ship. To do so, you'll use a low-pass filter in the track Equalizer to eliminate the high-frequency sounds just as a door would do in the real world.

How Do Physical Barriers Affect Sound?

Imagine you are in a pitch-black room. Somehow, you can still recognize whether the person speaking to you is facing you or has her back to you. You can also determine the difference between the speech of someone within your room, and the same person at the same volume level and distance positioned on the other side of a door, window, or wall. How do we know? Our minds can interpret these differences based on the clarity of the sound and the strength and range of audible frequencies without any visual clues.

In the real world, the best sound quality and clarity occurs when a sound source is in close proximity to the listener. The clarity of a sound source diminishes with a change in source position and orientation, or the placement of a physical obstruction between the source and the listener This aural compromise occurs because such changes alter the balance of certain sound frequencies: The greater the obstruction or positional change, the more limited the high and mid-high frequencies that reach the listener.

Why don't we hear the high frequencies through walls? In a nutshell, high-frequency sound waves are relatively short, and therefore more easily absorbed by solid materials. Low-frequency sound waves are much longer and require more power to become loud enough for human ears detect. As a result, lower frequencies not only have the power to reflect and bounce around a room, they can also resonate within and through walls, doors, and windows.

If you've ever been in a hotel room or apartment with noisy neighbors, you know it's the booming low to low-mid frequencies of their music, television, and shouting that penetrates your walls to drive you crazy.

1 Play the clips in the A5 Finished track to hear the finished footsteps with the EQ's low pass filter applied.

   Impressive, right? It absolutely sounds as if the footsteps are coming from a long room on the other side of a sealed door in a space ship. (You probably noticed that one of the footsteps clips was split into short sections, each containing a single footstep, which makes it easier to sync the individual steps to picture.)

Now, it's your turn to recreate that sound by applying the 6-band Equalizer to the A4 track. First, let's streamline the interface.

2 Select the A4 track, and mark the first clip for looped playback.
3 If necessary, hide the Effects Library and the Inspector.
4 Show the Mixer. In the A5 channel strip, look at the effects slots and EQ graph to see that both the Reverb effect and EQ are applied and active.
5 In the A4 channel strip, double-click the EQ control to open the EQ window.

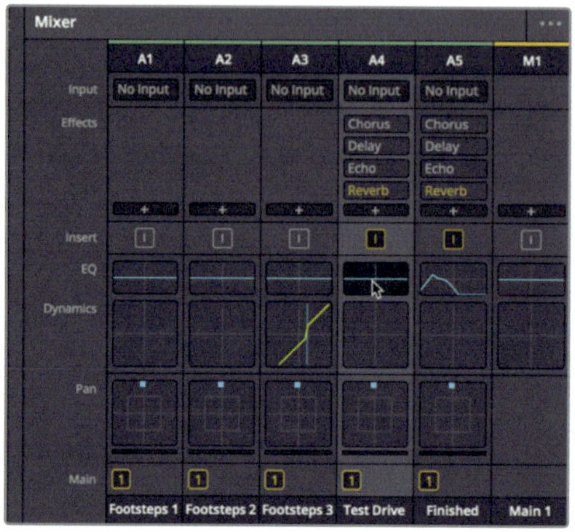

The Equalizer window opens for the Test Drive track.

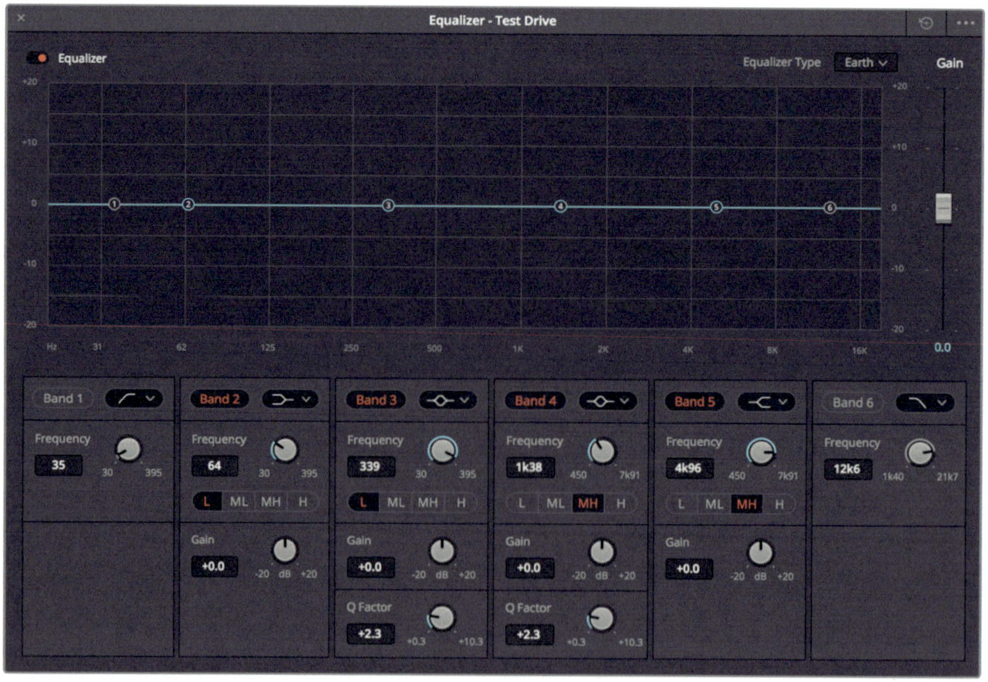

This EQ window is very similar to the 4-band EQ that you used in the Inspector to detect sibilance in Lesson 6. However, the track EQ includes six bands of equalization with graphical and numeric controls for attenuating (lowering) and boosting (raising) ranges of audio frequencies on each track. Bands 1 and 6 are primarily used as high- and low-pass filters. The middle bands control the Low (L) Mid-Low (ML), Mid-High (MH), and High (H) frequencies.

6 In the Equalizer window, click the Band 6 button to turn it on.

7 Start looped playback. Slowly drag the Band 6 handle to the left until the lower part of the low-pass filter curve reaches 1k.

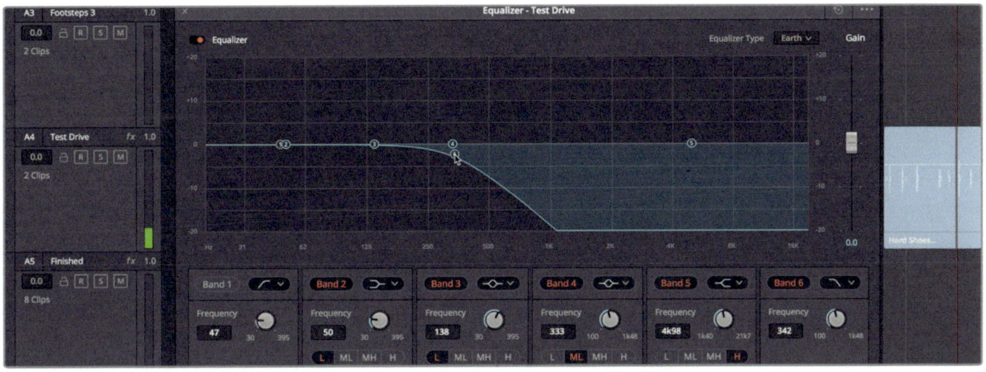

The footsteps sound as though they were limited by a physical barrier of some kind.

8 Click the A5 track header to open that track's Equalizer window.

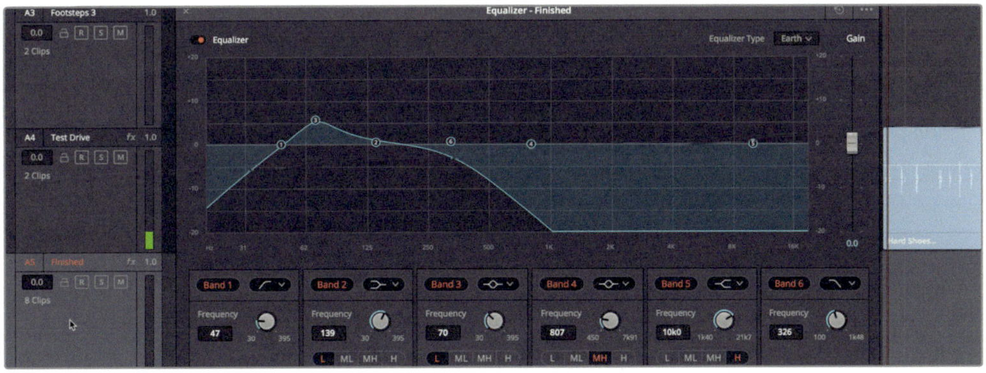

As you can see, the A5 track includes a high-pass filter (Band 1), and Band 3 has been boosted at around 70 Hz. Let's do the same to the A4 track.

9 Select the A4 track. In the Equalizer window for the Test Drive track, click the Band 1 button to turn it on. In the graph, drag the Band 3 handle up to the +5 dB line (one line above 0). In the Band 3 controls, drag the Frequency control to the left until it reads 70 Hz.

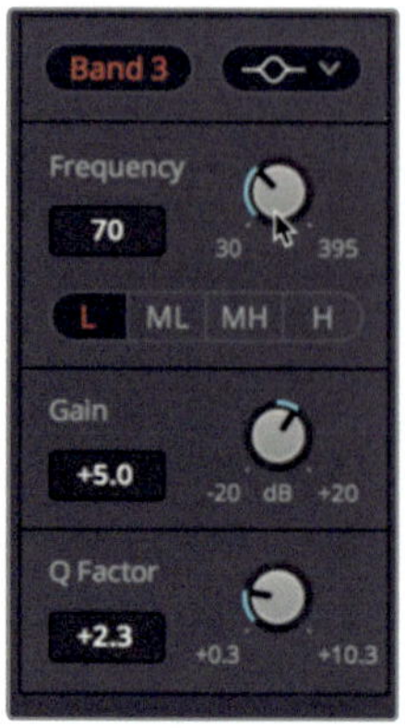

10 Stop playback. Close the Equalizer window.

NOTE Filtering out all the high frequencies will also diminish the volume level. You can always raise the output level (gain) of the EQ with the Gain slider on the EQ dialog. You can also raise the level of the clips in the timeline, or track in the Mixer, as needed.

Now the EQ graphs for the A4 and A5 tracks match.

11 Hide the Mixer and stop playback.

This technique of applying a low-pass filter to limit frequencies can be used for dialogue, music, sound effects, and anything else that needs to sound obstructed in your soundtrack.

This ends your exploration of sound design with footsteps. Let's take a moment to examine a finished mix to see and hear how the techniques you've learned in this lesson can be combined to create finished effects.

NOTE Frequency filters don't have to be extreme to be effective. For example, in a scene in which a character turns his back to the camera, the high frequencies would naturally be limited. So, to simulate this, you would simply apply a high-frequency shelf to reduce the high to mid-high frequencies, or a low-pass filter to eliminate them. In the Band 6 filter type pop-up menu, you can change from a low-pass filter to a high-shelf filter.

Working with Foley Sound Effects

Now that you've explored different sound design techniques for making sound effects fit the space within the scene, let's step back and look at a few advanced techniques for adding Foley sound effects to the timeline. In this set of exercises, you'll work with several professional Foley sound effects from the Fairlight Sound library to add Foley sounds to a scene. In the first exercise, you'll set a sync point for precision alignment. Then, you'll explore the Foley Sampler plug-in with which you can assign sound effects to MIDI keyboard keys and record them to the timeline in sync with the picture.

Using a Sync Point for Precision Alignment

When working with sound effects, there are different methods for syncing them to the timeline. Sometimes you base your sync on the head or tail of the sound clip. Other times, there is a specific frame in the clip that needs to perfectly align with the picture. In that case, you can use the playhead in the Sound Library to set a sync point that will align with the playhead in the timeline when you audition the clip.

For this example, you'll see a clip where Philip is digging around in a bin, with barely any sound to go with the clip. Therefore, you'll need to find a Foley sound effect in the Sound Library that will work. Once you find the right clip, you'll analyze the audio waveform and the video clip to determine the best sync point. Keep in mind that the audience's mind will believe what they hear, as long as it seems to be in sync with the picture. Hence, the need for using a sync point. Let's try it.

1 Open the 7 Foley SFX and Foley Sampler Start timeline.
2 Open the Viewer as a separate window and hide the Meters at the top of the interface.

3 Play the second clip in the A1 track.

The sound where Philip digs around in the bin looking for tools is a bit lackluster. Let's find a suitable sound effect that will enhance the drama of his search, as well as add another level of realism to the soundtrack.

4 In the Interface toolbar, click the Sound Library button to show the Sound Library.

5 Click the Database button to the right of the search field to show the Database pop-up menu. In the Database pop-up menu, choose > sound_fx.

If you recall, this is the database where you imported the sound effects in Lesson 1.

> **NOTE** If you did not follow all the steps to import a sound library in Lesson 1, please go to that section of the book and follow those steps before continuing to the next step.

Now, you can find a sound effect to audition in the Sound Library.

6 In the Sound Library search field, type backpack.

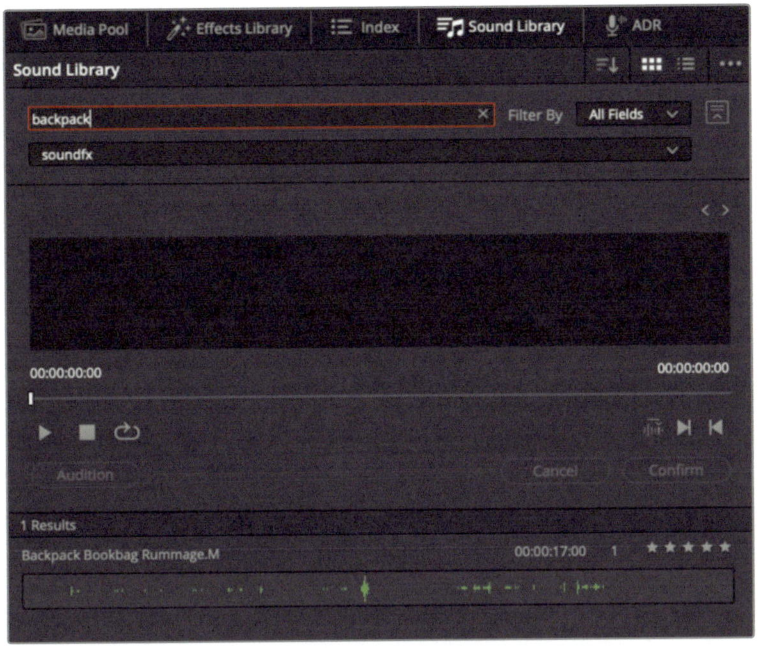

A foley sound effect titled "Backback Bookbag Rummage" appears in the results list. This is the sound that you'll use when Philip rummages around in the bins. The trouble is that the effect doesn't actually sync with his movement. No problem. In most cases, syncing just one specific moment of a clip with a generic sound effect will often sell the effect for the rest of the scene.

7 Select the sound in the list and preview it in the Sound Library. As you play the rummaging sound effect, watch the playhead move across the waveform, and listen for any distinctive moments in the clip.

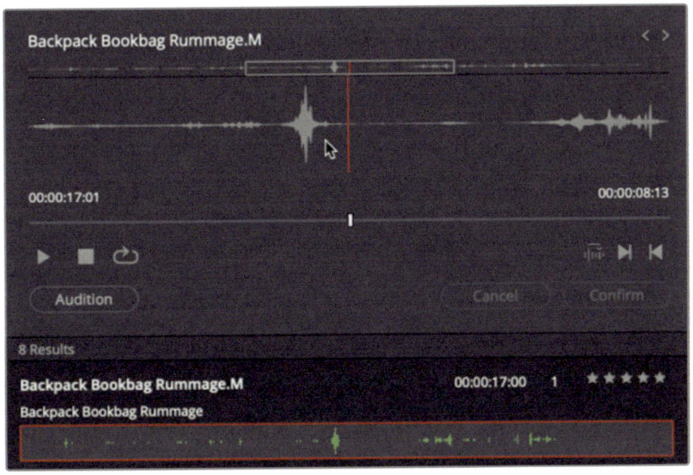

For the most part, it is just a lot of rooting and rummaging sound. That is, until the rather distinctive sound of a hand whipping out of the backpack. This is exactly the kind of Foley crescendo that can be synced to Philip's hand as he pulls it out of the first bin.

In a few minutes, you'll audition the sound effect to the A2 FoleySFX track. First, you'll need to set your sync point in the timeline with the playhead, as well as a sync point in the Sound Library with the preview playhead.

8 In the timeline, move the playhead to the blue marker above the second clip in the A1 track.

This marker position is where Philip's hand is almost all the way out of the bin.

9 In the Sound Library, use the JKL keys to move the playhead back and forth over the loudest part of the waveform near the middle of the sound effect. That is the section you will sync to the video. Refine the playhead position so the playhead is just to the right of the loudest part of the waveform.

10 Click the Set Sync Point button, to the left of the In Point and Out Point buttons, to set a sync point at the playhead position.

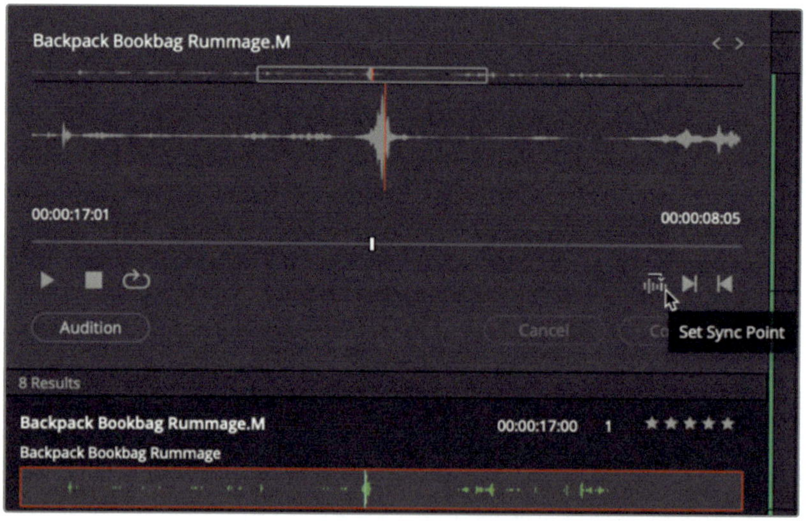

A green playhead appears over the waveform in the Results list to indicate where the sync point is currently set. You can set a new sync point anytime by simply moving the playhead to a new position over the preview clip's waveform and clicking the Set Sync Point button.

Before auditioning the clip in the timeline, it's a good idea to change the current In Point of the sound effect, because at the moment the sound effect is much longer than needed. Let's set an In Point three seconds prior to the sync point.

11 In the Sound Library, press Shift-Left-Arrow three times to move the preview playhead three seconds to the left. Then press I or click the In Point button.

12 In the timeline, select the A2 Foley SFX track, and make sure that the playhead is in position at the blue marker over the desired clip.

13 Click the Audition button to add the clip to the selected track, using the sync point and timeline playhead for sync alignment. Play the clip in the timeline to see and hear how well the Foley sound effect works with the video.

Pretty cool, right? It sounds as if the backpack rummaging sound was part of the recorded scene all along. Using this type of Foley fx sleight of hand (pun intended) is a great way to enhance your soundtracks. The volume level may be a little intense, but the alignment and content works great.

14 Feel free to finesse the sound effect clip's levels and timing to taste.

15 When you are finished, click the Confirm button in the Sound Library to commit the sound effect to the timeline.

Now that you know how to set a sync point in the Sound Library, let's move on to re-recording sound effects to the timeline with the Foley Sampler plug-in.

NOTE The next exercise requires DaVinci Resolve Studio 16 or higher to use the Foley Sampler to record Foley sounds in sync to the video using your mouse or MIDI keyboard. If you are working with the free version of DaVinci Resolve, you can still follow along with the lesson to learn how the Foley Sampler works. At the end of the lesson, instead of recording Foley samples to the timeline using the Foley Sampler plug-in, you can use other tools available in DaVinci Resolve to split a Foley sound effect clip into separate sample sections and edit them manually or with shortcuts to match the video.

Exploring the Foley Sampler (Studio)

The Foley Sampler plug-in is a built-in sampler FairlightFX plug-in that makes it easy to add sound effects that you want to play using a keyboard, pad, or other MIDI performance device connected to your computer, to add timed sound effects to sync with onscreen visuals. This powerful plug-in has also been designed to simplify the process of recording performed audio cues on the current track to which the sampler has been added.

In this exercise, you'll start by adding the Backpack Rummaging Foley sound effect to the Foley Sampler to see how it works as a sampled sound effect that can be performed with a MIDI keyboard or mouse. Then you'll add Foley footsteps to the Foley Sampler to demonstrate how to map additional sounds to the sampler. Finally, you'll learn how to record Foley Sampler sound effects into a track, in time to the footfalls onscreen.

The overall goal is to map Foley footstep sounds to the Foley Sampler so you can record them in sync with the video, and then apply sound design so the recorded Foley Sampler footsteps sound as though they are happening within the diegetic context of the scene.

1 Move the playhead to the beginning of the timeline and play the first two clips in the A1 track. During playback, pay close attention to Emiliana's footsteps because you will be re-recording them with the Foley Sample plug-in shortly.

2 In the Interface toolbar, click the Effects Library button to show the FairlightFX list.

3 Drag the Foley Sampler plug-in to the A3 Foley Sampler track header.

When you dragged the Foley Sampler to the A3 track, not only did the Foley Sampler window open, but it was also automatically patched as an input instrument ready for recording to the A3 track.

At this point, the Foley Sampler is ready to be used, but by default it has no samples loaded to play. Next, you'll add a sound effect from the Sound Library.

NOTE If you have a MIDI controller of some kind connected to your computer and properly configured, it will appear in the MIDI pop-up menu at the upper-right corner of the Foley Sampler window (next to the Keyboard button). Choose your device from this menu, and the Keyboard button will highlight to show it's enabled.

Adding Sound Effects to the Foley Sampler

There are three ways to add prerecorded sound effects to the Foley Sampler. You can drag and drop a sound effect from the Sound Library or Media Pool onto the Foley Sampler window. Another method is to click the Foley Sampler window's Option menu and choose one of the Add or Import sample options. Finally, you can simply drag an audio file in a supported format from the file system on your computer onto the Foley Sampler window. For this exercise, you'll use the first option and drag a sound effect from the Sound Library.

1. Open the Sound Library. The **Backpack Bookbag Rummage** sound effect should still be showing. If not, search for backpack to load that effect into the results list.
2. Drag the **Backpack Bookbag Rummage** sound effect from the results list to the Foley Sampler window.

Once you add a sound effect to the Foley Sampler, it is automatically assigned to the keyboard keys, starting at the C2 keys. If you look above the keyboard keys, you'll see a red bar indicating which keys the sound effect is currently assigned to. The first key plays the sound effect at its native pitch and speed. Subsequent keys to the right of the first key will play the effect incrementally faster and at a higher pitch.

3 Click and hold the first keyboard key (C2) on the far left of the Foley Sampler keyboard to play the sound effect in its native pitch and speed.

4 Click additional keyboard keys toward the middle and the far right to hear the difference in speed and pitch.

5 Right-click the C2 key to automatically play the full sample. Click the key again to deselect it.

Now that you know how to add a sample and play it using the keyboard keys, let's make a few modifications to the sample and mapping.

Modifying Sample Mapping

Although the default mapping works, it is important to be able to map the samples to the keys of your choosing. In this case, the backpack sound effect works well at its native pitch and speed and may also work a few keys higher for variety. However, it is unlikely that the very high and fast versions of this sample will be useful in the Hyperlight soundtrack. To modify the key mapping, you simply go to the Keys controls at the top of the Mapping panel.

At the top of the Foley Sampler, you'll find four panel buttons that can be used to show different panels of controls: Mapping, Sample, Level, and Filter. In this exercise, you'll work with the Mapping and Sample controls to remap the sample to the first twelve keys and change which part of the sample is played when you press a key.

The Low, High, and Centre control knobs determine which keyboard keys are assigned the sample.

1 In the Keys controls, drag the High knob toward the left to a value of 47. The red bar over the keyboard keys stops after one octave (12 keys), which is just before the C3 key.

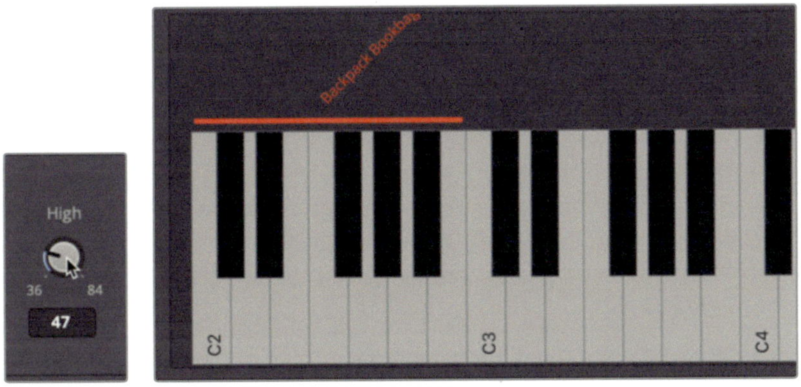

2 Change the Centre knob value to 41, which is the key in the middle of the mapped keys between the lowest key 36 and the highest key 47.

3 Right-click the forth white keyboard key from the left (beneath the name of the sample) to play the sample at its native pitch and speed. Feel free to click any of the other keys within the mapped area to hear how the sample has been remapped, with subsequent keys to the left playing lower pitch and slower, and keys to the right playing higher pitch and faster.

Next, you'll look at the sample and change which part of the sample is played when you press a key.

4 At the top of the Foley Sampler, click the Sample button to show that panel.

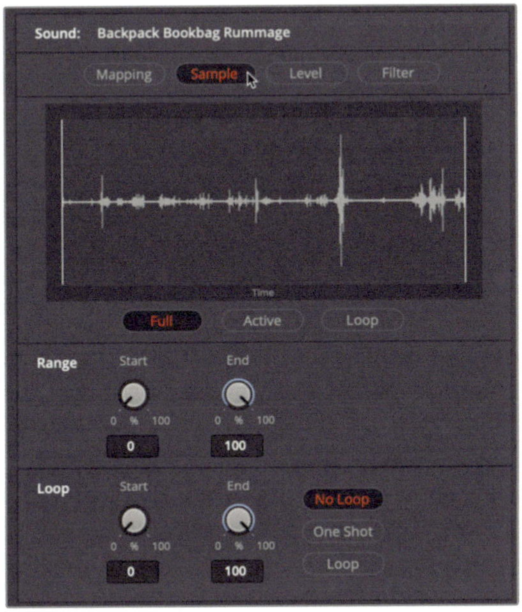

The Sample panel controls show the sample waveform and include controls for determining which part of the sample is played and whether it is played once or looped. White vertical lines on the far left and right side of the waveform indicate the Start and End points of the sample that are included when you play a key. Your goal for this exercise is to change the Start and End points of the sample so that it includes only the loudest section of the sample.

5 Right-click one of the keyboard keys beneath the red line to play the sample. As the sample plays, you can see the blue playhead move across the samples waveform.

6 In the Range controls, drag the Start knob toward the right until the white vertical line is just before the loudest part of the waveform (around 66.97). Then drag the End knob toward the left until the second white vertical line is just after the loud part of the waveform (around 72.62).

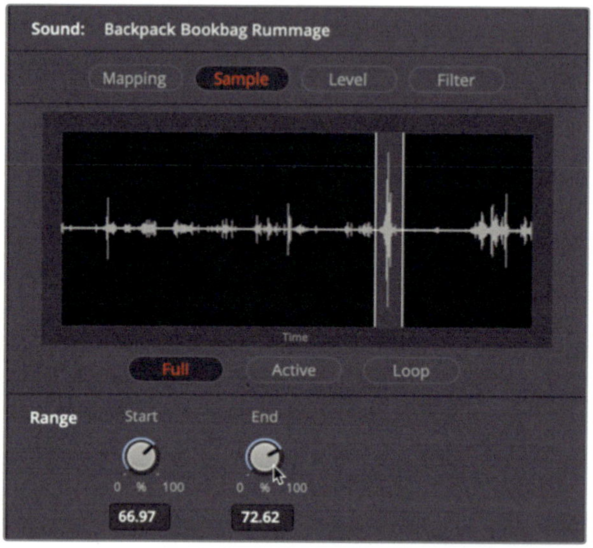

7 Click one of the mapped keyboard keys to play the modified sample.

As you have just demonstrated, it is easy to add and modify a sound effect sample in the Foley Sampler.

Manually Mapping and Splitting Additional Sounds

The Foley Sampler can accommodate additional sounds that can be mapped to different parts of the keyboard so they can be played and recorded in sync to the picture. For example, in the current scene with Emiliana performing maintenance on the ship, you could map sound effects for metal things being moved, buttons pressed, her hand grabbing things, and even her footsteps as she walks. To keep it simple, let's just add some footsteps. First, you'll need to set the Foley Sampler Keys controls to Manual mode. Otherwise, the sample mapping will override your previous mapping and automatically split the keyboard evenly between the previous sample and the new sample.

1 At the top of the Foley Sampler, click the Mapping button to show the Mapping panel.

2 In the Keys controls, click the Manual button.

3 In the Sound Library, search for the word *"catwalk"*.

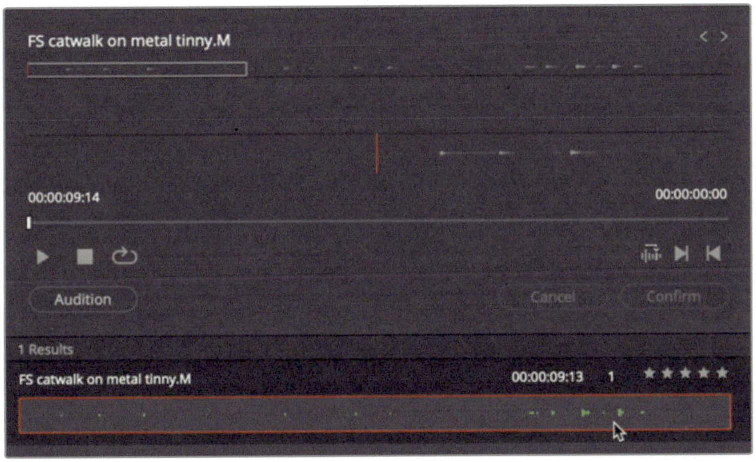

The Foley sound FS catwalk on metal tinny appears in the results list. This professional Foley sound effect is literally named and includes the initials FS for footsteps.

4 Drag the FS catwalk on metal tinny sound effect from the Sound Library to the keyboard area of the Foley Sampler. The new sample is mapped to all the keys to the right of the first sample in the keyboard.

5 Click a few keys mapped to the footsteps sample to hear what it sounds like.

Once again, each key incrementally plays the footsteps faster and at a higher pitch, which isn't really useful if you are recording footsteps to picture. What you really need is for each footstep to be mapped so that you can play them individually. You can do this by splitting the sample from the Options menu. When you split a sample, the sampler will automatically map the original full length sample to the first set of keys, followed by a spacer without sound, then individual split sounds effects are assigned to keys based on how many keys remain on the keyboard. That way the operator has the option to play the full sample, or individual split parts of the sample.

6 At the top of the Foley Sampler, click the Sample button to show the sample waveform.

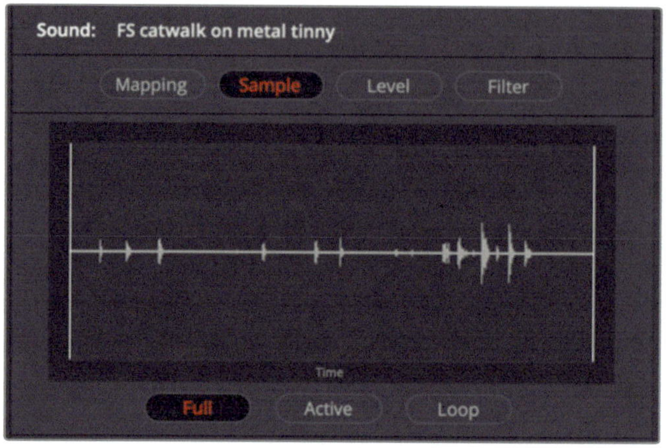

Here you can clearly see each individual footstep within the sample. You can also see that some footsteps are louder than others, especially toward the end of the sample.

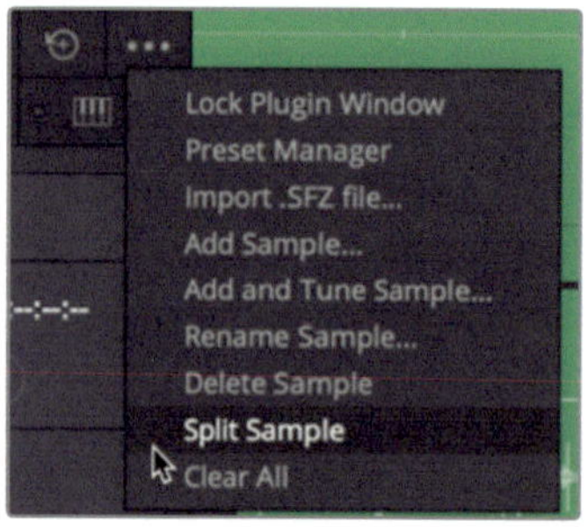

7 Click the Options menu in the upper-right corner of the Foley Sampler. This is where you'll find additional options to do things like Add, Rename, and Delete samples. In the Options menu, choose Split Sample.

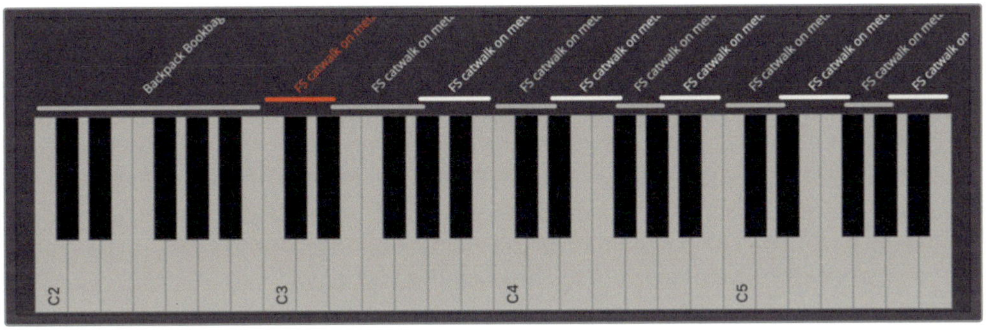

The individual footsteps are now mapped to keys in order from left to right. Let's look at the split samples to see the resulting waveforms.

8 In the keyboard, click the line above each of the FS catwalk samples from left to right to see the respective waveform in the sample waveform area.

As expected, the quietest footsteps are mapped to the earlier keys, and the loudest footsteps are mapped further down the keyboard, just like the original sample waveform.

Recording Foley Sampler Footsteps to Picture

With the keyboard mapped, it's time to move on to recording the foley samples to picture. Keep in mind that this takes some skill to master, so for the purposes of these exercises, just do the best you can. In this final Foley Sampler exercise, you'll start by setting a looped playback range so that you can practice performing footsteps via the keyboard to picture. Then you'll record a take in the timeline. To simplify the process, blue markers were added to the timeline to use a guide where the footfalls should occur.

1. Hide the Sound Library and Mixer if necessary to streamline your interface.
2. Use the zoom controls so you can clearly see the first two clips in the A1 track as well as the A3 track. You'll also need to position the viewer window and Foley sampler so that you can see the video and click the keys on the keyboard.
3. Set a range in the timeline that starts before Emiliana's last line in the first clip and ends before her last line in the second clip.

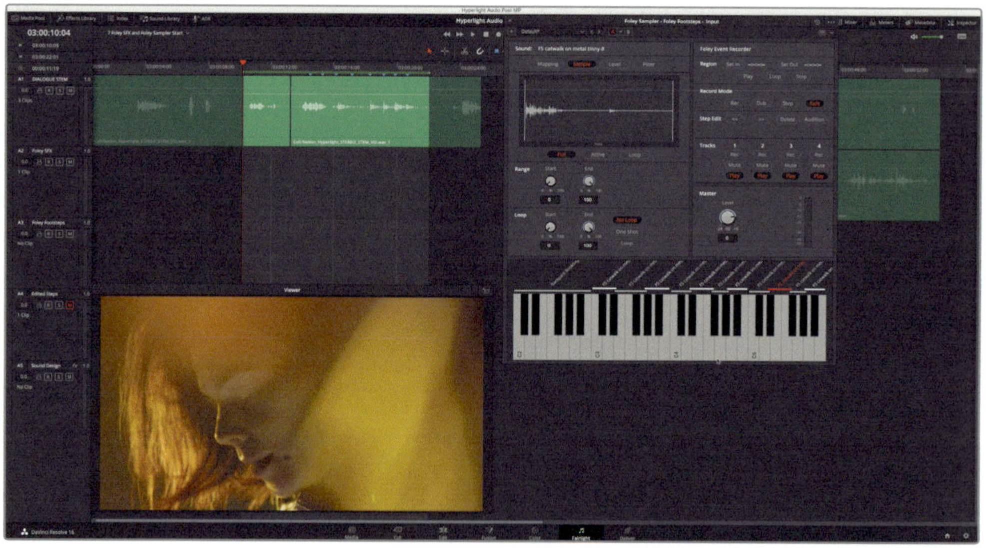

4. Start looped playback in the timeline and practice playing the samples using the soundtrack footsteps as a guide. You may notice that she takes a few steps during her first line in the second clip, "Who knows if any of them work?" that are not audible. Be sure to include those steps too.
5. Click the Arm button [R] on the A3 track header to arm the track for recording.
6. Move to the beginning of the range.

7 When you are ready to record a take, click the Record button in the transport controls above the timeline. Press the Spacebar to stop recording when you are finished.

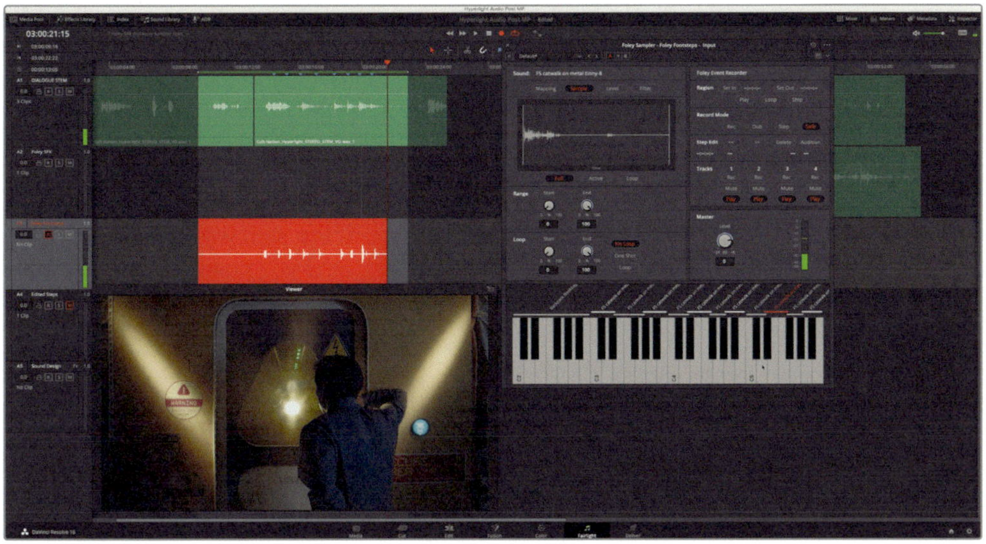

> **NOTE** If you want to try another take you can undo the previous recording or simply record another take on top of the first. Each take will be stored in the track in audio track layers with the most recent take on top.

You can record additional takes later. For now, let's move on to the last part of the exercise. Once you have recorded a take, you can easily split the clip with the Razor tool or shortcut and manually move or nudge the footfalls into position.

8 Close the Foley Sampler window and disarm the A3 track.

> **NOTE** If you are working with the free version of DaVinci Resolve, go ahead and use the clip in the A4 Edited Steps track to cut, copy, and paste footsteps where you'd like them in the track. When finished, move on to the next section.

Finishing the Foley Footsteps with a Little Sound Design

The last step for finishing the footsteps for this scene is to split the recorded footsteps when the scene goes from a shot of Emiliana inside her cavernous workspace to Philip watching her through the glass in the sealed door. To save time, track A5 Sound Design has already been set up with effects to alter the sound based on Philip's perspective on the other side of the sealed door. All you need to do is split the clip and move the footsteps down to the lower track.

> **NOTE** If you would rather try the sound design for the footsteps on your own, feel free to create a new track and move the footsteps to that track and apply whatever EQ and time-based plug-ins you think you need to achieve your goal.

1 In the timeline, move the playhead to the first frame of the shot where Philip is watching Emiliana through the glass in the sealed door. If you have a footfall landing on that frame, move the playhead right or left to finish the footstep waveform before splitting the clip.

2 Select the recorded footsteps clip in the A3 track and click the Razor tool in the toolbar (scissors)_or press command-B. You should now have two clips in the A3 Foley Footsteps track.

Now it's time to cut the second footsteps clip and paste it into the A5 track.

3 Use any method that you prefer to cut the second recorded footsteps clip in the A3 track and paste it into the same position on the A5 track.

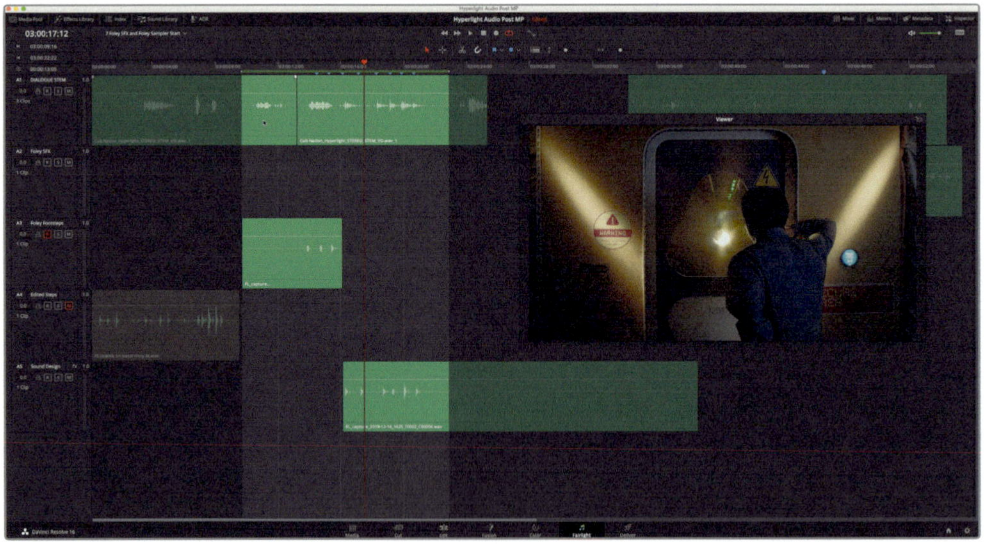

4 Play the finished scene and listen to the new footsteps and sound design as you watch the picture.

5 Make any edits or sound design decisions you feel are necessary to finesse the placement of the footsteps, their volume levels, and their perceived space within the scene.

6 To see and hear a finished example of the finished scene, open the timeline "7 Foley SFX and Foley Sampler Finished."

There you have it. You can now add Foley Sample recordings to your sound design skills. Don't worry if it isn't perfect, as long as you know how to perform this operation, you can use the Foley Sampler on your own soundtracks.

TIP You can save a Foley Sampler instrument with all the samples and mapping to the Preset Manager so you can use it again in other timeline tacks or even other projects on the same system. This is incredibly useful if you are working on a project with a lot of repetitive foley sounds, such as football tackles, helmets crashing, and crowd cheers for a football themed movie.

Analyzing Finished Foley Effects

The more you learn about sound design, the more techniques you will recognize in other soundtracks. Before recording some original sound effects in the next lesson, let's listen to some examples of Foley footsteps in the context of finished scenes. As you play these examples, listen to the volume level, reverb, delay, and other time-based effects.

1. Open the "7 Sample Sound Design Scenes" timeline.
2. Press Shift-Z to fit the clips to the timeline, if necessary.
3. Show the viewer. Resize and move it to the lower middle of the screen.
4. Solo the A1 STEREO MIX track. Move the playhead to the second-to-last blue marker in the track (03:04:08:14).

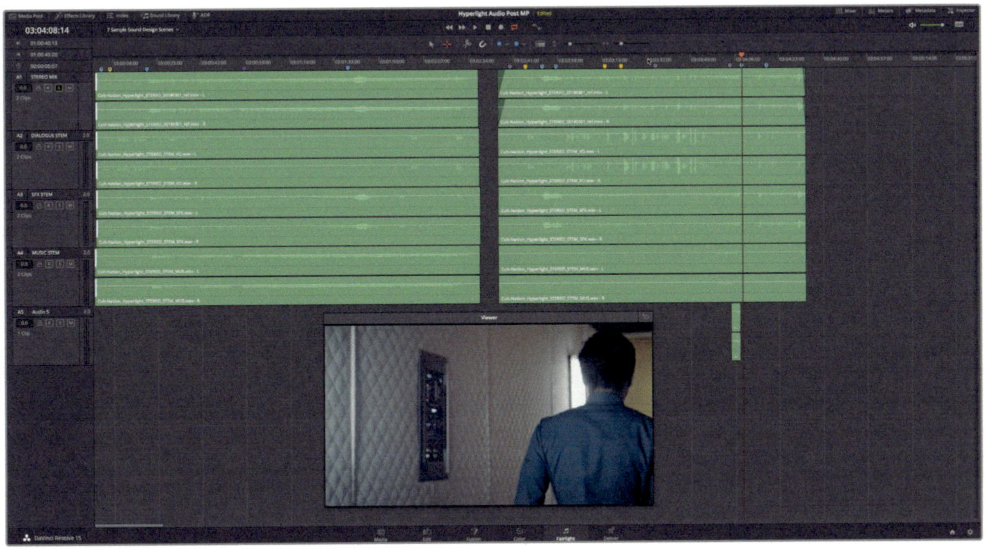

5 Start playback, and listen to the differences between Philip's footsteps as he walks down the padded corridor, and when he steps into a different space with different textile surfaces and reflective dimensions.

Could you hear the realistic change in footsteps between the two spaces?

Let's listen to one other example.

6 Move the playhead to the third yellow timeline marker (03:03:16:05).

7 Start playback, and listen to the techniques applied to the voices of the two characters. You'll hear examples of frequency cutting and reverb.

8 When you are finished, stop playback.

By now, your aural recognition of various sound design techniques should be more finely tuned as you hear them in context. Did you wonder how Philip's radio voice sound was achieved? Easy: by applying heavy low and high-pass filters, just like a real two-way radio.

In the next lesson, you'll apply what you've learned so far to create original sound effects.

Lesson Review

1 In which pages of DaVinci Resolve 16, can you reverse and change the speed of an audio clip?

- a) Media
- b) Edit
- c) Fairlight
- d) Fusion
- e) Deliver

2 Where do you turn on the Video and Audio scrollers in the Fairlight page?

- a) Timeline Options menu
- b) Inspector
- c) Sound Library
- d) Media Pool Preview panel

3 Which options are available to apply pitch changes to a clip? (Choose all that apply.)

- a) FairlightFX Pitch plug-in
- b) Right-click menu
- c) Pitch settings in the Inspector
- d) Clip EQ in the Inspector

4 Choose three time-based FairlightFX plug-ins.

a) Speed

b) Echo

c) Delay

d) Slow-Mo

e) Reverb

f) Flanger

5 To create the aural illusion that a sound is coming from the other side of a physical barrier, which tool would you use in Mixer?

a) EQ

b) Compression

c) Auxiliary Bus

d) Object-based pan controls

6 Where do you set a sync point on a sound effect clip?

a) In the Media Pool

b) In the Sound Library

c) In the Inspector

d) In the SFX Sync plug-in

7 True or False? When you drag the Foley Sampler plug-in to a track's header, the track is automatically patched with the Foley Sampler as the Input instrument.

8 Which methods can be used to add a sound effect to the Foley Sampler? (Choose all that apply.)

a) Drag a sound from the Sound Library to the Foley Sampler

b) Drag an audio file from the file system on your computer to the Foley Sampler

c) Right-click a clip in the timeline and choose > Add to Foley Sampler

d) Choose Add Sample in the Foley Sampler Options menu

Answers

1 b

2 a

3 a, c

4 b, c, and e

5 a

6 b

7 True

8 a, b, and d

Lesson 8

Creating Sound Effects

Where do sound effects come from? Some are recorded, some are purchased, and some are created from scratch. Regardless of their origins, nearly all sound effects need the added touch of the sound designer to help them fit the tone and aural texture of a scene. Transforming recorded sounds into realistic sound effects requires the active imagination and creative problem-solving skills of a sound designer

In this lesson, you'll apply skills you've already learned—along with new tricks, techniques, and creative plug-ins—to create original sound effects for the Hyperlight soundtrack.

Time

This lesson takes approximately 35 minutes to complete.

Goals

Creating Original Sound Effects

Do you remember the oscillator sounds that you recorded in Lesson 2? Well, it's time to put them to work to create four different sound effects for a heart monitor, tech background computer warble, sci-fi rotor-blades, and an interactive computer screen.

Let's start by opening up the finished Sound Design FX timeline that you recorded in Lesson 2. However, you'll work with a slightly modified version of that timeline with the clips already colored and markers placed to designate each of the recorded frequencies and noises.

1 Open the "8 Sound Design FX Recorded" timeline, and show the Markers index in list view.

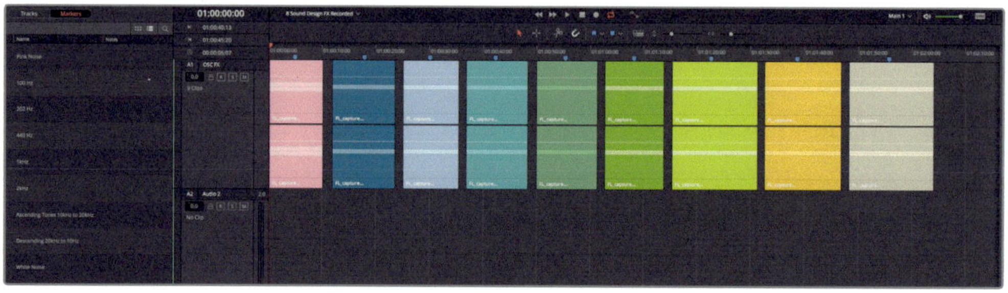

The marker names in the index show each type of recorded sound. If you recall, you have two ways to move the playhead from marker to marker. You can Press Shift-Down/Up Arrow, or in the index, you can double-click each marker frame.

> **NOTE** This timeline also includes titles for each clip that were created in the Edit page. So, you can use the viewer to see the frequency for each clip during playback.

2 Move the playhead from marker to marker, and play one second of each clip to refresh your memory of them.

To build the heart monitor sound effect, you could work within this timeline. However, as you already know, it's always best to build sound effects in their own timelines to contain the clutter and to keep them organized for use in multiple projects.

For this heart monitor sound, let's use the green 1-kHz clip for the sound effect. Also, select the My Timelines bin as the location for the timelines you create during this lesson.

3 Show the Media Pool, and select the Lesson Timelines > My Timelines bin. Hide the Media Pool.

4 In the index, double-click the frame for the 1-kHz marker. Right-click the green clip beneath the playhead, and choose Copy.

5 Choose File > New Timeline. In the New Timeline dialog box, name the timeline **Heart Monitor FX**. Retain the default settings for one Stereo audio track as in the following figure.

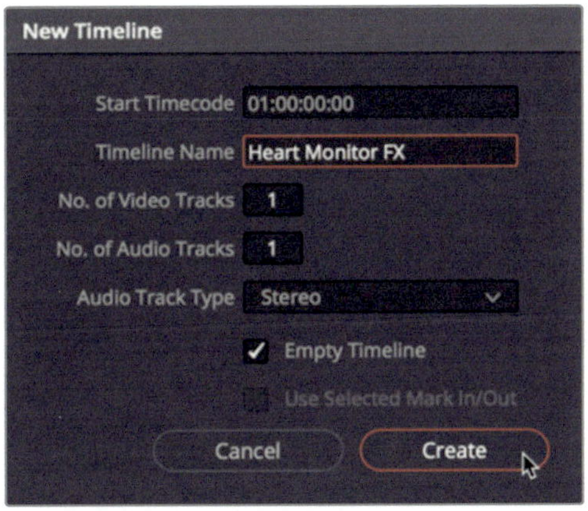

6 In the new Heart Monitor FX timeline, right-click the empty A1 track, and choose Paste.

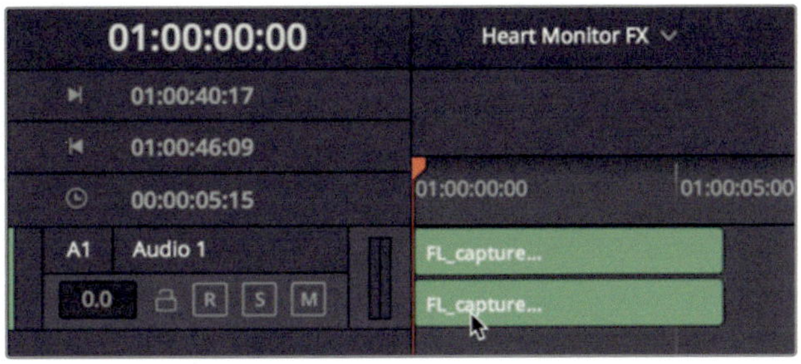

The 1-kHz clip is now ready to use in the Heart Monitor FX timeline.

7 Hide the index. Right-click the track header, and choose Add Track > Stereo.

Now the timeline contains two stereo tracks. You'll use the second track to build the heart monitor effect.

8 Zoom in on the tracks so that the clip in the A1 track fills about one-quarter of the timeline.

You have successfully copied one of your recorded effects into a new timeline. Now you are ready to build your first sound effect.

Designing a Heart Monitor Sound Effect

In this exercise, you'll design your own heart monitor sound effect. First, you'll make a short, five-frame version of the oscillator clip. Then, you'll adjust the clip's pitch to suit yourself. After the pitch is set, you'll copy and paste the clip at regular intervals to create a pattern. Finally, you'll bounce that pattern to a new track.

A seasoned sound designer could create this effect in about two minutes, including the recording of the oscillator sound. You will take only slightly longer than that because you already know most of the skills necessary to create the effect.

Let's start by splitting the clip to create a five-frame version. To do so, you'll use keyboard shortcuts that you already know.

1 Press the Down Arrow to move the playhead to the end of the clip. Press the Left-Arrow five times to move the playhead five frames to the left.

2 Press Command-B (macOS) or Ctrl-B (Windows) to split the clip at the playhead's position.

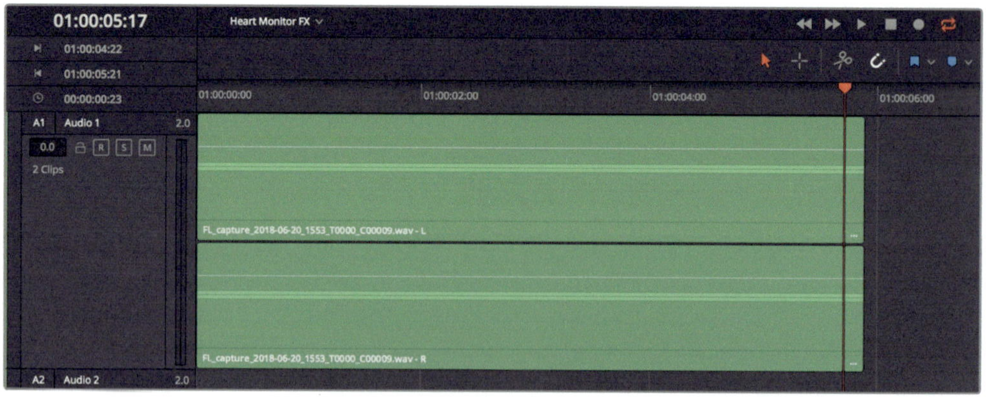

3 Right-click the short clip that you just created, and choose Copy.

4 Paste the copy into the A2 track. Mute the A1 track.

5 Mark the clip in the A2 track. Open the Inspector, and select the clip.

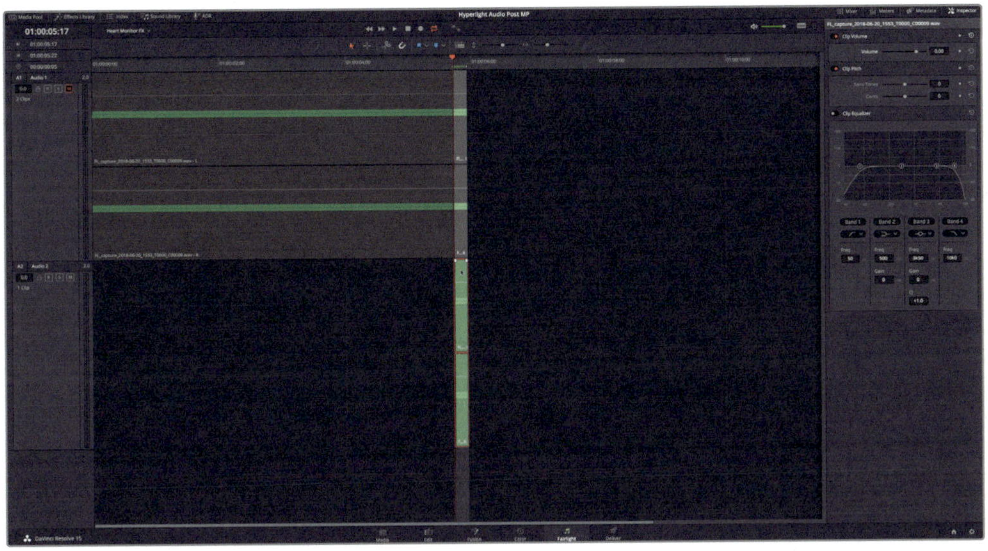

Next, you will choose the pitch for your heart monitor beeps. If this were a sound effect for a hospital on earth, the audiences would be expecting the standard tone they've heard hundreds of times on TV medical shows. However, because these Hyperlight scenes take place in a futuristic spaceship, you can use whatever pitch you'd like because the audience can't fact-check your authenticity.

> **TIP** In a dialogue scene, when choosing the pitch for an ongoing background effect–such as a phone ring, equipment alarm, or computer hum–it's best to select a pitch that doesn't conflict with the dialogue.

6 Start looped playback. In the Inspector, drag the Semi Tones slider to the left or right to choose a pitch for your heart monitor. A standard earth-based heart monitor is about -13 semitones lower than the 1-kHz clip.

Don't worry if you have trouble choosing a pitch. Just pick one. You can always change your mind later!

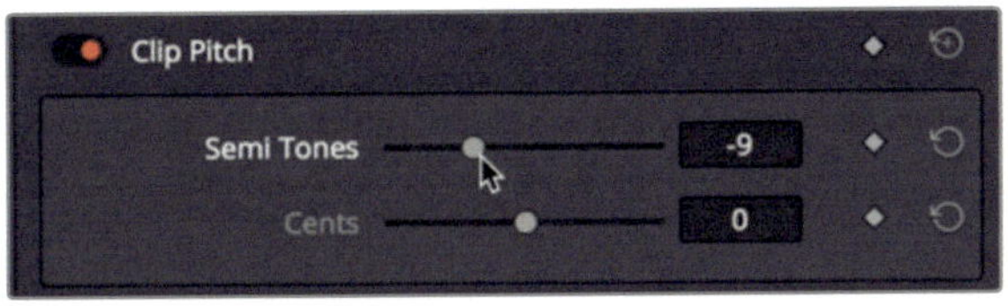

7 When you are finished, stop playback, and clear the play range.

Next, you'll select the clip, and paste it at one-second intervals to create a pattern.

8 In the A2 track, move the playhead to the last frame of the clip. Copy the clip.

9 Press Command- – (minus sign) in macOS or Ctrl- - (minus sign) in Windows to zoom out to see more of the timeline as you paste the clips.

10 Press Shift-Right Arrow to move the playhead one second to the right. Press Ctrl-V (macOS) or Command-V (Windows) to paste the clip. Repeat this step until you have 20 clips in the A2 track (for approximately 20 seconds of heart monitor beeps.)

11 Play the clips in the A2 track to hear your heart monitor.

It sounds great, but the volume level is a little low. In the track meters, you can see that the level is in the green and well below the middle of the meter. Let's raise the levels of all the clips simultaneously so they peak in the yellow range. While you have all the clips selected and visible in the Inspector, you could also change the pitch.

12 Drag the Selection tool across all the clips in the A2 track to select them.

13 In the Inspector, drag the Volume slider to the right to a value between 9.00 and 11.00. If you choose to change the pitch at this time, start playback, and adjust the Clip Pitch sliders for all the clips. Hide the Inspector.

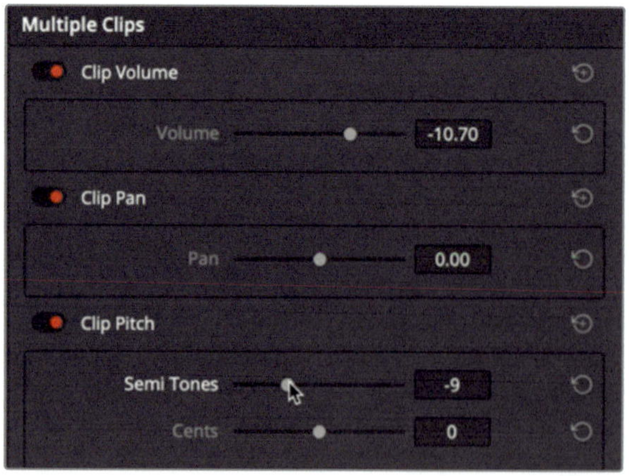

NOTE If the timing is off between any of your beeps, this is the time to fix it. This effect emulates a mechanical device that would keep perfect time.

Bouncing the Mix to a New Track

In Lesson 5, you bounced a selection to a new layer to create the room tone clip. In this exercise, you'll bounce the timeline's output to a new track in the same timeline. This is a common audio mixing technique used to create stems and consolidate multiple tracks into a single clip. In this case, you'll first mark the clips so you can loop them to make a longer sound effect, if necessary. Then, you'll bounce the play range to a new track.

1 Press Shift-Z to fit the clips horizontally in the timeline.

2 In the A2 track, move the playhead to the beginning of the first clip.

3 Press I to mark an in point.

4 In the A2 track, move the playhead to the first frame of the last clip. Press Shift-Right Arrow to move the playhead forward one second. Press O to mark an out point.

   Why the first frame of the last clip? To include the 5 seconds of silence at the end of the bounced clip. If you start and end the bounced clip with a beep, it will have two beeps back-to-back when looped or stitched together.

5 Choose Timeline > Bounce Mix to Track to open the Bounce Mix to Track dialog.

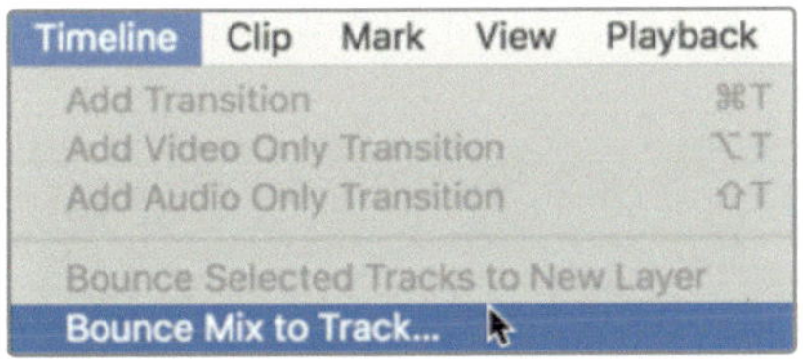

6 In the dialog, in the Destination track pop-up menu, choose New Track.

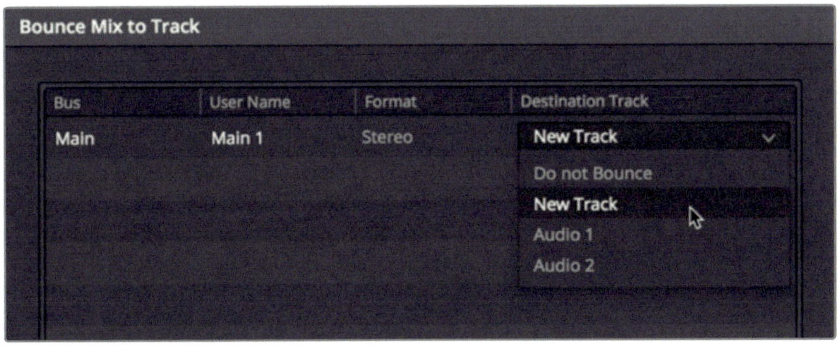

The Bounce Mix to Track dialog shows that the Main output bus, named Main 1, will be bounced to a new destination track.

TIP Before you bounce a mix to a track, it's a good idea to make sure that the appropriate tracks are soloed or muted. In this case, you are only bouncing the A2 track, so you want the A1 track to remain muted.

7 Click OK.

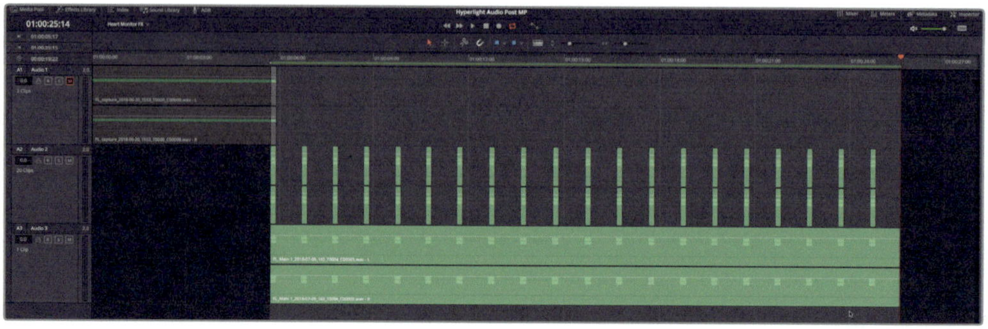

The new bounced heart monitor clip appears in a new A3 track. Notice that the bounced clip is exactly the length of the current play range. Your last step is to zoom out and move the newly bounced clip to the end of the current play range. By doing so, you'll have a full working timeline of the sound effect from start to finish.

8 Zoom out of the timeline horizontally. Drag the bounced effect clip to the right so it starts after the end of the play range.

9 Clear the play range.

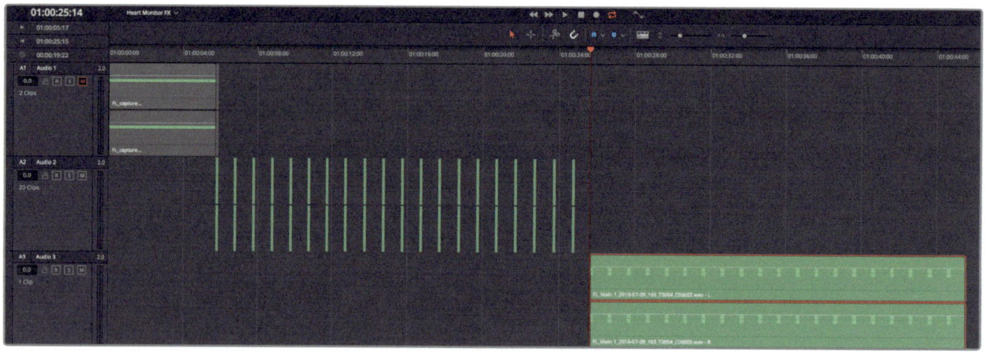

Your heart monitor sound effect is finished and ready to use in any of your other timelines.

Another Heart Monitor Version Please!

The original heart monitor effect you just made is pretty good. However, after a little research, it seems the beeps are a little too close together, even for a medlab in space. No worries. You can make another version in less than a minute. First, starting with beat number 2, select every other beat. Command-click (macOS) or Ctrl-click (Windows) the even-numbered clips to select them, and press D to disable them. The beats are now audible every two seconds. Mark a play range between the beginning of the first clip and the beginning of the last enabled clip. Bounce the mix to a new track. Move the bounced clip down the timeline to just past the first bounced clip.

When you are finished, you'll have two different heart beat monitor sounds to choose from, and the knowledge that in the future that you should do research before creating a sound effect from memory.

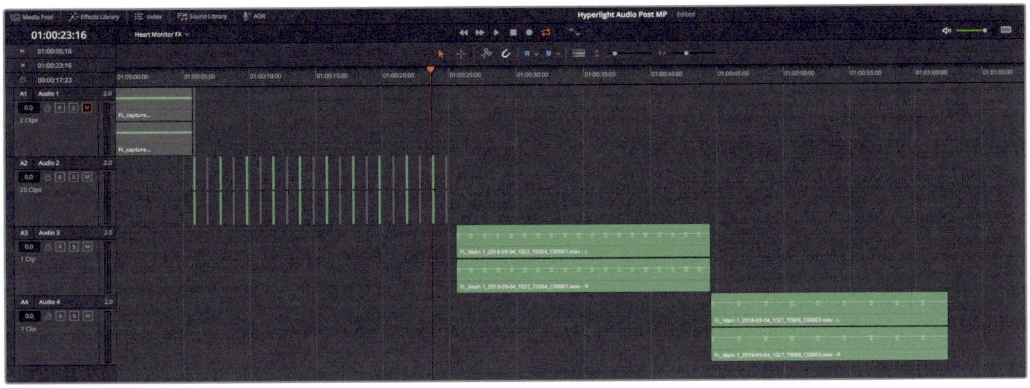

Transforming Noise into Atmospheric Sound Effects

To create dramatic atmospheric sound effects such as wind or warbling computer tones, you can use the Flanger. This plug-in mixes an identical copy of the signal while moving it back and forth, with modulation that constantly changes the amount of delay. In the following exercises, you'll start with a timeline clip that has the Flanger applied, and your goal is to create a cool warbling tech background sound for ADA.

Previously, you copied a clip into a new timeline to create the heart monitor effect. To save time, those steps have already been done for this example.

1 Open the "8 Computer Tech bg" timeline.

This timeline includes one track with a noise clip just like the one you recorded in Lesson 2. The initials "bg" in the timeline's name identify it as a background (atmospheric) sound effect or track.

2 Play the timeline to hear the finished effect.

You may not recognize this as a real-world sound, but you've probably heard hundreds of variations of this warble in science-fiction shows featuring high-tech devices.

3 Show the Inspector. Select the clip, and stop playback.

Clearly, this clip has some serious sound design upgrades, including a two-octave pitch change (-24), high- and low-pass filters to limit the frequency range, and a Flanger plug-in.

4 In the Inspector, click the bypass switches for the Clip Pitch, Clip Equalizer, and Flanger plug-ins to turn them off. But do not play the clip.

5 Click the DIM button to lower playback levels by 20 dB.

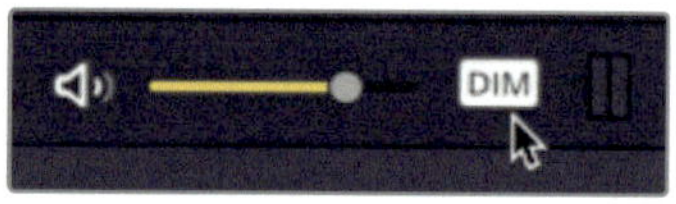

6 Play the clip.

Do you recognize the white noise sound? If you recall, you recorded it using the built-in oscillator as a source. It's also the sound a television makes when it loses signal.

7 Turn on the Clip Pitch, and play a few seconds of the clip.

8 Turn on the Clip Equalizer, and play to hear the changes in the sound.

9 Click the DIM button to return playback to the normal playback levels.

The result is interesting. However, it doesn't sound at all like a computer. Instead, it sounds more like the low roar of a distant ocean without the sound of crashing waves.

Creating a Computer Background Warble with the Flanger Plug-In

Now that you've heard the Flanger warble, you can explore the controls to create your own variation.

1 Start looped playback. Turn on the Flanger plug-in, and click the Customize button to show the Flanger window.

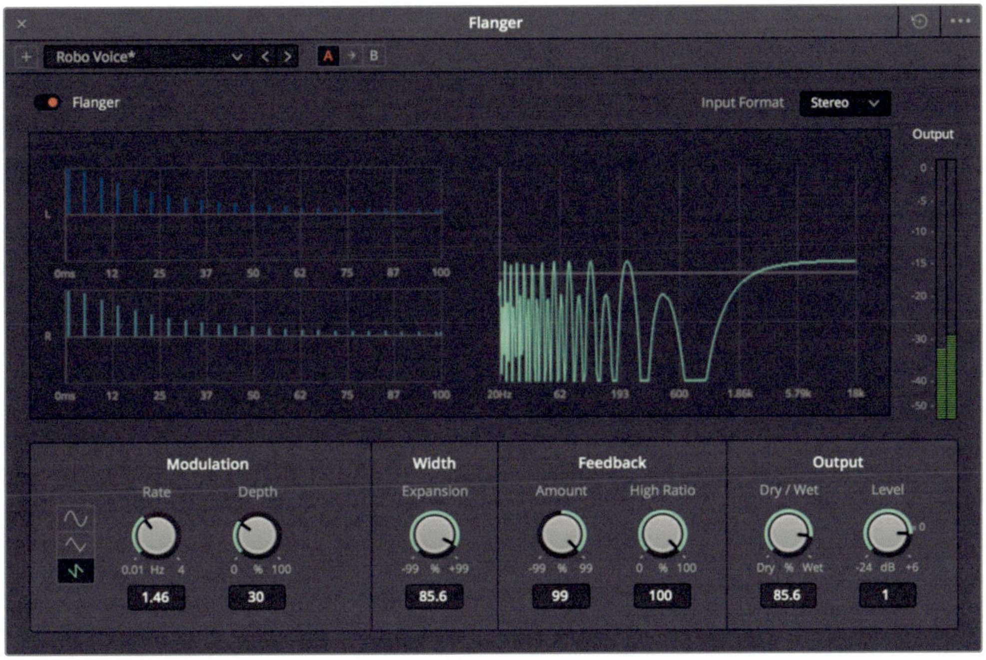

The Flanger window includes an animated graph that shows the results of the modulation parameters so you can see the warble added to the original signal. The current setting is a customized version of the Robo Voice preset.

Because this effect was applied to noise instead of sound with a fundamental frequency, some of the control parameters will have no effect on the signal. Let's build a computer warble sound from scratch.

2 In the upper-right corner of the Flanger window, click the Reset button to reset the plug-in.

To experiment with this creative plug-in, you'll start by adjusting the Dry/Wet control.

3 In the Output settings, drag the Dry/Wet control to the right to 100%.

Next, you'll use the Feedback controls to determine how extreme the flanging effect will be. The High Ratio control lets you filter the ratio of frequencies that will affect the feedback. This control won't be audible with a noise signal, so you'll have to try it another time.

4 While watching the graph of the Left (L) and Right (R) signals, drag the Amount control to the right to 99% to hear the highest amount of flanging.

The Amount control determines the percentage of the sound fed back to the delay line. Positive numbers add more of the Flange effect to the signal and subsequently raise the overall pitch of the warble, whereas negative numbers add the inverted Flange effect to the signal, thereby lowering the overall pitch. Increasing the value in either direction increases the audibility of the warble.

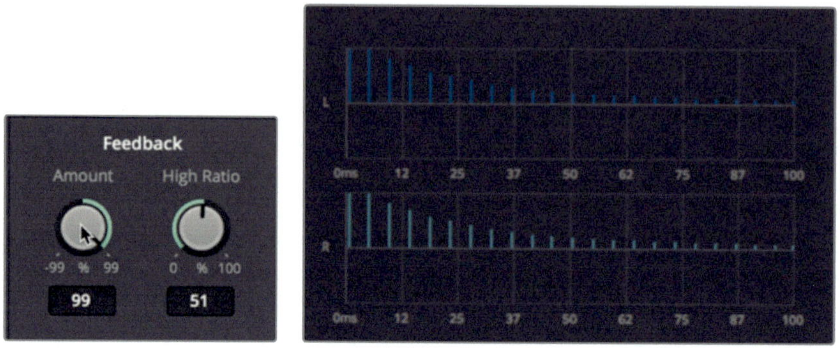

5 Drag the Amount control to the left to -99% to hear the highest amount of inverted Flanging.

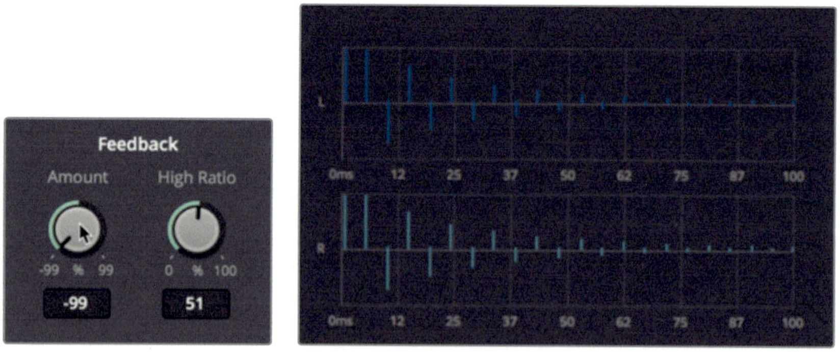

Next, let's jump to the Modulation parameters on the left where you can adjust the modulation shape, Rate, and Depth. Rate affects the speed of the low frequency oscillator (LFO), and Depth affects the length or definition of the warble. Low Depth values add only the beginning of the warble, while high Depth values include the entire warble from start to finish.

6 Drag the Depth control all the way to the left to hear only the beginning of the warble. Then, drag the control all the way to the right (100%) to hear the full warble effect. Set the Depth to around 25%.

7 Drag the Rate control all the way to the right for the fastest oscillator speed. Then, drag it to the left to around 0.40 Hz for a much slower "tech background" sound. Choose a Rate value that best sounds like the ongoing warble that you imagine as ADA's "pulse" for shots that show a computer screen or console. Don't worry about the pitch right now; you already know how to change that at any time.

The choices for the shape of the LFO affect the timing of the warble effect added to the Flanger sound : sine (smooth), triangle (jerky), and sawtooth (abrupt changes).

8 Click each of the Shape buttons to change the LFO modulation pattern. Choose a favorite.

Width has only one parameter, Expansion, which can set the differences in length for the left and right channels, along with the phase offset of modulators. In other words, this parameter helps make the flange pattern travel back and forth between the left and right channels.

9 Drag the Expansion control all the way to the right (99%) and all the way to the left (-99%) to hear the inverted effect.

With both extreme settings, the sound travels the "widest" distance between the left and right channels, which you can hear in your headphones or monitor speakers.

10 Drag the Expansion to a center (0%) position to hear a more centered signal. Then, add a bit of Expansion in either direction to suit yourself. Remember, this effect is ADA's "pulse," and will be associated with a ship-wide omnipresent computer.

11 Continue to experiment with the different settings until you have a tech computer background sound that you like.

When you are happy with the sound of your computer warble, you're ready to bounce the finished effect to a new track. First, it's a good idea to increase the clip's volume level until it peaks in the yellow in the track level meter.

12 During playback, drag the clip volume curve upward in the timeline until the clip starts to peak in the yellow in the track level meter (between 10.00 and 12.00 Clip Volume in the Inspector.)

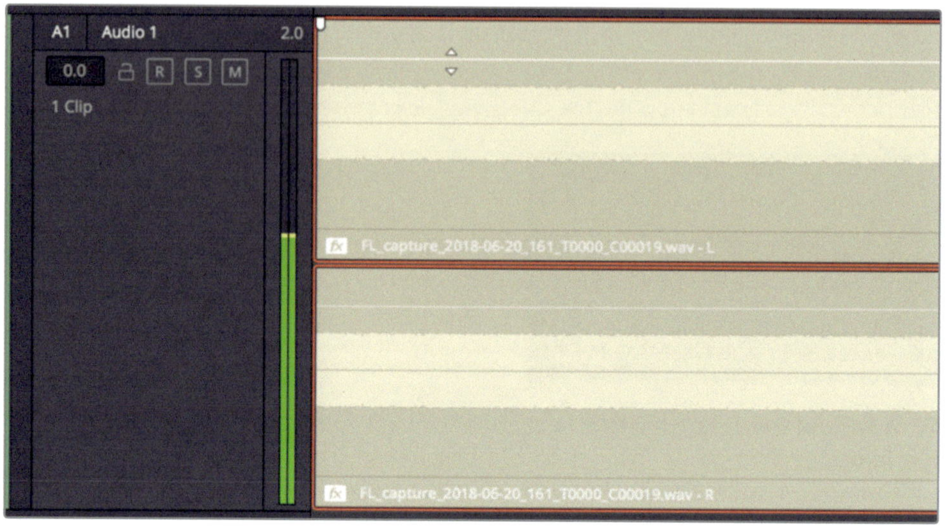

13 Stop playback. Choose Timeline > Bounce Mix to New Track.

14 In the Destination Track pop-up menu for Main 1, choose New Track. Click OK.

Your bounced mix appears in a new A2 track.

15 Solo the A2 track, and listen to the final rendered clip that includes the original noise signal, as well as pitch, EQ, and Flanger effects.

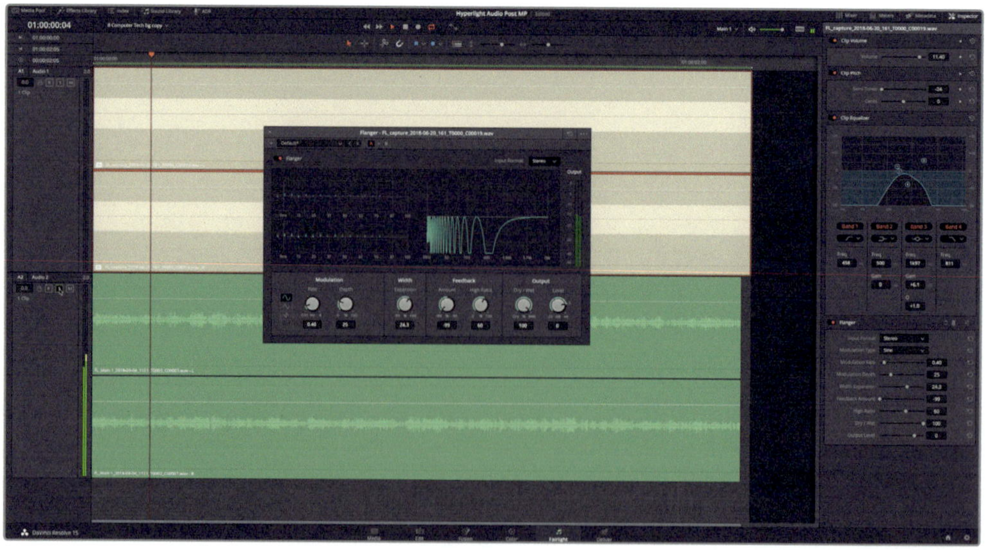

16 When you are finished, stop playback. Unsolo the A2 track. Turn off the Flanger plug-in, and close the Flanger window.

Using the right plug-in, some pitch controls, and a little EQ, you can turn white noise into a creative high-tech sound effect in minutes.

Exploring the 4-in-1 Modulation Plug-In

For creative sound design, such as transforming sounds into robotic voices, spaceship engines, or helicopter blades, the Modulation plug-in is your go-to Fairlight effect. If you enjoyed working with the Flanger and other creative plug-ins, you're in for a treat. Fairlight's Modulator plug-in includes many of the controls that you've used in other plug-ins; plus, as the name suggests, it also has expanded modulation controls. In fact, the Modulation plug-in lets you apply modulation to four elements at once, including the frequency, amplitude, tone, and rotation between the left, center, and right channels. If you're wondering what that means, let's check it out with presets. After you have heard the presets, you'll customize one to create a sci-fi rotor-blades sound effect.

1 Solo the A1 track.

2 From the Effects Library, drag the Modulation plug-in onto the clip in the A1 track. Hide the Effects Library.

The colorful graph shows the original signal in teal and the modulated pattern from the LFO in blue. The parameters below the graph affect the LFO, which in turn affects the sound of the signal.

3 Start playback to hear the default sound. In the preset pop-up menu, choose Auto-Pan.

   As you can see and hear, the Rotation parameters show that the sound is oscillating back and forth from the wide left to wide right channels.

4 Choose the Doppler preset to hear a doppler effect applied to the signal.

   If you've ever heard the sound of an emergency vehicle pass you with sirens wailing, or heard an airplane fly overhead, you've experienced a real-world Doppler effect. It's the perceived upward shift in frequency as a sound source approaches you and the downward shift in frequency as the source moves away from you.

   Just as in the real world, the Doppler preset applies the necessary change in pitch as the sound travels between channels.

Looking at the green modulation bars in the Modulation parameters, you can see that the Doppler effect preset requires modulation in all four of the control areas: Frequency, Amplitude, Filter, and Rotation.

5 Apply each of the remaining presets to hear the modulation effects on the noise signal. Some of the presets work much better on an actual audio signal, such as ADA's voiceover. As you go through each preset, which sounds the most like helicopter blades?

6 Choose the Tremolo preset.

Tremolo is a common effect used in music recording and stage performances that applies an oscillator to the amplitude (volume level). For this example, you will use the Tremolo preset as a starting point for the sci-fi rotor blades effect. At the moment, it sounds more like a fast-moving distant locomotive engine than rotor blades. Let's start by slowing the Rate (speed) of the oscillation.

7 In the Modulator parameters, drag the Rate control to around 3.40 Hz.

Next, let's give the sound a "voice" with some tone by applying modulation to the Filter and Tone parameters.

At the moment, the Tone control is set to 2 kHz, with a 0% Filter, which means no oscillation of tone between 10 Hz and 16 kHz.

8 Drag the Filter control to the right to increase the frequencies included in the oscillating pattern. Set the Filter level to around 50%.

9 Drag the Tone control all the way to the left and to the right to shift the filtered range to lower or higher frequencies. Set the Tone parameter to the highest level, 10 kHz.

Next, you'll increase the level of the Frequency parameters to increase the tonal effect on the original signal. Keep in mind that the original signal is white noise, so you won't hear much difference.

10 In the Frequency parameters, drag the Level control to 100%. In the Filter parameters, drag the Tone control to the left and right to hear the difference with the increased Frequency level. Set the Tone to around 1.4 kHz and the Filter to 100% to hear the full tonal range from 10 Hz to 16 kHz.

11 Lower the Filter level to 70% and raise the Tone to 10 kHz.

You've probably figured out that all these parameters affect each other to create an overall effect.

12 In the Amplitude parameters, raise the level to 100%. In the Amplitude effect, click the two Shape buttons to hear the effect of the different shapes and see their results in the graph.

Finally, let's add some movement between the left and right channels using the Rotation controls.

13 In the Rotation controls, drag the Offset control to the far left (-100) to hear sound only in the left channel, and drag to the far right (100) to hear sound only in the far right. Set the Offset amount to 0. Increase the Rotate value to oscillate the Rotation between the left and right channels. The higher the Rotation value, the wider the perceived movement of the sound between channels.

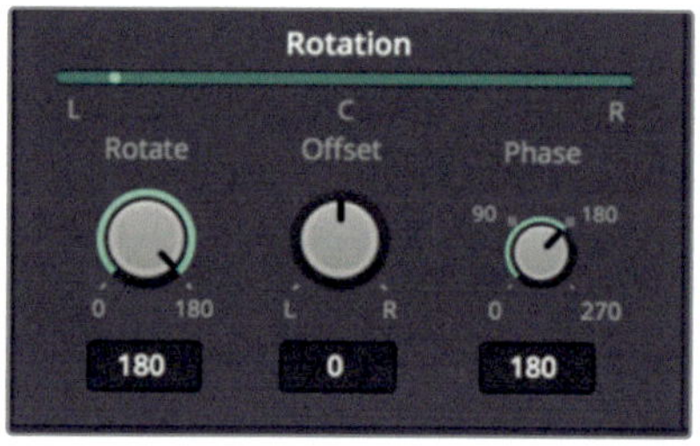

14 In the Modulator parameters, click each of the LFO Shape buttons to hear their effects on the signal. You'll see them update the graph.

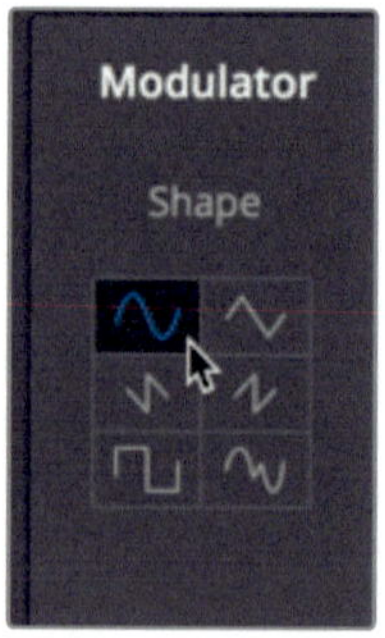

15 Continue experimenting with the controls until you have a sound that you like. If you don't like what you've created, reset the controls and start from scratch, or start with a preset.

16 When you like your new sci-fi rotor-blade sound, stop playback. Choose Timeline > Bounce Mix to New Track. In the Destination Track pop-up menu, choose New Track. Click OK.

17 Close the Modulation window.

You've created two original sound effects from a simple white noise recording. Keep in mind that plug-ins are creative tools, and the more you use them, the easier they are to work with. That said, look what you've accomplished so far in just one lesson. Impressive.

Analyzing Multilayered Sound Effects

You now have the basic editing and sound design skills necessary to start building multilayered sound effects. These effects aren't more difficult to create; they just require more steps. Think of making multilayered sound effects like making a multilayered dessert, such as a birthday cake. Doing so requires several steps: mixing the batter, baking the cake, making the icing, assembling the cake, and decorating the finished dessert. When you build multilayered effects, you have to gather sounds (ingredients), build each section and add effects, bounce elements to create a finished (baked) clip, and add final effects to create a polished result. Luckily, throughout this lesson, you've learned all the skills necessary to create exciting multilayered effects. Now, all you have to do is put those skills together. In this final exercise, you'll look at two finished examples of multilayered effects. Then, you'll create your own from scratch.

1 Open the "8 Respirator FX Build" timeline.

This timeline includes the elements to create a sci-fi medlab respirator sound to go with the heart monitor sound in the medlab scenes.

2 In the A9 Bounced Mix track, play the clip to hear the finished multilayered sound effect.

Pretty convincing, right? Now, move down the timeline to deconstruct the effect and see how it was made.

3 Show the Inspector and the Markers index in list view.

4 Move to the second blue timeline marker, 2nd Stitch with background. Play a few seconds of the green and pink clips.

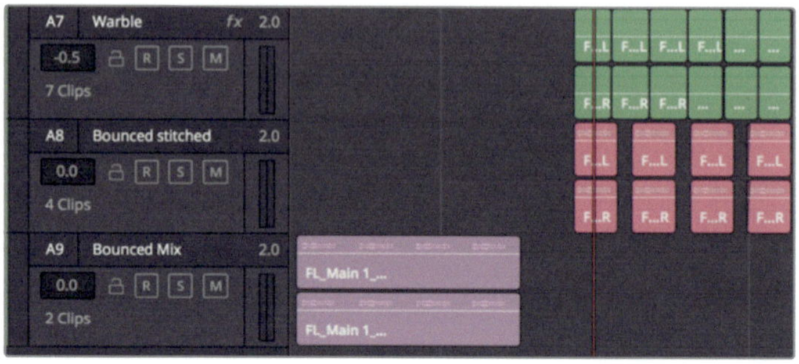

As the marker name suggests, these green and pink clips represent the final assembled effect before it was bounced to the rendered clip you just heard. The green clips should sound familiar; they are the flanged warble that you just created in the previous exercises. The pink clips are the bounced results of the individual components used to make the mechanical respirator sound.

5 Move to the next blue marker, and select the beige clip. Look at the Inspector.

This is the actual noise clip with the flanged warble that was copied from the previous timeline and pasted into this timeline. In the Inspector, you can see that because the changes and plug-ins were applied to the clip, they all were pasted with the clip into this timeline.

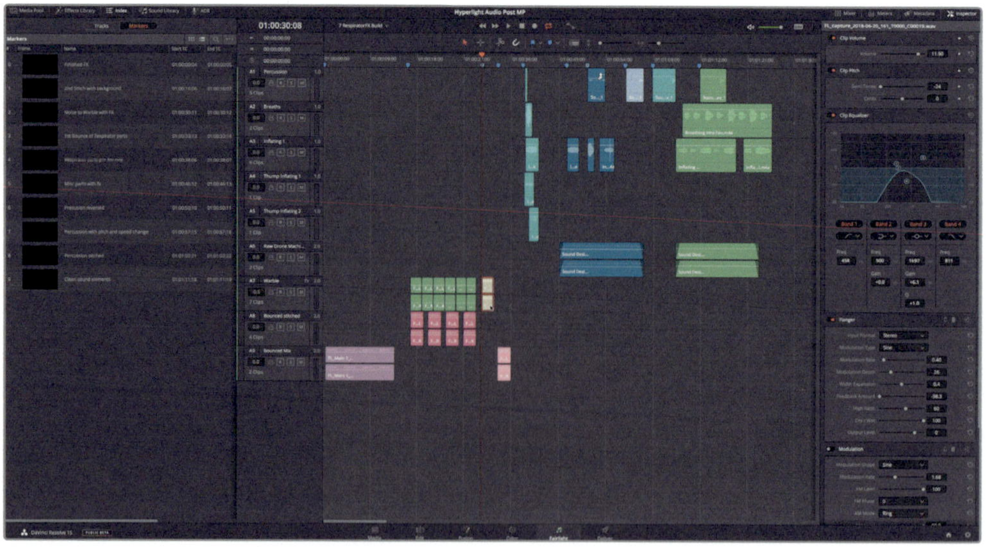

Move to the next blue marker, and play the pink clip to hear the bounced mechanical respirator sound.

6 Continue moving to the next marker to hear the individual clips on five tracks that were used to create the mechanical respirator sound. The track names indicate the original sound effects used to create the composite effect. Feel free to solo tracks to hear their individual clips.

The dark blue clips were sections that were split from the original recordings. Some have had effects or modifications applied. The green clips are the original sounds from the Media Pool.

Seize the Sound!

These inflating and breath sounds are not professional recordings. Instead, they are sounds of my 12-year-old daughter, Katie, inflating a beachball at the kitchen table that I recorded using a smart phone. I happened to hear her as I was passing by, and immediately said, "Wait, can you do that again slower so I can record it?" At one point, Katie's friend accidentally knocked over a cup while I was recording. That "klunk" sound ended up becoming the mechanical tick at the beginning of the respirator effect.

If you've ever used the Latin phrase "carpe diem" then you know it means "seize the day," or in other words, make the most of the day. In sound design, I have my own spin, which is "carpe canetis," or "seize the sound!" When you hear something interesting, record it right then and there with whatever device you have available! Even if you have only a smart phone, you have a way to record sounds on-the-fly. You can then bring them into DaVinci Resolve and experiment with them in the Fairlight page to see what they might become. If a sound works brilliantly, you can later recreate and rerecord it, if necessary. However, some of the most iconic movie sound effects were serendipitous discoveries of keen sound designers and sound effects editors who happened to hear something interesting in the real world and captured it for their projects.

Build Your Own Interactive Computer Display Sound Effect

In the real world, our interactive computer displays, such as the GPS screen in your vehicle, don't usually make beeps, chirps, and warbles as they show you information. However, as you know, part of the sound designer's job is to enhance a scene through sound. That includes adding tones, pulses, warbles, beeps, ticks, whooshes, and often alarm sounds to computer displays so that the audience "hears" as well as sees the results.

For this exercise, you will be on your own to create an exciting multilayer sound effect for ADA's interactive computer screen. Don't worry, you'll have an example for inspiration.

Start by opening the "8 Computer Ship Display" timeline. Show the viewer, and use the timeline options menu to show the video track.

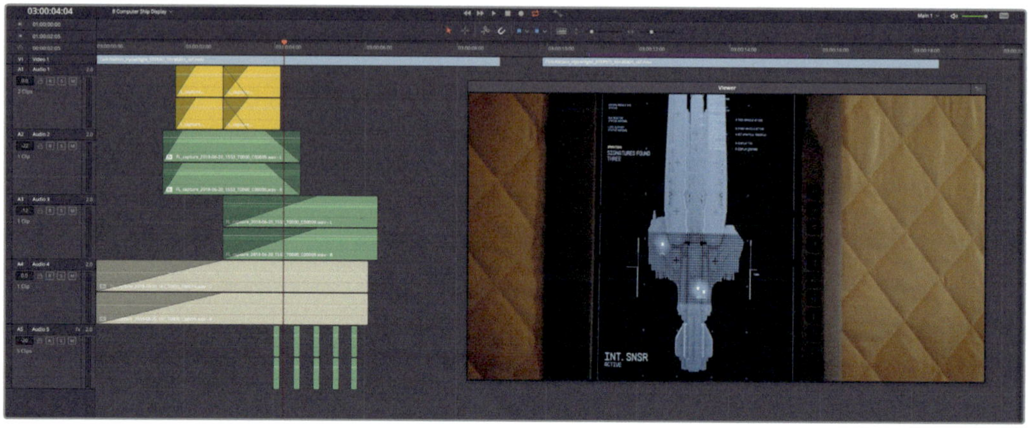

You'll see two identical video clips. The first clip has a multilayered sound effect that was created using only recorded oscillator sounds from the "8 Sound Design FX Recorded" timeline.

Your multilayer sound effect should include a warble for ADA's pulse, a searching sound to go with the moving search grid, a change to indicate the search has concluded, and an alarm or beeps to represent the results of the search (flashing dots).

You can copy and paste clips from any timeline in this project, although to really test your skills, it's best if you use only clips from the oscillator recordings, or sound effects that you created throughout this lesson.

Think like a sound designer. Push yourself to turn creative ideas into audible effects. Remember, you can change clip speed, reverse clips, apply pitch changes, add EQ, and insert plug-ins to craft your multilayered effects.

Have fun and enjoy the process!

In Lesson 10, you'll learn how to bounce your multilayered sound effect into both stereo and surround sound clips.

Lesson Review

1 When you copy a clip from one timeline and paste it into another timeline, what elements determine the track and placement of the pasted clip? (Choose two.)

a) Track number

b) Track selection

c) Markers

d) Play range

e) Playhead

2 What keyboard shortcut moves the playhead one second to the right (forward)?

a) Shift-1

b) Command-Right Arrow (macOS) and Ctrl-Right Arrow (Windows)

c) Shift-Right Arrow

d) Command-1 (macOS) and Ctrl-1 (Windows)

3 Which are two FairlightFX plug-ins that are useful for creative sound design can transform white noise into a computer warble or rotor blade sounds?

a) Reverb

b) Shusher

c) Flanger

d) Modulation

e) Hummer

4 True or False? It is a good idea to create layered sound effects in your primary working timeline to ensure that all of the elements remain in sync with the other tracks.

Answers

1 b, e

2 c

3 c, d

4 False. It's a good idea to create layered sound effects in separate timelines to allow more space for creative experimentation and creating multiple versions of the effect. Plus, doing so ensures that the individual effect timelines can be saved, organized, modified, and used in other projects.

Lesson 9

Mixing and Sweetening the Soundtrack

The last step of audio post involves enhancing and mixing the tracks to prepare them for output. Assuming that all other steps were completed prior to the mix, this process is fairly straightforward.

If you are a seasoned audio engineer or re-recording mixer, you already know how to transform a timeline full of tracks into a professional-sounding mix; you just need to learn how it's done using the tools in DaVinci Resolve 16. If you are an editor or are new to audio post-production, the next two lessons are designed to demystify the art of mixing sound by introducing you to the tools you'll need and how to use them.

Time

This lesson takes approximately 50 minutes to complete.

Goals

Mixing is an ongoing process. Every variable you change on every track can affect how other tracks interact. In other words, you have no set-it-and-forget-it in mixing. That's why you start small at the clip level by balancing levels within each track.

In this lesson, you'll explore the final stages of mixing including "sweetening" your soundtrack as you apply professional panning, equalization, and dynamic controls, to your tracks.

What Does a Re-Recording Mixer Do?

A *re-recording mixer*, also referred to as a *dubbing mixer*, is an audio engineer who specializes in post-production audio mixing. The name indicates that this mixer is not recording a live set, live audience, or a live performance. All the sound that he is mixing—including recorded dialogue, sound effects, and music for a soundtrack—has already been recorded and assembled into tracks. Although, a re-recording mixer may also work as the sound designer on lower-budget productions.

To build the final mix, re-recording mixers pre-mix the tracks to enhance the sound and reduce the number of tracks going to the final mix. The re-recording mixer carefully balances those tracks to create a final mix based on the sound designer's or director's audio goals. Along the way, the mixer records the output of the mix, or groups of tracks, back into the timeline into a new track. Hence the name re-recording mixer, because they continually re-record mixed sections of the timeline or the entire mix, back into the timeline to create finished stems or mixes. The mix must also comply with any regulations that enforce sound mixing standards for broadcast and theatrical distribution.

Although you can mix projects in a home office or recording studio, full-sized mixing or dubbing stages are preferred when mixing feature films destined for theatrical release. They enable the mixer to fully hear how the final mix will be experienced by audiences.

> **NOTE** The exercises in this lesson build from the tools and skills that you learned in previous lessons. If you skipped ahead to this lesson, you may need to review the previous lessons as a guide for skills and tools already covered.

Preparing the Project

In this exercise, you'll continue working with the scene that you've been editing and building since Lesson 2. For this lesson, you'll be looking at the meters often, so let's set up the interface accordingly.

1. Open the "9 Mix Scene Start" timeline.
2. If necessary, hide the Media Pool, index, and any other open panel on the Fairlight page. Hide the viewer so that it returns to the docked position in the Monitoring Panel.
3. Show the Monitoring Panel (Meters) and Mixer.

4 Press Shift-Z to fit the clips horizontally in the visible timeline area.

This version of the scene represents the culmination of your work throughout this book. In some cases, multitrack effects were bounced in their respective timelines and added to this timeline. Also, some of the track colors were changed to organize the tracks for mixing. Before you dive into mixing, here is a breakdown of the tracks you'll be working with in this lesson:

- A1–A4 are the dialogue tracks that you recorded, edited, balanced, and repaired in Lessons 2 through 6.

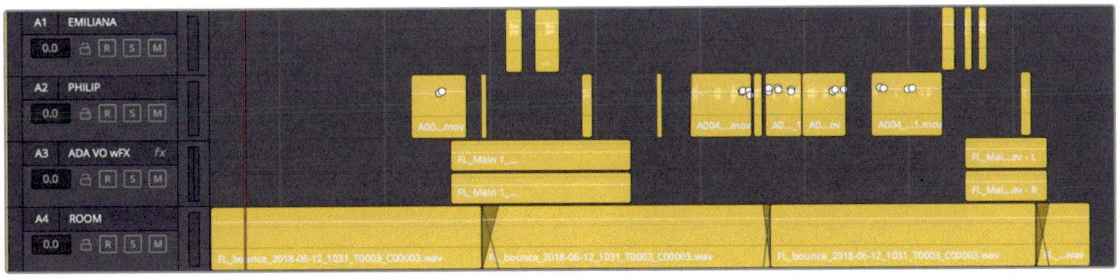

- A3 ADA VO wFX contains a bounced version of ADA's voiceover that you worked on in Lesson 7. Flanger and de-esser plug-ins were already added to this track to further enhance the sound.

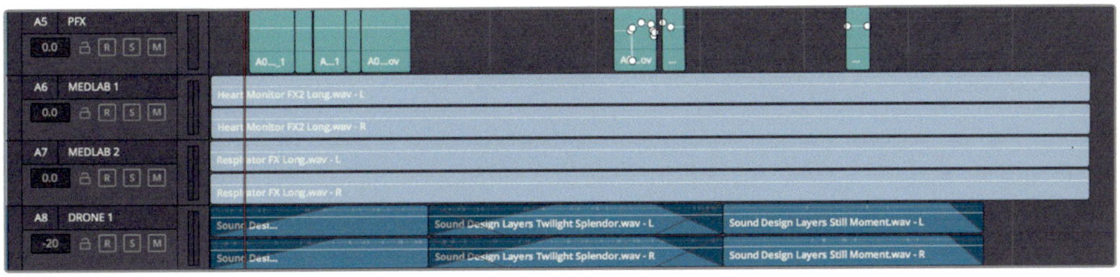

- A5 PFX contains all of the useful production sound effects (non-dialogue) that were recorded with the dialogue. It was split from the dialogue clips and moved to a separate track so it can be included with the other sound effects tracks in the mix.
- A6–A7 are the stitched and bounced versions of the heart monitor and respirator effects. They have been pitched up by +3 semitones in the Inspector.
- A8 is the DRONE 1 filler track that has been part of this scene since the initial edit. During this lesson, as the sound designer and re-recording mixer, you will determine whether you want to include it in the final mix.

> **NOTE** You can see the finished timelines for the medlab fx and ADA's VO in the timelines "8 MedLab FX Finished" and "7 ADA VO test Finished."

- A9 SFX contains the final mixed stereo sound effect stem that incorporates over 20 tracks of effects. Normally, the re-recording mixer would have access to the individual tracks, as well as a sub-mix and bounced stems of the sound effects.
- A10 MUSIC contains the final mixed stereo music stem with the original score.
- A11 SPACER is an empty 7.1 track used as a spacer to separate your primary mixing tracks from the reference tracks. Feel free to drag the bottom of the Spacer track header to change its height as needed.

- A12–A14 Reference tracks contain sections of the Hyperlight final mix and stems that can be used to set listening levels and compare mixing levels while you work. These tracks are muted.

NOTE If you are working in a quiet room with a good set of stereo audio monitors, you should be good to go. If you have been using your built-in computer speakers or an inexpensive single speaker up to now, you are well advised to use stereo headphones when performing these mixing exercises. Mixing with headphones is not ideal in most scenarios, but for learning purposes they will suffice.

Using Reference Tracks to Set Monitoring Levels

In previous lessons, you've learned some of the guidelines for balancing dialogue and sound effects. Those are important considerations, but stepping away from the clips and tracks for a moment, everything comes down to volume. Your personal preferences for volume levels may be completely different from your family and friends. That's why when you seize the television remote control, you usually change the volume as well as the channel. Regardless of your personal preferences, audio levels should be consistent from scene to scene and from one show to the next.

How do you do this? Meters are only part of the story. You still mix with your ears as well as the meters. So, before mixing a single track, you should set your computer output monitoring levels. Granted, you may be at home or in a classroom while stepping through this training guide, so you may not have the equipment to calibrate your listening environment—or even the speakers to listen to the output. And if you already have professional calibration gear and an awesome surround speaker setup at your mixing station, chances are your room is already calibrated.

NOTE The engineering details for calibrating your room go far beyond the scope of this book. Furthermore, the standards vary from region to region and are subjective to your equipment, room size, project type and delivery requirements. If you are considering a career in audio post production, by all means become an expert in the hardware, standards, and calibration. You can find reference details and sample wave test files online through professional organizations like the EBU, SMPTE, or ATSC, as well as some audio post facilities.

Regardless of your setup, for the purposes of this lesson, how loud should your monitoring output be? Basically, if your room, your computer output level, and your speakers are properly set up, then the levels should be comfortable to *your* ears while you mix. Loud sounds will sound loud, quiet sounds will sound quiet, and if something isn't working, you'll hear it.

One of the best ways to set your listening levels, is to use a reference track that contains professionally mixed content similar to your current project or scene.

Finding Useful Reference Tracks

In the real world, you'd need a time machine to reference a finished mix of your current project. Such is the power of training guides like this one. However, it is common for sound-mixers-in-training to use finished mixes as a guide to check levels and practice re-mixing a scene.

What should you use as a reference track? Something that is as close as you can get to the project or scene you are mixing. Of course, no two projects or scenes are the same, so when you are looking for reference tracks for comparison, here are some basic guidelines to follow:

- Compare apples to apples whenever possible. In other words, find a project that has the same delivery format. For example, if you are mixing a podcast, use a downloaded podcast as a reference. If you are mixing a film (even a short film) for theatrical release, use the highest quality theatrical mix you can of the same genre. Blu-ray and 4K DVDs often use the theatrical mix, or something close to it. You can patch a player to your system so you can hear the soundtrack, see the meters, and set your listening levels accordingly.
- If you are working on an episodic show, scripted or unscripted, use mixed elements from previous episodes as a guide.
- Choose reference scenes or sections that reflect the overall feel and sound experience (soundscape) that you are trying to convey in your mix. For example, if you are mixing an intense dialogue scene that involves a heated exchange ranging from hushed whispers to loud shouting, use a scene with the same dynamic dialogue range as a reference.
- Choose a reference soundtrack that is a good example of successfully balancing and mixing the subtle elements in the scene as well as the big moments. Pay attention to the background, ambience, and Foley sound effect levels, as well as the placement of the music and non-diegetic sound in the mix.

For this exercise, you'll use finished Hyperlight mix as a reference to set your listening levels. These can also be used later while mixing, to check your balancing and sweetening skills against the finished stems.

1. Adjust the vertical zoom on your timeline until you can clearly read the track numbers and names. Then scroll down to see the A12–A14 tracks.
2. Unmute and Solo the A12 MIX Reference track.

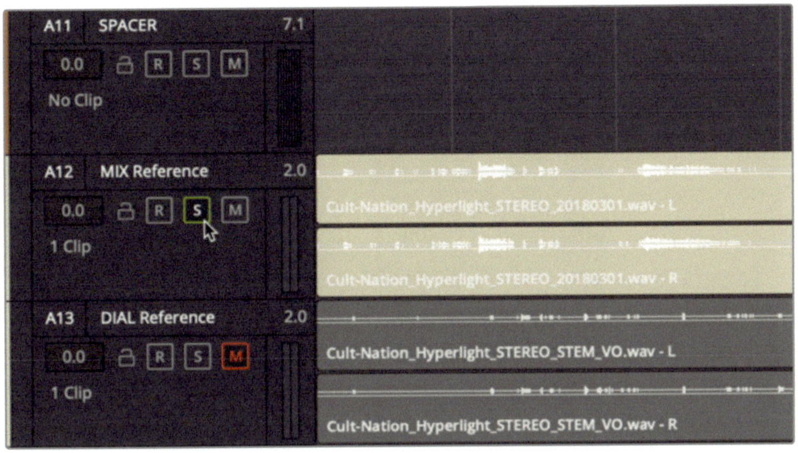

Because you will be focused on the meters for a few steps prior to adjusting the levels, let's use the DIM button to lower the playback volume by 20 decibels.

3. Mark a play range between 01:00:53:18 and 01:01:18:04. This section is a strong representation of dialogue plus music and background sound effects.

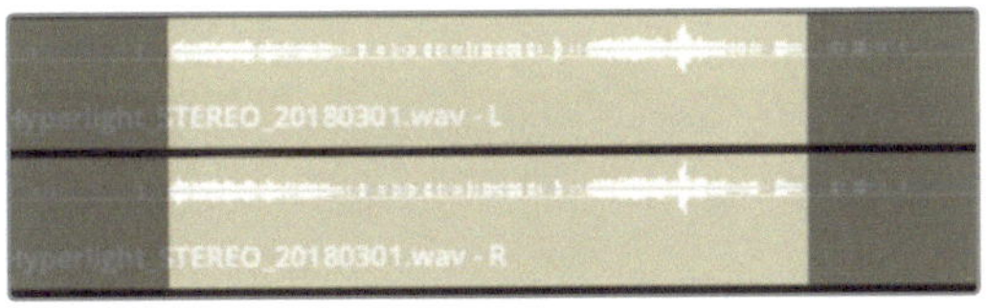

Keep in mind that this reference audio does not match the picture, so ignore the viewer and focus your attention on the meters.

4 Start looped playback of the A12 MIX Reference track. Look at all the meters in the A12 track header, Mixer, and monitoring panel.

Track header

Mixer

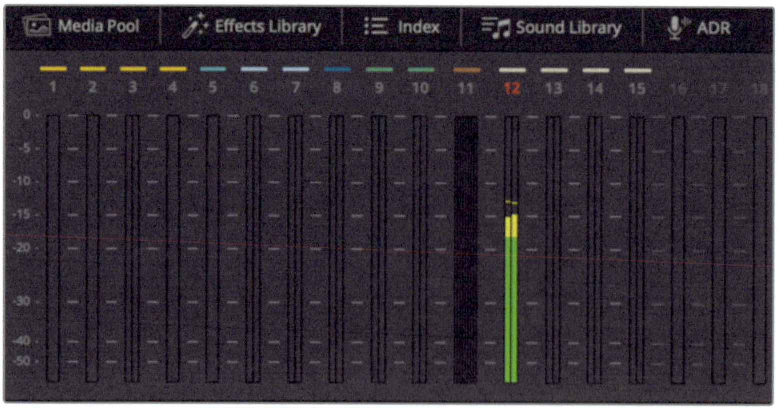

Monitoring panel

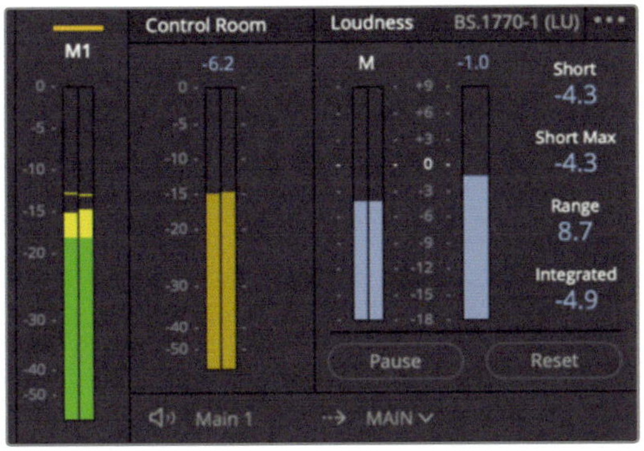

Monitoring panel

Seasoned audio engineers should feel right at home among these meters and controls. If you are new to audio post, you may find this setup as intimidating as a space shuttle cockpit. If so, relax. In the next few exercises, you'll discover just how useful and user-friendly the meters and mixing controls can be.

Keep in mind that all these meters are telling the same story but simply use different scales to do so. First, do you remember your target levels for dialogue? It's between -15 and -10 dB with the average at -12 dB. Also, if you recall, these target dialogue levels appear yellow in track meters, Mixer, and monitoring panel track meters. The dialogue in the reference track fits the target levels perfectly. Additionally, at powerful moments, music and sound effects swell to the same levels when dialogue is not present.

In the meters to the right of the monitoring panel, you'll see that each meter tells the same story: the reference track levels peak at the nominal levels between -10 and -15 dB in the Mixer and correspond with the Control Room (mustard colored) meter, and fluctuate around 0 (-23 LUFS) in the blue loudness meters. You'll work more with the Control Room and Loudness meters in a later lesson.

NOTE American loudness standards are actually -24 LUFS, which you can change in the project settings. Also, keep in mind that this is a stereo signal, just like the Main 1 output. However, single-channel (mono) pink noise is ideal for calibrating your various surround speakers and levels. So, you'll keep this as a stereo signal for now to set the level coming from the computer, but when you get to the panning section, you'll change the oscillator track to a mono signal.

Now that you are confident in the reference levels, you can adjust your monitoring levels.

In the master playback monitoring controls, click the DIM button to turn it off, and drag the fader to the far right position, 0dB. If you are monitoring through an external audio I/O device, adjust the monitoring levels accordingly. Continue looped playback.

5. While listening to the reference track, adjust the output level of your computer or speakers so that the levels are comfortable and clear for you. You can do so whether you are using headphones or external speakers. If necessary, you can also adjust the playback monitoring level.

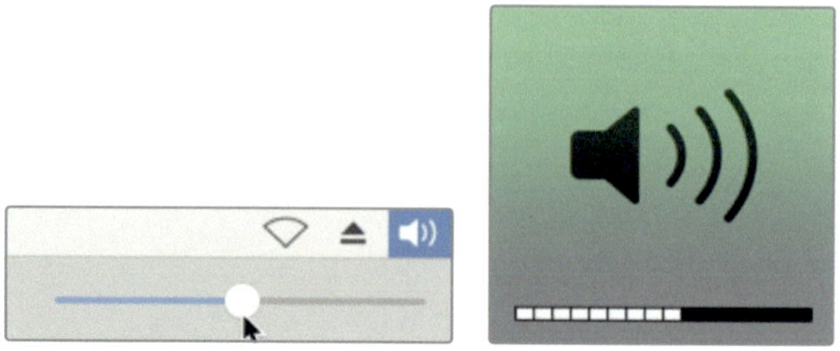

Let's check your setting with the actual dialogue tracks.

6. Clear the play range. Unsolo and Mute the A12 Reference track.
7. Solo tracks A1-A4. Move the playhead to around 01:00:22:00, before Philip's first line of dialogue, "ADA, identify the person in front of me."

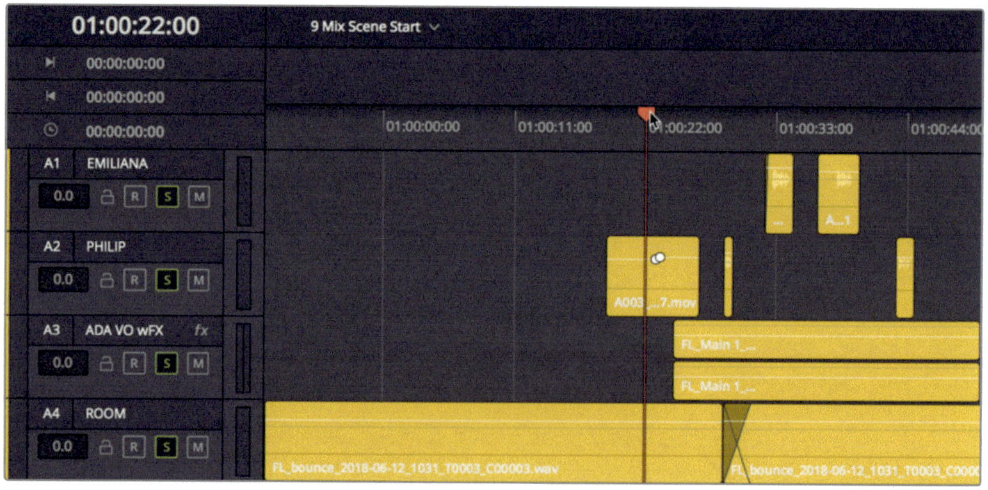

8. Start playback, and listen to the dialogue levels while looking at the meters in both the Mixer and monitoring panel.

   The levels should still average around -15 dB to with some louder peaks or quieter dialogue depending on the scene context. How did the levels sound to you? If they sounded too low, you may have set your computer output level a bit too low based on the reference track. Conversely, if the dialogue was too loud, you probably set the levels too high. Setting the best levels comes with practice. The important goal is to set a comfortable level for you.

9. Unsolo the dialogue tracks.

10 If you need to adjust your listening levels, unmute and solo the A12 track and adjust the levels now. When finished, mute and unsolo the A12 MIX Reference track.

11 Show the Tracks Index, and hide the visibility (eye icon) for tracks A11-A15. Make sure that all four Reference tracks are muted. Then close the Index.

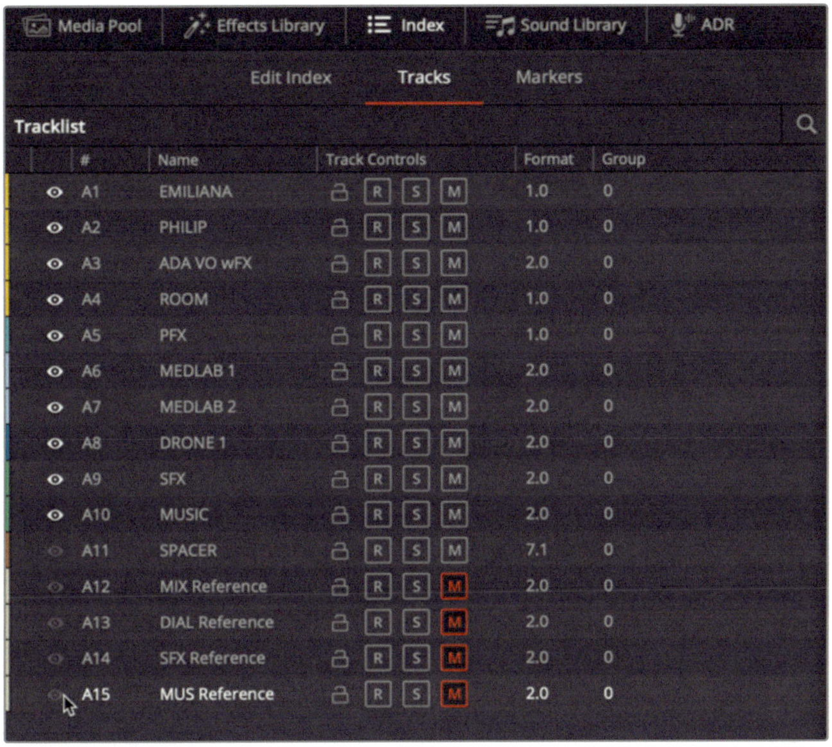

Once you have set up your monitoring levels for mixing, you must abide by the number one rule: Don't touch the levels again until you are finished with the mix. Really! Moving forward, when the tracks sound too loud or too quiet, they truly are. Fix them in the timeline or the Mixer, but don't touch the computer volume controls. Trust your ears as well as the meters.

Customizing the Interface

Now that the monitoring levels are set, let's resize the viewer in the monitoring panel. Because this timeline has only 15 tracks, you definitely don't need to see all the empty track meters at the top of the screen. Also, at any time, you can increase the height of the monitoring panel, while decreasing the height of the Mixer. In this exercise, you'll customize your interface for mixing. Then, you can make changes as you go to accommodate the mixing tasks at hand.

To resize the viewer or the monitoring panel, you can simply drag the viewer's boundaries.

1 In the monitoring panel at the top of the screen, move the pointer over the left edge of the viewer until it becomes a resize pointer.

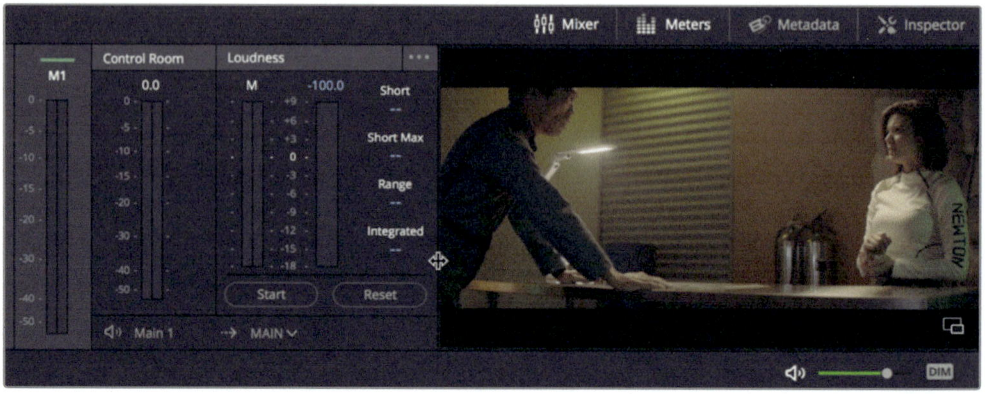

2 Drag the left edge of the viewer to the left until the loudness meters are to the left of the transport controls in the interface toolbar.

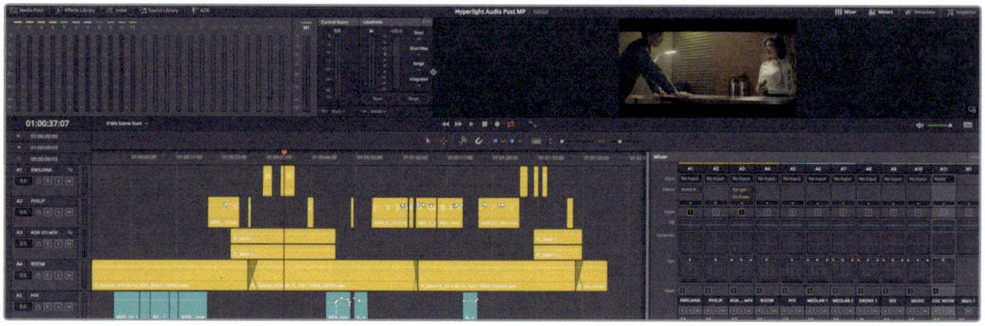

This customized layout has several advantages. First, the gray space around the viewer makes it easier to glance at, like a standalone window. Second, the loudness meters are much closer to the track meters in the monitoring panel, which makes it easier to see them all at once.

Next, you'll resize the monitoring panel in both directions to see how easy it is to quickly adjust the height of the monitoring panel or Mixer.

3 Drag the lower edge of the viewer down to increase the height of the viewer and monitoring panel. Stop dragging when the monitoring panel fills the upper-third of the screen.

This layout is great when you need a larger viewer, loudness meters, and overall track meters.

4 Drag the bottom edge of the viewer up as far as you can to minimize the height of the viewer and monitoring panel.

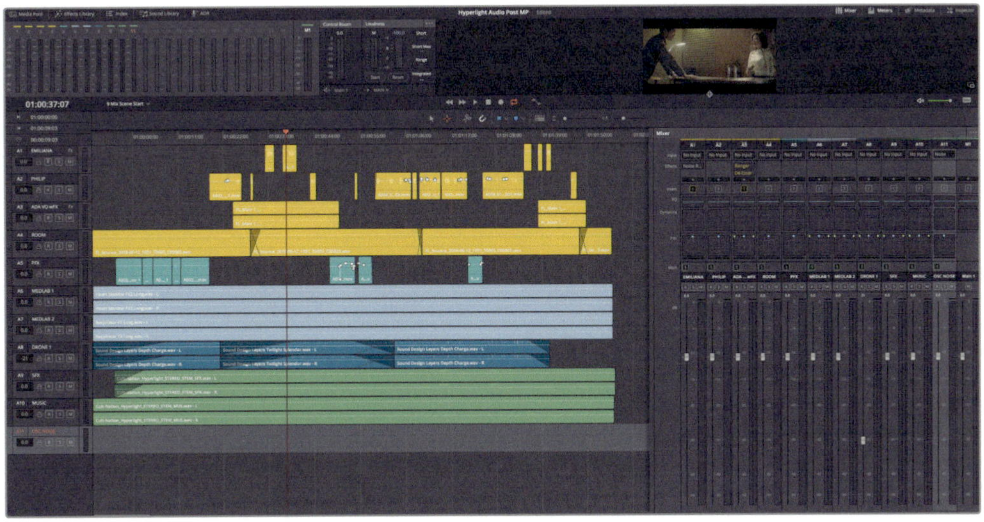

The trick to resizing your interface is to maximize the size of the elements you need as you need them.

Evaluating and Setting Initial Track Levels

The four fundamental elements of mixing are volume level, pan, dynamics, and equalization. However, each of these elements affects the overall sound at all times. In many ways, mixing a soundtrack is like cooking a single-pot dish. Many parts influence the final dish, so you continually mix, add more ingredients (tracks), taste (listen), and add spices (effects and EQ) until it's ready to serve.

The most apparent mixing element is the volume, so let's start with it.

In the Mixer, you can see that all the faders are at their default positions except for the A8 DRONE 1 track. This is the starting position for mixing. Assuming that all the clip levels are balanced within each track, you can use the mixing faders to adjust the levels within the mix. Keep in mind that the faders are only in their starting positions; their final positions will change constantly until the final mix is rendered in the Deliver page.

Let's play the first half of the scene and listen to the current levels of the tracks. At that time, if any track levels are obviously out of place, you can balance them in the track header or in the Mixer.

You have only one way to find out if the starting levels are balanced. First, as a re-recording mixer, you'll want all the faders in their default positions before you listen to the tracks. If a change in level is necessary, you'll want to be the one to do it.

1 In the Mixer, double-click the A8 DRONE 1 Fader to set it to its default position.

2 Play the beginning of the scene, and listen to the cumulative levels. During playback, listen for any tracks that seem too loud or distract from the dialogue. Also, listen for tracks that may need level changes over time.

3 When Emiliana says, "ADA, identify the person I'm speaking with." stop playback.

What did you think? You may have identified a few places that you must deal with before moving forward. Two of those are the medlab effects on the A6 and A7 tracks.

While the heart monitor and respirator effects sound good, especially in the beginning of the scene before the dialogue, they are way too "hot" (loud), especially when the medlab is offscreen and the characters are talking.

Also, the A8 DRONE 1 track is stepping over the music track and seems to be consistently too loud. No problem, these are all easy fixes.

NOTE Before working your way through this book, your first instinct may have been to turn up the levels of the dialogue tracks to compensate for the other louder music and sound effects tracks. If you found yourself reaching for the faders with an urge to turn down the medlab effects while listening to the scene, you have the reflexive aural instincts of a re-recording mixer!

Let's mute the medlab effects, and set the level for the A8 track during playback. After the A8 track level is set, you can tackle the medlab effects.

When mixing, the levels of background sound effects tracks are adjusted during playback in the context of other priority tracks. Because you are working on a digital audio workstation (DAW), you can also stop and adjust individual clips as necessary.

4 Mute the A6 and A7 tracks.

5 Select the A9 DRONE 1 track header to also select and brighten the corresponding channel strip in the Mixer.

> **NOTE** Keep in mind that Faders are the controls in the Mixer that control volume levels, whereas Fade handles are the white handles (dots) on the volume overlay at the head and tail of each clip in the timeline.

In the timeline at the first clip in the A8 track, you can see that the clip starts with a fade. So, wait until the playhead is beyond that fade-in before you adjust the track fader. Meanwhile, this is a great opportunity to listen to the drone track soloed during the fade-in to familiarize yourself with its sound.

6 Solo the A8 DRONE 1 track.

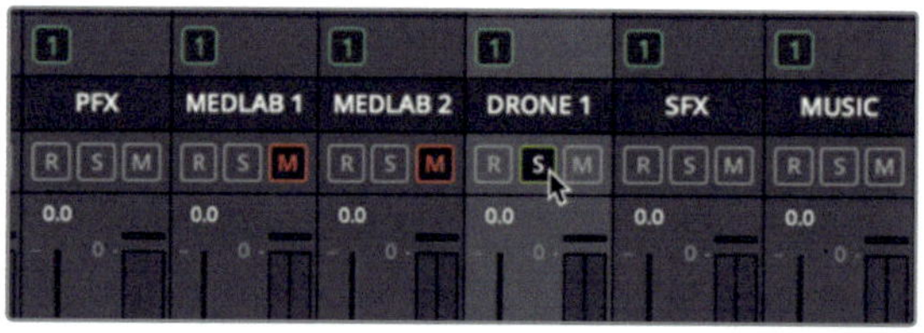

7 Start playback from the beginning, and listen to the A8 track soloed during the fade-in. Continue playback. In the Mixer, unsolo the A8 track, and drag down the A8 track fader to lower its overall level in the mix. Your goal is to lower the drone level until it is a subtle *bed* (a continuous sound) in the background. If you aren't sure which level to use, try a setting between -25 and -30 (around -15 dB in the track volume level field).

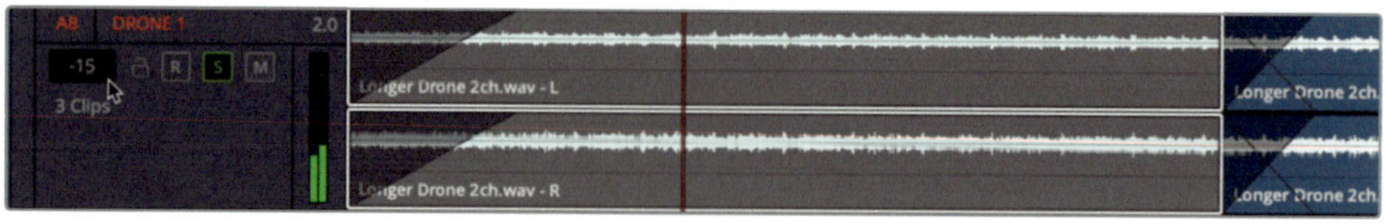

As you can see, it takes only a few seconds to set the faders and evaluate the tracks. Next, you'll balance the medlab effects tracks.

Finding the Right Balance between Two Background Tracks

The heart monitor and respirator sound effects are on separate tracks but represent one complex medlab machine. Your goal is to balance the two tracks to make them sound as if they are coming from one machine. You do not yet need to worry about their levels compared to the other tracks. First, just make them to sound great together. Once you find the right balance, you can use faders to lower their relative levels to fit into the overall mix. The best way to balance levels between two tracks is to do so during playback with the two tracks soloed.

1 Unmute and solo the A6 and A7 tracks.

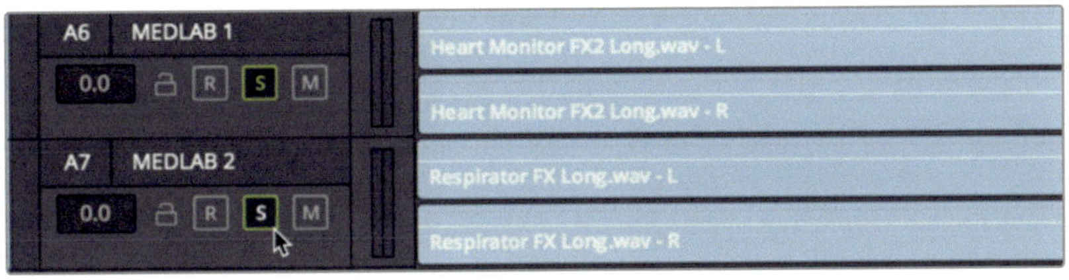

2 Select the A6 track, and then Command-click (macOS) or Ctrl-click (Windows) the A7 track to select them both.

3 Zoom in vertically on the selected tracks until you can clearly see the waveforms and volume overlays on the medlab clips. Then, deselect the tracks.

By zooming in on them, the tracks will be easier to work with while you change the individual clip levels. When balancing multiple sound effects clips into one effect, you should first determine which is the dominant clip, if any. As the sound designer, this choice is entirely up to you. In this example, you'll lower the respirator sound because it is more likely to detract from the dialogue. The heart monitor, on the other hand, can sell the idea of a medlab by itself.

So, let's leave the heart monitor clip's volume at the default level, and lower the respirator clip to about half the level of the heart monitor. To do so, you can drag down the clip's volume overlay while listening and watching the meters in the track headers, Mixer, or monitoring panel.

4 In the monitoring panel, drag down on the bottom of the viewer to increase the size of the viewer and monitoring panel, as needed.

5 Start playback from the beginning, and watch the video playback in the viewer and the meters for the soloed tracks.

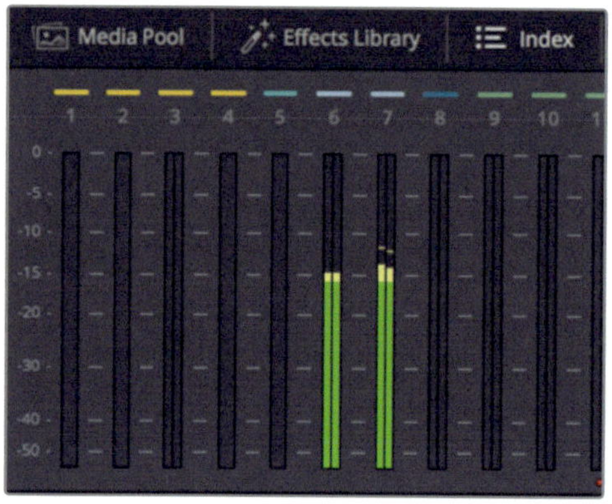

Both tracks average around -15 dB. So, to reduce the respirator level by half, you'll lower the level until it averages around -30 dB.

6 Continue playback. Drag the **Respirator FX Long.wav** clip's volume overlay down while listening and watching the meters until the average level of the clip in the A7 track is about half of the A6 track's level.

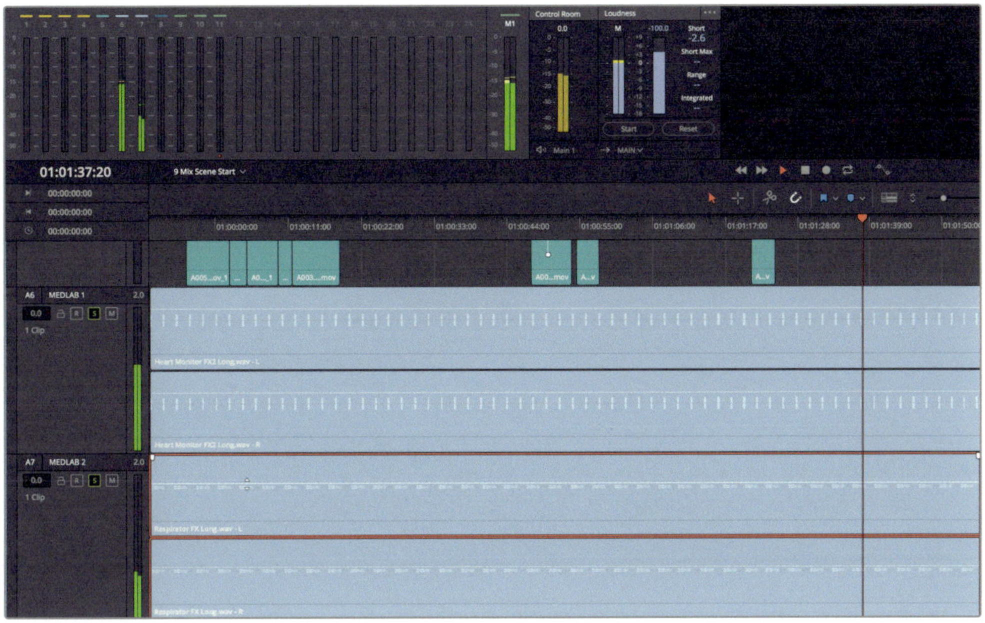

To make sure the heart monitor and corresponding respirator levels are working with each other, let's check them against the beginning of the scene where you see Philip in the medlab.

7 Start playback from the beginning, and listen to the combined medlab effects while watching the viewer. If the sound works with the picture and sells the idea of a real machine, your levels are good. If not, this is the time to fix them. Raise or lower the level of the respirator clip until you like the way it works with the heart monitor. When you are finished, stop playback.

Remember, sound mixing has a few delivery rules, but ultimately everything comes down to sound. Somebody must make the choices and decide if those choices work. For this lesson, it's all you. If it you think it works, move on.

Linking Tracks for Single-Fader Control

Now that the two different medlab clips sound great together, you need to adjust their levels relative to the mix. Trouble is, they are on different tracks. Because they are essentially one effect on two tracks, you can link them together to control them with a single channel strip in the Mixer. Linking tracks as a group has one prerequisite: the tracks must be single-channel (mono) tracks.

The waveforms in these clips show that the left/right channels in each track are identical.

In this exercise, you'll change the A6 and A7 tracks to mono and link them as a group for mixing. First, take a look at the Main 1 output levels for the two stereo tracks and compare that to the levels after they become mono tracks.

1 From the beginning, play a few seconds of the stereo medlab tracks, and in the Main 1 output meters, look at the output level of the cumulative tracks. Watch for the loudest levels in the yellow and in the peak indicators.

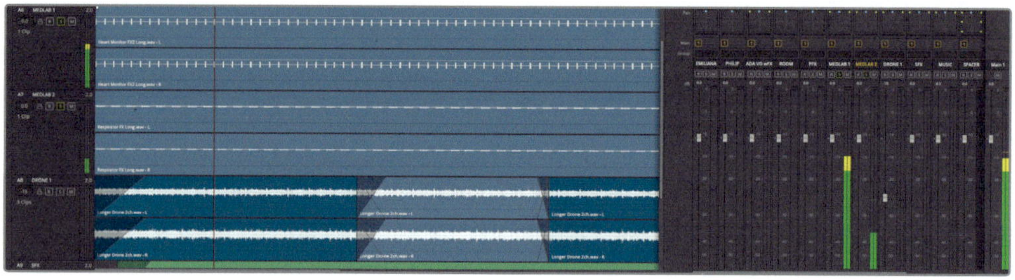

2 Right-click the A6 track header, and choose Change Track Type To > Mono.

3 Right-click the A7 track header, and choose Change Track Type To > Mono. Click OK.

4 Play the beginning of the timeline to hear the mono tracks played together. During playback, notice that the relative difference remains the same, yet the cumulative output level is lower.

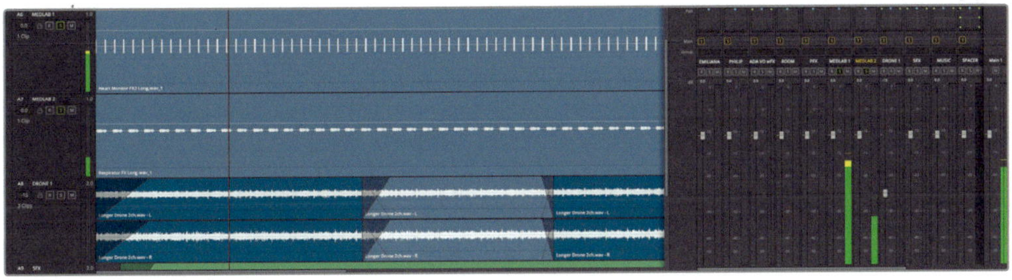

Why are the levels lower? To emulate the way human ears interpret sound in the real world, **pan laws** are applied to levels to compensate for the differences between mono signals panned to the center and stereo signals panned left and right. In short, a single-channel sound panned to the center is perceived to be lower in volume (approximately -3 dB) compared to the same signal playing equally from two stereo channels. Professional digital audio workstations, such as the Fairlight page, automatically apply the pan law standards to mono tracks.

NOTE You'll work more with track pan controls in the next section. This was just an opportunity to point out the level changes between the converted stereo to mono tracks.

Meanwhile, to link the two mono channels, you'll use the Link Group window.

5 Choose Fairlight > Link Group.

The Link Group window opens, displaying all the tracks in the timeline. If you look closely, you'll see that all the mono tracks are brighter to indicate that they are available for linking.

This window is primarily used to link stereo or surround sound channels that were imported into Resolve as separate mono channels. However, in this case, you'll link the A6 and A7 tracks to create a stereo track. Linking is non-destructive and doesn't alter the original files. Furthermore, you can unlink linked tracks at any time in the same window.

6 In the Link Group window, click the MEDLAB 1 and MEDLAB 2 buttons to select those tracks for linking.

7 Click the Link button.

In the Link Group window, the two linked tracks have a dark background and a bar connecting them to indicate that they have been linked.

8 Close the window, and look at A6 and A7 track headers.

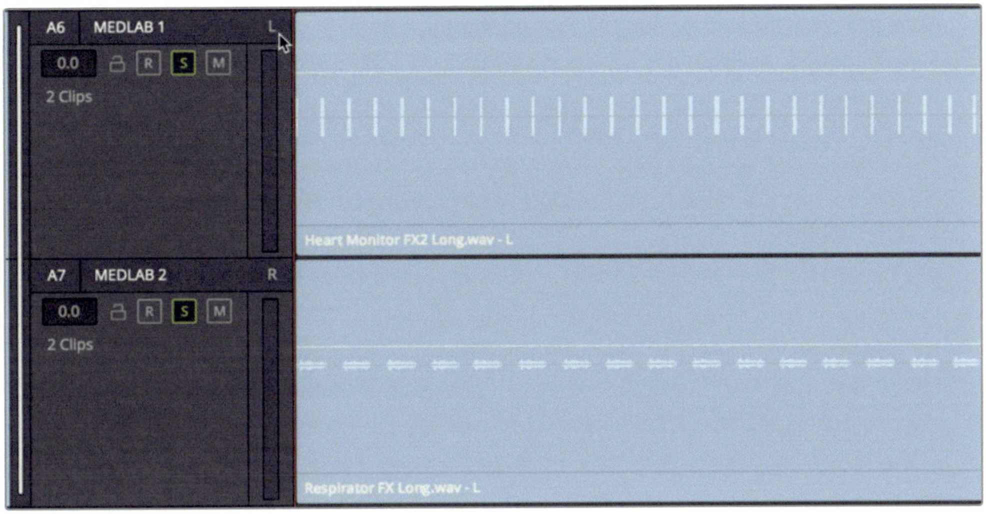

You can instantly recognize the linked tracks by the white vertical bar to the far left of the track headers, and the channel indicators that show that they are mapped as L (left) and R (right).

9 Drag the vertical zoom controls to resize the tracks until they all fit in the timeline.

10 Drag the left edge of the Mixer to the left to extend it to see all the channel strips.

The channel strips no longer include A7 because linked tracks are controlled by the channel strip for the first linked track, which in this case is A6 MEDLAB 1. The linked clips are also marked as such in the monitoring panel.

Now that you've linked the tracks, they should sound just as they did before, right? Never assume anything in audio mixing. In fact, you should get in the habit of auditioning changes in the context of the entire mix as you make those changes. Colorists have the advantage of instantly seeing every miniscule change they make. However, in audio mixing, you start playback and then rely on experience, your ears, and the meters.

11 Start playback from the beginning, and listen to the linked medlab tracks.

12 In the Mixer, lower the level of the A6 fader to around -25 dB, and stop playback.

As promised, you can now control the linked tracks using a single channel strip in the Mixer. However, you may have noticed a pronounced difference in the medlab sound effects during playback. Let's hear it in context with the other tracks.

13 In the Mixer, click the Solo button on the A6 channel strip to unsolo the linked tracks.

14 Play the beginning of the timeline, and listen to the medlab effect along with the other tracks. Stop playback before Philip's first line of dialogue.

Wait! What happened to the medlab effects? They used to sound like one machine in the middle of the room, and now they sound like two machines on opposite sides of the room. No worries, you just experienced the power of track panning, which you will explore in the next exercise.

NOTE Another method for controlling similar tracks with a single fader is to use a submix bus, as you will learn later in this lesson.

Placing Audio in the Panoramic Sound Field

Pan controls place a track's audio within a panoramic sound field. They enable you to compose the spatial arrangement of audio elements just as a cinematographer composes the visuals of a shot. You can precisely locate mono tracks to sound as if they come from an offscreen source, or virtually anywhere within the frame. DaVinci Resolve includes advanced pan controls in both the Edit and Fairlight pages that support 2D (stereo) audio, multichannel surround, and 3D (object-based) audio placement.

In this exercise, you'll use the pan controls to narrow the panoramic spread of the medlab effects. Then, you'll experiment with various panning options to widen the sound of ADA's omnipresent computer voice. Finally, you'll explore the surround-sound panning and 3D object-based panning. Let's start by using the A14 SFX Reference track's panning controls to compare stereo and mono tracks.

1. In the Index Tracklist, click the visibility icon for the A14 track. Then hide the Index.
2. Mark a range between 01:01:04:19 and 01:01:05:06. This is a short sound effects section that will work for this example.
3. Solo the A14 track and start looped playback. If the level of the reference sound effects track bothers you, feel free to click the DIM button for this panning exercise.

In the panning controls, the green dots represent the spread and positions of the speakers for each audio channel. The blue square is a handle that you can use to move the sound to a specific position in mono tracks, or between speakers in multichannel tracks.

Notice that the A14 track is in the default position with the left and right channels in the front (upper) left and right positions, and the blue handle is equidistant between them. During playback, the A14 meter and M1 output meters also indicate an equal amount of sound coming from the left and right channels.

4 Using the Mixer's pan controls, drag the blue handle for the 14 SFX Reference to the upper-left corner of the panning control in the control strip. Then, drag the 14 SFX Reference blue panning handle to the upper-right corner of the control to pan the sound fully to the right speaker.

Stereo track panned fully to the left speaker

Stereo track panned fully to the right speaker

In each example, you can hear the sound more prominently in either the left or right speaker or the headphone earpiece. While the track's meter stays the same regardless of the panning position, the M1 output meter shows twice as much sound coming from the left or right channel based on the panned position.

5 Drag the blue panning handle to different positions while listening to the output and looking at the meters.

No matter where the panning handle is placed, with these stereo pan settings, sound will always come from both the left and right speakers.

Now let's explore the same panning controls on a mono track.

6 Right-click the A14 SFX Reference track header, and choose Change Track Type To > Mono.

The A14 track's panning controls show only the blue handle with no specified speaker channels.

7 Drag the A14 SFX Reference blue panning handle to the upper-left corner, and then to the upper-right corner, to hear the sound travel from one extreme to the other. Repeat this pan while watching the M1 output meters to see the stereo output reflect the position of the single-channel panning handle.

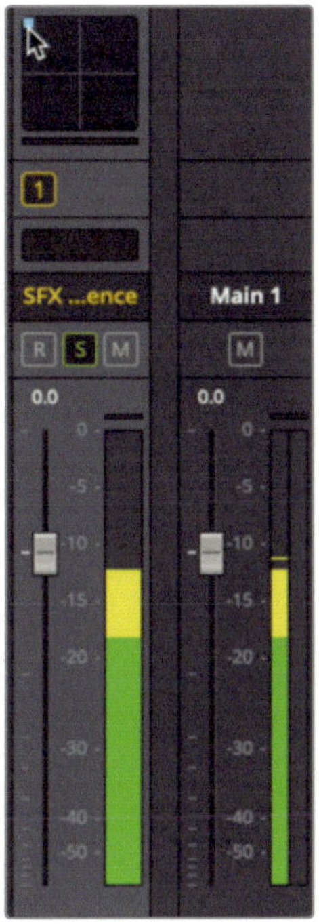

Mono track panned fully to the left speaker

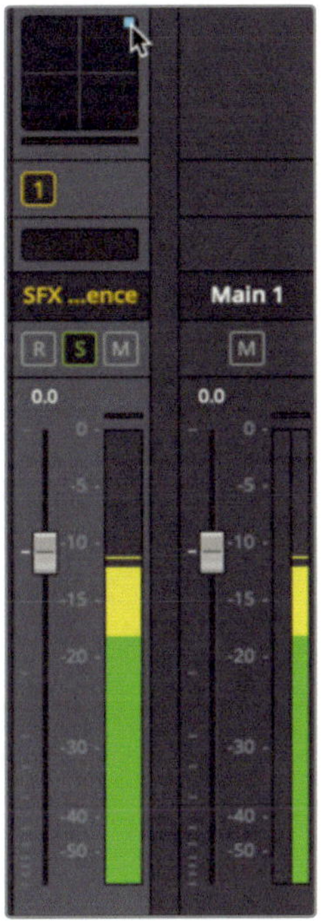

Mono track panned fully to the right speaker

When panning the mono track, you can move the single channel of sound fully to one speaker or the other or anywhere in between.

8 Clear the playback range. Then mute and unsolo the A14 SFX Reference track.

Clearly, the pan controls in the Mixer are a quick and easy way to pan your stereo and mono tracks. But what if you want a stereo track to sound more like a mono track?

Controlling a Track's Stereo Spread

In addition to panning sound between the left and right speakers, you can also narrow the perceived placement of the speakers and the size of the source. In this exercise, you'll narrow the distance between the two medlab sounds to regain the original illusion of a single machine. Let's solo the A6 MEDLAB 1 track and take a closer look its pan controls in the Audio Pan window.

1 In the Mixer, solo the A6 MEDLAB 1 track. Then, double-click the track's Pan controls to open the Audio Pan window.

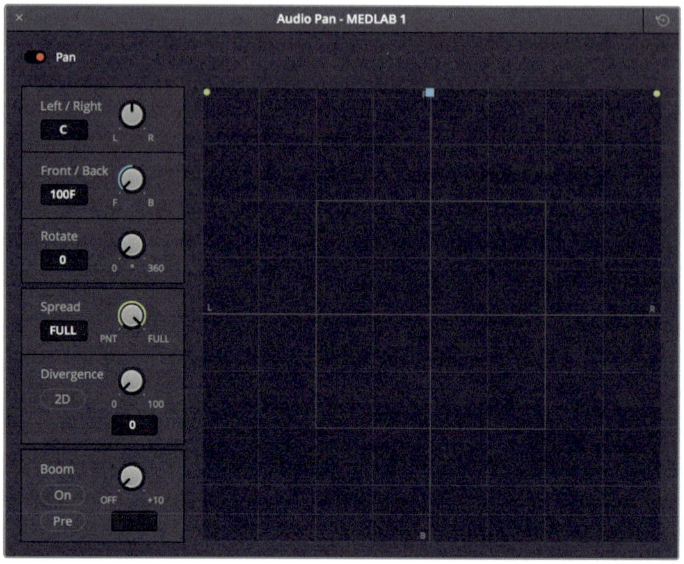

These Audio Pan controls work with either 2D (stereo) or traditional 5.1 or 7.1 surround-sound configurations, depending on your setup and project. The center of the graph represents the audience's perspective from which an audience can hear sounds placed in the surround sound positions around them. The top center position, which is the default position for the blue panning handle, is often referred to as the *phantom center* because it doesn't represent a speaker. Instead, it represents the illusion of a center speaker in a stereo mix. The small letters F, R, B, and L—positioned clockwise around the space in the top, right, bottom, and left positions—represent the Front, Right, Back, and Left positions in the panoramic field, respectively.

Where you place audio tracks within this panoramic space will reflect where the audience will locate each audio source. The Spread control is for linked sources. Divergence determines the spread of the audio signal to additional speakers in a stereo or surround mix, and Boom determines how much of a track's sound is sent to the low-frequency effect (LFE) speaker.

The A6 MEDLAB 1 track is currently panned to the front with the left channel going to the far-left front speaker position, and the right channel going to the far-right front speaker position. No wonder it sounds as if the respirator and heart monitor are on opposite sides of the room; that's exactly how the linked track is panned. Let's start playback and narrow the stereo spread to bring the sound of the left and right channels closer together.

2 Turn off the DIM button, if necessary.

3 In the Mixer, double-click the A6 fader to reset it to the default position.

4 Start playback from the beginning of the timeline, and listen to the current placement of the heart monitor to the left and the respirator to the right.

NOTE If you hear the heart monitor to the right and the respirator to the left, your headphones are on backward or your speakers are inverted.

5 Drag the Spread control counter-clockwise from the FULL position to the PNT (point) position. As you change the control, notice that the green dots representing the left and right channels move toward each other until they become a single point.

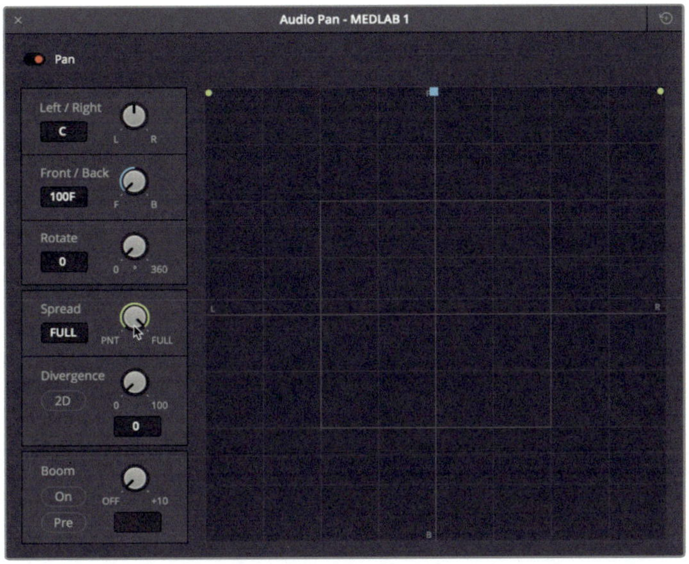

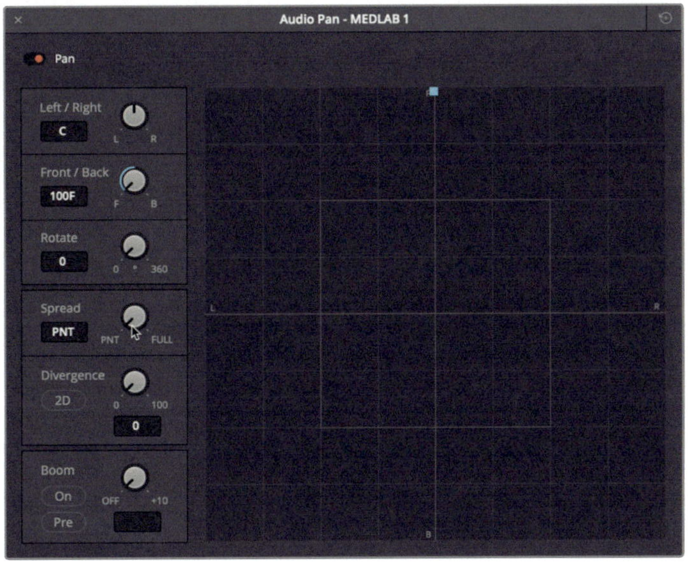

The sounds of the two effects now seem to be coming from the same position, just as they did when they were independent mono tracks. In fact, you can now position the sound using the blue pan handle, just as you could with a mono track.

6 Continue playback, and drag the blue pan handle anywhere within the panoramic field to hear the medlab effects move between speakers. Then, drag the handle to the center position on the panoramic graph. Use the Left/Right and Front/Back fields as a guide to see when the position is centered exactly (C).

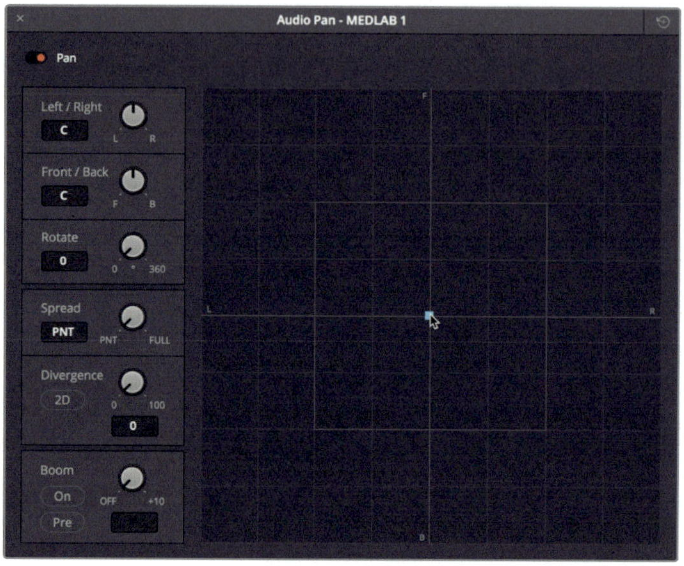

NOTE You may or may not hear the sound change from front to back when you are listening through stereo speakers or headphones. Also, if you are monitoring audio through an audio interface and Mixer that outputs only left and right channels, then you may not hear any output that is not panned to the front left or front right.

Next, let's use the Divergence control to increase the perceived size of the audio source.

7 Start playback from the beginning. Drag the Divergence control clockwise to around 20. The green line indicates the perceived size of the source sound.

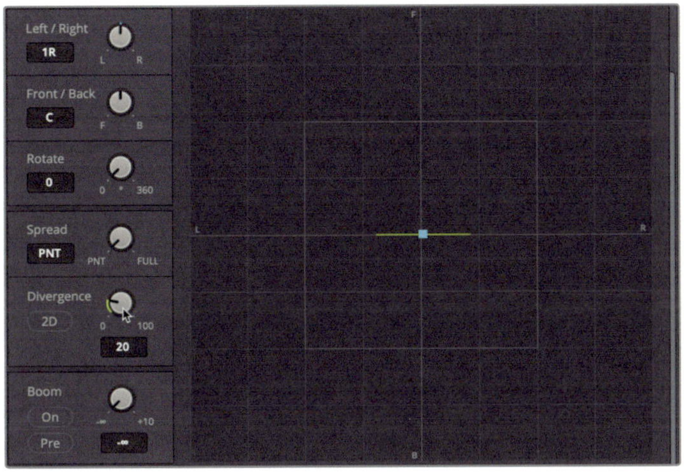

8 In the Divergence controls, click the 2D button to show a circle representing the perceived size of the source audio.

9 Increase the Spread to 6 to slightly separate the two sound effects.

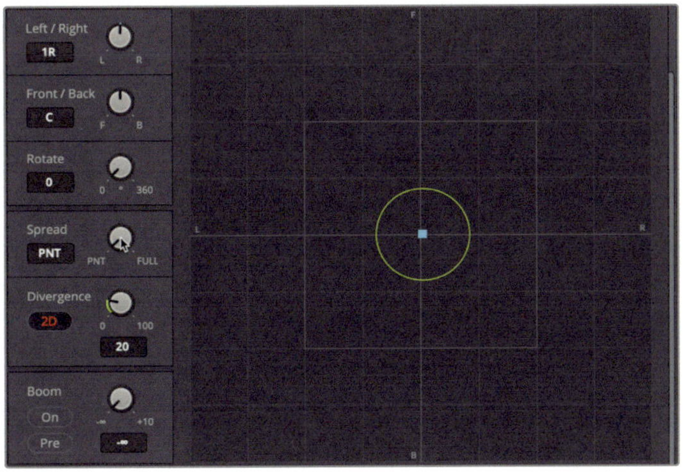

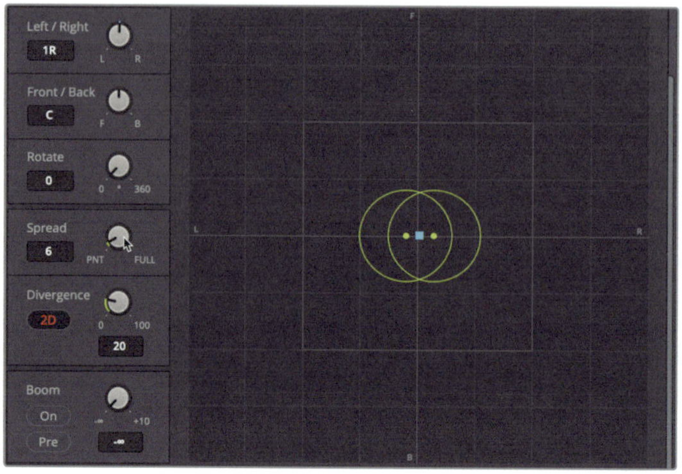

The Divergence and Spread adjustments scale the cumulative medlab sound effects to sound as though they are coming from the large medical device shown onscreen. Now, let's try setting the volume level in the context of the other tracks. For this exercise, focus on the sound of the medlab effects before Philip's first line of dialogue.

10 Close the Audio Pan window, and unsolo the A6 channel strip.

11 Play the timeline from the beginning, and adjust the A6 fader so the volume fits into the scene as a realistic background sound. If you aren't sure of the level, try setting a level between -25 and -30 in the Mixer. When you are finished, stop playback.

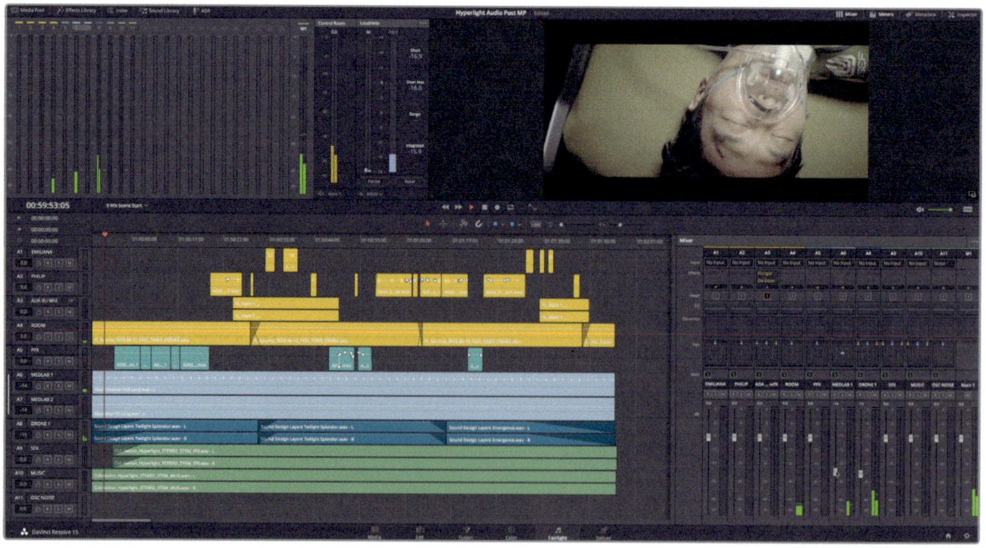

Wow! Goosebumps. Suddenly, the medlab looks and feels (and sounds) real. The medlab effects have come a long way since the beginning of this lesson. Finally, they are working in harmony with the other tracks and really "sell" the whole medlab location. And as you just witnessed, finding the right levels and track panning can make a huge difference in the impact of each track.

Exploring 3D Pan Controls

In addition to conventional stereo and 5.1- and 7.1-channel surround panning, DaVinci Resolve 16 includes 3D audio pan controls for object-based spatial audio positioning. These enable you to pan mono tracks to specific positions in 3D space for a more immersive audience experience.

Let's open the 3D pan controls for the A14 SFX Reference track to see and hear the 3D controls.

1 In the Mixer, Option-double-click (macOS) or Alt-double-click (Windows) the pan control for the A11 track to open the 3D pan controls.

Just like the standard surround panning controls, these controls include a panner viewer that displays a large 3D representation of the listener's perceived sound stage with a blue sphere representing the track audio being positioned within that sound stage.

To the right of the panner viewer are the 2D Front, Side, and Top panner controls for placing sound on specific axes.

- **Front panner** controls the horizontal left/right and the vertical up/down axis.
- **Side panner** controls the front/back and up/down axis.
- **Top panner** controls the left/right and front/back axis.

Let's solo the A14 SFX Reference track to hear it in the 3D panoramic space.

2 Unsolo the A6 MEDLAB track. Unmute and solo the A14 track to hear the SFX reference mix as a mono channel object. Do not start playback. Feel free to click the DIM button, if desired.

At the moment, the SFX reference sound is still in the front center position. To change the position of that sound in 3D space, drag the corresponding blue pan handle in the Front, Side, or Top panner controls.

3 In the Top panner, drag the blue pan handle down to move the track's audio toward the back and to the left of the 3D space. Watch the blue sphere in the 3D graph as it moves in the same pattern along the same path.

4 In the Front panner, drag the panner so that the sphere moves to the upper right and back of the sound stage. Keep in mind, the back of the stage is closer to you, the front of the stage is in the distance.

5 Experiment with the different panners to move the sound sphere within the sound stage. When you are finished, reset the pan controls and close the window.

6 Unsolo, mute and hide the A14 track.

As you can see, the Fairlight 3D panning controls make it easy to place sounds in 3D space. So, when your projects require object-based 3D panning, you know where to perform it. For now, let's return to the current mixing project and use a plug-in to control a track's stereo spread.

NOTE DaVinci Resolve 16 and higher also includes a Space View Scope that can be used to help visualize your 3D immersive track. You can find the Space View Scope in the Fairlight menu under the Immersive submenu.

Dialogue Is "King," Even When Panning Dialogue Tracks

Sound effects and music tracks can be panned anywhere in the panoramic field except the front center. That position is reserved for the most important tracks: dialogue.

The rules for panning dialogue are different than panning sound effects and music. Remember, dialogue is "king" in audio post-production. And the dialogue tracks get royal treatment to ensure that they are always heard front-and-center with all other tracks surrounding them. This translates to always panning the dialogue to the front center. That's also why most dialogue tracks are mono. Whether you are mixing in stereo or surround sound, you have plenty of panoramic space beyond the center to place the other tracks.

Controlling Stereo Spread Using the Stereo Width Plug-In

In the previous section, you used the audio pan controls to change the position and panning of a linked stereo sound effect. In this exercise, you'll listen to Emiliana and Philip's mono dialogue tracks that are dutifully panned to the front center position, and then apply a Stereo Width plug-in to ADA's track to create a more omnipresent sound quality. That's right; there's room for some dialogue panning creativity, as long as the final mix pans the dialogue *stem* or dialogue *submix bus* to the center channel.

1 Solo the dialogue tracks A1 through A4. Play the dialogue exchange starting with Philip's first line and ending with ADA's line, "I cannot explain this discrepancy" (01:00:22:00 - 01:00:47:20). Listen to the panoramic placement of the tracks.

Could you hear the difference in panning between the human voices and ADA's voice? Emiliana and Philip's voices seem to come directly from the middle of the screen (Front Center), whereas ADA's voice has a much wider sonic spread.

In the Mixer, the A1 and A2 tracks are mono and panned to the front center position. The A3 track is stereo with a full left to right spread.

> **NOTE** Although ADA's voice was recorded in mono, the doubled voice effect was finished by bouncing the tracks to a single rendered stereo clip.

Instead of using the panning controls to change ADA's stereo spread, let's use the Stereo Width plug-in. Because this is a mixing lesson, instead of dragging the effect from the Effects Library, let's use the effects slots in the A3 channel strip.

2 At the top of the A3 channel strip, in the Add Effect pop-up menu, click the Add Effect button (the + symbol), and choose Imaging > Stereo Width.

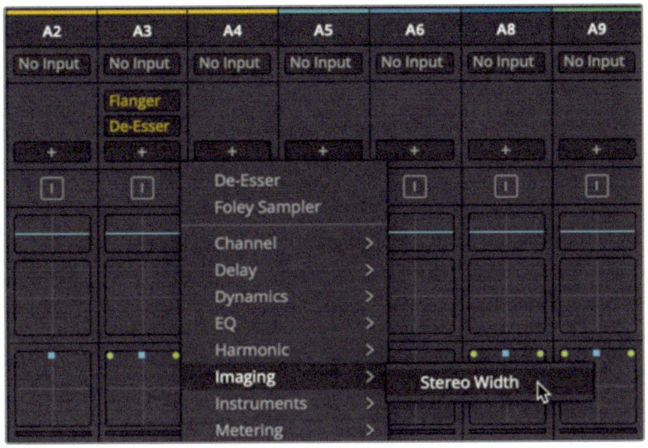

The Stereo Width plug-in is added to the A3 Effects slots along with the Flanger and De-Esser plug-ins. The Stereo Width plug-in window opens.

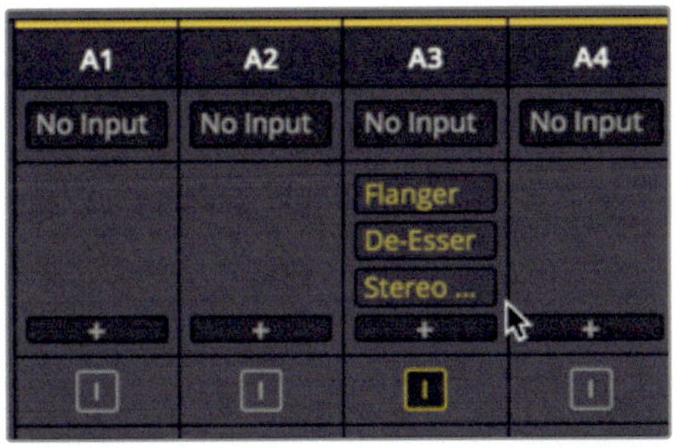

You can apply the Stereo Width plug-in to any stereo track and quickly adjust the stereo spread from mono to stereo, to super wide, or anywhere in between. Diffusion adds complexity to the output signal, and Sparkle increases the high-frequency gain.

3. Mark the first clip in the A3 track. If necessary, turn on the looped playback.
4. Start looped playback, and drag the Width control counter-clockwise to the far left (mono) position. Then, drag the Width control clockwise to the far right (wide) position.

Mono width

Super wide width

You can clearly hear the difference between the mono and super wide settings. The super wide setting sounds interesting, but it is spread so far that it may become a little distracting in the dialogue scene. Let's narrow the width so that ADA sounds as if she is coming from multiple built-in speakers without a specific point of origin.

5 While watching the viewer and listening to ADA's voice, slowly drag the Width control counter-clockwise toward the mono position. If you aren't sure what level to set, try a Width of around 0.50.

6 Stop playback. Unsolo the dialogue tracks and play that section to hear the dialogue panning along with the other tracks.

7 Close the Stereo Width window.

ADA's stereo panning works great to distinguish her disembodied voice from the two human characters in the room.

NOTE A spirited debate rages among audio professionals about which comes first: dynamics or equalization? This is a bit of a chicken-and-egg scenario because both processes effect the sound and each other. For this lesson, you'll address equalization first, but only because it makes sense to modify ADA's voice after adjusting the panning.

Things to Know about Dialogue Equalization

Audio frequencies are measured in Hertz (Hz) or thousands of Hertz, or *kiloHertz* (kHz).

Low numbers represent low frequencies, and high numbers represent high frequencies.

When you change the amount of gain in a frequency range, you either *boost* (increase) or *attenuate* (reduce) the volume level of the frequencies within that range. Filtering out a specific frequency or a narrow range of frequencies is referred to as *notching* or *cutting*.

The audible frequency range for the average human is between 20 Hz and 20 kHz (or 20,000 Hz). These frequencies are divided into three main categories:

- Bass 50–250 Hz
- Midrange 250–2300 Hz
- High 2500–20,000 Hz

The fundamental frequencies for human voices sit right in the middle of the audible frequency range and vary between men, women, and children:

- Men 80–160 Hz
- Women 165–255 Hz
- Children 250–300 Hz

Applying Equalization to Dialogue Tracks

Equalization (EQ) controls manipulate specific frequencies to shape or enhance the overall sound and are just like working with color, saturation, and hue in color correction. For example, the human voice is based on a fundamental frequency shared by millions, but additional frequencies add tonal qualities to "color" each voice and make it unique and recognizable. The primary function of equalization is to lower frequencies that detract from the voice and boost frequencies that clarify the overall sound.

If you recall, the Fairlight page includes a 4-band *parametric equalizer* for each clip in the Inspector, and a six-band parametric equalizer on each track, which is the perfect tool for enhancing and "sweetening" audio tracks during the mix.

In the following exercises, you'll look at two EQ techniques during mixing. First, you'll use the 6-band EQ on ADA's track to diminish her voice to sound as if it is coming from speakers in the ship. Then, you'll apply EQ to Emiliana's track to "sweeten" her voice.

1. Select the A3 ADA VO wFX track to brighten it in both the Mixer and timeline.

   Although selecting tracks isn't a necessary step, it does make the tracks and clips easier to spot while you work.

2. Solo the A3 ADA VO wFX track. If necessary, set a play range around the first clip in the A3 track.

3 In the Mixer, on the A3 Channel strip, double-click the EQ area to open the Equalizer window. Move the Equalizer window to the timeline area to the left of the Mixer.

4 In the Equalizer window, click the Band 1 and Band 6 buttons to turn on the high-pass and low-pass filters.

Your goal in the next step is to carve away the high and low frequencies to limit the bandwidth of ADA's voice until she sounds more as if she is coming from speakers in the ship. Instead of doing this by the numbers, you'll use your ears to determine how slight or extreme to make the effect. The quality and quantity of the speakers in the ship would also impact the sound of ADA's voice, so use your sound designer's imagination and the Mixer to create your desired effect.

Keep in mind that vocal clarity depends on the frequencies around 100 Hz–300 Hz and intelligibility between 1 kHz–3 kHz. Both clarity and intelligibility are necessary for great-sounding dialogue. You can carve away one end or the other for effect, but just keep in mind that carving both ends too far may make it difficult for the audience to understand the spoken words.

5 Start looped playback, and listen to ADA's voice while you drag the low-pass handle (6) to the left, and the high-pass handle (1) to the right, limiting ADA's vocal frequencies until her voice sounds like it is coming from speakers.

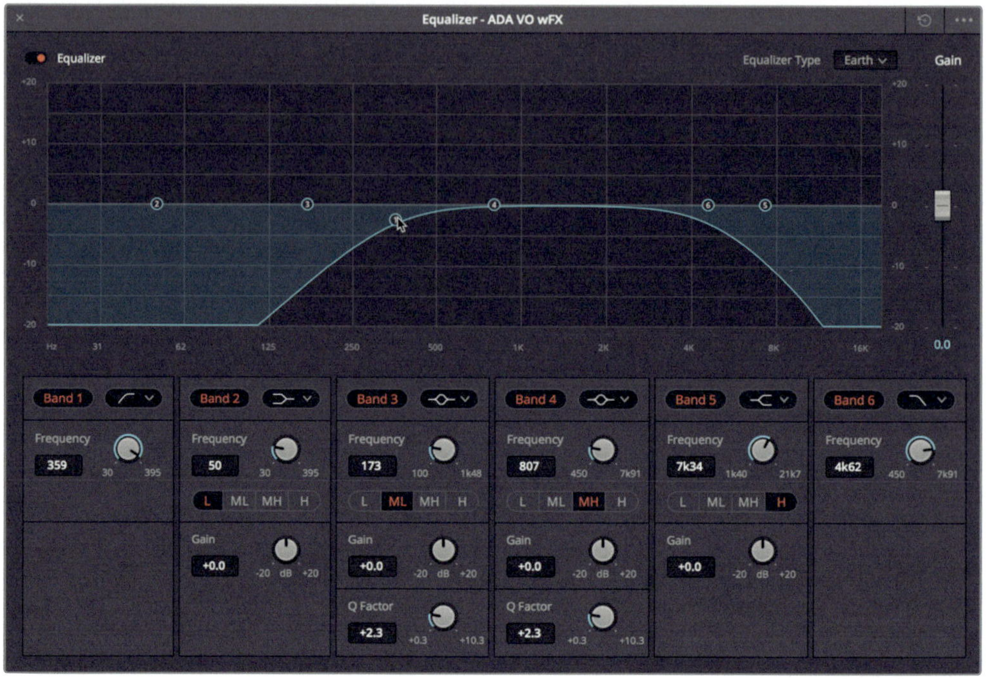

6 Unsolo the A3 track, and listen to it in the context of the other tracks. Make additional adjustments to the high-pass and low-pass handles to get the effect you want. When you are satisfied with the results, stop playback, close the Equalizer window, and clear the play range.

7 When you are finished, close the window.

Dialogue Frequency Level Enhancing and Troubleshooting

100 Hz–300 Hz	Too low sounds thin, too high loses clarity
200 Hz–500 Hz	Just right sounds warm, too high can sound boxy
250 Hz–750 Hz	Potentially muddy if too high
600 Hz–1.1 kHz	Potentially nasal/honky if too high
1 kHz–3 kHz	Adjust for intelligibility
3 kHz–6 kHz	Add presence
5 kHz–8 kHz	Adjust sibilance (esses)
9 kHz–15 kHz	Adjust sheen/sparkle
10 kHz–20 kHz	Adjust breathiness/airiness

TIP A common rule of thumb when adjusting dialogue EQ is to cut narrow and boost wide. Also, a little EQ goes a long way, so when possible, make subtle changes (no more than 2–3 dB) just as you would when adjusting hue on the color wheel.

Sweetening Dialogue with EQ

As you've just discovered, stripping the high and low ends of a voice recording is pretty easy. In contrast, actively listening, delicately cutting, and selectively boosting frequencies to improve a voice takes patience and practice. In this exercise, you'll apply three different bell curves to the EMILIANA track to find and reduce detracting frequencies, and then selectively boost frequencies to enhance the voice.

Every voice is different, so no single EQ setting will magically improve all your dialogue. However, you can tweak three key areas that will usually improve the overall sound of dialogue.

First, you'll start with a high-pass filter to roll off all the bass frequencies below 85 Hz. Next, you'll sweep the midrange frequencies around 250 Hz. This is the beefiest range of most voices and can often become boxy, muffled, or muddy. Attenuating a wide range of midrange frequencies, even by just a few decibels, can really warm up a voice and improve its clarity. Finally, you'll enhance the high frequencies to add vocal presence, intelligibility, and sparkle by applying a nice wide boost around 3–5 kHz.

TIP You should first find and cut, or attenuate, the frequencies that detract from the dialogue, and then boost the frequencies that sweeten the voice.

1 Mark the first clip in the A1 EMILIANA track.

2 In the Mixer, double-click the EQ display for A1 EMILIANA to open the Equalizer window.

3 Start looped playback.

NOTE If a track is distracting you during looped playback, such as the medlab effects in A6, feel free to mute that track while you sweeten the dialogue.

4 In the Equalizer window, turn on Band 1. Listen to Emiliana's dialogue while you drag the Band 1 handle right to around 85 Hz. You can watch the Band 1 Frequency field to see the current position of the handle.

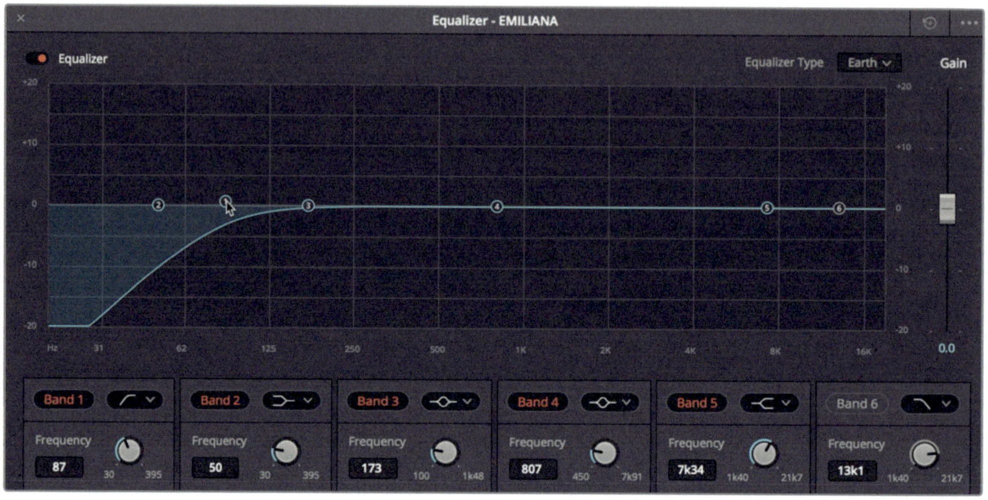

You probably won't hear any dramatic changes to the dialogue track by rolling off the bass frequencies, and some re-recording mixers are against the idea entirely. However, unless the voice you are working with is heavy on the bass end, you're unlikely to find any voice-enhancing frequencies below 85 Hz.

You can see the tracks frequencies for yourself with the help of a FairlightFX plug-in that you've already used. In fact, DaVinci Resolve 16 offers two different FairlightFX plug-ins that can show frequencies in real time: Frequency Analyzer and Noise Reduction. Since you are familiar with the Noise Reduction plug-in, and it shows a vivid depiction of the dialogue frequencies, let's use that one for this exercise.

5 In the Mixer, in the A1 EMILIANA channel strip, click the Add Plug-In button (the + symbol), and choose FairlightFX > Noise Reduction. Move the Noise Reduction window above the Equalizer window so that you can easily see both graphs at the same time.

Why are you adding the FairlightFX Noise Reduction plug-in to a track in the middle of this EQ sweetening session? Because this plug-in includes a powerful frequency analyzer that displays the track's frequencies during real-time playback. If you look closely, you'll see that both the Noise Reduction and EQ graphs show the same frequency range from 20 Hz to 20 kHz.

If you recall, the purple line in the Noise Reduction graph represents the noise profile, whereas the white line displays the live frequency analysis. If you reset the noise profile, no frequencies will be removed from the signal until you click the Learn button. In this case, you will just reset the noise profile so that you can focus on the white graphical representation of the vocal frequencies.

6 In the Noise Reduction presets menu, choose Reset Noise Profile.

Now you are free to use the frequency analyzer as a guide while sweetening your tracks.

7 Start looped playback, and watch the graph in the Noise Reduction window to see a live frequency analysis of Emiliana's voice.

As expected, the meatiest part of her vocal lives between 200–500 Hz.

8 Turn off the Noise Reduction plug-in, and close the window.

> **NOTE** You can leave the Noise Reduction plug-in applied to the track while you finish the sweetening in case you wish to see the frequency analyzer again. When you are finished with the EQ, you can delete the plug-in from the track.

Next, you can start sweeping Band 3 around 250 Hz with a wide bell curve to hear what happens if you boost or attenuate around that range. As you sweep, exaggerating the height of the bell curve by at least +10 or -10 dB will help you determine how much adjusting the 250–500 Hz range affects this voice.

9 Start looped playback. Drag the Band 3 handle up to around +10 dB and listen to Emiliana's voice as you sweep the frequencies between 250–500 Hz. Experiment to determine if tweaking within this range improves her voice or amplifies irritating vocal

qualities. Drag the Band 3 handle below the 0 dB line to around -10 dB and sweep the same frequency range. When you are finished, stop playback.

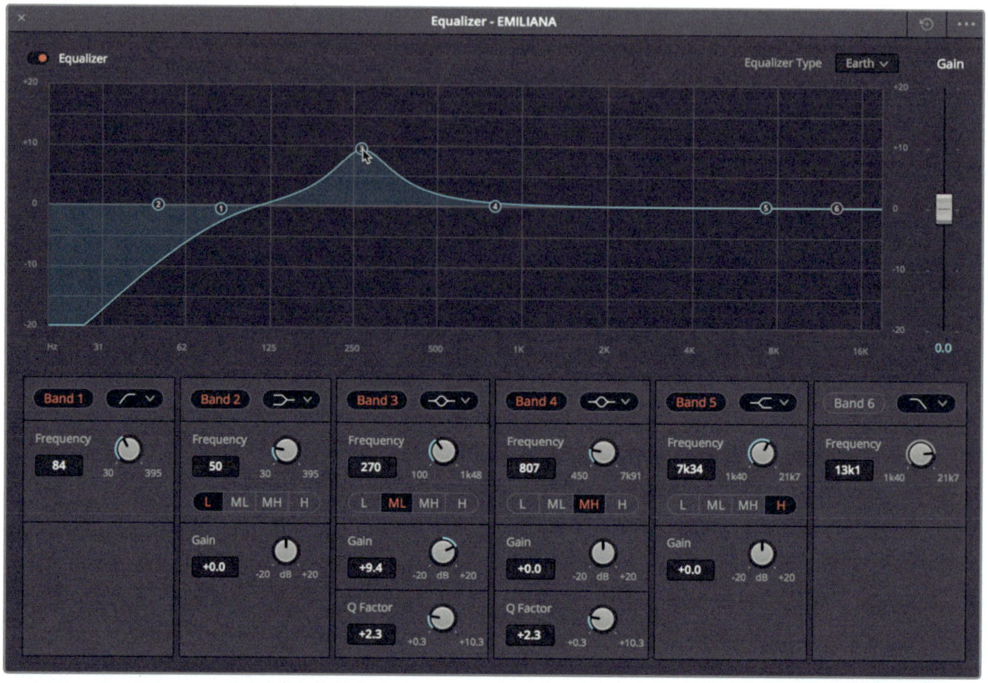

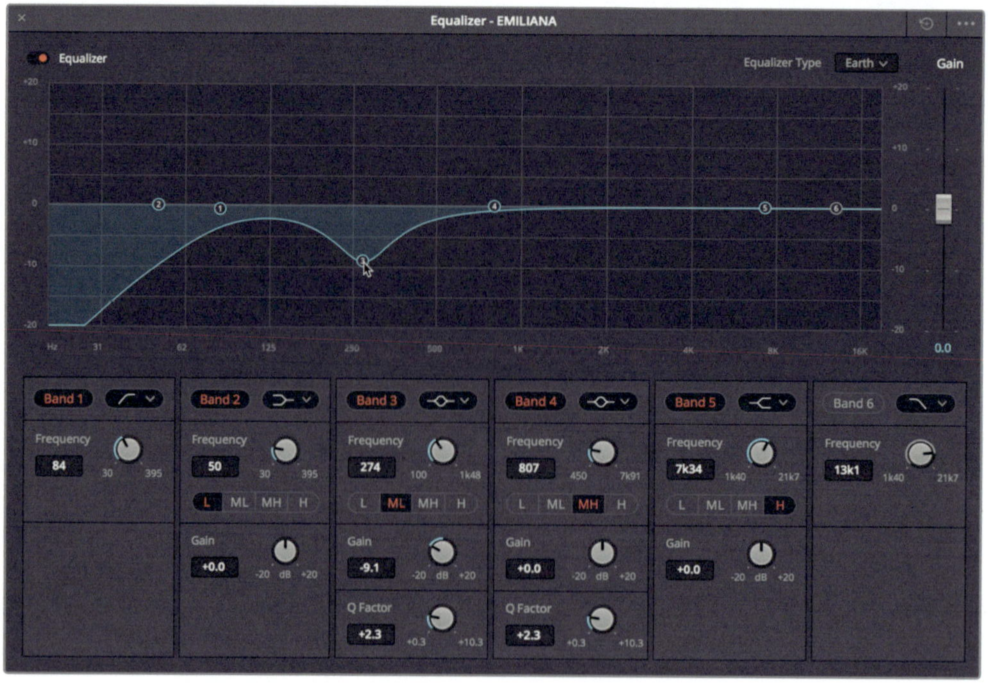

You probably discovered that boosting her vocal frequencies between 350 Hz–450 Hz brought out the most unflattering qualities. Also, cutting those frequencies too much made her voice sound thin and lifeless.

10 Start playback, and listen to Emiliana's voice as you drag the Band 3 handle down by around -2.0 dB at 350 Hz to attenuate that frequency range.

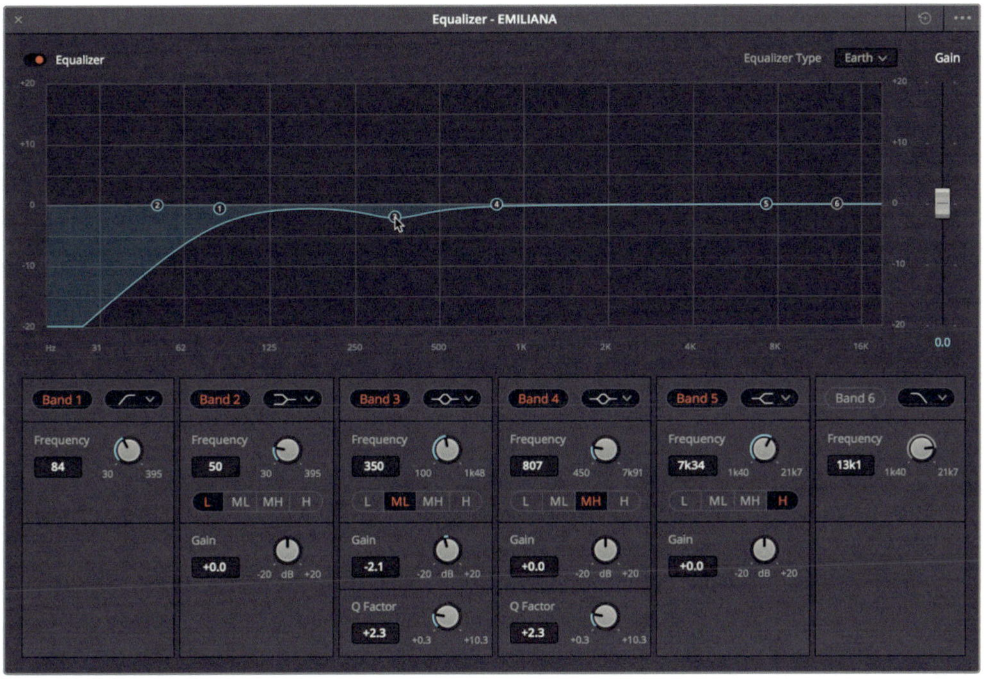

So far, so good; her voice sounds great. Now, let's boost the high end with a wide curve by 1–3 dB right around 3–5 kHz to brighten up the vocal presence and intelligibility.

TIP The secret to finding the sweet spot for vocal presence is to sweep for sibilance, and when you find the annoying esses and whistles, back off by 1–2 kHz and then reduce that amount to accentuate just a touch of brilliance in the voice.

11 Start looped playback, if necessary. Sweep the Band 4 bell curve between 3–8 kHz and listen for the sibilant esses and whistles. Sweep slightly toward the left until the sibilance is gone, and then lower the curve to around 1–3 dB. Trust your instincts and your ears. However, if you aren't sure, try boosting by +2 dB at around 3.70 kHz.

12 During playback, toggle the Equalizer Bypass button off and on several times to hear the difference in Emiliana's voice without and with EQ applied.

13 Stop playback, close the Equalizer window, and clear the play range.

14 If necessary, unmute the A6 MEDLAB 1 track.

TIP Training your ears to recognize subtle dialogue differences takes time.

Though subtle, the EQ improves Emiliana's voice significantly. This is another reason to separate each character's dialogue to its own track during editing so you can not only balance the volume levels but also apply equalization to the entire track.

Controlling Dynamic Range

The difference between a good soundtrack and a fabulous, professional-sounding soundtrack often comes down to dynamics control. Well-balanced dialogue tracks with modest compression have more consistently strong levels, and therefore stand out more in a mix, thereby leading to a better experience for the audience. Keep in mind that the entire audio post-production process is about delivering that awesome audience experience. Viewers don't care how it's done, they just want it to be incredible.

If you recall, the difference between the loudest and quietest levels in a track is its *dynamic range*. If you have ever worked with a waveform or Parade scope in the Color page, controlling a track's dynamics is very similar to adjusting the white and black levels of a clip. Just think of white as the loudest and black as the quietest.

In this set of exercises, you'll apply compression to two dialogue tracks. First, you'll apply heavy compression to ADA's track to emulate the processed yet consistent levels that commonly suggest computer-generated voices. Then, you'll apply standard compression to Emiliana's track to help her voice stand out in the mix.

1. Solo the A3 track. In the A3 track, set a play range around the first clip.
2. In the Mixer, on the A3 Channel strip, double-click the Dynamics area to open the Dynamics window. Move the Dynamics window to the timeline area to the left of the Mixer.

The Fairlight page Dynamics controls include four tools for modifying a track's dynamic range. So far, you have used only the Gate to eliminate low-level noise.

The *Compressor* is the most common control and is used to narrow the dynamic range by lowering the highest peaks and bringing them closer to the lowest peaks. It's a great tool for bringing out some of the quieter spoken words, while simultaneously lowering the level of the loudest words.

In this exercise, you'll apply heavy compression and boost overall gain in the Radio track to further suggest an acoustically constricted radio sound. Adjusting the frequencies was just the first step. Compressing the dynamic range will further modify the computer's voice to emulate the under-the-hood vocal processing in computer speakers and two-way radios.

TIP Many devices—including smart phones, intercom systems, and two-way radios—use heavy compression to limit the dynamic range so that each spoken word can be heard. Obviously, the results don't follow natural volume fluctuations in vocal patterns but are a convenient enhancement to the voice amplification process.

The first step for working with the Compressor is to assess the dynamic range for your track without compression applied. You can use the handy Input meters in the Dynamics window to see the current track levels. In natural speech, the dynamic range between two syllables in the same word can be high, so pay attention to the levels for each syllable, as well as for words and phrases.

3 Start looped playback. In the Dynamics - ADA VO wFX window, watch the Input meters to determine the lowest level (solid green) and the highest peak (yellow). When you are finished, stop playback.

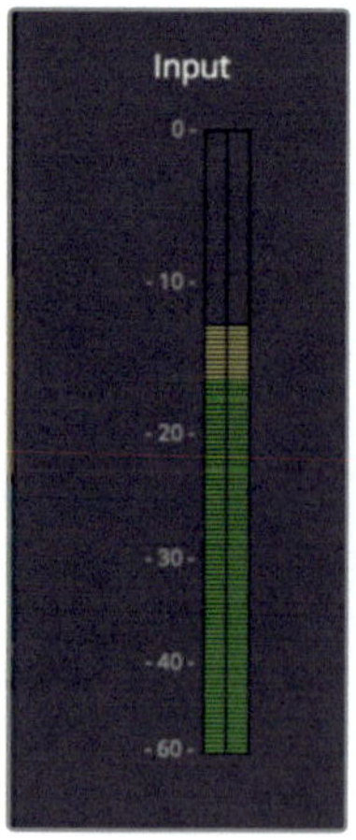

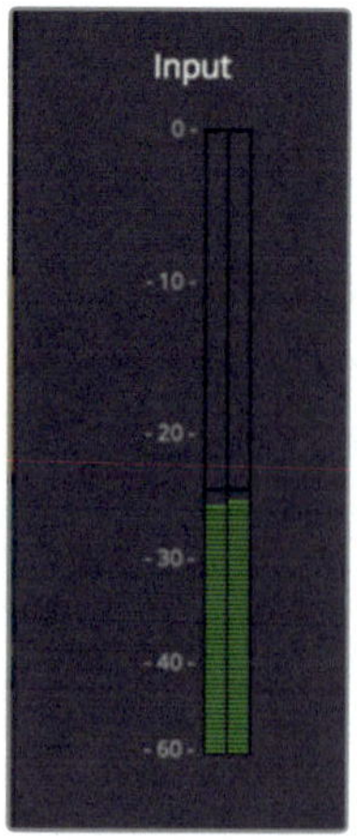

The lowest level was around -25 dB, while the highest was around -12 dB. This track has a whopping 13 dB dynamic range! If you recall from when you were balancing dialogue tracks, the lowest dialogue level for a whisper is -20 dB, and a loud voice averages around -10. That's only a 10 dB difference between a whisper and a loud, booming voice. Clearly, the dynamic range in this track is far too wide.

> **NOTE** If you're wondering why ADA's levels are so much lower than the original tracks that you balanced long ago, it's because all the plug-in FX, EQ, panning, and processing has taken a cumulative toll on ADA's levels. So, while working with the Compressor, you can also boost the gain to return that output to nominal levels.

Because ADA is supposed to have an artificial voice, the dynamic range should only be around 2–3 dB. You can compress that dynamic range with—you guessed it—the Compressor.

4 In the Dynamics - ADA VO wFX window, click the Compressor button to enable the Compressor controls.

The default Compressor settings include a Threshold of -15 dB and a Ratio of 2.0:1. The Threshold indicates where in the dynamic range you want to start compression. So, in this example, any peaks over a Threshold value of -15 dB will be compressed by a 2:1 ratio, which means that for every two decibels increase over -15 dB, only a one decibel increase will be permitted. When a track has a wide dynamic range, you can apply a higher ratio value, such as 5:1, so that for every 5 decibels of increase over -15 dB, only a 1 dB increase will be permitted. The heavier the compression, the more likely you will start to hear it in the sound quality. However, because this scene features a computer voice emitted from multiple speakers, the more compression, the merrier.

5 Start looped playback.

The lowest levels are between -25 and -30 dB, which is lower than the current Threshold.

6 Drag the Threshold knob to the left to lower it to around -25 dB. Then, drag the Ratio knob to the right to raise it to a 12:1 ratio. As you apply the compression, listen to the change. Watch the Output meter to see the compressed dynamic range.

The compression clearly worked, because the clip now has a dynamic range of around 3 dB. With this much compression applied at -25 dB, notice that the graph of the upper end of the decibel curve (light green) is nearly horizontal, as are the peaks.

Let's boost the gain to more clearly hear the heavy compression and make up for the loss of gain caused by all the other effects and processing. The Make Up gain fader lets you raise the overall sound level coming from the track's Compressor. You can still adjust the track's overall output level independently using the track fader in the Mixer.

7 Drag the Make Up gain fader up to +8.0 to raise the output level peaks to around -12dB.

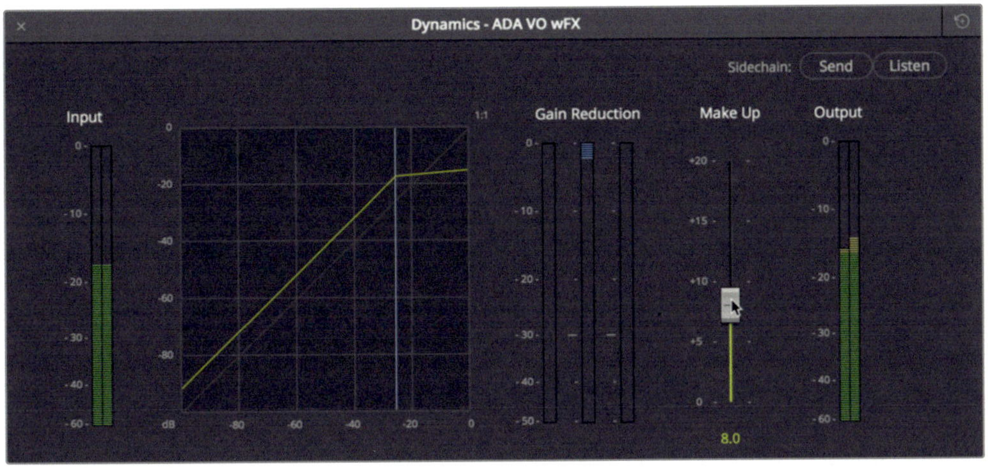

8 Unsolo the A3 track. Play the timeline to hear ADA's compressed voice along with the other tracks. Toggle the Compressor button off and on to compare the voice without and with the compression applied. When you are finished, move the Dynamics window to the lower-right corner of the screen, and clear the play range.

This was another successful exercise in voice manipulation. Although this is an extreme example of compression and EQ, it should pave your way to experiment with heavy compression when warranted in future projects.

> **NOTE** The Dynamics window always updates to show the selected track. To save time during mixing, instead of closing the Dynamics window, you can simply select the next track that you need to work on.

Compressing Dialogue Tracks

In this compression exercise, you'll use similar methods to apply compression to Philip's track. This track was edited and balanced prior to mixing, so its dynamic range should be much less than ADA's track. Keep in mind that the reason for applying compression is to limit the dynamic range of a track so that everything in your mix sounds balanced and even.

Also, when compressing standard dialogue, you want to keep the Ratio in the 2:1 to 3:1 range. Doing so will keep the compression subtle while effectively creating a more balanced dialogue track.

Let's start by finding the high and low average levels in Philip's track. To do so, you can mark a play range around a group of clips.

1 Solo the A2 PHILIP track. Press R to enable range selection mode, and drag a play range around the four consecutive clips in the middle of Philip's track (approximately 01:00:58:10 to 01:01:20:22).

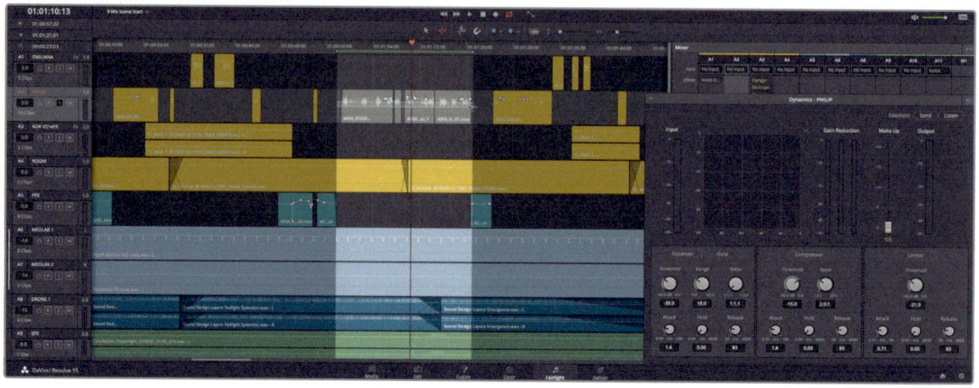

Dragging the play range in the A2 track automatically selected the track and updated the Dynamics window accordingly.

2 Start looped playback. In the Dynamics - PHILIP window, watch the Input meters to determine the lowest and highest peaks.

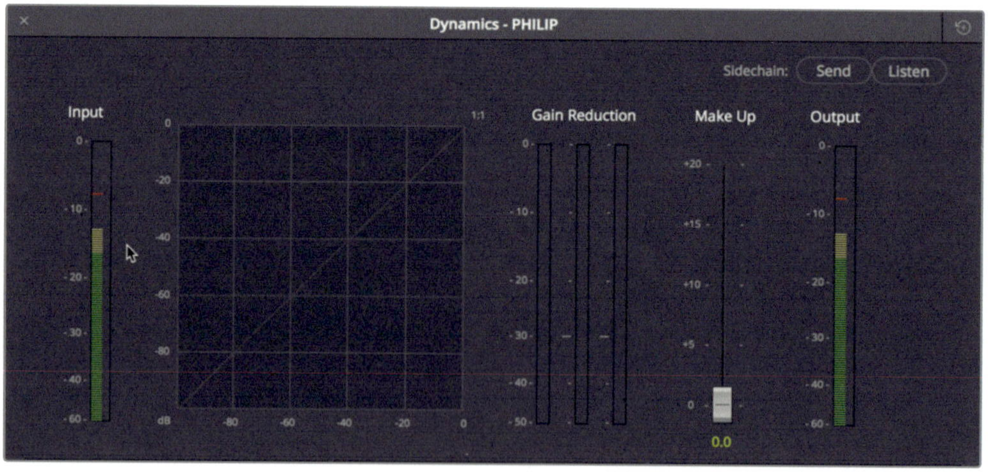

Even though Philip's clips were balanced, his peaks range from +5 to -15 dB. Although this is just a section of conversational dialogue with natural fluctuations in level, some light compression can rein in the excessive peaks and narrow the dynamic range.

3 Click the Compressor button to enable the Compressor controls. Look at the Output levels during playback.

The default compression settings are working on this track. However, the dynamic range is still a little high for conversational dialogue. Let's lower the Threshold, raise the Ratio slightly, and raise the Attack level to give the Compressor a faster attack time when it encounters levels above the threshold. Remember, the compression Threshold value is the level at which the Compressor kicks in; so, if you have strong dialogue tracks, you can play it a little lower to rein in more of the louder peaks. In most cases, you'll set your dialogue compression Threshold between 10–20 dB.

4 Unsolo the A2 PHILIP track. Start looped playback, if necessary. In the Dynamics window, set the Compressor Threshold to around -20 dB, Attack to 20 ms, and Ratio to around 3.0:1. Close the Dynamics window.

5 Clear the play range, and play the scene from Philip's first line of dialogue (around 01:00:22:00). Listen to the cumulative results of your dialogue sweetening and dynamics processing. When you are finished, stop playback.

What a difference a little mixing TLC can make on the dialogue tracks. All three characters voices really stand out in the mix now.

Saving and Applying Track Presets in the Presets Library

DaVinci Resolve 16 includes a Presets Library that you can use to save track settings and apply them to another track, timeline, or project. In this exercise, you'll save the settings from the ADA VO wFX track so you can apply them to another timeline containing an ADA track.

1 Choose Fairlight > Presets Library to open the Presets Library window.

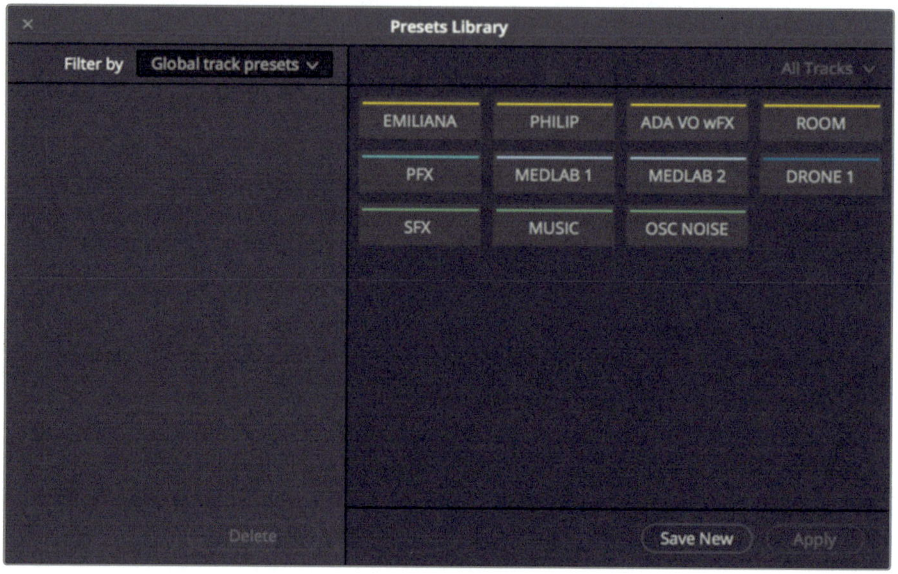

You'll see a list of tracks in the current timeline.

2 In the Presets Library tracks list, select the ADA VO wFX track.

3 In the upper left of the Presets Library, click the Filter By pop-up menu to see the presets options.

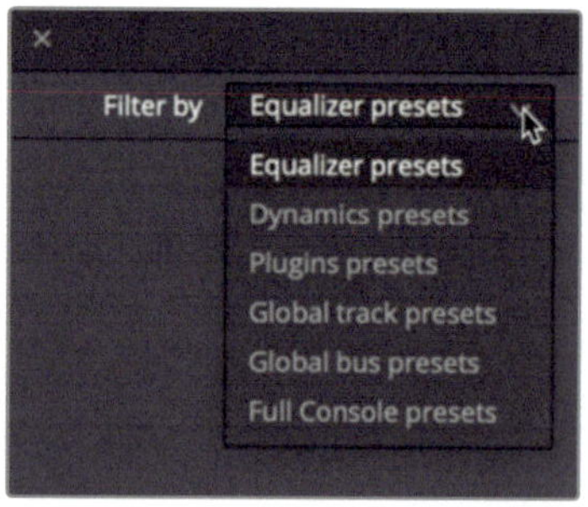

In the pop-up menu, you can save Equalizer, Dynamics, Plugins, or Global track presets such as volume, panning, and track color. For this example, you'll create a Global preset for ADA's computer voice.

4 Choose Equalizer.

5 Click the Save New button to open the Create Global track preset dialog. Name the preset **ADA computer voice**, and click OK.

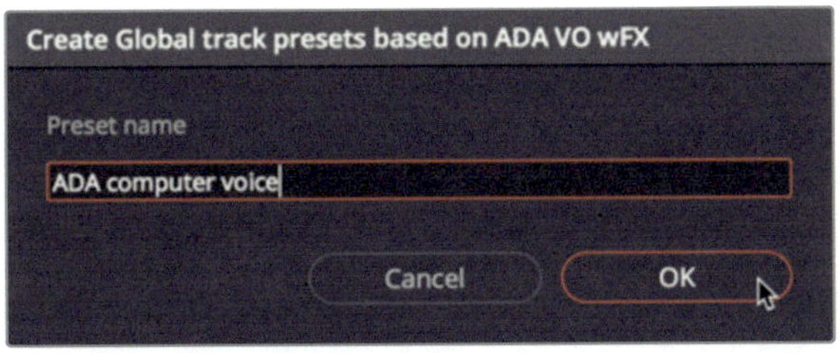

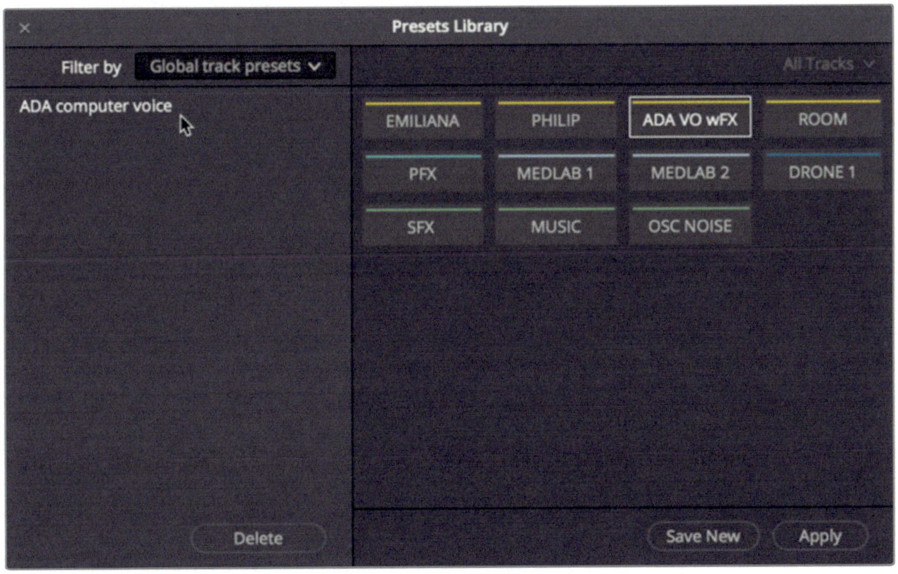

The ADA computer voice global preset appears in the list to the left of the Presets Library. Now, you can apply that preset to any other track in this or any other timeline. Let's open another timeline containing ADA's voice and test this new global preset.

6 Open the timeline, "9 Mixing Add Preset."

This is an alternate version of the previous timeline in which the A3 ADA VO track has no plug-ins, EQ, or Dynamics applied.

7 Click anywhere in the Presets Library window to update it to the new timeline.

8 In the Presets Library, select the ADA VO track. Select the ADA computer voice Global track preset, and click Apply

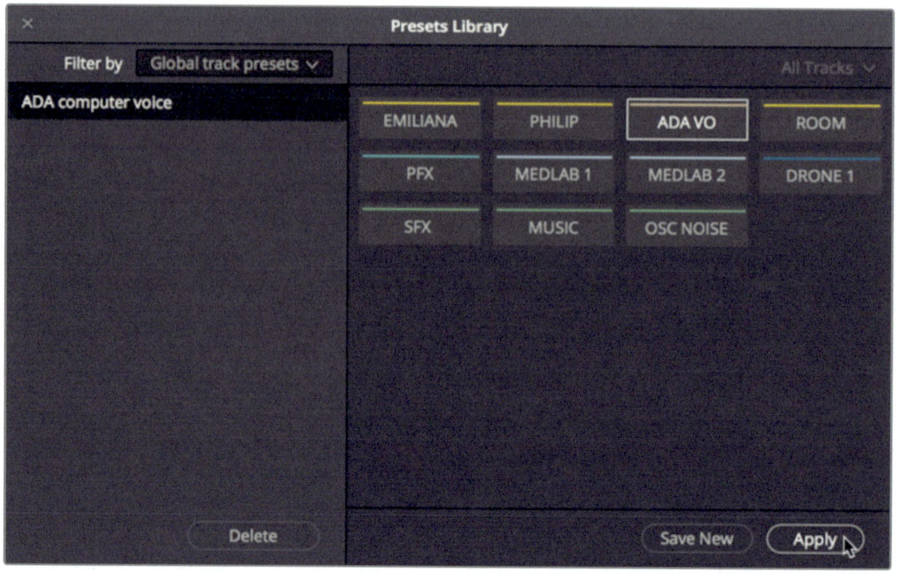

All the settings, including track color, are applied to the ADA VO track.

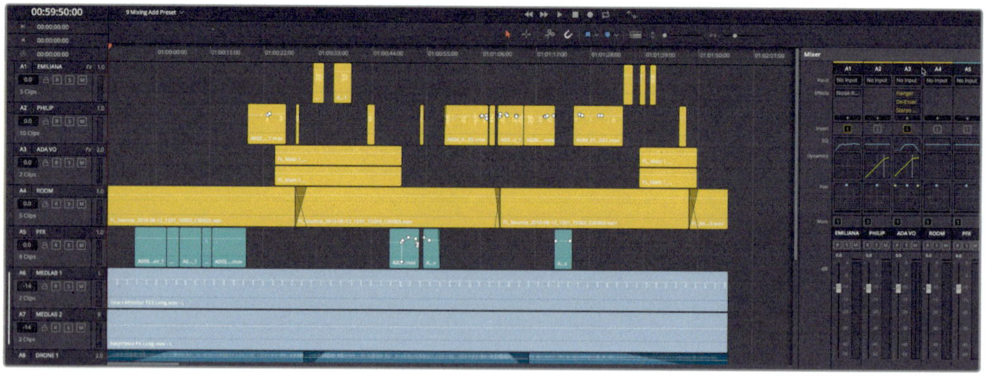

9 Close the Presets Library.

As you can see, the Presets Library is a powerful way to save and apply your track throughout the post-production process.

Using Sidechain Compression to Automatically Dip Music Levels

Previously, you have applied keyframes in the timeline to dynamically control volume levels of individual clips within a track. You also used the compressor to control the dynamic range of an entire dialogue track. In addition to standard compression, the Dynamics window also offers sidechain compression, which lets you use the signal from one track to control the amount of compression that is applied to another track. Sidechain compression is often referred to as *auto-ducking* because it automatically "ducks" or attenuates the level of one track based on the content of another. One of the most common uses of sidechain compression in dialogue scenes is to use the dialogue tracks to control the compression on the music tracks.

In this exercise, you'll work with a different scene from Hyperlight that has a heavy music bed under the dialogue. First, you'll apply compression to the dialogue tracks. Then, you'll set up the sidechain compression on the Music track so that the music level is reduced whenever there is dialogue spoken in the dialogue tracks. Let's get started.

1. Open the "9 Auto Ducking Start" timeline.
2. If necessary, show the Monitoring panel (Meters) and Mixer.
3. Play the timeline and listen to the competing music and dialogue levels.
4. In the Mixer, open the Dynamics window for the A1 EMILIANA track and turn on the Compressor.

   For this exercise, you'll use the default compressor settings on the A1 and A2 tracks.
5. On the right side of the Compressor settings, click the Send button to send the levels of this track to the compressor on another track.

6 Repeat steps 4 and 5 for the A2 ADA 2 track.

7 Click the header for the Music track to select that track and update the Dynamics window to the newly selected track.

8 Turn on the Compressor. Click the Listen button to automatically feed all "sent" levels to the sidechain of the Compressor on this track.

9 Set a range around the first dialogue exchange between Emiliana and ADA that includes Emiliana's first line, "ADA where am I?" and ends after ADA's line, "Philip's cryopod has lost power; his oxygen level is critical."

10 Start looped playback.

11 During playback, lower the Threshold and raise the Ratio. Continue adjusting these controls until the music dips sufficiently beneath the dialogue levels and you can clearly hear every spoken word. Don't worry if the music seems to cut in and out. You'll finesse those levels next.

If you aren't sure of what values to set, try a Threshold around -30 and Ratio of 4:1 or 5:1.

12 Expand the range from the beginning of the first yellow clip in the A1 track to the beginning of the fourth yellow clip. The expanded play range will make it easier to smooth the transition between the music with and without dialogue.

The music is now compressed to be less loud during dialogue. Now it's time to adjust the Attack value to determine how quickly the music level is reduced, raise the Hold value so the music level doesn't fluctuate wildly, and increase the Release value so the volume adjustment won't end too abruptly. These controls are easier to use during playback when you can hear them in action.

13 Start looped playback and change the Attack value first to determine how quickly (by milliseconds) the music level is lowered when dialogue levels are detected.

14 Set the Hold value to smooth how long the level stays consistent.

15 Set the Release value to determine how gradually the music level will return to the uncompressed level.

You should hear that whenever there is dialogue in the A1 or A2 tracks, the music volume automatically lowers, or "ducks."

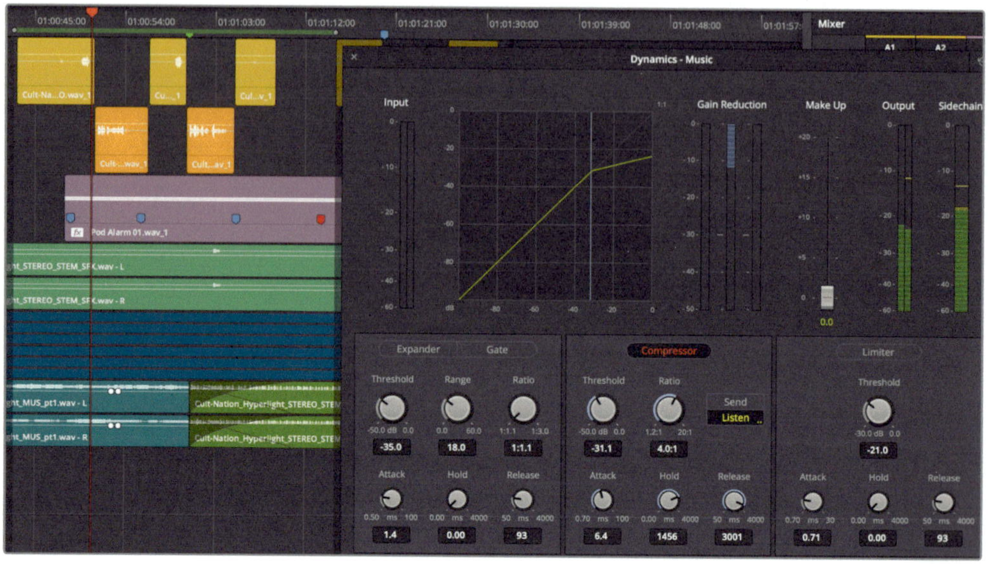

16 When you are finished, clear the range and play the entire scene from the beginning to hear your ducking in action. Feel free to re-open the Dynamics controls for the Music track to further modify the levels.

These controls take practice, so the more you use them the easier it will be to set up and apply sidechain compression to your own projects. However, as you can hear for yourself, this is an awesome way to quickly "duck" your music levels beneath your dialogue tracks. If you would like to open a finished example of this scene with Sidechain compression applied, open the timeline "9 Auto Ducking Finished." As a result, you should hear that whenever speech plays, the music volume is automatically lowered. Adjust Threshold and Ratio to set the amount of this gain reduction, and adjust Attack, Hold, and Release to set how quickly gain reduction responds to changes in the Sidechain level (as defined by the Send levels).

More Sweetening and Compression Please!

Open the timeline "9 Mix Add More Sweetening."

Now that you've applied EQ to Emiliana's track and compression to Philip's track, it's time to finish the dialogue sweetening. Take a few minutes to add a little EQ to Philip's track and some compression to Emiliana's. Feel free to proceed on your own without any guidance. Or, you can apply the methods that you used earlier for the other character. Your final goal is to deliver great-sounding dialogue in your soundtrack—even if one of the characters sounds like a computer.

Lesson Review

1 True or False? You should balance the volume levels of individual clips in the timeline before adjusting track levels in the Mixer.

2 Where do you set and balance overall track levels?

a) Mixer

b) Track Level window

c) Monitoring panel

d) Control room

e) Loudness controls

3 True or False? It is possible to use the controls in the Audio Pan window to make a stereo track sound like a mono track.

4 4. In the Fairlight page, how can you link two mono tracks as a single stereo track that uses a single fader control?

a) Channel mapping controls

b) Right-click the track headers, and choose Link Tracks.

c) Link the tracks in the Link Group window.

d) Right-click the tracks in the Mixer, and choose Link Group

5 What is a common reason for applying a compressor to a dialogue track?

a) To make noise sound like a computer warble

b) To change the signal-to-noise ratio (SNR)

c) To reduce the dynamic range between the loudest and quietest peaks in the track

d) To create the aural illusion of a physical barrier

6 How can you use sidechain compression (auto-ducking)?

a) To remove bird sounds in your dialogue tracks

b) To automatically stitch room tone for dialogue tracks

c) To use the dialogue track to control compression on a music track

d) To compare volume levels between tracks

Answers

1. True
2. a
3. True
4. c. Link the tracks in the Link Group window, available in the Fairlight menu.
5. c
6. c

Lesson 10

Simplifying the Mix with Buses and Automation

The goal of mixing is to balance each track level so they all sound good as a whole. You do so by making subtle changes to those levels or by combining similar tracks into submixes to make them easier to adjust with a single fader. Automating the changes to your faders and other parameters while you work allows you to free up your hands and focus in each pass of the mix. The final master mix must sound great and meet delivery standards for loudness. Fortunately, the Fairlight page includes everything you need to make sure that the levels are on target.

In this lesson, you'll explore some of Fairlight's advanced mixing tools like busing and automation to simplify the mixing process and deliver a cinema-quality soundtrack!

Time

This lesson takes approximately 45 minutes to complete.

Goals

Are Other Sound Mixers Working on the Same Project?

In large-scale Hollywood films and shows, several other mixing teams often work in parallel prior to the final mixing phase. Those Mixers may include a dialogue mixer, mixers for ensemble pieces, and sound effects mixers. Like the re-recording mixer, their jobs are to balance and sweeten (enhance) the sound and levels of their respective tracks and narrow them down to a few well-balanced tracks, submixes, or bounced submixes for the re-recording mixer to blend with the other sound elements.

Creating an Auxiliary Reverb Bus

This scene is coming along well. The tracks sound great, the levels are pretty good, and it's time to add a little reverb to the whole scene so that the dialogue and sound effects in the room sound as if they are in a real room. Previously, you applied plug-ins to individual clips and tracks. What would you do if you had multiple tracks that needed the same effect at the same time? Instead of adding the same plug-in to every track, you can simply send the tracks to the plug-in via an auxiliary (aux) bus. Doing so saves processing and time.

In the next series of exercises, you'll create an auxiliary bus with reverb for the dialogue and medlab effects.

DaVinci Resolve includes a powerful busing architecture available in the Fairlight menu in which you can create multiple main buses, submix buses, and auxiliary buses. In this exercise, you'll create an auxiliary bus for the FairlightFX Reverb plug-in and send the dialogue and medlab tracks to the aux bus to add reverb.

1 If you didn't complete all the previous mixing exercises, open the 9 Mixing Buses Start timeline to catch up.

2 In the Fairlight page, choose Fairlight > Bus Format to open the Bus Format window.

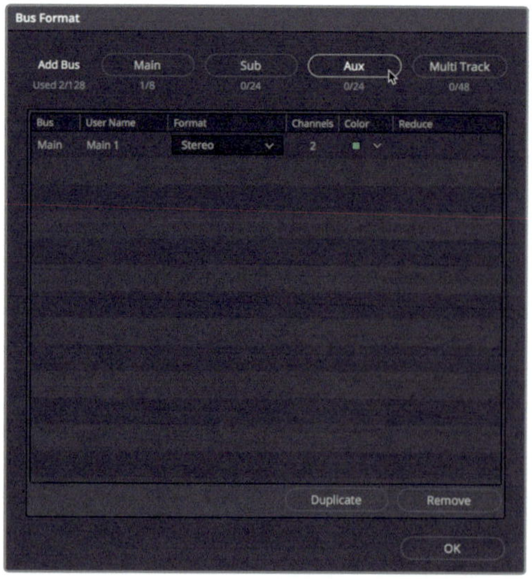

In the Bus Format window, you can assign tracks to one of four types of buses. The project currently contains one bus: the default main bus named M1. The Format and Channels columns indicate that this main bus is in stereo with two channels (left and right).

3 Click the Aux button to add a new auxiliary bus to the list.

The new aux bus appears below Main 1 in the Bus Format pane.

Let's rename the aux bus and change its color. Because most of the tracks are mono, you can leave the Format set to mono. If you need to change the Format from mono to stereo, you can do so at any time in the Bus Format window.

4 In the User Name column, double-click the Aux 1 name, and enter **VERB** (an abbreviation for reverb).

5 In the Color column, change the VERB color pop-up menu to beige.

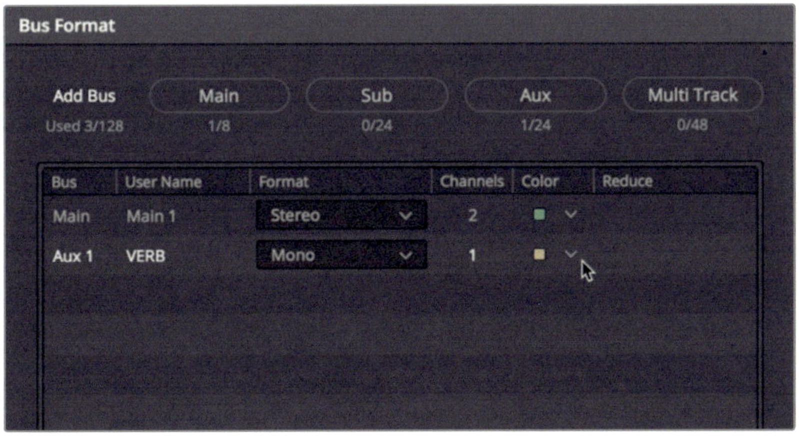

6 Click OK to close the Bus Format window. Expand the Mixer to see the new Aux 1 VERB bus to the right of the M1 output. Notice that the new aux bus—labeled Aux 1—is also to the right of the M1 meter in the monitoring panel.

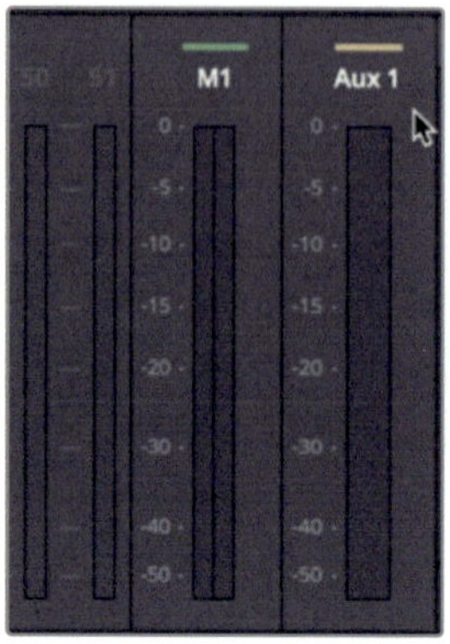

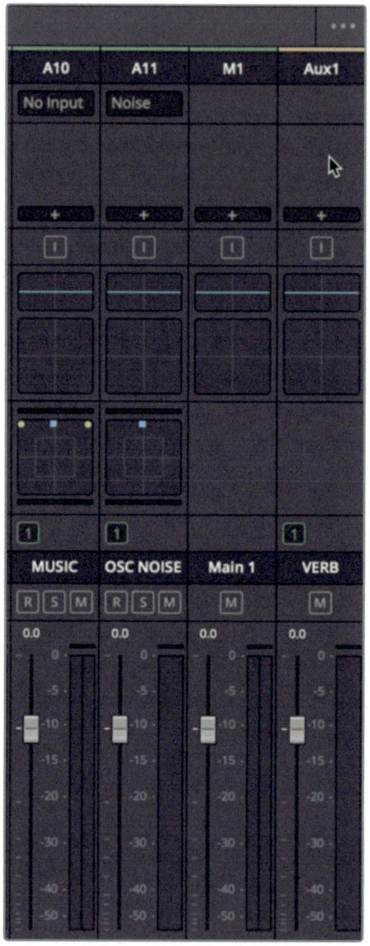

You are all set up with the aux bus. The next step is to add the reverb plug-in.

7 In the Mixer, on the Aux 1 bus, click the Add Effect button, and choose FairlightFX > Reverb.

The Reverb window opens, and the Reverb plug-in appears in the Aux 1 effects slots.

8 In the Reverb window, in the Preset pop-up menu, choose Cathedral. Close the Reverb - VERB window.

Obviously, the Cathedral preset is too much reverb for this scene, but it will work great for demonstrating the auxiliary reverb. You can always change the reverb settings later. The last step is to send the signal from the tracks to the aux bus.

Sending Track Signals to the Aux Bus

At the moment, the aux bus is configured with the Reverb plug-in ready to go, but you still have to send the signal from the tracks to the bus. This is one of those times when audio mixing terms and functions are one and the same. That's right, you are about to employ the use of *sends* to do exactly what their name implies. If you look closely at the Mixer, you'll find the aux sends between the Dynamics and Pan controls.

1 In the Mixer, in the A1 channel strip, double-click the Aux send bar to open the Auxiliaries - EMILIANA window.

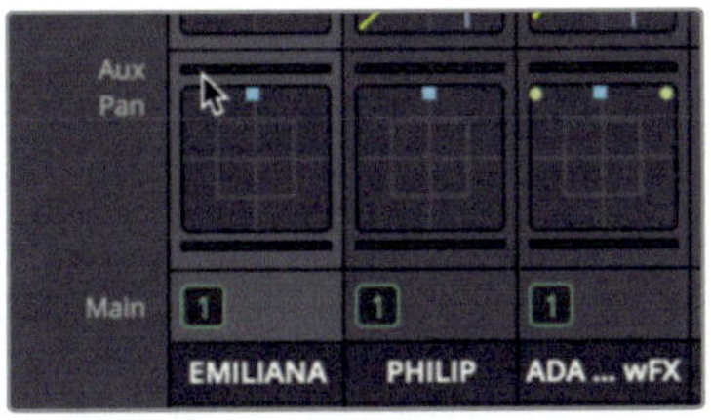

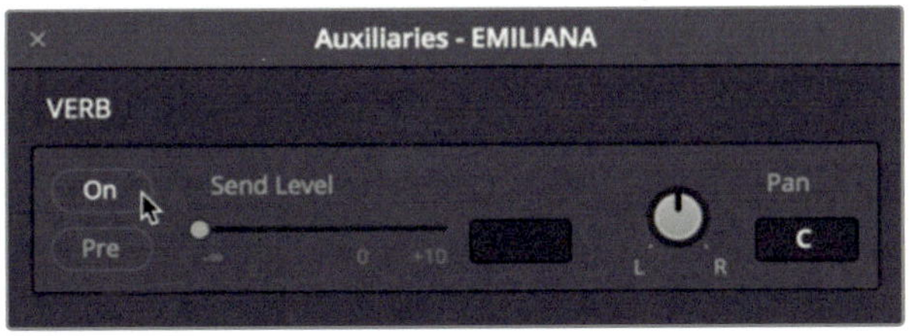

2 In the Auxiliaries - EMILIANA window, click the On button to turn on the send.

The Send Level adjusts the amount of signal sent from the selected track to the aux bus.

3 Drag the Send Level handle to 0.0 to send the full signal to the VERB bus.

The send level is also visible as a green bar in the Aux sends area of the channel strip.

4 Play the first clip in the A1 EMILIANA track to hear it with the aux VERB effect applied.

Emiliana's voice has definitely picked up the reverb. In fact, it's so wet that her voice is practically drowning in reverb. No worries, you'll fix the reverb levels in a minute. First, let's send the other tracks.

5 In the Mixer, double-click the Send bar for the A2 track. Turn on the VERB send, and set the Send Level to 0.0.

6 Repeat step 5 on tracks A3 and A6. When you're finished, close the Auxiliaries window.

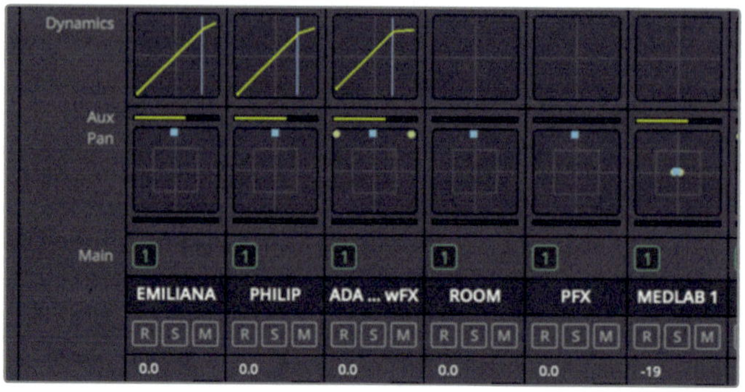

The last step is to use the fader on the Aux 1 bus in the Mixer to raise or lower the amount of reverb applied to the sends. This is a fantastic way to control effects on multiple tracks over time. Keep in mind, any changes you make to the aux bus levels will apply to all the incoming sends. If you want to vary the amount of the effect on a specific track, lower the Send Level on that track's send.

7 Move the playhead to the beginning of the first clip in the A2 PHILIP track.

8 Start playback, and listen to a few seconds of the dialogue with the heavy aux reverb applied. Then, slowly lower the Aux 1 bus fader to decrease the amount of reverb until it is barely noticeable in the mix. Stop playback.

As you just demonstrated, auxiliary buses are a quick and easy technique for applying effects to multiple tracks.

Customizing Auxiliary Bus Effects Parameters

Now that you've seen and heard the Cathedral reverb preset with the tracks, let's change that Aux VERB bus preset to apply a more realistic sound. To customize an effect applied to a track or bus, you just hover your pointer over the effect in the Mixer and click the Customize button to open the corresponding plug-in window.

1 In the Mixer, hover the pointer over the Reverb effect in the Aux 1 bus to see the available options buttons, including Bypass and Customize.

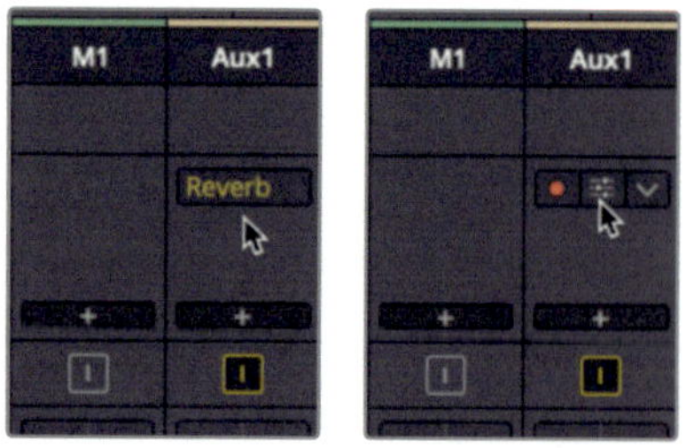

2 Click the Customize button to open the Reverb - VERB window.

3 In the Reverb window, click the Reset button to reset the plug-in to the default settings. Feel free to adjust the reverb settings to suit yourself.

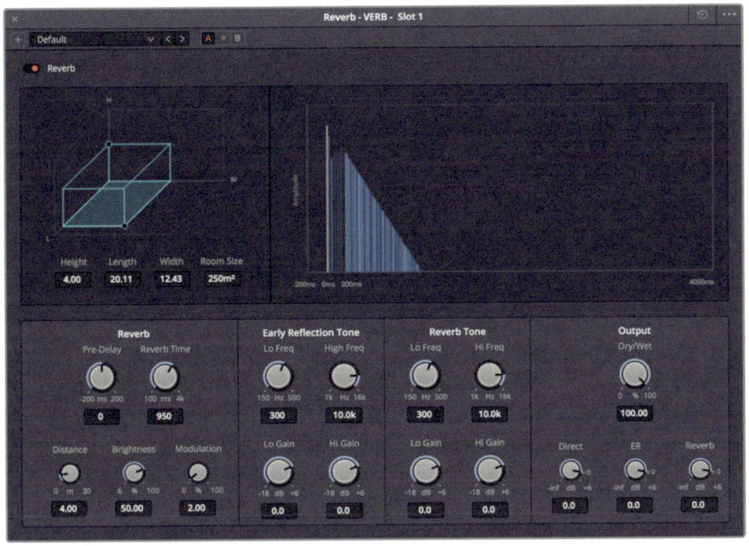

The default reverb settings more closely match the room in the scene.

4 Close the Reverb window. Play the scene from the beginning to hear how it sounds with the reverb applied to the dialogue and medlab tracks.

5 Continue playback and adjust the Aux VERB fader, if necessary, to find a realistic reverb level for the scene.

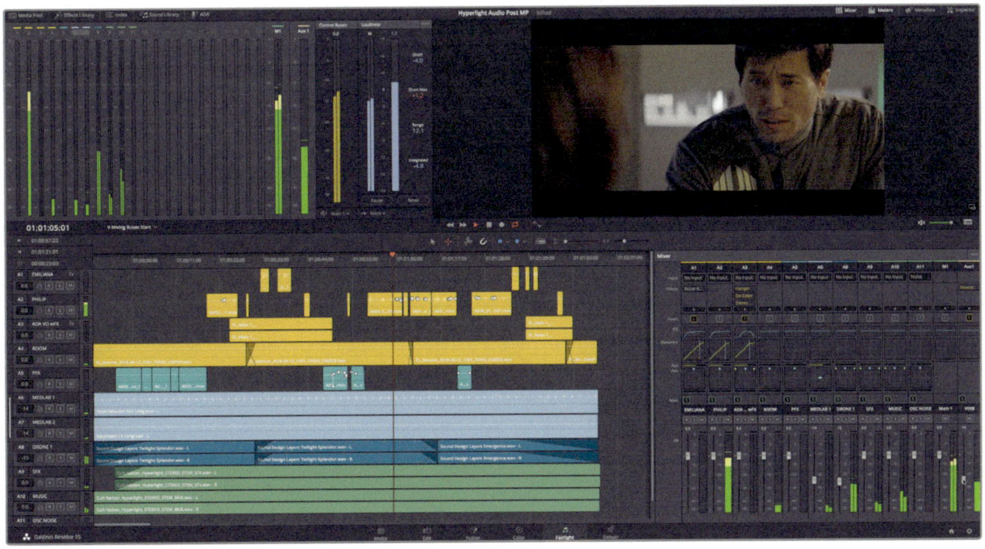

It's truly amazing how the cumulative mixing techniques in this lesson, plus a hint of reverb, elevate this soundtrack.

Simplifying Mixing Using Buses

You already know how to balance individual clip levels, but if you have numerous similar tracks that are already balanced—such as dialogue, sound effects, or music—you can combine them and send each of their output signals through a bus to create a *submix*. As the name suggests, a submix lets you independently mix a subset of the tracks.

Submix buses are signal paths, or *vehicles*, that send the signal from multiple tracks to a new channel strip in which the sum of the signals can be controlled as a single track.

In this exercise, you'll create submix buses for the dialogue, music, and background FX tracks.

1 In the Fairlight page, choose Fairlight > Bus Format to open the Bus Format window.

2 Click the Sub button three times to add three new submix buses to the list.

The three sub buses (submixes) appear in the Bus Format pane below Main 1.

Let's name the sub buses and change their formats and colors based on the tracks they will contain.

In the User Name column, double-click the Sub 1 name, and enter **DIAL**. Then, change the name of Sub 2 to **MUSIC** and Sub 3 to **BG FX**.

3 In the Format column, change the Format of the Music and BG FX subs to Stereo. Leave the DIAL sub Format as Mono.

Finally, let's change the colors of each submix.

4 In the Color column, change the Sub 1 DIAL color to yellow, the Sub 2 bus color to navy, and the Sub 3 bus color to pink.

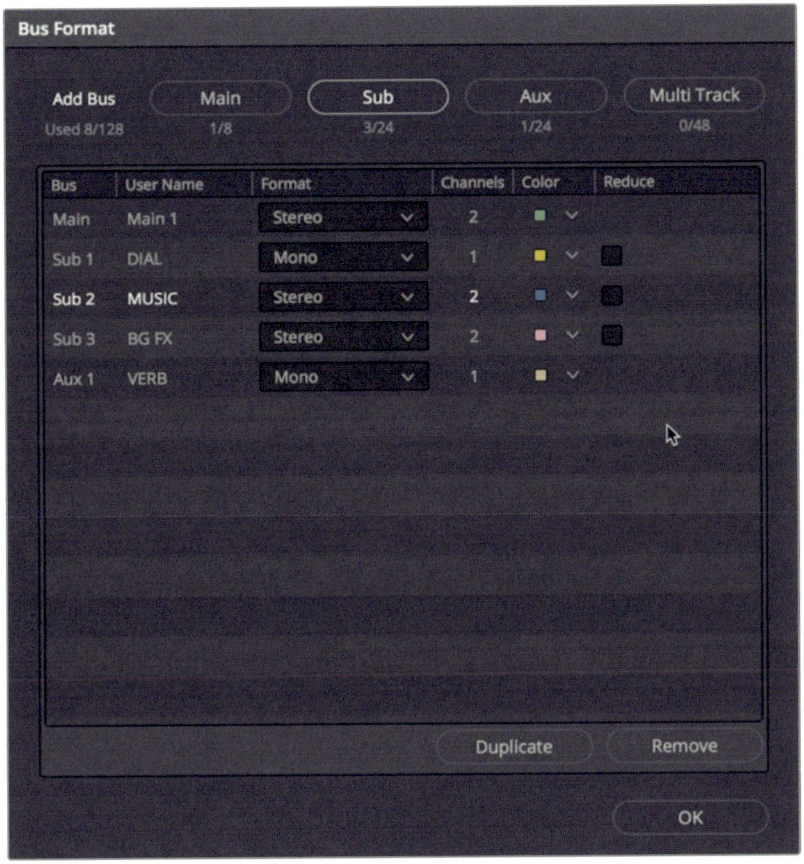

5 Click OK to close the Bus Format window.

The new sub bus names and colors also appear in the Mixer and the Meters.

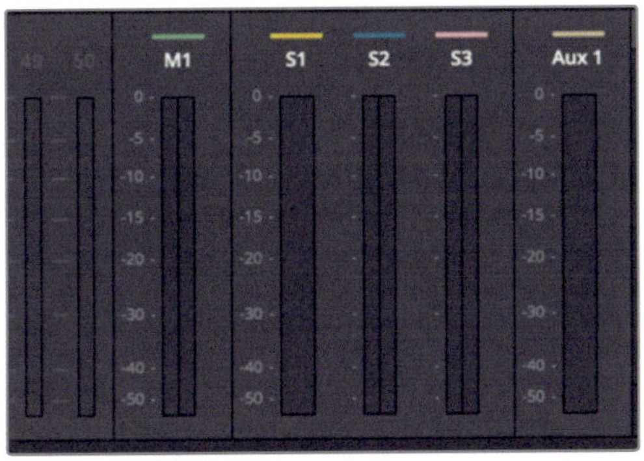

You are all set up with submixes. The next step is to assign the tracks to go into those submixes.

Assigning Tracks to Submix Buses

You must now assign the tracks that you want to send to each bus.

1 Choose Fairlight > Bus Assign to open the Bus Assign window.

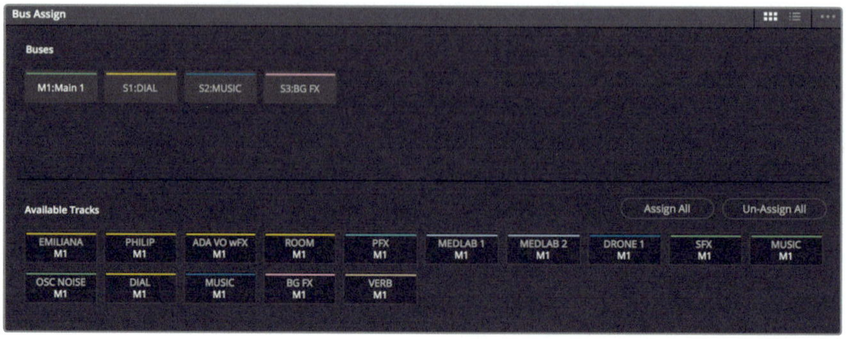

The Bus Assign window has two sections. At the top of the window, you'll find is a list of all available Buses. The rest of the window shows the available tracks. In the Available Tracks area, initials under each track name indicate the current assignment for each track. Below each track name, an M1 indicates that the track is assigned to the Main 1 output.

To better understand the bus and main architecture, let's clear all the current track assignments and reassign them. To do so, you'll select each bus individually, and click the Un-Assign All button.

2 In the Buses area in the Bus Assign window, click the M1:Main 1 button.

The name of the selected bus highlights to indicate it is selected.

3 In the Available Tracks area, click the Un-Assign All button.

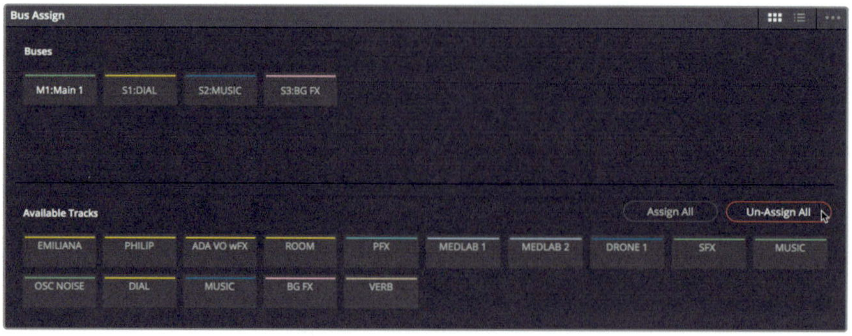

All the tracks are removed from the M1:Main 1 output bus. Remember that only tracks assigned to a Main output are audible. So, you'll eventually need to reassign these tracks.

Now that the tracks and buses are unassigned, you can manually select each bus and assign tracks to it.

4 In the Buses area, select the S1:DIAL bus. In the Available Tracks area, select the EMILIANA, PHILIP, ADA VO wFX, and ROOM tracks to assign them to the S1:DIAL bus.

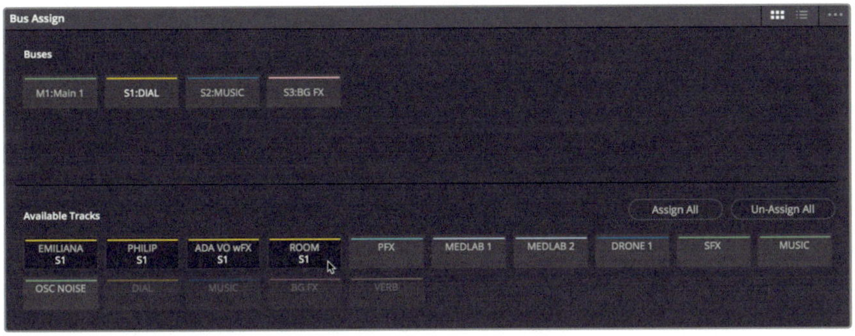

5 Select the S2:MUSIC bus, and then select the DRONE 1 and MUSIC tracks to assign them to that bus.

6 Select the B3:BG FX bus, and assign the MEDLAB 1 and MEDLAB 2 tracks.

Next, you'll assign the remaining unassigned tracks to the M1:Main 1 output. They represent the PFX and SFX tracks in this case. The PFX are the production effects included with the sound effects mix. The SFX track is the actual stereo mix from the finished soundtrack. The levels of these two tracks can be adjusted using the standard faders in the Mixer. You'll also need to assign each of the three submix buses to the M1:Main 1 bus so that the audio output of those buses will be audible through the Main output.

7 Select the M1:Main 1 bus, and assign each of the unassigned tracks to it, as well as the DIAL, MUSIC, BG FX, and VERB buses.

8 In the Bus Assign window, click the Save button.

All the tracks are assigned to the main output or a sub bus, so you can see and hear them in action in the Mixer.

Working with Buses in the Mixer

The middle section of the Mixer always shows your main and submix buses, and the current signal flow of every track so you know where each is assigned. In this exercise, you'll play through the timeline to verify that all your tracks and submixes are playing properly. Then, you'll try out your new submix faders.

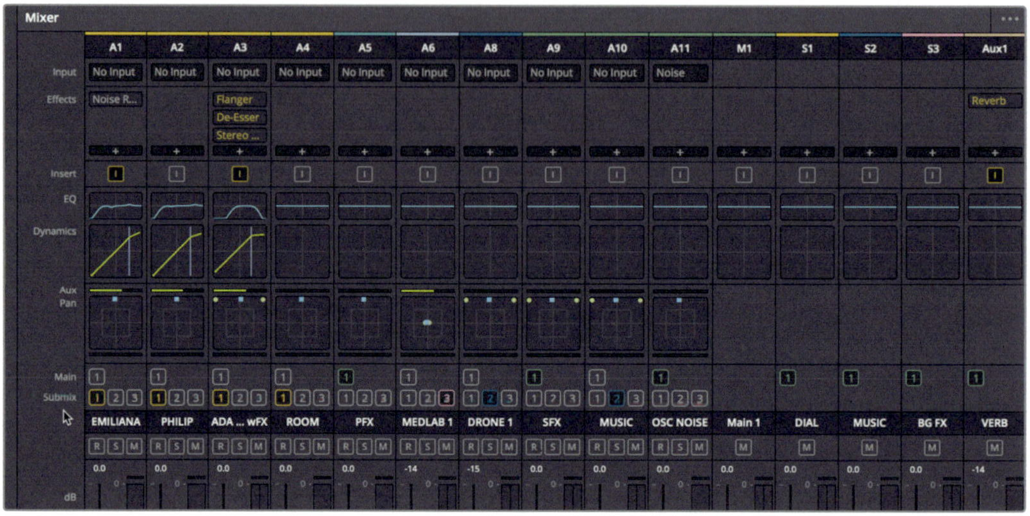

TIP You can expand or compact the Mixer at any time by dragging its left edge. The buses are always visible to the right of the Mixer.

1. Drag the left edge of the Mixer to the left to expand it, if necessary.
2. Press Shift-Z to fit all the timeline clips horizontally in the visible timeline.
3. Play the project from the beginning, and listen to the current mix.

   The mix should still sound pretty good. You haven't actually changed any of the levels; you've just altered the way that you will control them. You can continue to balance clip level curves and individual faders as usual.

   The BG FX tracks containing the medlab effects seem a little distracting when the dialogue starts. Also, the Music and Drone sound a bit low at the beginning. You can emphasize the dialogue by lowering the BG FX submix fader after Philip says his first line and the medlab is no longer in the shot.
4. Start playback, watch the scene, and move the S3 fader down until the track sounds less distracting (about -28 dB). Also experiment with the music levels in the beginning and return them to their original levels during the dialogue.

Now you know why submixes are a Mixer's best friend. It is much easier to move one fader that lowers multiple tracks than to manually lower multiple separate faders on-the-fly.

TIP Re-recording mixers working with large mixing consoles are typically moving hardware faders during playback and can maneuver their hands around the controls like a keyboardist playing a church organ. When you are working in a software-only environment, you must do the best you can using a mouse or track pad. You can always record automation to your tracks or submix buses to help simplify the final mix. Also, numerous hardware mixing consoles, including the new customizable Fairlight consoles, are compatible with DaVinci Resolve and its Fairlight page. You can find a list of compatible mixing hardware and the Fairlight consoles at www.blackmagicdesign.com.

Automating Track Changes

Adjusting faders and other controls during a mix is part of your job. You could try making all the changes on-the-fly during your final mix, or alternatively record the parameter changes as automation. In this exercise, you'll use Resolve's automation controls on the Fairlight page to record volume changes to the medlab tracks over time. You could record automation using either the submix fader or the A6 MEDLAB 1 track fader to achieve the same results. However, recording automation to the track fader gives you the added bonus of a visible automation curve in the timeline.

You'll find the Automation button at the top of the Fairlight page, next to the transport controls.

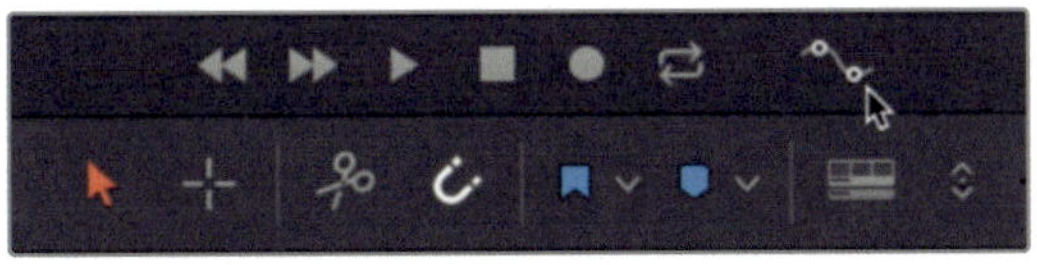

NOTE If you didn't complete the steps to create submixes, open the 9 Submix and Automation timeline to catch up.

1 Click the Automation button to open the automation toolbar.

The automation toolbar includes buttons that control every available option for setting up and recording automation for your mix. The buttons are organized in groups from left to right.

You can record automation data in either Write or Trim mode. Write mode records absolute changes to controls, whereas Trim mode records relative changes that increase or decrease levels that are already recorded.

The **Touch mode** settings determine what happens when you begin automation. When the touch mode is turned off, no automation is recorded. Touch includes two automation recording modes:

- **Latch mode** begins recording as soon as you touch a control that is set for automation, and continues to record automation after you release the control.
- **Snap mode** begins recording automation when you touch a control, and stops recording when you release the control. Additionally, once you release the control and automation stops recording, the physical control snaps back to the starting value or position where it was at the start of the automation recording. For example, if you are recording fader automation with a starting position of -12dB in Snap mode, you can record new fader automation as long as you are holding the fader. Once you release the fader, the automation recording stops and the fader snaps back to the -12dB position.

The **On Stop** controls determine what happens when you release the control you are recording:

- **Event** maintains the most recently recorded automation value and overwrites other previous levels for that parameter until the playhead reaches the start of the next recorded data on the track.
- **Hold** maintains the most recently recorded value and deletes all previously recorded data for the rest of the track.
- **Return** creates a ramp from the automation value you just recorded to the previous values on that track.

Let's record volume automation for the A6 MEDLAB 1 track so the medlab effects are more prominent at the beginning of the scene, and then gradually diminish when the medlab is no longer visible onscreen. The medlab sound effects are a reminder for the audience that the other Philip is nearby, so keep the medlab audible but not overpowering. Rather than giving you specific numbers and levels to set here, trust your ears and your own judgement to decide when to dip the medlab effects and when they should be more prominent in the mix.

> **TIP** Before recording automation, it's always a good idea to set the initial track levels and practice changing levels on-the-fly.

2 In the Mixer, double-click the BG FX submix fader to reset the levels.

3 In the timeline, select the A6 MEDLAB 1 track to select both linked MEDLAB tracks.

4 Zoom vertically to increase the height of the selected tracks.

With the larger track height, you can more clearly see the Automation button and Automation curve pop-up menu in the track header.

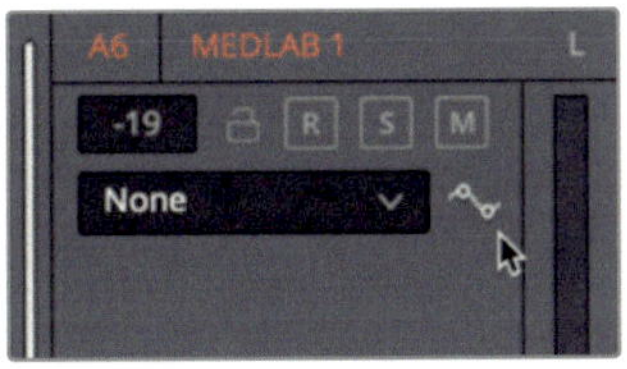

Let's set the initial level for the medlab effects.

5 Start playback from the beginning of the scene, and set the A6 track level for the medlab effects. If you aren't sure of the level, try around -10.

Notice the current fader position so that you can return the fader to that position after your practice run.

6 Make sure that the touch automation controls are off. (The Touch Mode Off button will be red.)

7 Start playback from the beginning of the scene, and in the channel strip, adjust the A6 fader throughout the scene. Experiment with different levels until you find values that work best for the highest and lowest levels.

8 Move the playhead to the beginning of the track, and set the A6 fader to the starting level (around -10).

9 In the Touch controls, click the Latch button. In the enables section, click the Fader button. Also, ensure that automation is in Write mode, and On Stop is set to Hold.

10 Click the A6 fader once.

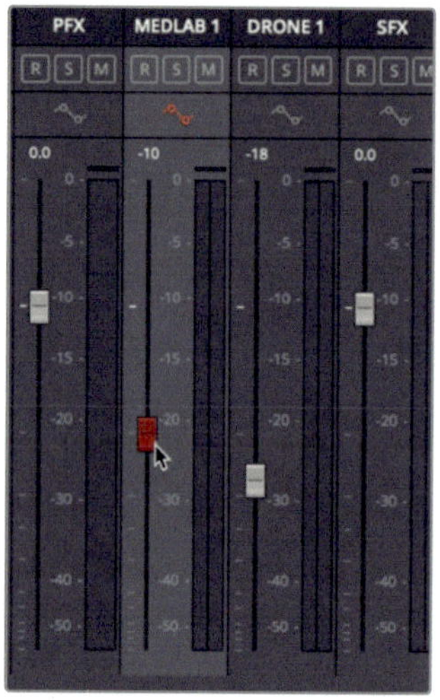

The fader turns red to indicate that it is enabled for automation recording. You'll see that the Automation button on the track header also turns red.

11 In the A6 track, in the automation curve pop-up menu, choose Fader Level so you can see the automation curve as you record automation.

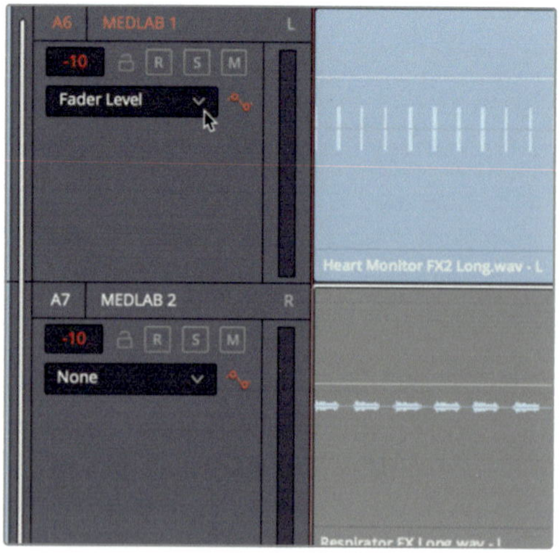

12 Start playback from the beginning, and lower the fader just as you practiced in step 5. When you are finished, stop playback, and release the fader.

Notice that when you stop recording, the fader turns green to indicate that the control is automated. Also, you'll see the green fader level automation curve on the A6 track.

13 Set the Touch mode to off.

> **TIP** When you are working with automation, it's a good idea to turn off the Touch mode when you aren't actively recording. Otherwise, you could accidentally record or overwrite automation during playback. Also, you can click the automation enable switch to the left of the automation toolbar to disable or enable all automation that is applied to a timeline.

14 Play the scene from the beginning, and watch the A6 fader as it moves automatically to follow the recorded automation.

Once you have recorded automation, you can write new automation over the original, or edit the automation curve using the Select Range tool, which is available only when the automation controls are active.

15 In the timeline toolbar, click the Select Range button.

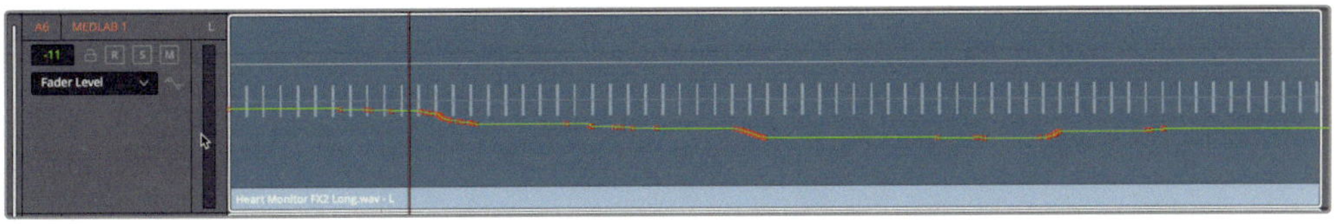

The Select Range button turns red, as do the level changes in the Fader Level curve in the timeline.

16 Using the Select Range tool, in the timeline, drag a range around one of the red sections of the curve.

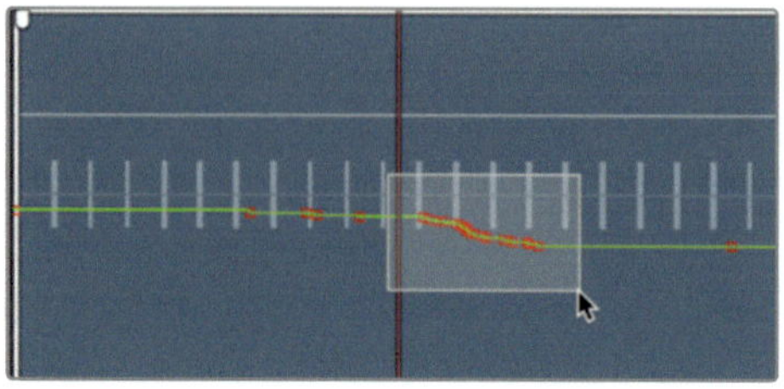

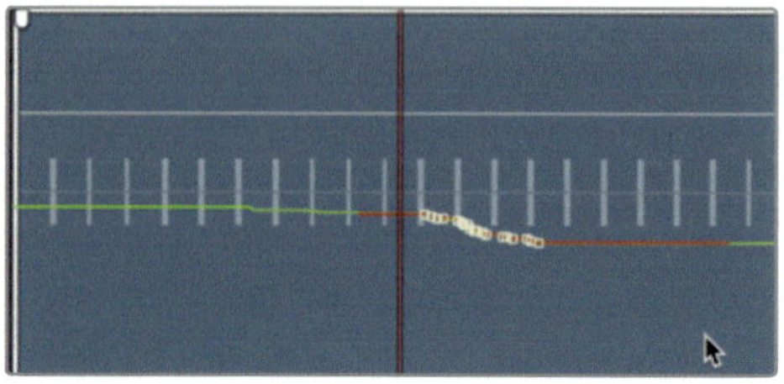

The changes in level become control points that you can edit individually or as a group. You can even delete all the control points in a selected range. For now, let's go back to the standard Selection tool to deselect the control points.

17 Press A, or click the Selection Mode tool, to return to standard selection mode.

Now that you have seen how easy it is to automate fader levels on a track, let's automate some other parameters.

Automating Pan Controls

Automation goes far beyond adjusting volume levels. In the next set of exercises, you'll take automation to the next level by automating the panning of the MEDLAB tracks at the beginning of the scene to follow the respirator mask onscreen.

Just like before, you'll be able to see the associated automation curve for Pan in the timeline track.

> **NOTE** If you did not complete all the previous steps, feel free to open the timeline "10 More Automation Start."

1 In the A6 MEDLAB 1 track header, change the Automation curve pop-up from Fader Level to L/R Pan.

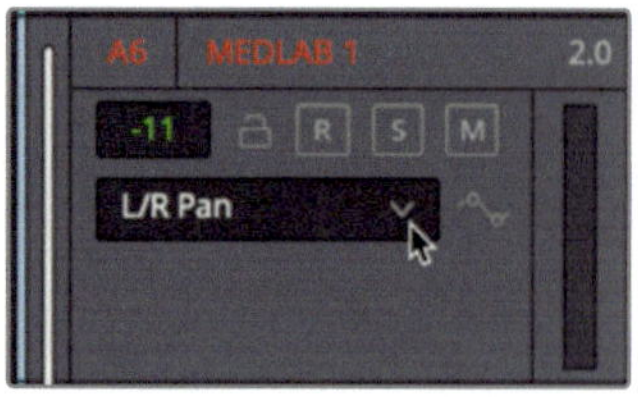

2 In the Mixer, double-click the A6 Pan control to open the Pan window.

Automation can be recorded with or without the playhead moving. In this case, let's record the panning during playback. First, let's watch the beginning of the scene.

3 Play the scene from the beginning and pay close attention to Medlab-Philip's position when he is onscreen, and the direction of the other character's eye lines when they are looking his direction when he is not onscreen.

Your goal is to automate the panning to simulate the movement of Medlab-Phillip's position onscreen. Once the scene evolves to Philip and Emiliana talking back and forth without seeing Medlab-Philip in the background, you can resume the Center pan position for the medlab sounds for the rest of the scene.

To expand your automation horizons a bit in this exercise, this time you will set the On Stop mode to Event. First, you'll record the center position at the point where you want the center panning position to remain for the rest of the scene. Then, you'll go back to the beginning of the scene and record the pan movement for the early part of the scene.

4 In the Automation Toolbar Enables section, disable the Fader button and enable the Pan button. Then, enable Event in the On Stop controls.

5 Move the playhead to the red timeline marker. This is where you want the panning to be in the center position and remain there for the rest of the scene. To record this automaton point, or Event, you can use the Pencil tool.

6 In the automation toolset, select the Pencil tool.

7 Make sure the panner is set to the center position. Then, double-click the L/R Pan automation curve on the A6 track at the playhead position with the Pencil tool to record the current pan data to the new control point.

8 Move the playhead to the first frame of the timeline and click the Automation write button on the A6 track header.

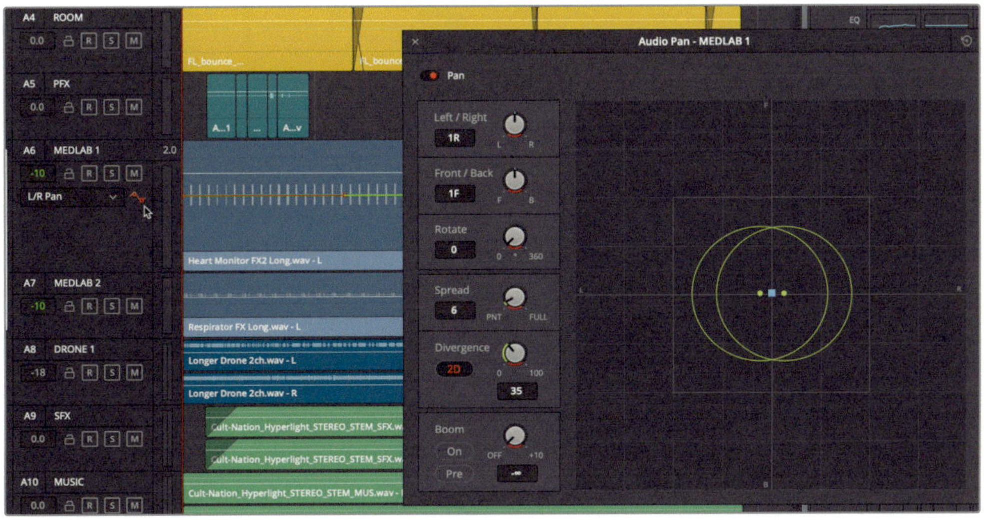

All the controls on the panner turn red to indicate that any changes will be recorded as automation data. You'll also notice that the first part of the L/R pan graph is orange.

When automation is set to Event on stop, the graph changes to orange from the playhead to the next event (control point) in both directions. This will isolate the new pan automation data you are about to record to the orange section of the graph.

9 Move the panner to the front right position about one gridline from the right corner (approximately 75R, 100F).

The orange part of the graph turns red to show the new recorded panning position, and where it will end up when it gets to the next event.

10 Start playback and move the panner toward the left to follow the action onscreen. Once Medlab-Philip is offscreen, move the panner to -75L and keep it there until Philip crosses the room to talk to Emiliana. Stop playback with the Spacebar any time before the red marker.

When you stop recording, the graph turns green again. You'll also see red control points while the Pencil tool remains active.

11 Press A to return to the default Selection tool. Then play the beginning of the scene to see and hear your Pan automation. When finished, close the Pan window.

12 Change Touch Automation mode to Off.

> **NOTE** If don't like your automation and would like to try again, simply choose Edit > Undo, and then repeat the previous steps to record again. To clear all automation on a curve, move the playhead to the first frame. Change the On Stop mode to Hold, and set the parameters to settings you want for the entire track. Press the Spacebar twice to start and stop automation recording. The parameter you set was recorded for the duration of the track.

Next, you'll automate the Reverb on the VERB Aux bus.

Automating Plug-Ins

For a little added drama and realism to accentuate the distance between Medlab-Philip and his beeping monitor during Philip and Emiliana's conversation, let's add an Echo plug-in to the medlab tracks and automate the wet/dry control on the fly during playback. When the camera is near Medlab-Philip, the Echo plug-in will be dry, without echo. When the camera is far from the medlab and the medical equipment, you will increase the wetness of the effect to exploit the distance between them.

Let's start by adding an Echo plug-in to the A6 track in the Mixer.

1 In the Mixer, A6 channel strip, click the Add Effect (+) button and choose Delay > FairlightFX > Echo to add the Echo plug-in to the track.

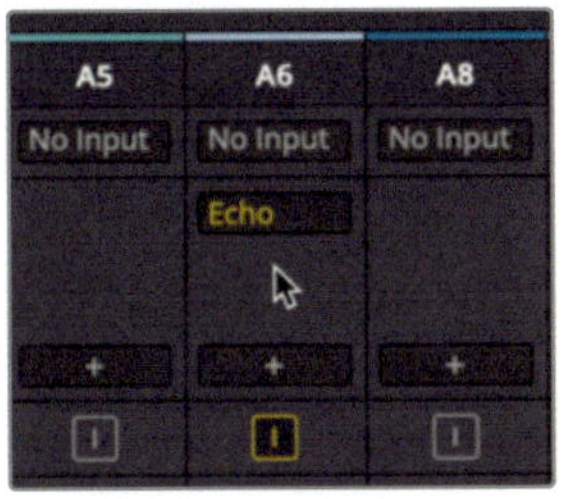

2 In the Echo - MEDLAB 1 window, change the Default preset to Large Hall.

Before automating the Dry/Wet control on the plug-in window, let's hear how it sounds dry and at around 30% wet.

3 Move the playhead to the beginning of the timeline.

4 In the Echo window, Output controls, drag the Dry/Wet control all the way to the left (0) for a dry signal, without any Echo effect applied.

5 Start playback and change the Dry/Wet control from Dry (0%) when the camera is in the medlab near Medlab-Philip to around 20-30% when the camera is near Philip and Emiliana across the room away from the beeping medlab machines. Anytime Medlab-Philip is in the foreground onscreen, the Echo effect should be dry. Once the scene settles at the table with Emiliana and Philip (red marker), you are finished.

Practice a few times. When you are ready to record automation, move on to the next step.

6 In the A6 MEDLAB 1 track header, in the Automation Curve pop-up menu, choose Plugins > Echo > Dry/Wet to show that curve in the track.

7 In the Automation Toolbar, enable Plugins and set Touch mode to Latch and On Stop to Hold.

8 Move the playhead to the beginning of the timeline. Turn on the Automation Write button in the upper-right corner of the Plugin window. Then, turn on the Automation Write button on the A6 MEDLAB 1 track.

All the controls in the Echo window turn red to show that they are armed for recording automation data.

9 Drag the Ouput Dry/Wet knob to the Dry position (0%).

10 Start playback and record the changes to the Dry/Wet parameter. When you finish (before the red marker), release the knob and press the Spacebar.

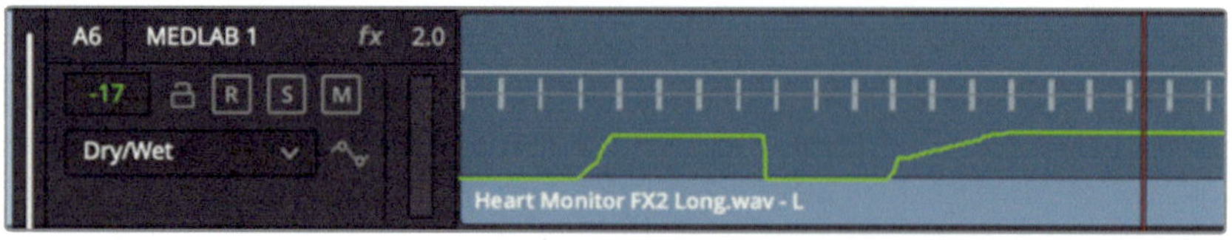

11 Close the Echo window and set the Touch mode to Off. Then deselect the Pan and Plugin enable buttons.

12 Play the beginning of the scene to hear your Volume, Pan, and Plugin automation in context. If you didn't complete all the automation steps and you'd like to hear the finished track, open the timeline 10 More Automation Finished.

Excellent work! This is just scratching the surface of what you can do with automation to bring a soundtrack to life. Hopefully, you will now feel confident enough to try it on your own.

NOTE You can also turn on Preview mode in the Automation toolbar, which lets you audition changes to automated parameters without writing new data or overwriting existing automation data.

Beware of Ear Fatigue!

Ear fatigue is a phenomenon that listeners experience after prolonged exposure to constant audio stimulus. Common symptoms include ear discomfort, an inability to focus or distinguish between similar sounds, and tiredness.

As you know, your ears are a vital part of your complex auditory system. They are always working and sending information to your brain to process. Our brains, in turn, attempt to tune out unnecessary background noises while consciously processing priority sounds such as human speech. When you focus attention on a specific sound, you are consciously overriding your brain's subconscious sound dampening and sifting processes.

Now imagine sitting in a theater, watching loud action movies and focusing on every element of the soundtrack for 8 to 10 hours a day, five days a week. Such is the life of a Hollywood re-recording mixer. Audio pros learn to recognize ear fatigue and know when they need to step away for a bit. In the early days of sound mixing, the necessity to change film reels created built-in breaks throughout the day. However, with the advent of digital audio and high-end mixing stages, prolonged and sustained exposure to loud audio is a daily occurrence. Whether you are a seasoned pro or a headphone-wearing student in a classroom, when you notice your mind drifting away from your work, feel uncomfortable pressure in your ears, or you are suddenly tired—take a break! If you are an audio post supervisor, film producer, or medical professional who doesn't believe in the ear fatigue phenomenon, feel free to spend a day in a theater watching eight hours of action films non-stop or sit near the mixing console during an eight-hour rock concert.

Creating Additional Output Buses

Every audio signal in the timeline flows from the track to the main output. Whether you are listening to speakers or headphones at your workstation, you are hearing the main output every time you play your timeline.

The current timeline has one main output, three submix buses, and sixteen individual tracks. However, depending on the delivery requirements for your final project, you may need to create additional main output buses in multiple formats. In fact, most professional projects require several versions of your mix, such as stereo and surround.

Furthermore, when dubbing a program into another language, you need to provide an M&E (music and effects only) version of the mix that is devoid of all intelligible onscreen dialogue so that it can be replaced by actors speaking their lines in another language.

You can create and assign main buses just as you created submix buses in the Bus Format and Bus Assign windows.

In this exercise, you'll create a 5.1 surround main output and an M&E main output based on the current mix. Because the M&E mix will be identical to the Main 1 mix without dialogue, you can simply duplicate the Main 1 with all tracks and subs identically patched and make any changes to the copy. As for the 5.1 main output, you'll create that one from scratch.

1 In the Fairlight page, choose Fairlight > Bus Format.

2 In the Bus Format window, select Main 1, and click the Duplicate button.

The duplicate main appears below Main 1 with the Bus assignment of Main 2 and the User Name Main 1.

3 Click the Main button once to create a new main output bus for the 5.1 surround main.

The two new mains (Main 2 and Main 3) appear in the Bus Format pane below Main 1.

4 Name the Main 2 bus **M&E** and the Main 3 bus **5.1**.

Although the default setting for the Main 1 output format is Stereo, you can change it to a different format at any time, or create multiple mains for each of your delivery formats. Digital Cinema Projection (DCP) delivery, for example, requires a minimum of three channels (LCR) or a maximum of 16 audio channels for multichannel surround sound.

Because you probably are not doing this lesson with a surround sound setup, you can create the 5.1 surround main with the understanding that you may not be able to hear all the channels in the surround format.

5 Change the Format of the Main 3 bus to 5.1. Then, change the Color to purple to distinguish it from the other stereo mains.

You now have three Main outputs—labeled M1, M2, and M3—to the right of the Mixer and on the monitoring panel.

6 In the Bus Format window, click OK.

7 Choose Fairlight > Bus Assign.

Now, you'll assign tracks to these main outputs. Let's start with the M3:5.1 bus. Later, you can set up the M2 bus on your own.

8 In the Buses area of the Bus Assign window, select the M3:5.1 bus. In the Available Tracks area, select the DIAL, MUSIC, and BG FX buses. Finally, select each of the tracks that are mapped to the M1 output: PFX, SFX, and OSC NOISE.

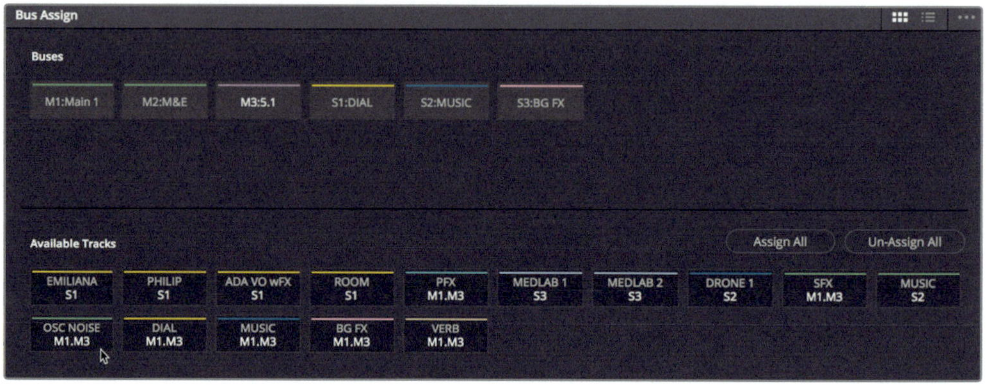

9 Click Save and close the Bus Assign window.

Your new 5.1 and M&E output buses are set up. If you are wondering why you included the OSC NOISE track with your 5.1 main, it's so you can use it test your surround channel mapping in the next lesson.

Monitoring Multiple Outputs and Buses

In the monitoring panel, the Control Room settings let you determine which outputs or submixes are audible in the Control Room during your mix. The default is always the Main 1 output. To monitor a different main output or bus, you choose from the pop-up menu below the Control Room meters.

In this exercise, you will change the Control Room monitoring to the 5.1 output. Play the timeline once, and then return to the Main output. Keep in mind that your Control Room monitoring can be changed anytime.

1 Below the Control Room meters, in the Control Room monitoring pop-up menu, choose 5.1.

2 Play the beginning of the timeline to monitor the 5.1 main output.

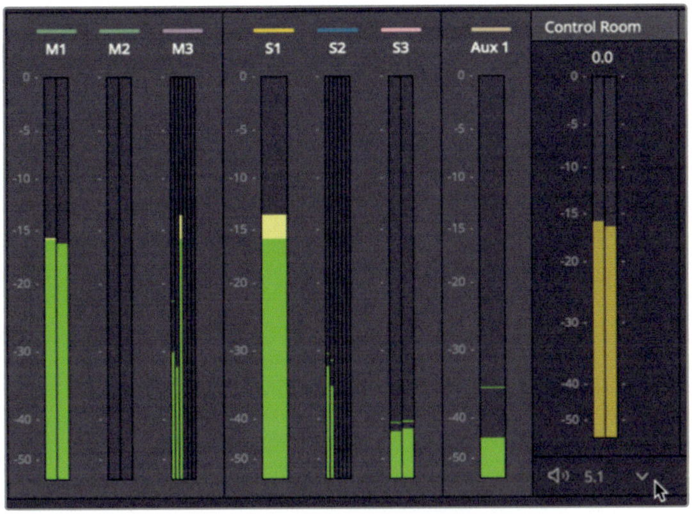

Notice that six meters are under the M3 main, but only the first three channels have a signal. That's because this isn't a 5.1 surround mix, and no signal is going to the LFE (Low Frequency Effects), left surround, or right surround channels.

No problem. Let's change the MUSIC submix Format to 5.1 surround. Then, you'll change the pan settings for the DRONE 1 track to place it in the LFE, left surround, and right surround positions.

3 Choose Fairlight > Bus Format. In the Bus Format window, change the Sub 2 MUSIC bus Format to 5.1. Click OK.

4 In the Mixer, double-click the A8 DRONE 1 pan controls to open the Audio Pan window.

5 Start playback from the beginning. In the Audio Pan window, drag the blue pan handle down to the rear center position which will, in turn, place the two green stereo channels in the left surround and right surround positions. As you drag the handle, watch the meters on the M3 main output to see them reflect the audio panned to the surround channels.

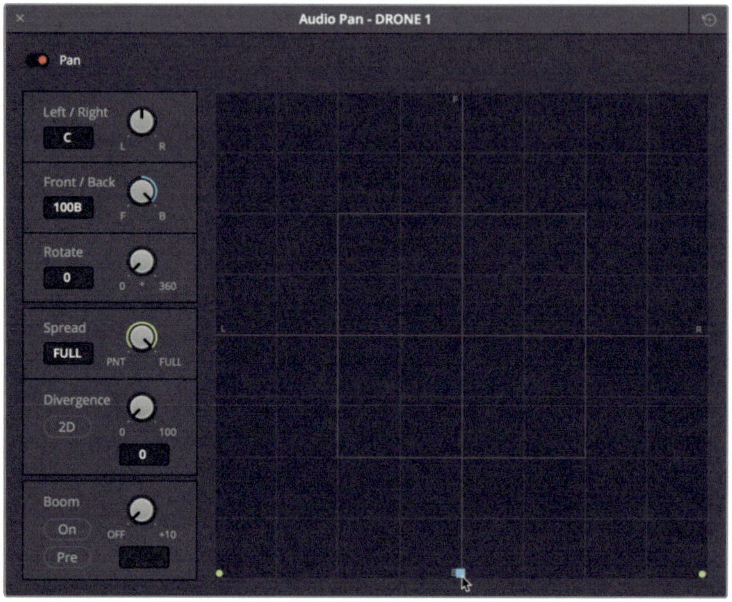

Now a signal is going to all but the LFE channel. To send a signal to the LFE channel, you use the Boom controls in the Pan window.

6 In the lower-left corner of the Audio Pan window, click On to turn on the Boom controls.

The dial in the Boom controls is the track send that lets you determine how much of the track's signal is sent to the LFE.

7 Set the send control to around -20.

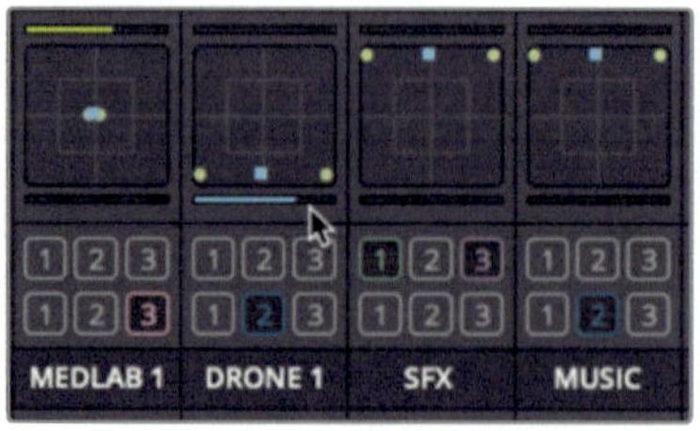

Notice that a blue LFE send bar is now present below the track's Pan controls in the Mixer to indicate the amount of signal that is being sent from the track to the LFE.

Close the window. Start playback from the beginning to see all six surround channels in action in the M3 meters.

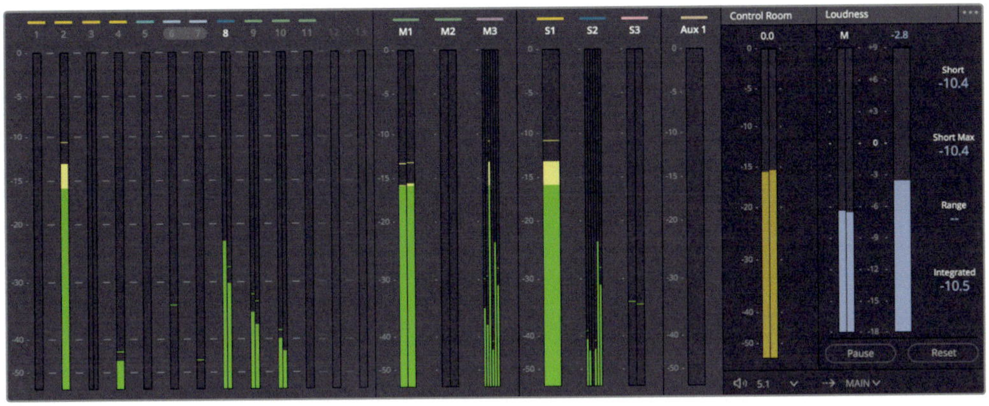

If you continue watching the M3 meters, you'll see that the third channel (Center) is panned to that channel because of all the dialogue.

8 Stop playback.

9 Return the Control Room monitoring to Main 1.

> **TIP** When you are finished, it's a good idea to always reset the Control Room monitoring to the Main output because the Control Room settings are global. Otherwise, if you leave the Control Room set to Main 3 and open a timeline that does not have a Main 3, you won't hear any sound until you go to a timeline with a Main 3 and reset the pop-up menu to Main 1.

Setting up and patching multiple mains is easy once you understand how to create and assign buses in Resolve.

Assigning the Music and Effects Tracks

Now it's up to you to think through the M&E output to make sure all the correct tracks or submixes are included. Your assignment is to open the Bus Assign window, select the M&E Main, and assign the submixes and tracks that do not contain any dialogue tracks. (Hint: Since room tone is considered part of the dialogue track, it should not be included in the M&E mix.)

After you have assigned those tracks, change the Control Room pop-up menu to monitor the M&E output, and listen to the scene to make sure no dialogue is present. When you are finished, set the Control Room to Main 1.

Working with Bus Tracks in the Timeline

With DaVinci Resolve 16, you can now view any of your buses as tracks in the timeline. Bus tracks let you record, edit, and view automation curves for buses just like standard tracks. In fact, you show and hide bus tracks, just like standard tracks, in the Tracks Index. This feature is available only while the Automation tools are showing.

In this exercise, you'll open a different version of this mix with fader automation already applied to some of the buses.

1. Open the timeline, "10 Bus Tracks Example."
2. Hide the Media Pool and Effects Library if showing.
3. Show the Tracks Index to see all the timeline tracks in order from top to bottom. In this case, all tracks are visible in the Trackslist.
4. In the timeline, click the Automation button to show the automation toolbar.

   With the automation toolbar showing, you'll also see all the buses including mains, submixes, auxiliary buses, and VCA groups.

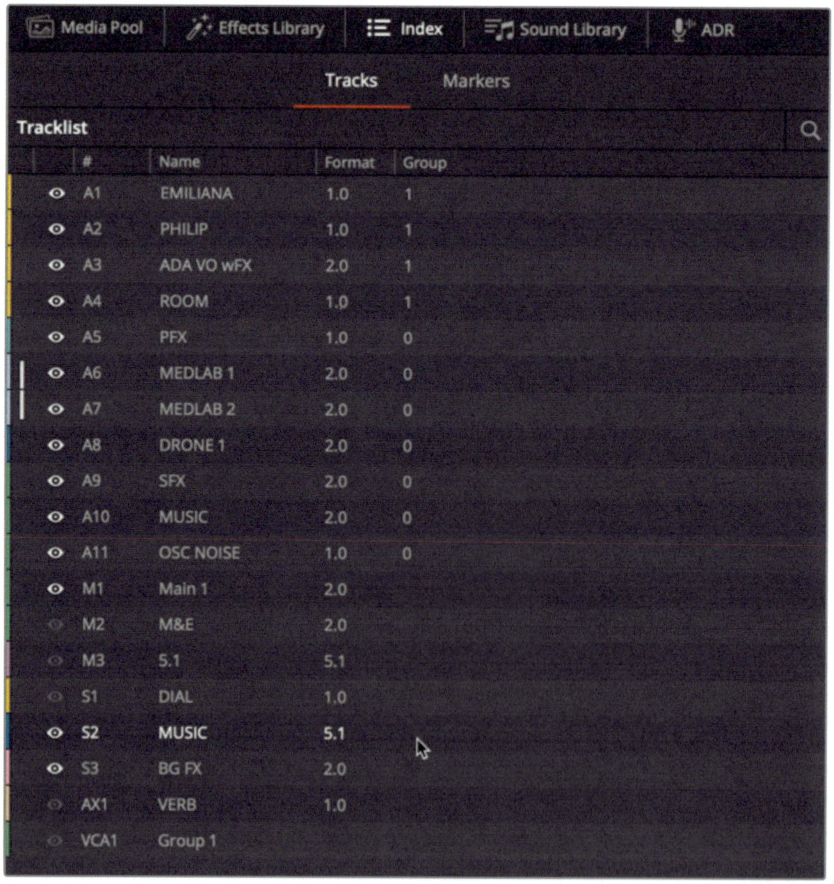

Here you can see that the Main 1 (M1), MUSIC submix (S2), and BG FX submix (S3) have their visibility icons enabled and are therefore showing in the timline.

5 In the timeline, scroll down to see the three bus tracks at the bottom of the timeline starting with the M1 Main 1.

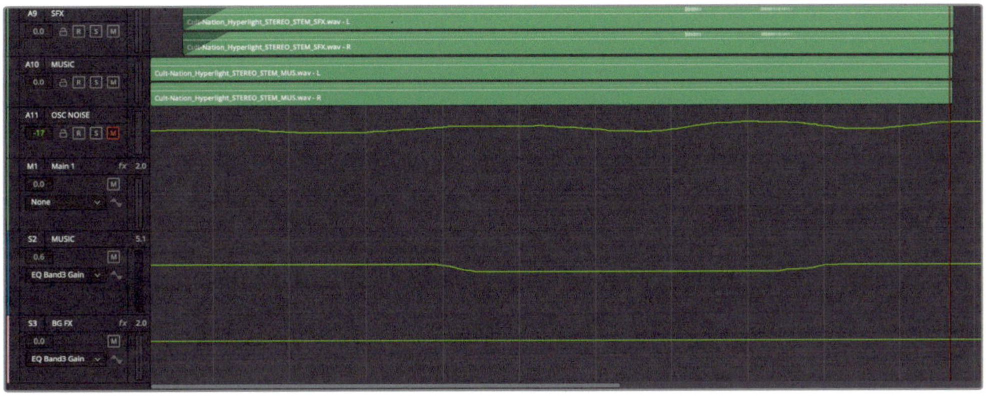

Notice that the S2 and S3 submix bus tracks show automation curves for the EQ Band3 Gain. This type of EQ automation is called *EQ carving* and is common for dipping or carving the mid-range EQ frequencies in music and background tracks that compete with important dialogue frequencies.

Coping and Pasting Automation

If multiple tracks or bus tracks need the same automation, you can copy and paste automation data from one track to another via the Fairlight Automation menu.

In this exercise, you'll copy the EQ automation curve from the S2 bus track and paste it on the S3 bus track. To copy and paste automation, you first need to set a range in the timeline.

1 In the timeline, mark a range from the first frame to the last.

2 Select the S2 MUSIC submix bus track header to select that bus track.

3 Choose Fairlight > Automation > Copy.

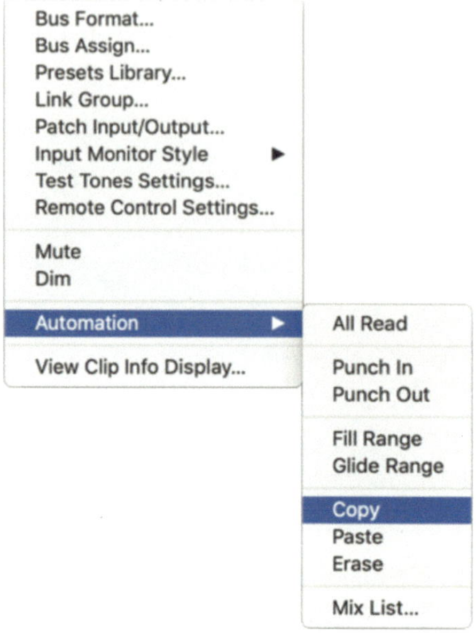

4 Select the S3 BG FX bus track.

5 Choose Fairlight > Automation > Paste.

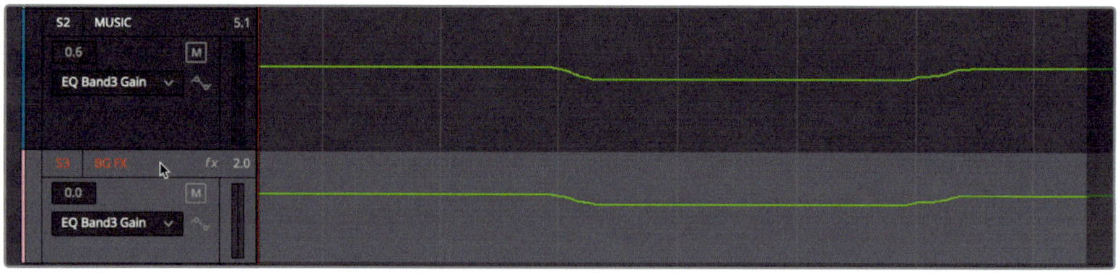

The S3 BG FX bus track now has the same EQ Band3 Gain automation curve as the S2 MUSIC bus track.

Showing the Loudness History Curves in the Main 1 Bus Track (Studio)

If you are working with DaVinci Resolve Studio, you can now show loudness curves in the Main 1 bus track so that you can easily spot exactly where loudness issues occur and fix them.

1 Move the playhead to the first frame of the timeline. Then, if necessary, show the Automation controls in the timeline and enable visibility for the MI Main 1 bus track in the Trackslist.

2 In the timeline, increase the height of the M1 Main 1 bus track until you see the Loudness History controls in the track header.

3 Switch on the Loudness History controls to see the Loudness history graph in the timeline. Then check the Integrated, Momentary, and Short Term options to see all three loudness metering graphs.

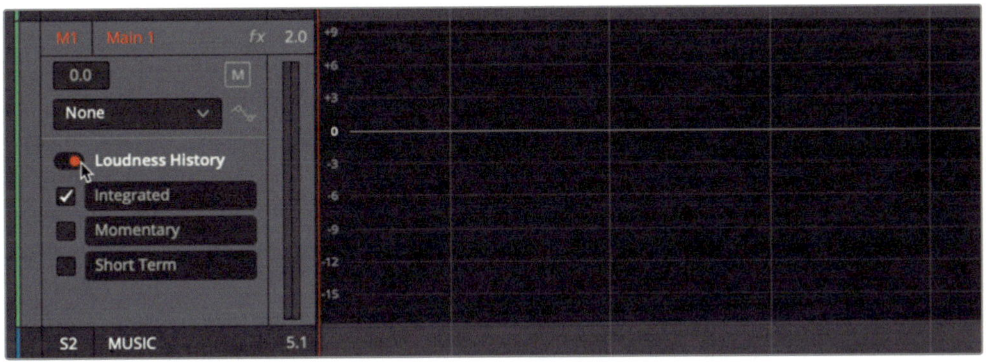

To see the Loudness history in the graph, you must start the Loudness meters.

4 In the Monitoring panel, click the Loudness meters Options menu and choose a loudness metering standard for your region. If unsure, choose BS.1770-4, which is the most current loudness metering used worldwide that employs algorithms to measure audio program loudness and true-peak audio levels.

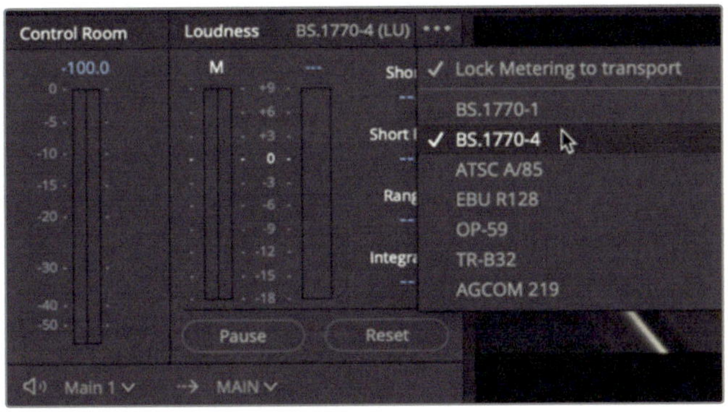

5 On the Loudness meters, click Reset and Start.

6 Start playback. The loudness metering colors and values are drawn in the graph as you play.

These three Loudness curves follow the same color indicators as the loudness meters. If the program level exceeds the -23 LKFS loudness standard, that part of the graph shows in red along with the meters. If the loudness levels are within tolerance (+/- 2 dBFS), the graph and the meters will display yellow.

> **NOTE** The Loudness History graph shows the visual history of the program loudness in the timeline. You cannot edit the loudness history curves. If there is an issue with the loudness levels, you must fix the volume levels in the tracks and buses using standard mixing and automation workflows.

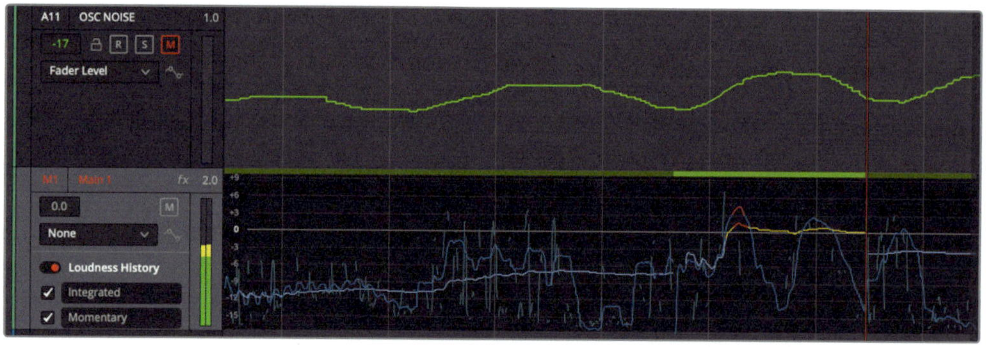

7 When you are finished looking at the loudness history graph, hide the Tracks Index and Automation controls in the timeline.

Now that you know how to show and use timeline bus tracks and the Loudness History graph in DaVinci Resolve, you can employ these powerful mixing and finishing tools in your own projects.

Massaging the mix

If you are up to a little more mixing, this is a great time to play the timeline from the beginning and follow all your new mixing instincts to adjust levels, enhance the EQ of a track, record automation, or simply experiment with different settings and listen to the outcome.

Open the timeline, "10 Mains and Subs Finished," or the previous timeline you were working with and take a turn at massaging the mix.

Congratulations! You have finished the scene and it's ready for delivery. It may not be perfect, but it has come a long way since the beginning of this lesson. And you are now armed with a new set of Fairlight mixing techniques—including EQ, dynamics control, auxiliary buses, automation, and submixes—that you can apply to your own projects.

Lesson Review

1 What type of bus can you use to apply one plug-in, such as Reverb, to any track in the timeline?

a) Main

b) Effects

c) Sub

d) Aux

2 Where do you access the Bus Format and Bus Assign windows?

a) Mixer

b) Monitoring panel

c) Fairlight menu

d) Auxiliary controls

3 Which controls enable you to automate faders, pan controls, and other parameters?

a) Automation

b) Animation

c) Auxiliary

d) Agnostic

4 True or False? You can apply automation to the Loudness History curves in the Loudness History graph.

5 To see bus tracks in the Tracks Index and timeline, what other controls must be showing?

a) Mixer

b) Automation Controls

c) Effects Library

d) Bus Format window

Answers

1 d

2 c

3 a

4 False

5 b

Lesson 11

Finishing and Delivering Tracks

The final phase of audio post-production is finishing and delivery. This process includes ensuring that your final soundtrack meets delivery standards, and that the files are rendered in the correct format. You also need to know how to bounce or render out smaller elements, such as your multilayered sound effects, to multiple formats.

In this lesson, you'll start with the multilayered computer ship display sound effect, and prepare it for bouncing as a stereo and a 5.1 sound effect. Then, you'll take those skills to the next level and bounce stereo and surround sound stems in your primary timeline. Finally, you'll explore DaVinci Resolve 16's professional finishing tools and techniques to deliver tracks as finished stereo and surround sound mixes.

Time

This lesson takes approximately 35 minutes to complete.

Goals

Panning and Bouncing Multilayered Sound Effects

Previously, you bounced multiple clips into a stereo file to create room tone and the heart monitor effect. In this exercise, you'll bounce the finished multilayered computer ship display effect into a stereo clip and a 5.1 surround sound clip. The trick to bouncing tracks into a multichannel track is to first set the main outputs for the format that you want to bounce, and then pan each track to the correct speaker channels prior to bouncing. Let's start by panning the tracks for bouncing a two-channel stereo clip.

Before bouncing or delivery, it's always a good idea to remove fades and extend the background element of your sound effect. Doing so gives the sound effects editor more editing options. This practice is similar to including an additional one-second duration to either end of an animated graphic.

First, on the A4 track, you'll need to remove the fade at the beginning of the background clip. Then, extend the clips on A3 and A4 by two seconds.

1 In the A4 track, drag the fade handle at the head of the clip to the left to remove the fade.

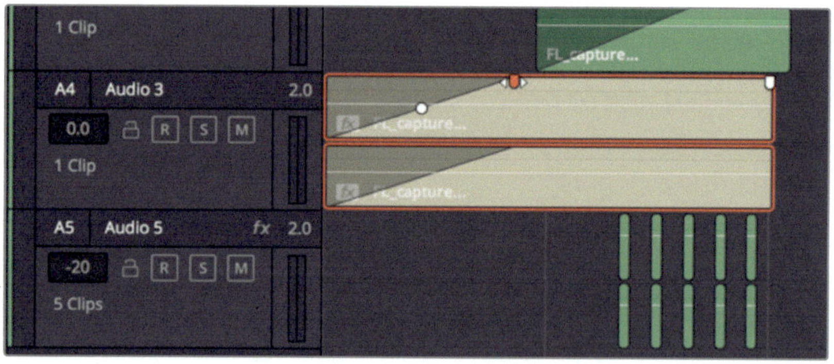

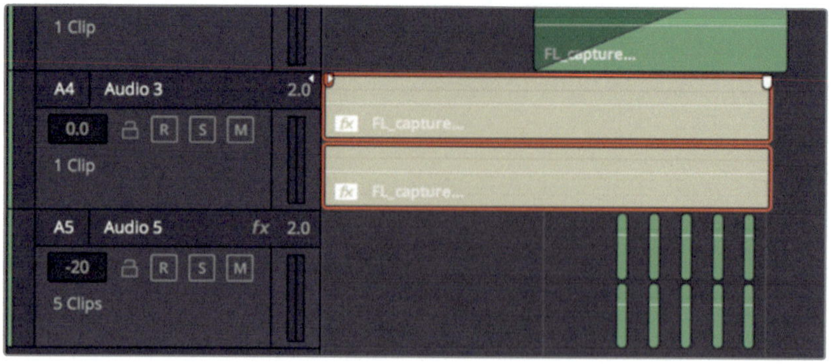

2 Move the playhead to the end of the clip on the A3 track. Press Shift-Right Arrow to move the playhead one second to the right.

3 Drag the tail of the clip on the A3 track to the playhead to extend it by one second. Then, extend the clip on A4 by the same amount.

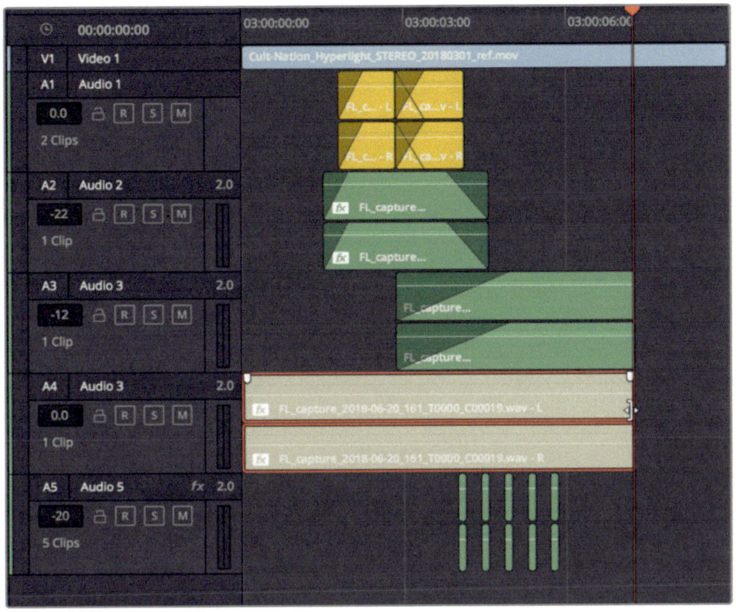

4 Mark the clip on the A3 track.

The ship display is ready for the next step: panning.

Spreading Tracks in the Stereo Field

Now, you'll pan the tracks to spread the sound throughout the panoramic stereo field. If you are a music producer, you are very aware of how much room you have in the stereo field to place sounds, much like placing musicians and instruments on the stage. For a sound effect such as this computer display, its panned position in the mix is based on its visual position in the shot. For a wide shot with a small computer display in the background, you would probably mix the sound effect as a mono or stereo element with a narrow spread to match the distant position from the camera and the visually perceived "size" of the sound source. In this example, however, you have a hero shot of the computer that fills the entire screen, so the mix must do the same for its sound.

A panoramic photograph is an image that includes everything you can see without turning your head. A panoramic stereo field is everything you can hear from the far left to the far right without turning your head.

By default, all stereo tracks start with the pan position set to left/right (0), or center. Stereo tracks panned to center are played at equal levels out of both speakers, which leaves the sonic impression that the sound is directly in front of the listener in the middle of the L/R audio space. Be aware that this audible "phantom center" is not the same as panning a mono track to the center channel speaker in a multi-speaker configuration.

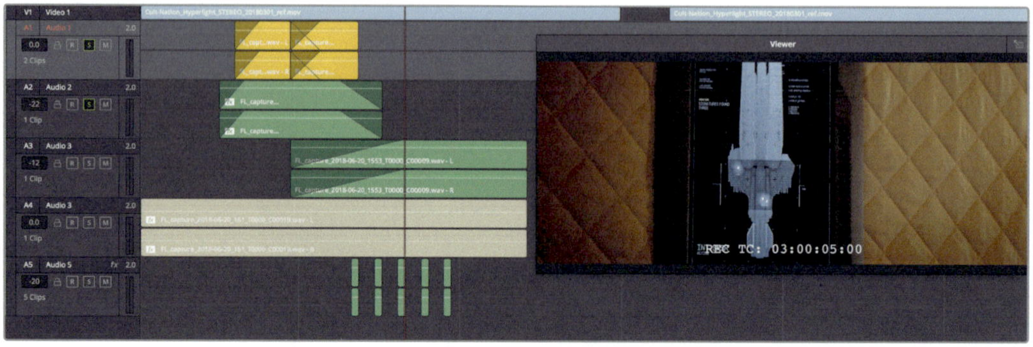

Let's vary the stereo spread by adjusting the panning controls for tracks A1-A5, each of which represents a different element in the sound effect. A1 is the *searching* sound. A2 is the *thinking* sound. A3 is the *finished search* sound to replace *thinking*. A4 is the background warble, or ongoing *pulse* of the computer; and finally, A5 represents the *they are here* beeps that match the blinking lights on the display.

1 Play the effect through once and listen to how it sounds.

  With all tracks identically panned, your ears and brain tell you that the sound is a single object. However, because the associated image could be 20-feet tall or larger on a movie theater screen, let's give the sound a larger dimension to match.

2 In the Mixer, in the A1 track, double-click the pan controls to open the Audio Pan - Audio 1 window. Set the Spread to 30.

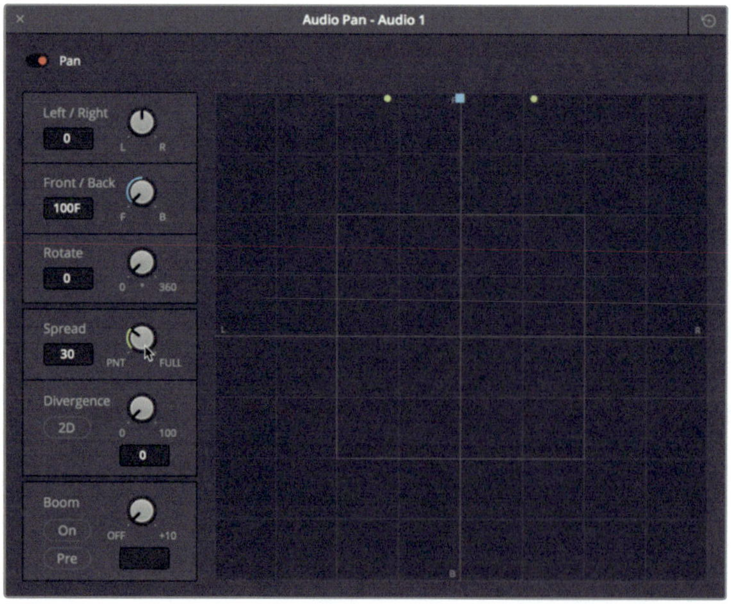

3 Set the following values for the remaining tracks:

- Tracks A2 and A3 - Spread 60
- Track A4 Panned 0 with Spread at 100 (Full)
- Track A5 - Spread 10.

If you are listening with headphones, you should easily be able to hear the distribution of sounds within the stereo field.

4 Close the Audio Pan window.

5 Play the timeline from the beginning to hear the sound effect fully spread throughout the panoramic stereo field.

Now that the tracks are panned, you can prepare them to bounce.

Bouncing to a Stereo Track

You should check three things prior to bouncing tracks. The first is to determine what format you want to bounce to. The second is to select a destination for the bounced file in the Media Pool. And, finally, you should check the output levels. As you already know, bounced files go to the same location as your recorded files, the current capture location on your system. But where do they go in the Media Pool? Just as with recordings, they go to a selected bin. Let's take a moment to create a bin to contain this effect. Then, you can check the levels and bounce the tracks.

1 Show the Media Pool.

2 In the Media Pool bins list, right-click the SFX for Audio Post bin, and choose Add Bin.

3 Name the new bin **Computer Ship Display 2ch**, and select it.

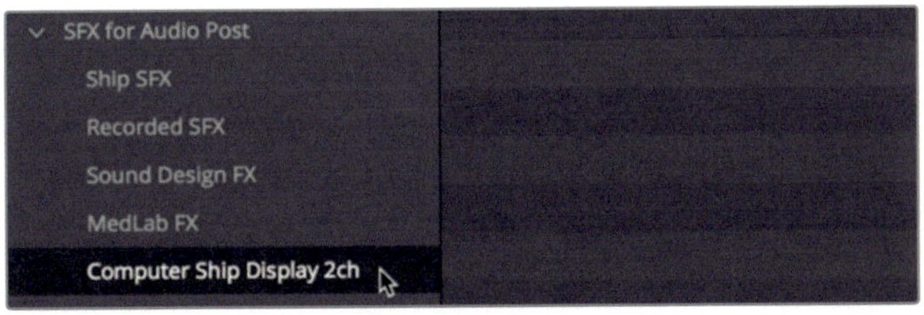

The 2ch stands for 2 channels, or stereo. Now, you should check the levels during playback. Your goal is to bounce a strong signal with a nominal average level (-12 dB). In this case, the *searching* sound on the A1 track should average around -12 dB.

NOTE You could add a compressor to the main output to limit the dynamic range between elements, or leave it as is. For this exercise, the output sounds fine with its full dynamic range.

4 Mark the clip in the A4 track.

5 Start looped playback, and watch the Main 1 output levels. In the Mixer, set the Main 1 fader so that the strongest part of the signal (yellow) averages around -12 dB.

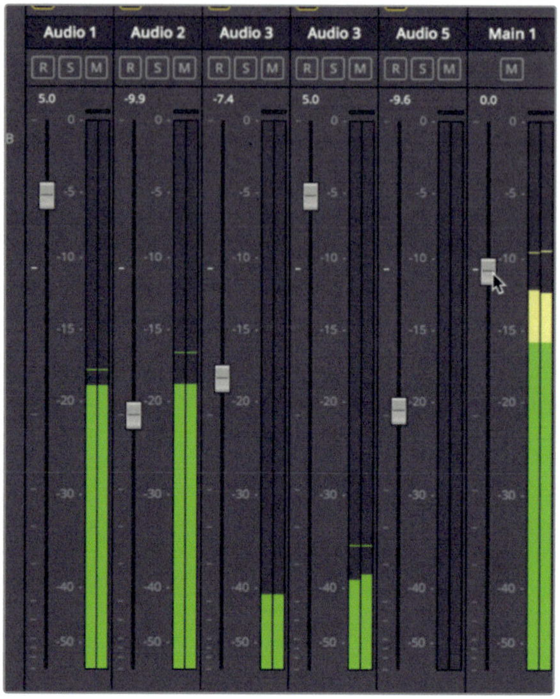

The two channels of audio in the main output meter indicate you're monitoring a stereo output so you needn't perform additional steps to convert the sound to stereo.

6 Choose Timeline > Bounce Mix to Track.

7 In the Bounce Mix to Track window, in the Destination Track pop-up menu, choose New Track. Click OK.

You have a finished stereo clip in both the timeline and the selected bin in the Media Pool.

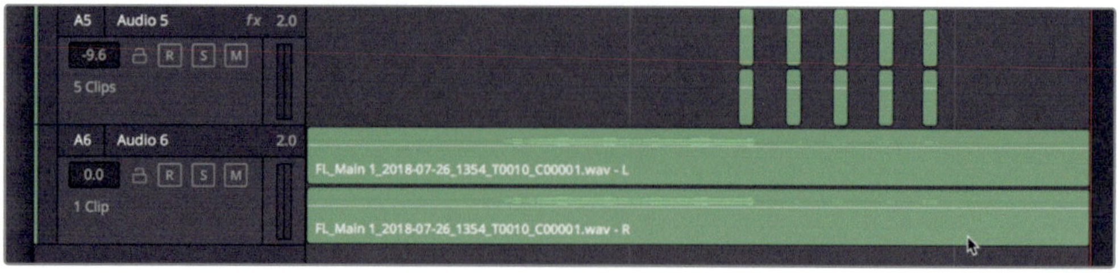

8 In the Computer Ship Display 2ch bin, select the clip to see it in the preview player. Play the clip once in the Media Pool preview player.

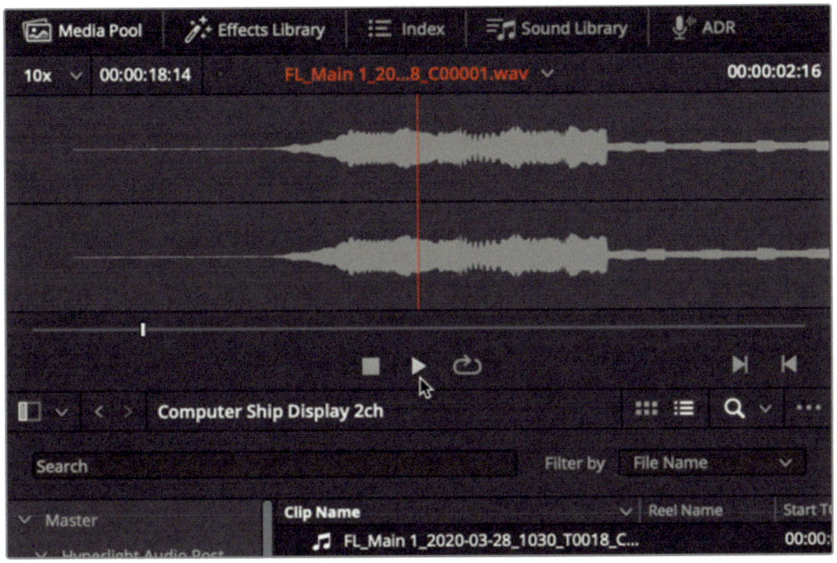

TIP To locate the bounced file on your system, you can right-click the clip and choose Reveal in Finder. The bounced clip in the Media Pool contains a single track with two channels, left and right. The bounced media will become two separate mono files with the same names except for their ending which adds _C00001 and _C00002.

One bounce down, one to go.

Bouncing to Multichannel Surround Formats

The only differences between bouncing an effect as stereo and bouncing it as 5.1 surround are the main output format and how the tracks are panned. In this exercise, you'll create a second main output bus for 5.1 surround, and change the track panning and LFE to take advantage of all six surround channels. Let's repeat the previous methods with some added steps to create a six-channel surround clip.

1 In the SFX for Audio Post bin, create a new bin called **Computer Ship Display 6ch**.

2 Choose Fairlight > Bus Format. In the Bus Format window, click the Main button.

3 Name the Main 2 bus **5.1**. Set the Format to 5.1, and change the Color to green. Click OK.

4 In the Mixer, open the pan controls for the A1 track. Increase the Divergence to around 50 to spread the sound between the left and right channels.

5 Set the A2 and A3 Tracks to Spread 100 (Full).

6 For the A4 background computer sound, drag the blue pan handle down to the 100B (back) position. Turn on the Boom, and set the send value to -10.

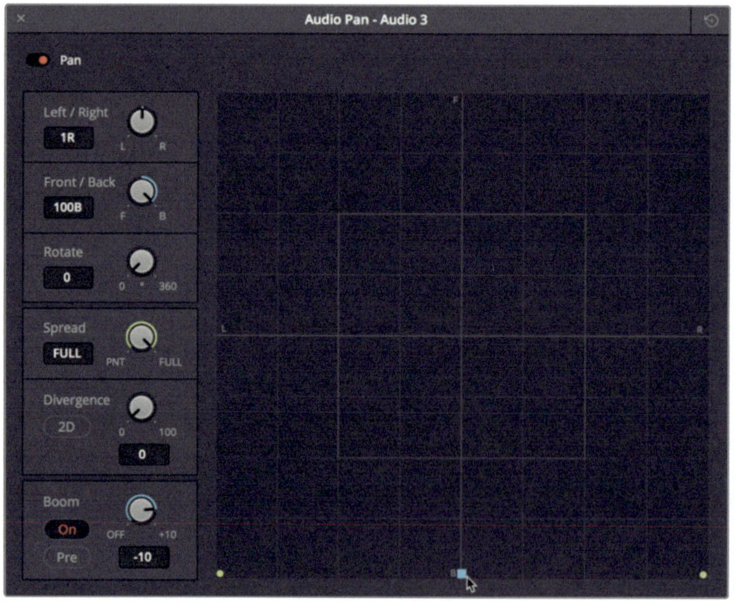

7 In the A5 track, drag the pan handle to the surround center position, set the Spread to 50, and the Divergence to 50.

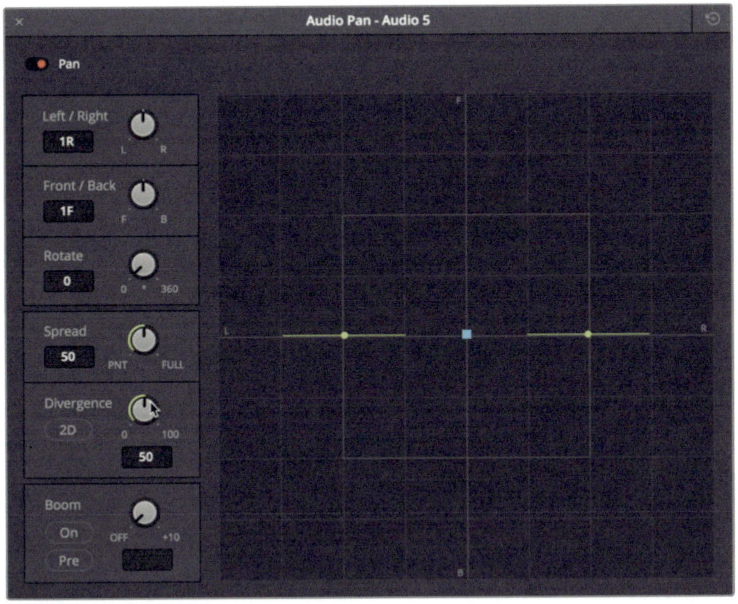

8 Close the Audio Pan window.

9 In the Mixer, Click the Main 2 button for each track to patch the output to the Main 2 bus.

10 Mute the A6 track that contains the bounced stereo clip.

11 Play the timeline to hear the surround channel panning and see the M2 channels in the meter.

12 Mark the clip in the A6 track to set a play range to bounce.

13 Choose Timeline > Bounce Mix to Track.

The Bounce Mix to Track window displays both Main outputs.

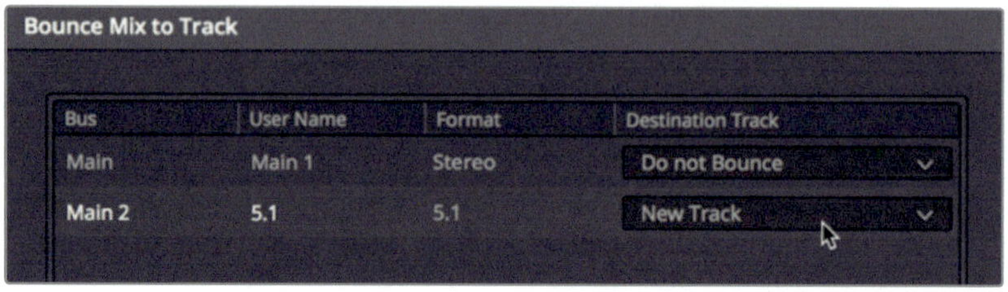

14 Set the Main 2 5.1 Destination Track to New Track. Click OK.

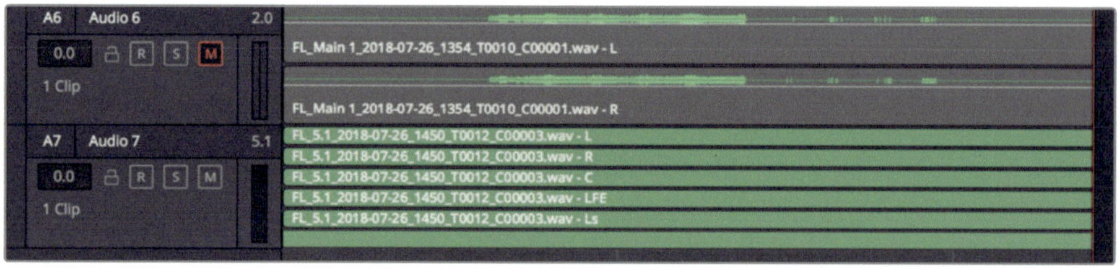

The six-channel clip appears in the A7 track and the Media Pool.

15 In the Media Pool, select the new six-channel clip, and preview it.

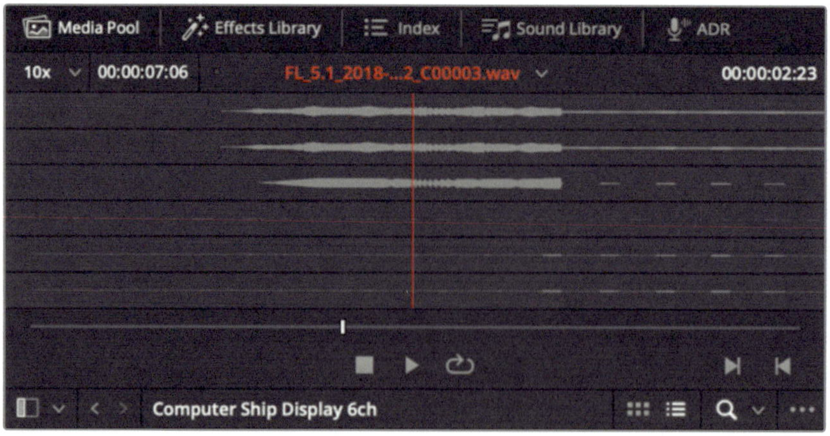

Now that you've explored multichannel bouncing for a sound effect, let's look at bouncing the dialogue tracks in your mix.

Bouncing Dialogue to One Track

In addition to using submixes for simplified mixing, another option is to bounce all the dialogue tracks, or the dialogue submix, to a single track as a *stem*. In this exercise, you'll create a mono dialogue stem for the scene you mixed in Lesson 9. This time, instead of bouncing to a new track, you'll create the track ahead of time, above the other dialogue tracks.

1 Open the "10 Mix for Stems" timeline.

  This should look familiar because you spent quite a bit of time mixing it in the previous lesson.

2 Right-click the A1 track header, and choose Add Tracks. In the Add Tracks window, choose the following settings:

  - Number of tracks 1:
  - Insert Position > Above First Track
  - Audio Track Type > Mono.

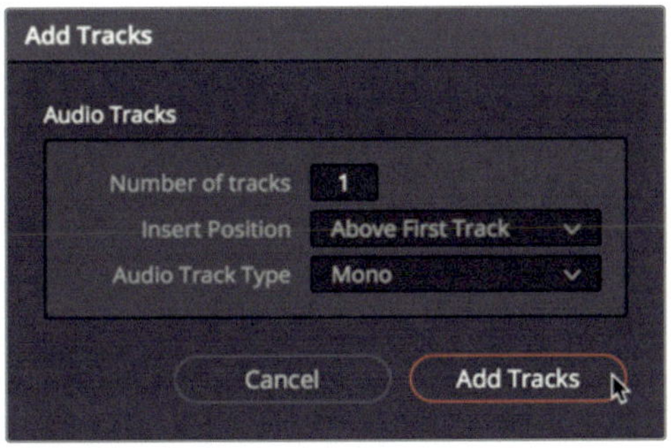

3 Click Add Tracks.

4 Name the new A1 track **DIAL**.

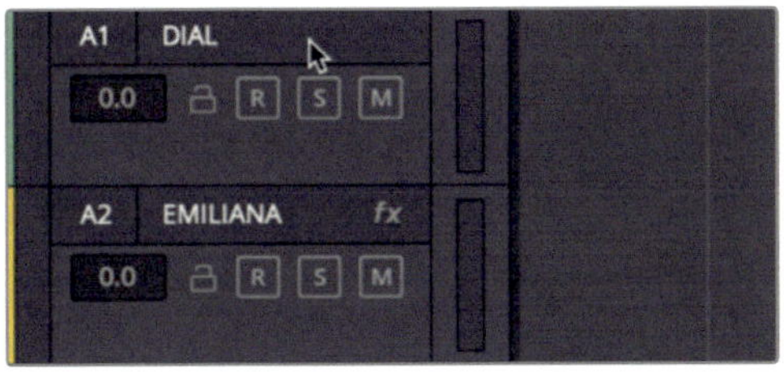

5 Set a play range from the first frame to the end of the last frame of audio.

  You could solo the dialogue tracks and bounce the main output to the mono DIAL track. Or you could simply bounce the DIAL submix to the DIAL track. For this exercise, let's use the latter method.

6 In the Media Pool, create a bin called **My Bounced Mixes**. Select the bin, if necessary, and hide the Media Pool.

7 Choose Timeline > Bounce Mix to Track.

The Bounce Mix to Track window includes options to bounce from any bus.

8 Set the Destination Track for the Sub 1 DIAL bus to the DIAL track.

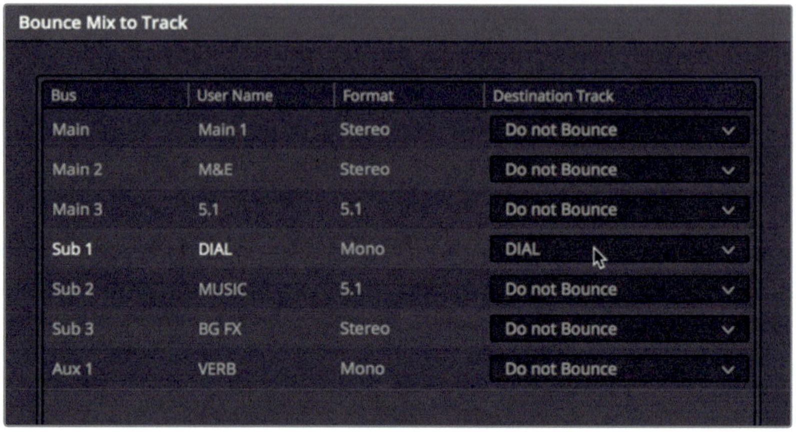

9 Click OK to bounce the DIAL submix to the DIAL track.

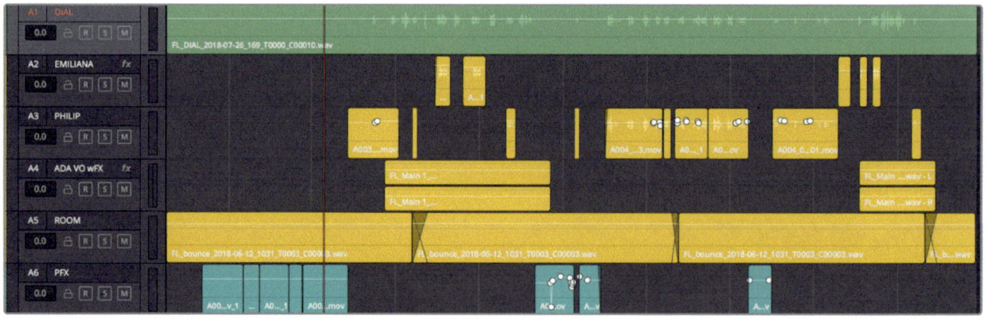

Done! With a little knowledge of buses and mixing, you were able to bounce four tracks into one. Let's play it back to audition the results.

TIP Changing the color of clips and tracks is very useful for organizing, marking, and maintaining your audio post-production workflow. However, when you get down to the final mixed elements, it's traditional for some mixers to leave them native-audio green to indicate that they are the final version.

10 Mute tracks A2 - A5. Play the scene from the beginning, and listen to the bounced dialogue with the other tracks. Enjoy! When you are finished, stop playback.

Don't worry if the levels aren't perfect at this time. The important thing is that throughout this book you transformed the rough dialogue from the edit through the entire audio post-production process, resulting in a final mix. If you have any issues with your levels or the bounced mix, you can always make changes and bounce again later because the process is easy when you understand how it all works. You're now ready to produce the finished mix.

Delivering the Mix

Assuming that all the levels in the meters are great, it's time to deliver your mix, whether it is your final stereo mix, stems, or just an intermediate timeline such as your sound effects. Whatever type of deliverable you need to output, the Deliver page of DaVinci Resolve 16 is where you'll do it.

1 Click the Deliver button to open the Deliver page.

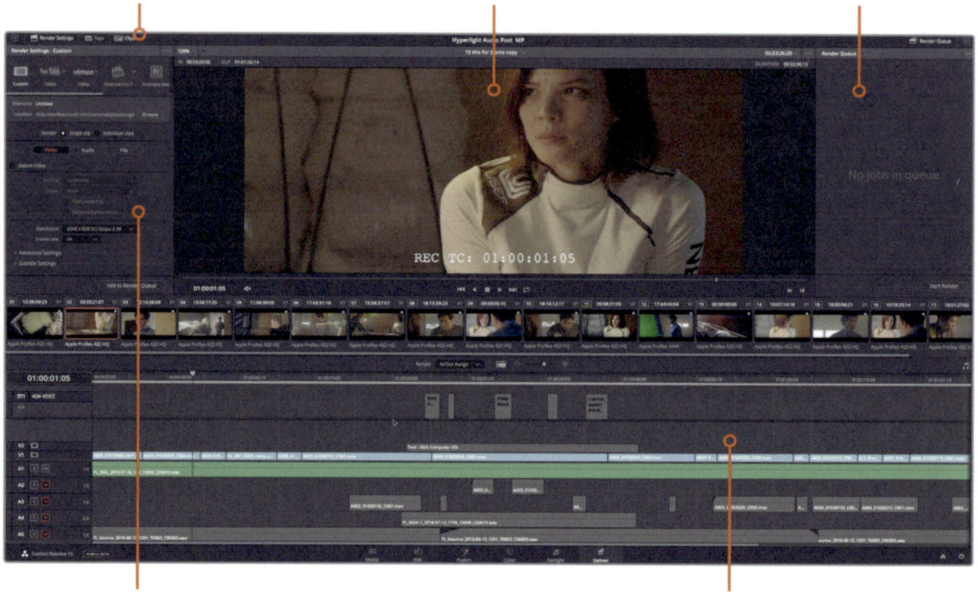

2 In the upper left, in the Render Settings, click Audio Only. Also, choose Single Clip, if necessary.

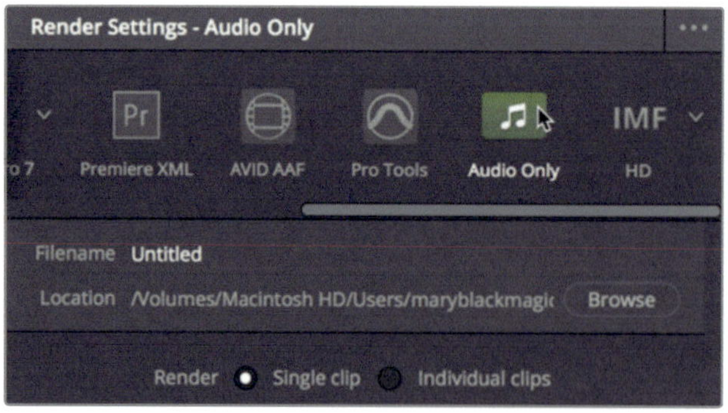

When rendering a mix, you will always want to output a single file for your project.

3 In the Export Audio tab, in the Format menu, choose Wave.

You can choose whatever format you need on your projects, but for this one, you'll export in a standard Wave format.

4 In the Codec menu, choose Linear PCM, if necessary.

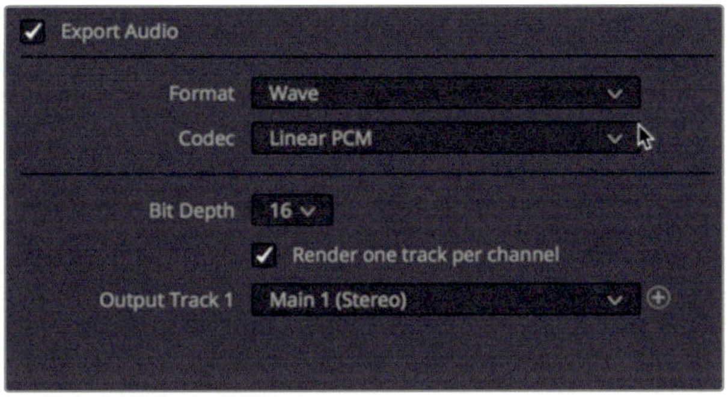

The current Bit Depth is 16, which is fine for this lesson. You can change it to 24- or 32-bit audio files at any time depending on your delivery requirements.

Below the Bit Depth is a very important checkbox because it determines whether your mix will be rendered as a two-channel stereo file or as two single-channel mono files for the left (L) and right (R) channels.

For this example, let's render a stereo mix.

5 Deselect the "Render one track per channel" option, if necessary, so you will output a single, stereo file.

6 Click the Output Track pop-up menu to see the multiple available options.

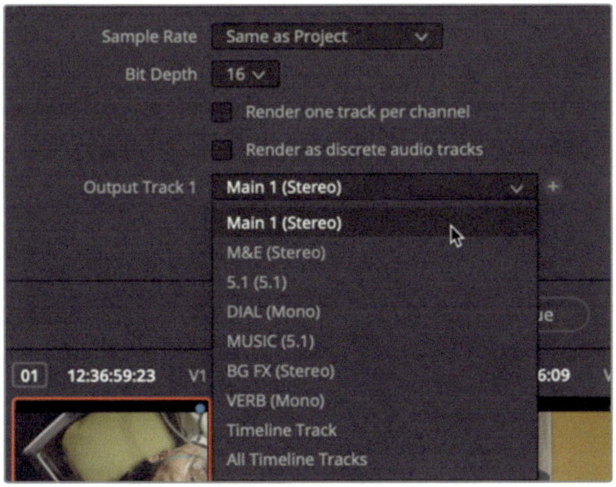

Here you can select any of the buses, a timeline track, or all timeline tracks. With the All Timeline Tracks option, the channel mapping of the rendered clip will match the entire timeline all in one clip.

7 In the Output Track 1 menu, choose Main 1 Stereo.

TIP To simultaneously output additional buses or tracks as part of the same rendered file, click the Add button (+) to the right of the Output Track pop-up menu, and choose another output.

Next, you'll need to add a custom name and browser location for your file.

8 Click the File tab, and add a custom name as you would for any other output. In this case, let's name the file **HL_Scene4 _2ch _mix**.

Finally, you need to select a destination for the mixed file.

9 Click the Browse button to change the location of your mixed file to the Desktop folder. Create a new folder on the desktop called **My Mixed Stems**. Click OK.

10 When you are done selecting your drive destination, click Add to Render Queue.

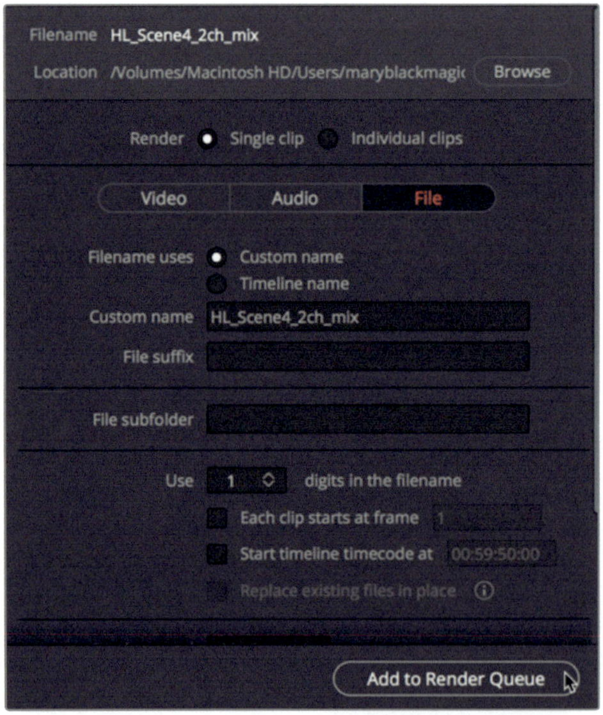

11 In the lower-right corner of the Render Queue, click Start Render.

12 When your file is finished rendering, head to your Desktop to locate and audition your finished stereo mix.

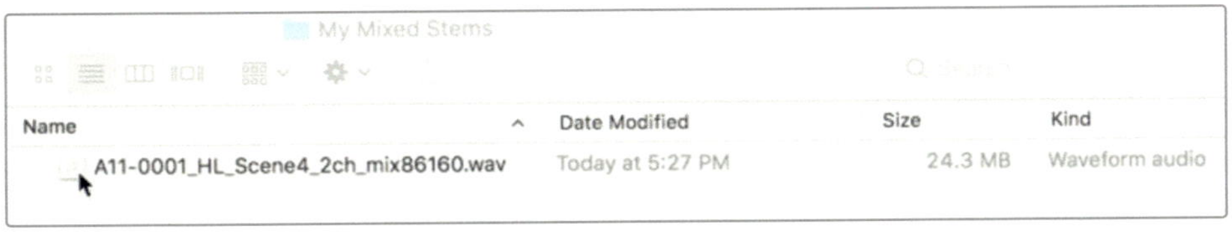

As you can see, the file-naming conventions added some information to the filename, including the render date. Once the file is rendered to the desktop, you're free to name it anything you want.

Audio Finishing with Mixed Stems

Now that you can bounce tracks in the Fairlight timeline and render stems and mixes in the Deliver page, you've reached the end of your audio post-production journey. All the audio mixing is complete, the files are rendered, and you're poised to put it all together and deliver a finished movie. Whether the audio stems were created in DaVinci Resolve 16 or another audio application, every project finished in Resolve reaches this final stage when the completed video, color correction, and compositing meets the finished soundtrack for a final delivery.

In this exercise, you'll open the finished scene with all its finalized sound elements, check the loudness, and pan the mono tracks to prepare them for delivery.

1 Press Shift-7 to return to the Fairlight page.

2 Open the timeline, 10 Finishing. You can use the Zoom sliders to fit all the clips horizontally and vertically in the timeline.

This timeline includes the finished mixed stems from the audio post team at Cult Nation Studios for the Hyperlight soundtrack. The tracks are as follows:

- A1 - Stereo Mix
- A2 - VX Stem (stereo dialogue stem)
- A3 - SFX Stem (stereo sound effects stem)
- A4 - MX Stem (stereo music stem)
- A5 - Audio 5 (channel 1 of a 6ch 5.1 surround sound mix)
- A6 - Audio 6 (channel 2 for a 6ch 5.1 surround sound mix)
- A7 - Audio 7 (channel 3 for a 6ch 5.1 surround sound mix)
- A8 - Audio 8 (channel 4 for a 6ch 5.1 surround sound mix)
- A9 - Audio 9 (channel 5 for a 6ch 5.1 surround sound mix)
- A10 - Audio 10 (channel 6 for a 6ch 5.1 surround sound mix)

Each of these track groups fulfills a different delivery requirement. They are all in the same timeline at the same time, but they aren't meant to be played all at once. Let's start at the top with the A1 - Stereo Mix track.

> **NOTE** Many post-production mixers use VX, SFX, and MX as shortcuts for the voice (dialogue), sound effects, and music stems.

3. Solo the A1 - Stereo Mix track. Resize the viewer in the monitoring panel so you can clearly see the video playback while listening to the audio.
4. Start playback, and watch the scene with the final mix. When you are finished, stop playback.

   You probably noticed that many sound design and mixing techniques were used to achieve this final mixed soundtrack. And, of course, ADA's voice turned out great when performed by a professional actress. In addition to listening to the finished mix, you also need to check the loudness levels.

Monitoring Loudness in Your Mix

Now that you are ready for final output, you need to monitor the output levels. In the past, broadcast standards were based on the highest peak level in a soundtrack. As long as audio content did not exceed that level, it would pass quality control. That led to commercials that applied heavy audio compression to narrow their dynamic range and sustain all voiceovers at the maximum allowable (and overbearing) level.

To even the acoustic playing field, new loudness standards were introduced that apply to all broadcast programs (television and radio), regardless of the length or type of program. In North America and parts of Asia, an integrated target of -24 LUFS (Loudness Units Full Scale) is the norm for broadcast content, whereas in Europe the target is -23 LUFS. (Theatrical films, trailers, and streaming videos have different standards.)

DaVinci Resolve 16 includes two types of monitoring panel meters to measure such standards. *Peak meters* are traditional Root Mean Square (RMS) meters available for every track and bus that uses a decibel scale, and *Loudness meters* measure the loudness in your program based on the loudness unit scale (LUFS). There are also presets available in the Loudnes options menu (...) where you can choose the latest metering BS.1770-4 as well as international Loudness standard for the meters.

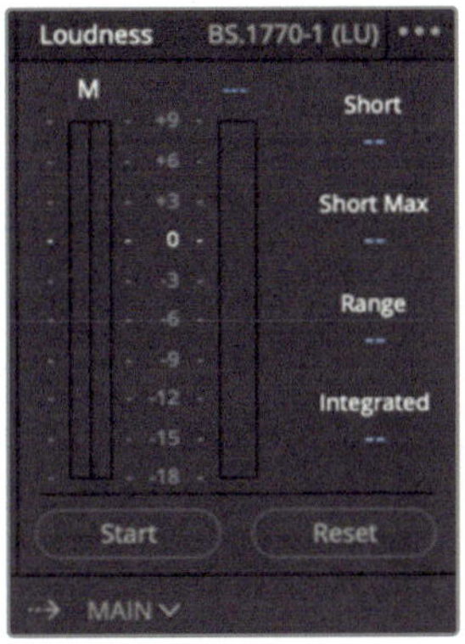

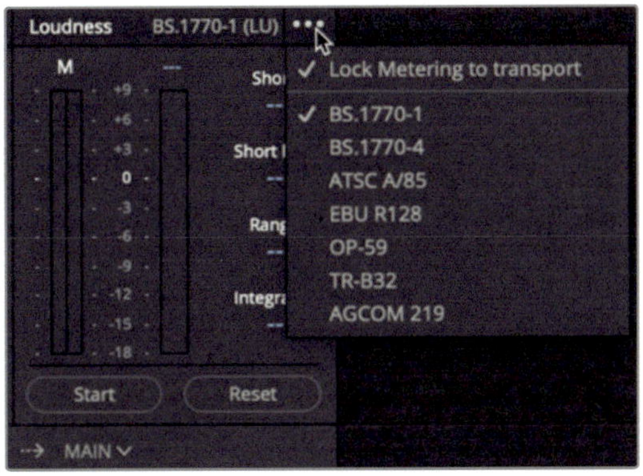

You can use the Loudness meters to measure the perceived loudness over the entirety of your program to verify that the levels of your final mix meet the current loudness standard, ITU BS.1770.

The Fairlight page Loudness meters include a numeric display that outputs a variety of useful loudness measurements, with the most important being the Integrated level.

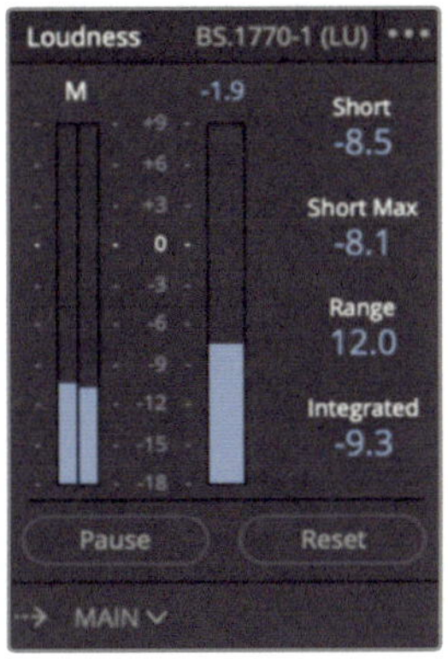

- M displays the momentary loudness unit at the playhead location.
- The loudness units meter displays the sum of all channels for the duration of playback. The numeric value at the top is the max LU value over that range.
- Short displays loudness over a 30-second range.
- Short Max displays the played range's maximum true peak level.
- Range displays the dynamic range of loudness in the entire program.
- Integrated displays the average loudness for the entire played range. This is the value that targets either -23 LUFS or -24 LUFS as required by broadcasters and configured in your Project Settings.

In this exercise, you'll monitor the integrated loudness display to ensure that your project audio comes within +/- 0.5 of the target -23 LUFS.

The integrated number is measured over the entire start-to-stop range as you play it.

1. At the bottom of the Loudness meters, click the Reset button to clear all current measurements.
2. Move the playhead to the beginning of the scene.
3. Under the Loudness meters, click the Start button to enable the analysis.
4. Press the Spacebar to play the timeline. Watch the integrated loudness display during playback to see how closely to 0 this stereo mix measures on the loudness scale (-23 LUFS).

   The Loudness meter is a solid blue color that measures from -18 to +9, unlike decibel meters that show from -50 dB to 0 dB across their green, yellow, and red ranges. The Loudness meter is designed in this manner because -23 LUFS is equal to 0 on the Loudness meter scale. When you monitor levels with the Fairlight Loudness meter, your goal is to keep the final level near 0 and the integrated level to +/- 0.5 .

> **TIP** If you are delivering content for a region that requires an integrated loudness level other than -23, you can change the target loudness level in the Project Setting's General options.

Although loudness units and decibels express audio measurements differently, they have a 1-to-1 relationship, which makes it easy to adjust faders when targeting the integrated loudness level. For instance, if the integrated loudness display shows +2 LUFS after you play through your program, you can use the master fader to lower the decibel level by -2 dB to reach the targeted 0 value in the Integrated display.

Because these are finished mixed stems, their Loudness levels should be good to go.

5 When the scene finishes playing, look at the Integrated value (yellow) on your Loudness meter.

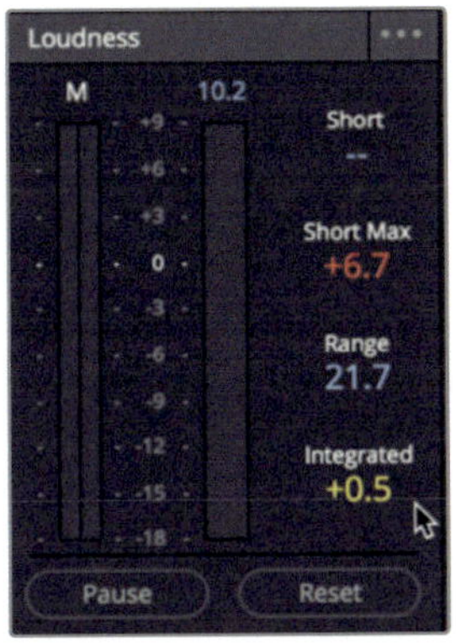

These levels are just within the loudness standards with an Integrated value of +0.5.

6 Reset the meter.

7 Unsolo the A1 track.

Panning Multichannel Mono Tracks

You'll now turn your focus to the last six tracks in the timeline, which contain the surround sound stems. Earlier in this lesson, you panned tracks to create a surround sound effect. When panning tracks to create a dimensional aural soundscape, the sound designer and mixer have plenty of room to be creative. However, when the final mix is rendered, the master audio channels must be mixed to the proper channels, or they won't be reproduced properly by the speakers. Worse, the show won't pass quality control because the audio channels are wrong. That's why the channels are all clearly labeled in the Audio Pan window.

Let's zoom in on the tracks to see their current channel configurations, and then look at the pan controls to move them to the correct output channels.

1 Solo tracks A5–A10. Zoom in on the timeline tracks so you can clearly see the headers for those tracks.

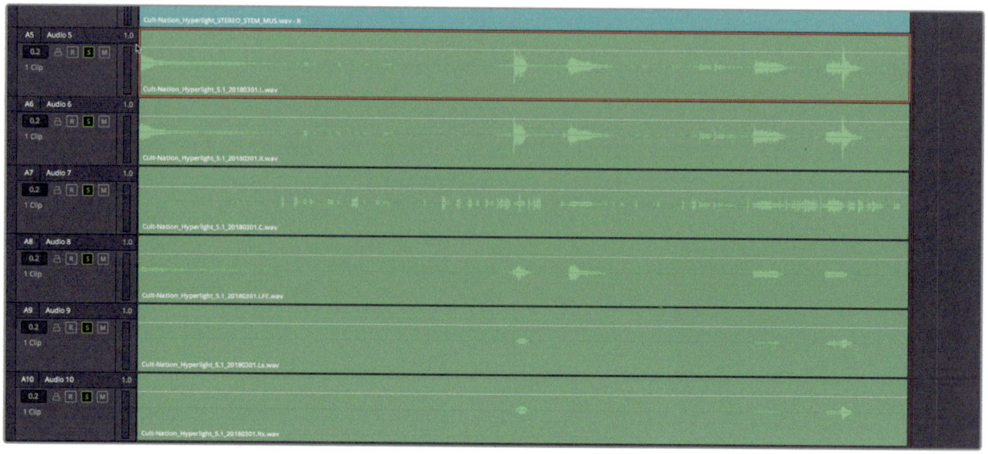

In the upper-right corner of each track header, you can see that the current channel mapping is 1.0 to indicate one mono channel. Also, the Mixer shows that each of these mono tracks is currently panned to the Center channel. As you know, in surround sound mixing, the center channel is reserved for dialogue and LFE only.

You could manually pan each mono track to the proper channel in the Mixer, or you could use the handy Link Group window to link all six tracks to a single fader. Let's try plan B.

2 Choose Fairlight > Link Group to open the Link Group window.

If you recall, the Link Group window highlights all mono tracks in the timeline. In this case, all six of the 5.1 mono stems.

3 In the Link Group window, select the Audio 5 track, and then select the Audio 10 track to select all six tracks.

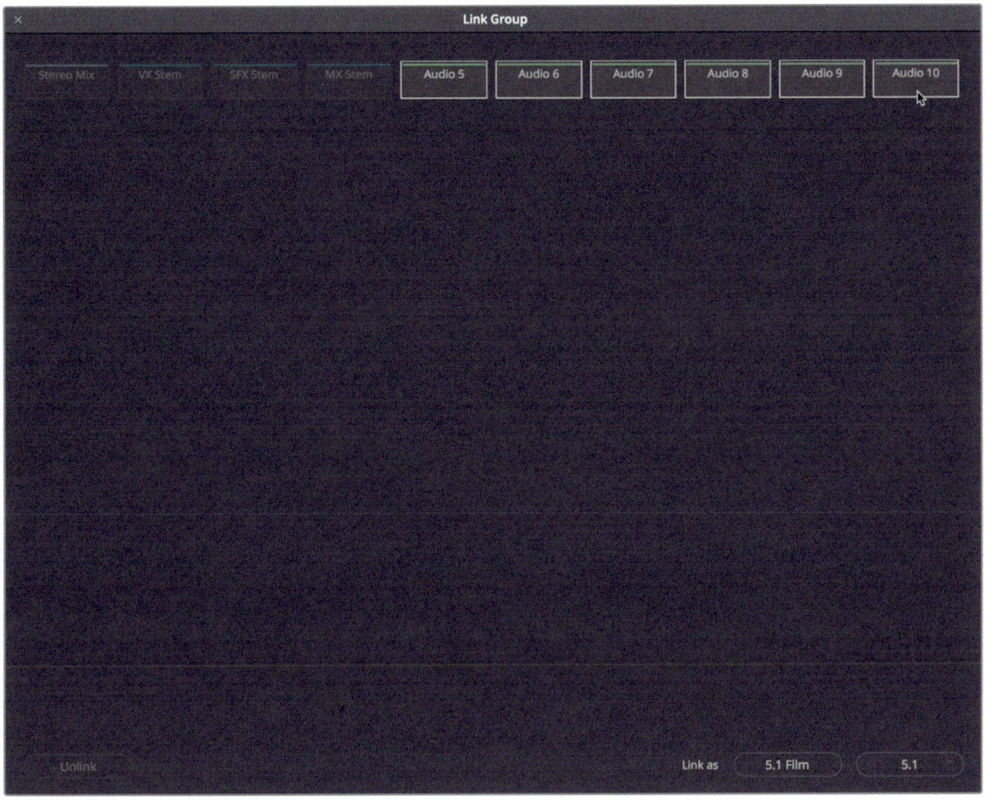

At the bottom of the window, you'll see options to Link as 5.1 Film or 5.1. The difference between these options is that 5.1 Film configures to Academy standards, whereas 5.1 configures to a standard six-channel output.

4 Click the 5.1 button, and close the window.

Two things just happened to the track panning. First, all the track headers now indicate the channel panning for each track. Second, all six tracks appear under one fader in the Mixer with perfect 5.1 panning.

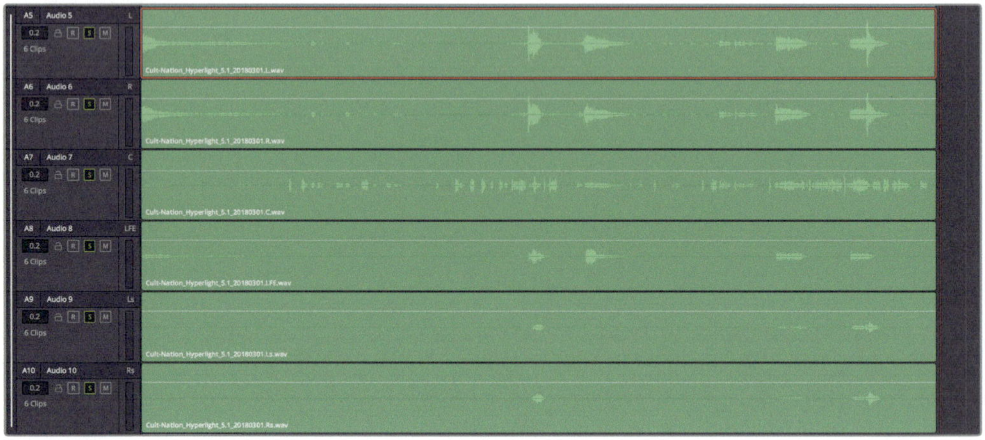

You can now head to the Deliver page to render out the final program with any combination of these audio tracks, stems, or mixes. But wait, there's one more thing that's totally awesome about finishing in Resolve's Fairlight page. What if the deliverables require separate panned mono tracks instead of a linked group represented by a single 5.1 track? No problem.

5 Choose Fairlight > Link Group to open the Link Group window.

6 In the Link Group window, select the linked group, and click Unlink.

The tracks are separated into single-channel mono tracks, but with a twist. In the Mixer, take a look at the track panning.

Awesome, right? The individual mono tracks maintain the same surround channel panning that they had in the 5.1 linked group.

You still have one thing to do, which is turn on the Boom for the LFE track (Audio 8).

7 In the Mixer, double-click the A8 pan control to open the Audio Pan window.

LFE channels are panned to the center, just like the dialogue.

8 In the Audio Pan window, turn on the Boom, and drag the control to 0.

9 Close the Audio Pan window.

The tracks are now panned and the LFE channel is good to go.

NOTE With DaVinci Resolve 16.2 and higher, you can convert any multichannel track into a Linked Group of mono tracks. Just right-click the track header and choose Convert to Linked Group from the contextual menu. Once converted, you can adjust levels for each track independently, while maintaining single fader control of the group. You can even unlink the group in the Link Group window to turn the tracks into independent mono tracks with separate faders.

Explore the Mix!

You've finished your last official exercise in this book. Now you can kick back and play through the different stems to really appreciate all that you put into making this outstanding soundtrack.

When you are finished, feel free to return to any of the previous timelines and try your hand at finessing their editing, tracks, effects, mixing, and more.

Mix the Trailer Soundtrack!

Time to put your new audio post-production skills to the test. When you have a few minutes, open the Lesson 01 build soundtrack project, and mix the "3 GOW Trailer Basic Tracks Finished" timeline. If you recall, when you started Lesson 1, you were just learning the basics of audio editing in the Fairlight page. Now that you are a seasoned Fairlight user, listen to the "finished" trailer to be reminded of how much you have actually learned. In fact, you probably will have trouble listening to the timeline once through without feeling the need to dive in and start fixing levels. Of course, the actual trailer has an amazing soundtrack, so you can listen to the timeline "4 GOW Trailer Mix" for inspiration. Have fun.

Hopefully, stepping through the exercises in this book has demystified the audio post-production process. It takes years to gain the experience and the educated ears of a professional sound designer and re-recording mixer. However, if you have enjoyed any or all of these lessons, you are encouraged to spend as much time as you can exploring, experimenting, and working in the DaVinci Resolve 16 Fairlight page to hone your skills at crafting your own soundtracks.

Lesson Review

1 What factor determines the channel configuration of a bounced mix?

   a) Selected tracks

   b) The type of track that is bounced

   c) Bouncing matches the channel mapping of the A1 track.

   d) The channel mapping of the Main or Submix bus that is bounced

2 Identify one or more channels that are used for low-frequency effects (LFE) in a standard six-channel, 5.1 surround sound mix.

   a) Channels 1 and 2

   b) Channel 3

   c) Channel 4

   d) Channel 5 or 6

3 Where in the Mixer will you find the Boom settings for the LFE? (Choose all that apply.)

   a) Insert

   b) Dynamics window

   c) Aux Send

   d) Audio Pan window

   e) Effects window

   f) EQ window

4 True or False? In the Deliver page render settings, you should always render your mix as a single clip.

5 True or False? Linking six mono tracks as a 5.1 group will automatically pan them to the correct channels with a single fader control. Subsequent un-linking of that same group of tracks will retain each mono track's 5.1 channel panning.

Answers

1 d. A bounced mix follows the channel configuration of the main output or submix bus selected in the Bounce Mix to Track window.

2 c

3 d

4 True

5 True

Congratulations!

You have completed **Fairlight Audio Post with DaVinci Resolve 16** and are ready to explore more editing, visual effects, color grading, and audio mixing functionality using the additional certified books in this series. Completing all the lessons in this book has prepared you to become a certified DaVinci Resolve user. You can take the online exam by following the link below to earn your certificate.

We also invite you to become part of the DaVinci Resolve community by joining the web forum on the Blackmagic Design website. There, you can ask further questions about the creative aspects of editing, color correction, and audio mixing.

We hope that you have you have found DaVinci Resolve 16's professional nonlinear editing and world-class color correction tools to be intuitive to learn and a perfect fit to become the hub of your entire creative workflow.

Test your skills by taking the online assessment: **https://bit.ly/2M7kbLu**

Index

G

H

I

J

K

R

U

V

W

Z

About the Author

Mary Plummer is a professional video editor, audio artist, and composer with over 20 years of experience in video and film production. Mary is based in Los Angeles, California, where she lives with her husband, Klark Perez, and daughter Kathryn. She has worked on a wide range of productions from coast to coast, including shorts, PSAs, trailers, music videos, documentaries, and independent feature films. Between projects, Mary has written more than a dozen industry training books including *Apple Pro Training Series: GarageBand* (all versions), *Apple Pro Training Series: Soundtrack*, *Apple Pro Training Series: Soundtrack Pro*, *Apple Pro Training Series: Getting Started with Final Cut Studio*, and *Media Composer 6: Part 1—Editing Essentials* (Avid Learning). She has also created over 26 hours of online Logic Pro X videos.

Mary currently works for Blackmagic Design with their marketing and curriculum development team.

Made in the USA
Las Vegas, NV
04 November 2026

57424225R10312